Unity from Diversity

Ray W. Cooksey

Unity from Diversity

Pluralist Systemic Thinking for Social and Behavioural Research

Springer

Ray W. Cooksey
UNE Business School
University of New England
Armidale, NSW, Australia

ISBN 978-981-97-3461-0 ISBN 978-981-97-3462-7 (eBook)
https://doi.org/10.1007/978-981-97-3462-7

This Springer imprint is published by the registered company Springer Nature Singapore Pte Ltd.
The registered company address is: 152 Beach Road, #21-01/04 Gateway East, Singapore 189721, Singapore

Preface

This book is about the building of knowledge in social and behavioural spaces. Knowledge is the *unity* we seek and, given that social and behavioural research is a human endeavour focusing on human lives, there are *diverse and ever-evolving pathways* towards achieving that unity. If people were rocks or stars or molecules or animals, that diversity would be greatly reduced and there would be no need for a book like this. We could simply refer to any book on the methods and approaches used in the natural and physical sciences, which unfold under the pattern of guiding assumptions, generically known as 'positivism'. However, human beings are conscious, thinking, and feeling beings, with identities, beliefs, and agendas and, as such, there are many pathways for understanding their lives and experiences; researchers are human beings, but so are the beings whose lives and experiences they research. Some of those pathways are very old, some are rather newer and emerged as a counterpoint to the older pathways, and some pathways have only recently been developed in order to explore life spaces and experiences not previously encountered or acknowledged. Pathways will only ever yield partial glimpses into human life, so diversity serves to enrich, expand, and enlarge those glimpses in pursuit of more complete understanding. Researchers and the researched are far more connected than disconnected in this world and those connections are spread out through a network of interlinked complex systems.

Given this starting position for social and behavioural research, it is only natural to ask if there are such diverse pathways, how does one choose an appropriate pathway? The answer put forward in this book is *pluralist systemic thinking*. We start by conceptualising a research endeavour as a journey, undertaken by a researcher, in pursuit of specific goals. That journey unfolds within and is subject to potential influences of an array of contextual forces; those forces may constrain the researcher's journey or open up opportunities. Therefore, making choices about how to go about undertaking a research journey means understanding where one is coming from as a researcher, where they want to get to, in and for what contexts, under what constraints, and in light of opportunities pursued. The journey is not static or linear; it is dynamic and nonlinear. All of this means, that as researchers, we need to enact a mode of thinking that considers all of the potential pathways that could be followed (this is *pluralist* thinking) and makes choices amongst those pathways against the background of the myriad contexts that the research journey will be part of, is subject to influence by, as well as potentially able to influence. It is *systemic* thinking that allows us to understand those myriad contexts (as well as their interconnections and cross-influences); in fact, contextualisation is an essential and ongoing process for any social or behavioural researcher. When pluralist thinking and systemic thinking are synergistically harnessed, we have a meaningful way to make defensible choices for contextualising and completing a research journey. Furthermore, we will see that successfully completing a research journey means that the research is convincing to one or more audiences. Although researchers will make choices to enhance their chances of achieving high quality research outcomes (and, in this book, we offer meta-criteria to assist the researcher in achieving such outcomes), the ultimate arbiter of researcher quality is not the researcher themselves, but the audiences and stakeholders to whom and for whom the research might be of interest and impactful, which is why *convincingness* is essential. No single pathway for building knowledge can claim to be the best or the one and only. That arrogance is what tends to feed disciplinary preferences and biases. Instead, we ask researchers to hold their predispositions, personal and disciplinary preferences, and comfort zones at bay, while they work through the diversity of available choices. Two things will matter most: choose what will make the research convincing and choose what will make the research journey feasible/viable to complete. One skill will be essential to develop: learning to be *flexible and adaptable*, in the face of emergent constraints or unexpected opportunities.

This book should not be considered an introduction to social and behavioural research methods. Rather, it builds on a platform of prior knowledge/experiences of researchers but brings into focus ways of thinking and a diversity of possibilities they may not have encountered before. Thus, the proper place for this book is as a value-add to initial research training and experiences. It can also be a value-add for experienced researchers looking to break out of the bonds imposed by disciplinary constraints and expectations. This book is all about breaking down boundaries and opening up experiences. It does not provide recipes for success or step-by-step guidance. Rather it is intended to rattle comfort zones, question the status quo, and build a mindset for harnessing diversity in service of unity. For many, this will be a challenge, because it is hard to overcome one's own predilections, preferences, and past training experiences in pursuit of research outcomes. However, I hope that, in the shorter term, researchers will find their research experiences to be greatly enriched and perhaps more convincing, after reading this book, and, in the longer term, the cumulative nature of knowledge building in human systems will be better served.

I must state for the record here that many of my pluralist systemic ideas, issues, and concepts I discuss in this book were first put forward, in a more fragmented, partially developed, and less well-organised form, in the chapters I wrote for Cooksey and McDonald (2019, namely Chaps. 3, 9 through 12, 14, 18 through 22). In this volume, I sought to build, reorganise and integrate my thinking with respect to those ideas, as well as incorporate new concepts and ideas, into a coherent pluralist systemic framework, centred around seeking unity from diversity. Thus, while my discussions and narratives here have drawn upon, in part, my work in those earlier chapters of

Cooksey and McDonald (2019), they have been substantively reshaped, condensed, or expanded, as appropriate, as well as refocused on social and behavioural research in general. Where I have adapted, extended, or revised a figure/diagram or a table from my work in Cooksey and McDonald (2019), I have appropriately cited it in this volume. It would be fair to say that my earlier ideas and concepts were in a nascent fragmented form in my chapters of Cooksey and McDonald (2019) and, in this volume, they matured to become fully realised and integrated.

About the Author

I am a white/Caucasian male who was born into a military family in California in the USA. My early childhood involved multiple postings in different states, and included one 2.5 year posting in Ankara, Türkiye, in the early 1960s (I was in primary school at the time, and it was my first exposure to cultural diversity). I developed an interest in psychology in my senior year of high school and went on to earn B.Sc., M.Sc., and Ph.D. degrees in psychology at Colorado State University. My Ph.D. was in Environmental Social Psychology, with a heavy background in experimentation and quantitative methods. While I was a graduate student at CSU, I took a Philosophy of Behavioral Science elective unit, and my mind was blown. There, set before me, were a number of other pathways for carrying out behavioural research (e.g., interpretivism, constructivism, verstehen, critical social science); I had only ever experienced one pathway in my training in psychological research, the 'positivist' pathway (or hypothetical-deductive science), which eschewed qualitative data in favour of quantitative data and statistical analysis. Also, while in graduate school, I encountered the work of Egon Brunswik (1930s, 40s, and 50s) and, once again, my mind was blown. There was a psychologist who actually thought like a systems theorist, espousing 'heresies' such as psychologists not being able to fully understand and generalise what they learn unless they also understand the real-life contexts in which the people they were studying were situated, and before generalising (the nomothetic imperative), researchers must first understand what happens at the level of the individual (idiographic appreciation). This was my first encounter with an imperative to contextualise aspects of one's research.

When I graduated with my Ph.D., I spent a year with the Bendix Corporation, doing ergonomic design work in the coal mining and environmental industries. This exposed me to the realities of connecting academic knowledge with practice-based knowledge as well as exposing me to the necessity of making design trade-off decisions in the face of constraints (such as engineering or workplace limitations). After Bendix, I accepted a lectureship at the University of New England in the Faculty of Education, teaching research methods. Thus began my first interdisciplinary experiences. My teaching had to meet the needs of a variety of educational disciplines, so I had to adapt to teaching a more diverse array of approaches, including some that my colleagues absolutely thought were rubbish approaches. I persevered and established a reputation as open-minded, adaptive, and responsive to novice researcher needs. I developed relationships with research colleagues that provided me with more diverse research experiences including encounters with qualitative data and non-positivist patterns of guiding assumptions. I also began supervising students who were not from my discipline areas, and in some cases, not from my cultural background. I have learned from all the students I have taught and supervised as well as from the colleagues and mentors I have had the good fortune to work with.

In the early 1990s, I encountered chaos theory and complexity theory, both of which spoke to the interconnectedness and complexity of systems and began to apply those concepts to my disciplinary work in decision-making research as well as in my research methods teaching. At the same time, I spent 4 years in the Psychology department at the University of New England, then moved into a new business/management school at the same university, where I remained until my retirement in 2014. My career also involved several senior leadership roles, where I played a role in establishing or modifying institutional policies governing aspects of research (experiences which acquainted me with the importance of institutional research policies, the opportunities they can provide, and the constraining influences they impose on researchers). My research as well as my teaching began to reflect pluralist systemic thinking and I began to insert my thinking into my research training and supervisory work. In 2011, a colleague, Prof. Gael McDonald, and I published a book entitled *Surviving and Thriving in Postgraduate Research*, whose chapters were based on questions postgraduate students would ask us at yearly doctoral training workshops for the Australian and New Zealand Academy of Management. Those questions had a common feature; they remained unanswered by the research training courses they were undertaking at their own universities. It was in that book that I actively implemented pluralist systemic thinking to influence the content of my chapters. When we wrote the second edition of the book in 2019, my thinking had expanded further to encompass Indigenous/First Nations perspectives and arts-based perspectives. My involvement in developing the Ph.D. Innovation at the University of New England has enriched my perspectives on critical engagement, working with multiple stakeholders, and sharing research outcomes with academic as well as industry/professional/practitioner audiences.

Given my personal/professional history, I can say with some certainty that this book has been in gestation since around 1979; I just hadn't realised it until much more recently. I decided to write this book in order to focus directly on the value and utility of pluralist systemic thinking for making convincing choices from a diverse landscape of possibilities in pursuit of the unity of knowledge. In a way, it is the capstone achievement of my career and all that came before it.

Personal Statement with Respect to Indigenous/First Nations People

It is important that readers know my positioning with respect to Indigenous/First Nations people and research, especially as I have endeavoured to devote much more attention to it in this volume. I wrote this book on Anaiwan country in the Northern Tablelands of New South Wales, Australia; in fact, my entire academic career has unfolded on Anaiwan country. As I am a white American expat, I cannot claim to have any connection to or special knowledge of the Indigenous people of Australia, although I acknowledge that I do come from a culture whose nation building efforts were achieved only after displacing, corrupting, oppressing, enslaving, and/or destroying the First Nations before it. My positioning with respect to Indigenous/First Nations people and research is one of resonance with the imperative to decolonise research, to free Indigenous/First Nations research from the strictures, assumptions, and restorying imposed by western colonising cultures. My hope is that I have been able to show proper respect for and acknowledgement of the importance of, and on-going advancements in, Indigenous/First Nations research perspectives and approaches.

Acknowledgements

I would like to acknowledge some very important mentors and colleagues, people whose influences on my professional development, career trajectory, and achievements are directly or indirectly reflected in this book:

- the late Mr. James Stewart (the social science teacher who stimulated my interest in psychology when I was a senior at Mitchell High School in Colorado Springs);
- the late Professor Egon Brunswik (University of California at Berkeley, influential via his published works);
- Assistant Professor David Crocker (philosophy professor at Colorado State University who taught Philosophy of Behavioral Science);
- Professors Ross J. Loomis and Terry L. Dickinson (my Ph.D. co-supervisors at Colorado State University);
- Craig S. Hartley (ergonomics engineering colleague at the Bendix Corporation, with whom I designed an adjustable seat and an operating cab for underground coal mining machines);
- the late Professor Emeritus Kenneth R. Hammond (University of Colorado, world renown judgment and decision-making researcher, who greatly influenced my thinking about decision research and mentored me through writing *Judgment Analysis: Theory, Methods, and Applications*);
- Professor Peter Freebody (now affiliated with the University of Wollongong, with whom I produced many research outcomes in the education discipline);

- Associate Professor Richard Gates (retired from the University of New England, with whom I produced several chaos/complexity-oriented research outcomes);
- Professor Ingrid Moses (former Vice-Chancellor of the University of New England, under whose leadership I worked to develop or update research and research training policies for the university while I was Chair of the Academic Board);
- Professor Tom Maxwell (University of New England, with whom I developed the concept of the Ph.D. Innovation); and
- Professor Gael McDonald (formerly of RMIT Vietnam, who was a colleague of mine in the Australian and New Zealand Academy of Management (ANZAM) for many years and with whom I wrote two editions of *Surviving and Thriving in Postgraduate Research*).

Over the years, I have had the good fortune to supervise a diverse range of domestic and international postgraduate students in many disciplines and, invariably, I learned as much or more from them as they did from me.

I could not have completed this book (or any others for that matter!), without the steadfast love and support of Christie, my wife of 47 years and counting.

March 2024

Emeritus Professor Ray W. Cooksey
UNE School of Business
University of New England
Armidale, NSW, Australia

Acknowledgement of Country

I acknowledge the traditional custodians of the land on which I live my personal, family, and professional journey as an outsider and guest, *the people of the Anaiwan nation*. I want to pay my respects to Elders past, present, and emerging and celebrate the diversity of Indigenous peoples, whose stories and songlines traverse the millennia connecting land, water, life, culture, community, and cosmos, I also acknowledge and pay my respects to Aboriginal and Torres Strait Islander people who are reading and experiencing this book.

Contents

Chapter 1
Pluralist Systemic Thinking for Researchers

Abstract This chapter conceptualises research as a nonlinear journey influenced by a myriad of potential contextual forces. That journey is merely one facet of a researcher's life, comprising experiences that occur at specific times, places, and circumstances; experiences that should be regularly recorded to provide a source of information for continual reflection. The contextual complexities surrounding the journey emerge because the researcher is human (influenced by their own contexts and identity as a researcher) undertaking research on, with, or about other humans (influenced by their own contexts, identities, and roles) and/or the handiworks they produce. Importantly, the researcher is not the sole owner, shaper, and experiencer of their research journey, others will exert influence and share an interest in its outcomes. Achieving convincing research outcomes benefits from pluralist systemic thinking: *pluralist* - envisaging multiple potential pathways the research could follow; *systemic* - understanding contextual influences, feedback, and feedforward. Importantly, the researcher is not the arbiter of what counts as convincing, that role resides in the various audiences/stakeholders with and for which the researcher aspires to share research outcomes. In this light, planning plays an important yet limited role, because some influences, constraints, and opportunities cannot be anticipated at the outset.

1.1 Prologue

In this book, we will be talking at length about a more holistic way of thinking about and undertaking quality social and behavioural research and identifying/understanding where that research sits in the larger scheme of activities and learning relevant to human life and its myriad contexts (issues I first discussed at some length in Cooksey, 2001, with respect to human behaviour in general). We consider research to be a distinctly human activity, where one or more human beings adopt and enact the role of researcher and other human beings take on the role of participants or sources of information for learning or produce/have produced artefacts of interest for learning. Research is inherently a learning activity, one intended to discover, produce, and share knowledge and understanding and perhaps use that knowledge

R. W. Cooksey, *Unity from Diversity*, https://doi.org/10.1007/978-981-97-3462-7_1

for innovation, change and transformation, within one or more specific contexts. Since social and behavioural research involves people studying people or artefacts people have produced, we cannot ignore the potential impacts that our humanness and experiences may have on how research is conducted and what impacts it might have on the lives of not only those people in the researcher role, but the lives of any others involved in, with an interest in, or potentially affected by the research. People who adopt a researcher role cannot simply ignore or discard the potential impacts that their own background, humanness, and experiences may have on their research. Equally, they cannot ignore or discard the potential impacts that their research may have on others and the influences that others may have on them (whom we will call *stakeholders*). There will always be stakeholders relevant to any social and behavioural research journey: people/communities, organisations, cultures, societies who stand (or perceive/fear/ anticipate) to gain or lose something as a consequence of the research, either its conduct or its outcomes, and it is those stakeholders who must be convinced by what the researcher does and learns. Some stakeholder interests will be immediately engaged by the research journey prior to its commencement; others will have interests that emerge during the research journey or downstream once the research is completed, outcomes are being produced and shared, and impacts are being felt/realised. A critical subset of stakeholders may be in a position to exert power over the researcher (through control over access to resources, contexts, sources of data (we refer to this stakeholder subset as *gatekeepers*), or through political/social expectations that the research must yield something useful/innovative/ effective for informing policy, enhancing social good, facilitating technological or social improvement) with respect to what, how, where and why some aspect of the research must or must not be done.

To produce effective and convincing research requires a person to learn how to adopt and enact the role of researcher, how to contextualise the research they wish to carry out, how to anticipate and manage the potential forces that may attempt to influence the scope and shape of that research, how to effectively plan the research, and, most importantly, how to be flexible and adaptive in the face of unanticipated, unplanned, or emergent constraints, obstacles, or opportunities, all without suffering substantive damage to convincingness and credibility. Research emerges from a long and interconnected series of choices made by the researcher and other interested and relevant stakeholders, in and from specific contexts and under dynamically changing circumstances. We will find it useful throughout this book to metaphorically characterise any specific instance or project of social or behavioural research as a *journey* of learning and discovery undertaken to achieve one or more valued research outcomes in one or more specific contexts. It is a journey instigated and travelled by those adopting a researcher role and potentially involves or implicates many other people along the way, some of whom may be taking on specific roles in the research (e.g., stakeholder, participant, gatekeeper, supervisor, funder, innovation adopter). It is a journey over which those adopting the researcher role don't always have complete control in the all-too-frequent circumstances where stakeholders exert their power to either directly or indirectly influence how that journey should unfold. In short, what we are saying is that social and behavioural research is a contextualised human activity that is lived and experienced just as any other journey in life. The research journey, however, must be actively managed so that the researcher's goals can be achieved despite any shocks and surprises that may emerge along the way and, at a minimum, without adversely affecting the lives of others. The journey itself is all about making choices, some at critical junctures (i.e., *critical leverage points*), to source, gather, and process information needed to convincingly produce/build/apply meaningful knowledge.

Every research journey is embedded within a complex array of contexts that is generally not static but whose aspects are dynamically interconnected and continuously evolving and changing. Poth (2018, p. 13) was aware of such complexities when she wrote that "a complexity-sensitive approach guides mixed methods researchers' responses to emergent and dynamic realities under conditions of complexity while engaging in the research process". Furthermore, Poth argued that a complexity-sensitive approach can stimulate adaptation and innovation of research practices. The research journey is seldom a purely linear one; it will not necessarily travel in a direct series of steps from start to finish. Rather the journey tends to be nonlinear and messy, punctuated by bumps in the road such as dealing with unexpected constraints, obstacles and events, experiencing periods of doubt and uncertainty, having to manage relationships with colleagues and stakeholders and perhaps other researchers in a team, having to deal with existing as well as emergent complexities in the research landscape, responding to political manoeuvring, experiencing lapses in motivation and focus, encountering blind alleys and information voids, managing the research in conjunction with other events in one's life and career, experiencing excitement and enlightenment, suffering disappointment and discouragement, having to manage movements back and forth between contextualisation, data, and interpretation and so on. Sanbonmatsu et al. (2021), for example, recently argued that contextual complexity can have dramatic effects on theorising, sampling, measurement, method choices, research practices, analytical strategies, generalisations, and acceptance by the academic community in quantitative psychological research, creating shocks that can produce adverse impacts on research rigour and quality. Sometimes in research, it can be more a case of one step forward, two steps back. Contributing to knowledge and learning is no small achievement; it takes training, thinking, time, resources, perseverance, engagement, commitment, flexibility and adaptability, resilience, and occasionally a bit of luck. How the research journey unfolds is determined by choices that are made along the way. When we speak of pluralist systemic thinking, we are talking about ensuring that such choices, from the widest set of possibilities, are made in ways that are productive, viable, and feasible for achieving research goals under constraints while remaining contextually sensitive, responsive to stakeholders, and convincing for one or more audiences. We are talking about understanding that there may be no single right way to pursue the intended research objectives; multiple potential trajectories may be available to choose from. Finally, we are talking about realising that planning will only get research so far, the rest must emerge as the journey unfolds in real time and space. Part of this realisation is the knowledge that researchers may not enjoy total control or say over their journey's trajectories and outcomes; some choices may be made by other stakeholders in the research.

One important practice for a researcher to adopt is maintaining a journal as a record of their research journey. The research journal is multidimensional in focus and dynamic in nature. Through judicious and focused recording throughout their research journey, the researcher can capture and reflect upon all aspects of their decision making, contextualisation, learning, personal as well as professional observations, thoughts, and reflections with respect to the research journey. Reflexive learning can also be recorded where the researcher's beliefs, ideas, and conceptualisations are compared with those of others, including significant stakeholders; changes that stem from such reflexive considerations can also be recorded. The research journal provides a platform for recording the researcher's evolving pluralist systemic thinking and how such thinking gets translated into concrete research intentions, choices, actions, and outcomes. Via their journal, a researcher can retrace/recall where they came from, what they aspired to achieve, and how they got to where they ended up. It is an invaluable resource to draw upon when making choices to frame, scope, configure and implement research activities and producing outcomes for sharing/communicating what has been learned.

1.2 Systems of Human Existence

As human beings, we comprise, inhabit, influence, and are influenced by multiple mutually embedded and interconnected systems. A system involves a network of relationships, linkages, feedback loops, and feedforward influences that operate at some level of existence from the microscopic to the cosmic. If a system is open, learning, change, and transformation are possible through responsiveness and adaptation to *feedback* (where current actions yield information that stimulates learning and change) and effective use of *feedforward* (where learning provides information that guides subsequent actions). If a system is closed, it cannot learn, change, or transform because it is unaware of or unresponsive to feedback and cannot generate feedforward (i.e., cannot innovate). Any research we conduct must acknowledge, tacitly or explicitly, the multiple systems we inhabit as this will help us to properly contextualise what we learn. It is in our best interest, as researchers, to assume, think, and behave as if the systems we inhabit are open rather than closed so that the learning and impact potential of the research we conduct can be maximised and so we remain alive to the possibilities as well as risks that we can both influence and be influenced, by such systems. Of course, as with all assumptions, the open systems assumption may be erroneous in certain cases, creating additional challenges to overcome, especially if our goal is to create change.

With respect to human behaviour, we can identify at least 15 different systems that are relevant to human existence (see Table 1.1 which displays these systems along with exemplars of the aspects of human existence they may encompass). Social and behavioural researchers may choose to investigate behaviours, experiences, and/or relationships within any one or more of these systems. As we will later see, these systems can also potentially serve as contextual sources of influence on the researcher and their research journey. These systems interconnect and form synergies and interdependencies with each other in complex and often unpredictable ways. Since researchers are human beings, they are also embedded within these systems, and it takes focused effort and mental energy to stand at least somewhat apart from their embeddedness to carry out convincing research. We need to cultivate systemic thinking to understand the connections, relevance, and potential influences of these systems on a particular research journey. For example, researchers can view these systems and their potential relevance to their research through various contextual lenses: a personal experience/relevance lens, one or more temporal lenses (historical; present day; future-oriented), and/or lenses external to but available or provided for the researcher (e.g., community, cultural, or legal frameworks). We also need to cultivate an open mental attitude and learning stance (i.e., pluralist thinking) to embrace the diversity of potential ways that a research journey may be conceived and undertaken and make reasoned choices in light of that thinking. Out of such diversity can emerge a unity of focus on learning and impact for discovering/building new knowledge and insights about the systems we inhabit; enhancing or modifying existing knowledge about the systems we inhabit; and productively sharing and using knowledge to change or improve the systems we inhabit. Thus, the central massage of this book is: ***unity is knowledge achieved by knitting together pluralist thinking to open diverse potential pathways to building knowledge and systemic thinking to contextualise and choose amongst those pathways and the potential outcomes they could yield.***

1.3 Unpacking the Research Journey Metaphor

Figure 1.1 (adapted and revised from Cooksey & McDonald, 2019, p. xi) provides a conceptual visualisation of the research journey, which unfolds against a complex contextual background. There is an important embedded conceptual metaphor displayed in Fig. 1.1, called the *'Data Triangle'*, where the researcher navigates their way through the central activities associated with sourcing, gathering, and analysing data. It is important to note that there are three distinct versions of any research journey: (1) the *anticipated/imagined/planned journey* which is often conveyed in a research, grant, or consultancy proposal; (1) the *actual journey* experienced by the researcher(s), which unfolds in real time and is influenced by a range of events and forces, both anticipated and unanticipated; and (3) the *reflective/reflexive journey* where the researcher revisits key decisions, perspectives, and essential aspects of their experiences and learning to assemble and convey a convincing story in or for a specific research outcome. Each version reflects a different way of negotiating the various processes and stages associated with the research. No matter which version is being considered, the metaphorical representation of the journal reflects some important systemic thinking features. The journey takes place against an amorphous and complex contextual background that can never be fully unpacked or understood, nor its influences fully predicted. This creates a fundamental tension in that the research

Table 1.1 Identifiable systems of human existence (list should not be considered exhaustive)

System	Illustrative aspects
Material	External world of objects and things and their relationships and dynamics; space and time; environment; land, sea, and air; astronomical/astrophysical; physical, biological, chemical, geographical, geospatial, climatic, energy, agricultural, ecological conditions, relationships; events and resources
Technological	Machines and other hard and soft technological tools and innovations for supporting/mediating human thought, creativity, and action; facilitative platforms for supporting infrastructure, engineering, communication, transportation, commerce; support for and/or replacement of aspects of living or non-living systems; artificial intelligence and simulation; ergonomics (relationships between technology and humans)
Conceptual	Ideas; concepts; judgment and decision making; language(s); imagination; visualisation; innovation, creativity, and intuition; education, learning, training and teaching; knowledge building and sharing; theories, philosophies; logic and reasoning; mathematics; simulations and games; research paradigms and patterns of guiding assumptions
Personal	Personal physical, psychological, and emotional health and well-being; internal/mental world views; physical/cognitive/ emotional development as a person; personal beliefs, attitudes, and values; preferences, perceptions; motivations; cognitions; memories; indigeneity/ethnicity; personal, social and professional identity(ies); physical, mental, and social aptitudes, abilities, skills, educational background, and learning; strengths, limitations, weaknesses and capacities; aspirations; ancestry/ family/cultural history, and socioeconomic background; occupational/career/professional status and position; personal institutional and community affiliations
Biological	Physiological, chemical, neurological, biological and microbiological processes associated with human life and with human-non-human relationships/interactions; ecological relationships; evolution; genetics, genetic modifications, and mutations
Health	Human physical and mental health and well-being; medicine; doctors; nursing; counselling, therapeutic, and surgical interventions; management of quality of life; diseases, illnesses, diagnoses, and treatments; emergency support
Social	Interpersonal/group development, relationships, and networks; parents and family (immediate and extended); friends, peers, colleagues, and mentors; relationships w/ Elders and leaders; communication patterns and content
Community	Collectives of people, united by geographical, ethnic, cultural, religion, common experiences, common suffering, generation, sexuality/sexual and other types of identity, political preferences, teams, sports and leisure activities and/or other factors; local government; groups, gangs and factions having specific identities; Indigenous/First Nations peoples; demographics
Institutional/ organisational	Businesses and organisations; non-government organisations; national and multinational corporations; governments; councils; towns and cities; educational and training institutions; schools; universities; industries; hospitals; geospatial links and relationships

(continued)

Table 1.1 (continued)

System	Illustrative aspects
Political	Power dynamics and distribution; historical patterns and geospatial links; respect and reciprocity; domination and colonisation; governance and social policy; equity and inequity; marginalisation and oppression of different groups; social justice, discrimination, systemic advantage and disadvantage; international relations
Legal/ethical	Legal systems and laws; courts at multiple levels; regulatory systems; conception and administration of justice; policing and criminal behaviour; rules of evidence; burden of proof; societal/institutional codes of ethics and professional standards
Economic	Commerce and trade; monetary and financial resources, systems and dynamics; labour and production systems; marketing, purchasing, and consumer relationships; work and leisure; agribusiness; demographic movements
Artistic	Artistic expressions/outcomes of human creativity; storytelling; dance, performance, poetry, photography, writing, presenting, drawing, painting, carving, sculpting, graffiti, social media; multi-media
Cultural	Customs and traditions; language(s); communities; states, nations and countries; concepts of knowledge building and sharing; rituals; symbols; totems; ideologies; geospatial links; culturally-linked values and ethics; practices; hierarchies of knowledge and power; stories, songlines, talking circles, and other Indigenous/First Nations traditions and practices; practices for knowledge sharing
Spiritual	Cosmos; heaven; universe; religions; creation; inclusivity/exclusivity; holistic connectivity, relationships between living and non-living; relationships between beliefs, actions, and consequences; how life is/should be conducted; relationships with supreme beings; how life begins, flows, dies; enlightenment; relationships with other human systems

journey must be contextualised yet there will be limits on the extent to which it can be contextualised, leaving the researcher to continually grapple with the question 'have I contextualised my research sufficiently enough to be convincing?'.

The typical linearised/idealised journey is represented by the flow of solid green arrows between research stages from top to bottom of Fig. 1.1. However, this idealised linear path is seldom congruent with how the journey is experienced. When a researcher conceptualises their journey, they are thinking forward by anticipating and planning (especially prominent in *anticipated/imagined/planned* version of the journey), and as their *actual* journey unfolds, creating feedforward responses and adaptive choices to feedback they have received (symbolised by the dashed red arrow flowing downward). However, as their *actual* journey progresses, the researcher will also be thinking 'backward' reflectively and reflexively, reacting to feedback signalling that things may not be working as they were intended or that something they have done has created unintended consequences, and thereby revisiting/reconsidering/recontextualising choices that were made earlier on (symbolised by the dashed purple arrow flowing upward). As well, thinking 'backward' (in single

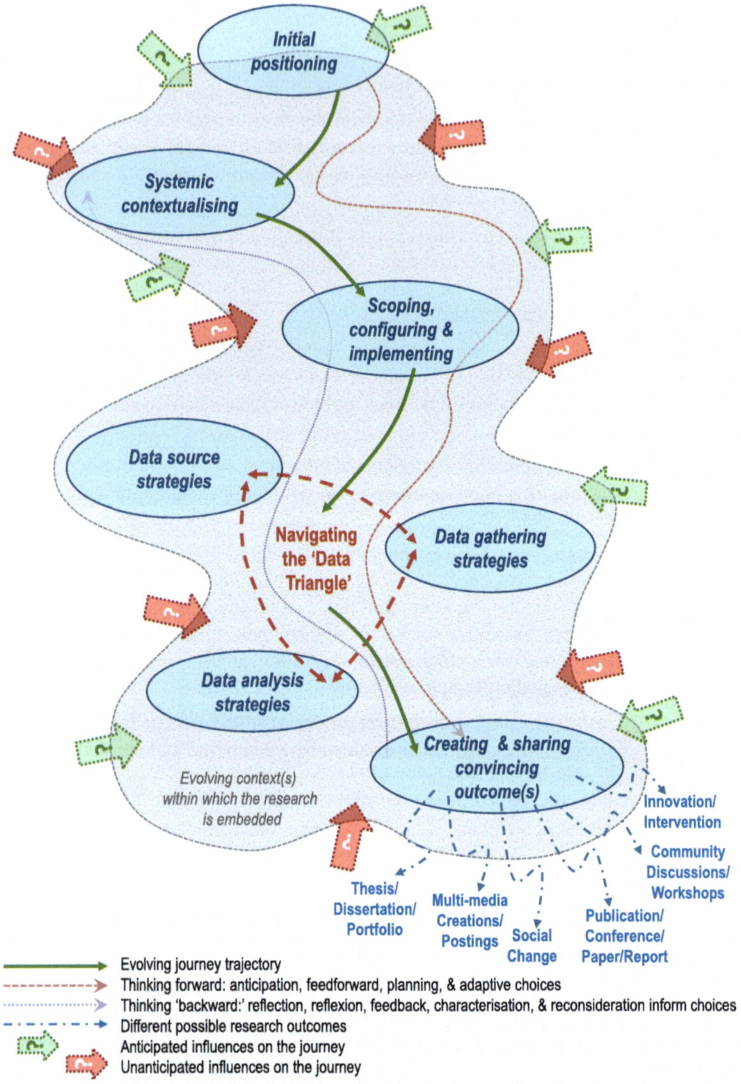

Fig. 1.1 Conceptually visualising the research journey

quotes to avoid suggesting actual time travel), reflection, and reflexion become especially prominent in the *reflective/reflexive* version of their journey as the researcher contextualises and makes sense of what they have and have not learned and where they stand with respect to that learning, in light of what they have done during the process of creating and sharing their research outcome(s). As the journey unfolds, some influential contextual forces may be anticipated by planning and steps put in place up front to mitigate or cope with that influence (the green arrows with the question mark). Other influential contextual forces may not be anticipated or planned for, instead emerging as circumstances change during their journey (the red arrows with the question mark). These forces often create the demand for a trade-off decision to be made on the fly by the researcher—an adaptive response.

Initial positioning: The research journey starts with the researcher working out their initial positioning for the research. Prior to undertaking a research journey, a researcher needs to think broadly and critically about their motivations, their skills, strengths and weaknesses, choices of and requirements for the specific research program to which they will commit, their needs (including developmental and skill enhancement needs), intentions and goals, the opportunities and constraints they can see ahead, the research context(s) they will need to work within, and the needs/expectations of other relevant stakeholders in the research. Many of these considerations are bound up in the researcher's identity as a researcher and part of initial positioning is seeing where and how their research identity fits with the journey they intend to take. This may also involve understanding where tensions might need to be dealt with between their researcher role and identity and other roles and identities they have in other facets of their life. The researcher will also need to cultivate and draw upon support systems, peer networks, resources (time, space, finance, human), and coping strategies to help them manage progress through their journey.

The research journey needs to be planned as far as possible. Researchers need to marshal resources, skills, and capacities to support their impending endeavour. Maintaining a detailed journal is an essential aspect of a social or behavioural research journey, beginning with thoughts on initial positioning at its genesis point. The journal is used to record key facets of the research journey, including ideas, reflections, contextual notes, encounters with and learning from prior research and knowledge building efforts as well as significant stakeholders and gatekeepers, and the choices and adaptations they made along the way. Throughout the journey, it is important to maintain a healthy respect for the fact that effective planning will only go so far. There will be things over which researchers have little or no control and things that they cannot plan for, or must wait to plan for, downstream. Flexibility and adaptability are therefore critical skills for researchers to develop. Changing or evolving circumstances as well as emergent feedback during the research journey may require revisiting, rethinking, making new choices, adapting, sacrificing, or augmenting some aspect of the research. In short, trade-off choices will likely be demanded at some stage, which may alter aspects or even the entirety of the journey's trajectory, and the researcher needs to be prepared for making some of those tough choices.

Systemic contextualising: To be convincing, research must be effectively and as fully contextualised as possible (i.e., understanding the who, what, where and why with respect to the research and where and how it fits with the research of others and with the needs of relevant stakeholder individuals and groups). This is where systemic thinking plays a critical role. The research needs to be positioned in terms of overall intent, purposes, goals, objectives, and the questions, propositions, and/or hypotheses to be addressed. Researchers must appropriately position not only themselves (using knowledge of their own context(s) and researcher identity to help make choices as to where they are coming from) but also potential and actual research 'participants' (considering their context(s) with respect to where they are coming from and what they bring to the research) as well as relevant stakeholders in the research and gatekeepers who may control access to something the researcher requires. Part of contextualising and positioning research involves making very clear decisions about the pattern(s) of assumptions to be adopted to guide the research (often, yet unproductively, referred to as 'paradigm' choices). To be effective and to have impact and contribute theoretically and/or practically, research must be convincing to relevant stakeholders and interested parties/audiences. This means that all aspects of the research should reflect close attention to research quality and stakeholder needs. Pluralist thinking plays an important role here, keeping the field of potential choices as diverse as possible and harnessing contextual knowledge to help narrow and focus those choices. All these considerations come together to help inform the researcher's choice of frame—a holistic way of describing researcher intentions.

It is important to recognise that research quality is itself a contextual consideration and can be addressed at two levels: criteria specifically relevant to an adopted pattern of guiding assumptions and meta-criteria. Meta-criteria are especially useful in helping researchers engage pluralist systemic thinking to plan and implement their research as well as to evaluate it. Contextualisation and positioning serve to shed light on fleshing out the amorphous light-grey shaded area in Fig. 1.1 which then helps shape subsequent choices made throughout the journey. That is, early contextual considerations may influence, constrain, or open up downstream choices and, equally, downstream (i.e., emergent) contextual considerations may cause the researcher to reflect upon and perhaps recast, qualify, or better understand earlier choices—clearly signalling the nonlinearity of a research journey.

Scoping, configuring and implementing: The researcher needs to scope their research journey by progressively refining and focusing the choices they make in light of research contextualisation, choice of pattern(s) of guiding assumptions, and framing. At this juncture, the researcher develops a clear picture of what they want to achieve and how they want to achieve it. They will have been mentally navigating through the landscape of methodological choices for the 'Data Triangle' while thinking forward and anticipating/planning their journey. Pluralist systemic thinking is critical to engage in this process especially as the landscape continually evolves as a consequence of needs-based researcher, technological, and social/cultural creativity and innovation. To move forward, the researcher needs to configure their data sourcing and data gathering activities so that they can effectively achieve their intended goals and address their research questions/propositions/ hypotheses. Configuring research involves deliberations on types of data sources, forms of data (quantitative/qualitative/hybrid) to be gathered or generated, and specific data gathering strategies to be employed. Each specific combination chosen comprises a 'Method Unit' (MU). MUs form the basic building blocks for any research configuration. Each MU is accompanied by its own set of procedural considerations that must be planned, fleshed out, and resourced before implementation. Implementation occurs during the *actual* version of the research journey.

Data source strategies: This forms one vertex of the 'Data Triangle' in Fig. 1.1. To focus data gathering activities, the researcher will need to use sampling and/or selection processes to facilitate accessing or connecting with their data sources, in a manner consistent with their adopted pattern(s) of guiding assumptions and intended research goals. Data sources may be human or non-human (e.g., texts/documents, YouTube segments, movies, policies, paintings/drawings, cultural artefacts, secondary databases, animals) in nature. Data source choices may emerge from anonymous probabilistic (e.g., random or systematic) sampling using numerical identifiers or be specifically identified on practical, intentional, and/or contextualised grounds. If the data sources/participants to be chosen are human, then their ethical right to give informed consent to participate overrides the researcher's desire for them to participate, which means that participation is their choice, not the researcher's. This then becomes one of the aspects of the research journey that the researcher cannot completely control. If the data sources to be chosen are non-human, then the researcher must assemble any information they can to ensure they understand the context, gatekeeper(s), nature, original purpose, and quality of whatever data source they access (nowadays, this sort of information tends to be characterised as *meta-data*).

Data gathering strategies: This forms the second vertex of the 'Data Triangle'. Data gathering/data creation strategies need to be implemented in a manner consistent with the researcher's adopted pattern(s) of guiding assumptions, paying close attention to relevant research quality criteria. The researcher wants any data gathered or created to be directly relevant to their research goals. Many data gathering strategies can work synergistically together to enrich the learning they can achieve and, where appropriate, the participant involvement they seek. For social and behavioural research, data gathering strategies can be categorised as those that facilitate connecting with people, those that facilitate exploring the handiworks produced by (or valued by) people, and those that help the researcher to, in many cases, structure the experiences of people. Some are oriented toward gathering quantitative data and others toward qualitative data (hybrid or merged data forms are also possible to gather). Mixed method practitioners argue that both forms of data are needed to convey a convincing story, but this thinking is too limiting. Gathering the same form of data using different data gathering strategies can be equally convincing. Here, we reinforce the value of pluralist thinking rather than just mixed methods thinking.

Data analysis strategies: This forms the third vertex of the 'Data Triangle'. Data that the researcher has gathered (or created) need to be analysed in a manner consistent with their adopted pattern(s) of guiding assumptions as well as the MU configuration they have implemented. The researcher's goal is to build meaning from the data so that they can convey the stories they reveal relevant to their research goals and questions/propositions/hypotheses. Descriptions, social constructions, patterns, connections, relationships, dynamics, and changes over time may emerge from appropriate analyses; models may be tested, and theories constructed. While analyses of quantitative, qualitative, and hybrid/merged data typically draw upon very different procedures, the end goal is the same, namely, to discover what the data are telling the researcher and to create displays and narratives that convey those stories to others.

If the stories that arise from quantitative, qualitative, and/or hybrid/merged data can be integrated to highlight points of convergence, divergence, qualification, or deeper understanding, this can add to the convincingness of the research.

It is important to note that (1) relevant characteristics of contextualisation and positioning, including relevant ethical and cultural considerations, underpin every vertex of the 'Data Triangle' and (2) depending upon the research frame, chosen pattern(s) of guiding assumptions, and MU configuration, the researcher may have to navigate/cycle through all vertices of data triangle many times before they can build convincing meaning from their data.

Creating and sharing convincing outcome(s): Researchers can produce different research outcomes to achieve different kinds/levels of impact (see the blue dot-dashed arrows at the bottom of Fig. 1.1). A research outcome typically involves creating an integrated *reflective/reflexive* story about the research journey that is convincing when shared with one or more relevant audiences, with respect to both the quality of the work that has been done and the value of what has learned. For research that has an applied or transformational intention, a research outcome may also be a concrete series of, or recommendations for, actions or an innovative product or process. Research outcomes may be prospective (e.g., an *anticipatory* journey embedded within a thesis, dissertation, or portfolio proposal, a research grant proposal, consultancy proposal, planned innovation, or policy intervention) or retrospective in focus (e.g., a *reflective/reflexive* story embedded within a thesis or dissertation, a portfolio, a journal article or conference paper, technical report, presentation, community discussion, multi-media event, realised innovation, policy or process change, creative outcome). Meta-criteria for research quality should prove especially useful in helping the researcher to produce a convincing outcome. Desired impact will differ depending upon the research outcome being produced. For example, for a proposal, the immediate impact is that the researcher is given the go-ahead, for example, to conduct their research or undertake a consultancy. For a thesis, dissertation, or portfolio, the immediate impact is that the candidate convinces examiners or their supervising committee that they have created important value for learning and deserve the award of the degree they are enrolled in. For an applied or transformational research outcome, the immediate impact is that the researcher convinces relevant audiences/stakeholders that they have created important value for learning as well as demonstrated/facilitated change or impact in some social, professional, community, or industry practice or process. For a multi-media event, impact may be visible in terms of number of reads/downloads as well as, perhaps, tangible influence on some social, community, or policy process. Some research outcomes may require a more personal/hands-on and contextualised approach, such as offering a community workshop or meeting for sharing research learning, and the impact could be reflected in attendance numbers, interaction quality and quantity, and perhaps invitations for extended work. This can be an important vehicle for achieving reciprocity in research, by returning value to the community in which the research was undertaken (especially important in Indigenous/First Nations research). For a conference paper, the immediate impact would be acceptance for a presentation or discussion at the conference and the downstream impact might be enhancement or enrichment of the researcher's professional network because of the value for learning they have generated for future potential research and collaboration. For a journal article, the immediate impact is acceptance for publication and the downstream impact is creating value for learning for other researchers, as they cite and build upon the research. For an innovation, the immediate impact may be adoption and the downstream impact may be enhanced community or organisational benefit. Irrespective of the particular research outcome being created and shared, both *thinking 'backward'* (to contextualise, interpret, and quality what has been learned as well as what limitations and opportunities accompany what has been learned (in short, telling the story effectively to an intended audience)) and *thinking forward* (to anticipate what the implications are for further research, who might benefit from the research, or what might be done in the future) are essential.

1.3.1 Recording the Research Journey

As we mentioned earlier, a valuable social and behavioural research practice is to maintain a research journal that records the entire research journey (see, for example, Banks-Wallace, 2008; Clarke, 2009; Glaze, 2002; Hayman et al., 2012; Nadin & Cassell, 2006; Meyer & Willis, 2019; Ortlipp, 2008). This is especially important if one engages in pluralist systemic thinking as the complexities of such thinking make it difficult to remember the myriad paths it can take, particularly across a span of several years. A research journal serves as a record of decision making throughout the entire journey (Ortlipp, 2008, discusses the importance of this particular use for a research journal in qualitative research; Carcary, 2009, p. 11, refers to this as maintaining a "research audit trail"). A well-maintained research journal will provide the researcher with a rich and unique source of textual and visual data that are analysable, which means that the research journal can serve as a qualitative data source in its own right (Glaze, 2002; Lamb, 2013). As well, a research journal provides an invaluable memory and reflection/reflexivity aid for producing specific research outcomes. If the researcher also uses their journal to record data, such as participant observations, interview details and contextual notes, important conversations (e.g., with mentors, colleagues, stakeholders, and gatekeepers), it essentially takes on the role of field notes.

A research journal can be characterised as a type of autoethnographic reflective/reflexive research activity providing a critical historical account of the research journey (see Hughes & Pennington, 2016). It records an evolving journey that is lived by the researcher, shaped by stakeholders and gatekeepers and other forces, experienced by research participants, and ultimately shared with others. Amongst other things, a research journal can be a vehicle for:

- recording the story of the journey;
- exploring the influence of various contextual forces, pressures, and expectations;
- setting out relevant contextually-situated positionings to be adopted or understood;

- tracing the evolution of research goals, objectives, questions, propositions, and/or hypotheses to be addressed;
- arguing for the pattern(s) of guiding assumptions and research frame to be adopted and why;
- recording the choices/decisions made along the way and why;
- detailing the obstacles and constraints encountered and the opportunities that were pursued as well as spelling out how they were dealt with;
- identifying and contextualising key stakeholders, noting how their expectations and potential influences were perceived and managed;
- identifying and contextualising key gatekeepers along with strategies for gaining access to desired data sources or other research-relevant information, including a record of promises made and assurances given in exchange for such access;
- recording the trade-off choices (and accompanying reasoning) made to render the research feasible and realistic to achieve, deal with emergent constraints, or take advantage of an emergent opportunity;
- critically reflecting on what has been learned throughout the journey, including from prior research and knowledge building efforts; and
- recording, in a separately identifiable way, personal observations, thoughts and reflections about the research journey as it unfolds (e.g., emotions felt, reminders to check things and follow-up, preliminary hunches or interpretations, relationships developed, thoughts about how the research is progressing and what might need to be changed, reactions to what is being learned, other knowledges experienced/encountered as the journey unfolded, reflexive considerations of researcher perspective with other perspectives).

Apart from providing a permanent record of the research journey, the process of writing and maintaining a research journal helps the researcher to implement reflexive practices (Le Gallais, 2008):

- reflect on and more deeply critically consider the evolution of their research using notations that clearly record contextual as well as prior research observations as well as any thoughts that help them to situate their research more clearly;
- reflect on patterns or critical events associated with their life more generally (day-to-day as well as longer-term) and on how they are influencing their research thinking and activities;
- critically analyse and reflect on research outcomes (e.g., published and unpublished research) produced by others that help the researcher to build the platform and context(s) for their own research;
- clarify the researcher's own role(s), expectations, and other aspects of their researcher identity, which becomes very important when positioning, planning and configuring the research, choosing data gathering strategies, and identifying data sources;
- clarify the roles and expectations of participants and relevant details about other data sources in the context of the research, which also becomes important when working through the processes of gaining ethics approval (where required) for the

research and negotiating access to research settings, research participants and, possibly, important contextual documents;

- monitor where the researcher is in their own research cycle(s), including milestones, accomplishments, inspirations, and dialogues with colleagues, peers, mentors, and other sources of advice and information, all of which are things that can help maintain motivation throughout the research journey and, if the journal is properly chronologically maintained, the researcher can review the status of their research at any point in time;
- realise that there are potentially many different paths to achieving research goals, and that negative outcomes aren't necessarily fatal – something that will be reflected in the on-going record of the decisions and research trade-offs the researcher makes (including the limitations imposed or created by their choices), blind alleys identified, bumps in the road, unforeseen events, obstacles, unexpected outcomes from analyses, emerging opportunities, and changes of perspective or direction pursued as a result of what the researcher is learning;
- focus attention on making careful and defensible choices with respect to the constraints the researcher confronts, intended starting and ending points for the research, pattern(s) of guiding assumptions and frame(s) the researcher has adopted and why, strategies and approaches they decided on or rejected, form(s) of data needed to address the research questions, and planned strategies for approaching the analyses of data – all of which can help the researcher to continually reflect on and perhaps re-appraise the choices they have made along the way and the reasons behind them;
- reflect on and critically consider emerging analysis results, interpretations and conclusions as a path toward achieving a convincing story, i.e., what the researcher is learning from the data;
- reflect on specific observations and reflections amassed during data gathering activities to help contextualise each data gathering event;
- access and maintain a ready source of qualitative data to reflect on and analyse in support of the research, i.e., the contents of the journal can be analysed as easily as participant responses to survey questions or participant statements obtained through an interview; and
- construct a more convincing anticipatory story of the research journey (i.e., a research proposal) and/or a more convincing and holistic retrospective account of the research journey and the learning it yields in whatever research outcome the researcher is producing (Engin, 2011, argued that this is an essential part of the scaffolding process for building up the researcher's own knowledge).

Daily recording in a research journal is a useful habit to cultivate but requires discipline to maintain the momentum. By maintaining this discipline, the researcher can minimise the influence of memory on their entries. For some researchers, daily journalling may not suit their personal style or circumstances, preferring instead to record entries on a more episodic basis (e.g., weekly or when specific events occur). Recording journal entries should be an activity done with a minimum of distraction in order to maximise the likelihood and accuracy of recall, the goal is to minimise interference with memory processes. Journal recording may work best in concentrated bursts because, once the researcher starts recalling the day's events relevant to their research and recording things about those events, other associations, recollections, and ideas will begin to emerge and they want to be able to capture these as well (a useful but more challenging aspect of pluralist systemic thinking as well as the rich associative nature of the human memory system). Even if the researcher is already part of the way through their research journey, it is not too late to start building up a research journal. Of course, this will be a retrospective task up until the time journalling commences and such retrospective journalling will be subject to memory limitations and potential distortions. This retrospective process can also work well if the researcher has gaps in their journal where they either forgot to make some entries or did not have access to their journal to make the entries.

Structuring the Journal

A research journal should have a consistent structural format, either using a formatted paper template or an electronic format supported by a specific journalling software program or app. For a journal to support critical reflection, reflexivity, and researcher learning, there should be, at minimum, two main sections: a section for recording narrative notes and observations (the 'objective'/factual story/data about what happened and what was done, e.g., what happened and in what context?, what was said or done and in what context?, to whom?, by whom? where?, when?) and a section for recording personal notes, ideas, thoughts, reflections, reflexive thinking (the subjective' story about how the researcher thought and felt about what happened and what was done and what has been learned as a consequence, e.g., why?, what did it mean?, so what? how did it make me feel? how can I explain this? what did I learn? what do I do now?). The distinction between these two sections is not too dissimilar from what Ballenger (2004, pp. 146–154) termed the 'double-entry journal' which helps the researcher to keep the record of what happened disentangled from the record of feelings, thoughts, and reactions to what happened (which is also where deductive, inductive, and abductive reasoning are utilised and displayed). In this way, the tensions and connections between the two parallel records are always available to be unpacked and explored.

The research journal is an expanding and ever-evolving record of the research journey. Used correctly and with discipline, it can map out the entire journey, including all its bumps, fits and starts. Probably the best mindset to adopt with respect to using what is contained in the research journal is to consider it as a distinct source of qualitative data. These 'data' may prove useful not only for helping the researcher to flesh out and enrich their research story, but also to qualify, amplify or modify their learning from other data sources (see Lamb, 2013, for a personal account of how keeping a research journal assisted in a number of these aspects of their research). Research journal entries that focus on prior research the researcher has read or encountered and theoretical frameworks, methodological approaches, constructs, ideas and gaps the researcher has identified can inform not only their research planning and configuration but also aid in producing the 'prior research/knowledge building' sub-stories for various research outcomes. They can help the researcher to clarify why they looked at certain material and how they responded to that material as well as tracing how prior research learning has influenced the shape and conduct of their research. Research journal entries that focus on contextualisations, pattern(s) of guiding assumptions, frame choices, research configuration, data source selection, data gathering strategies, strategies for gaining access and ethical clearance, and analytical strategies can inform the researcher's writing of 'navigating the 'Data Triangle' sub-stories in a research outcome and may help them to openly defend the choices they have made to the reader. This will add to the convincingness of the research. Additionally, such entries may help the researcher to appropriately qualify or contextualise their findings in 'learning from/meaning of data' sub-stories for research outcomes. Research journal entries that focus on any aspects of the research context itself as well as the role(s) that the researcher and research participants played in that context can help the researcher to contextualise their entire journey. This would certainly help the researcher to develop any conceptual framework for their study, defend any methodological tailoring they had to do, and qualify, amplify, or further characterise their results and their implications. If the research methodology involves detailed interactions with research participants and the researcher has used their journal to document details of those interactions, then those journal entries can help the researcher to tell a much clearer and more convincing story about what they have learned from such interactions.

In a more general sense, maintaining a research journal is an exercise in self-reflection. This process is termed *reflexivity* or 'reflection on their own data-making role' (Richards, 2020, p. 69) and is an analytical strategy advocated by many researchers who deal with qualitative data, including those employing 'grounded theory' (see Carcary, 2009: Charmaz, 2014, Chap. 2; Flick, 2023, Chap. 25; Nadin & Cassell, 2006; Ortlipp, 2008; Richards, 2020, Chap. 2). This is where the power of the research journal really lies. The researcher decides what is important to record and they record how they react to or think forward from what they have recorded. Thus, the journal evolves dynamically. For example, the researcher might identify and record areas of weakness in their skill set and record plans to undertake activities to strengthen those areas, thus, dynamically developing aspects of their researcher identity. Using the journal, the researcher may become aware of and record their thinking and logic about the pattern(s) of guiding assumptions they might want to adopt. The researcher might record their thoughts on things they have observed or heard and reflect on how they may or may not have influenced them. Throughout their journey, the researcher can record the decisions and trade-off choices they have made and reflect on their implications. The researcher can also record and reflect on limitations of their research and their impact on what they can say has been learned (or not learned) and how future research might address such limitations. Such reflexive thinking and self-reflection will be invaluable when producing a specific research outcome.

1.4 Researcher Identity

In addition to other identities which researchers may take on in their lives, they also develop an identity specific to the positioning, conceptualisation, and conduct of research; an identity that begins to take shape during research training years and continues to develop and evolve as research experiences accrue over time and circumstances (see, for example, discussions in Boyd & Smith, 2016; Castelló et al., 2021; Giampapa, 2011; Hamilton et al., 2012; Harvey, 2013; Mertkan & Bayrakli, 2018; Ponterotto & Grieger, 1999; Salmons, 2020; Uluğ et al., 2021). The researcher identity reflects who the person is as a researcher, how they think about research, why they do research, how they relate to contexts in which/for which/with which they conduct research, how they relate to others who do research, how they learn from research, as well as how they learn from their own research experiences. The researcher needs to understand their place within/relationship with the various systems of human existence when it comes to their role as a researcher. This understanding informs and underpins their researcher identity, and that researcher identity will influence the choices they will make when it comes to a specific research journey. Thus, a person's identity, as a researcher (distinct from but not disconnected from other roles they play out or take up in their lives), can constrain and/or liberate their thinking about and shaping of their research journeys and shape their learning about how to develop, extend, and evolve that identity so that the research they carry out is more likely to be convincing and impactful (overcoming their constraints by amplifying and expanding strengths).

Figure 1.2 displays a concept map of the issues and considerations associated with the development and maintenance of a researcher's identity. Researcher identity is strongly influenced by three central contextual forces: the researcher's own contextualised background (i.e., where they are coming from), their research-oriented capabilities and relationships (i.e., what they know and can/cannot do), and their capacities for reflexivity, reflectivity, self-awareness/learning, and adaptation (i.e., how they learn and change). These central forces can be further unpacked to reveal a myriad of complexly interconnecting influences. Researcher identity is fluid rather than static, which means it evolves and changes in response to/anticipation of changing circumstances, new information, and new relationships (Thomson & Gunter, 2011). Most importantly, when researchers first begin to develop their researcher identity, such development is strongly influenced by key relationships (such as with mentors/supervisors) as well as contextual circumstances surrounding the conceptualisation and conduct of research and acquisition of research skills and experiences. Through these interactions and early experiences, most researchers encounter the pressures and expectations associated with disciplinary norms (i.e., how research is typically conceived and conducted within an academic discipline) unless they are part of/involved in multidisciplinary training. The extent to which a researcher internalises disciplinary norms (symbolised by the shaded yellow enclosed area in Fig. 1.2 encompassing aspects where such norms can manifest their major impacts) will constrain their capacities for pluralist systemic thinking. It will require cognitive effort, reflexivity (critical introspection about the impact of disciplinary norms and the value of breaking their grip on researcher thinking and practices), concrete experiences, and an expanded circle of professional relationships for the researcher to break the shackles imposed by disciplinary norms. One should not underestimate the power of disciplinary norms to shape one's research identity nor the amount of effort it will require to question and move beyond them. This task is made even more difficult when we consider that various contextual forces, external to the researcher (such as supervisory and examination practices in postgraduate training programs, editorial practices for academic journals, guidelines for funding research, promotion practices/career expectations in universities, necessities for managing the identity tensions created by conducting insider vs outsider research (e.g., Aiello & Nero, 2019; Hayfield & Huxley, 2015; Kerstetter, 2012), demands for more research impact, critical engagement (e.g., Fear et al., 2006), participatory/community input (e.g., Muhammad et al., 2015), and meaningfulness/ usefulness), create a complex amalgam of pressures that pulls the researcher's identity in different perhaps contradictory directions.

Part of researcher identity involves achieving/enacting a balance in researcher emphasis on three different modes of logical reasoning: deduction, induction, and abduction (see Fig. 1.3; see also Åsvoll, 2014; Kennedy & Thornberg, 2018; Reichertz, 2004; Yu, 1994):

- *deductive reasoning* builds locally specific logical statements/propositions/ hypotheses from more general premises or a theoretical/conceptual framework (moving from general to specific);
- *inductive reasoning* builds logical extensions, inferences, or generalisations by pulling together learning from specific instances or cases to inform a more holistic perspective or set of inferences (moving from specific to general); and
- *abductive reasoning* builds up new and plausible explanatory statements/ propositions/ hypotheses/theories from observed patterns or outcomes, some of which may be unexpected or anomalous (moving from specific to new).

Yu (1994, p. 19), building upon the philosophy of science of Charles Pierce, characterised these three modes most succinctly: "abduction creates, deduction explicates, and induction verifies". It is through deductive reasoning that researchers derive hypotheses or testable explanatory statements from theoretical propositions and networks; it is through inductive reasoning that researchers build generalising statements or theoretical propositions from observations, and it is through abductive reasoning that researchers construct plausible accounts for observed patterns (including, importantly, unexpected, or surprising patterns or observations). Over-reliance on only one or two modes of logical reasoning risks leaving any researcher's account of outcomes from their research journey incomplete. This implies that the researcher's identity must evolve a workable balance between the three modes, a balance that will likely shift for specific research journeys. There is no set ordering to the use of these modes of reasoning, different patterns of guiding assumptions (to be discussed more fully in Chap. 4) invoke different emphases and sequencing

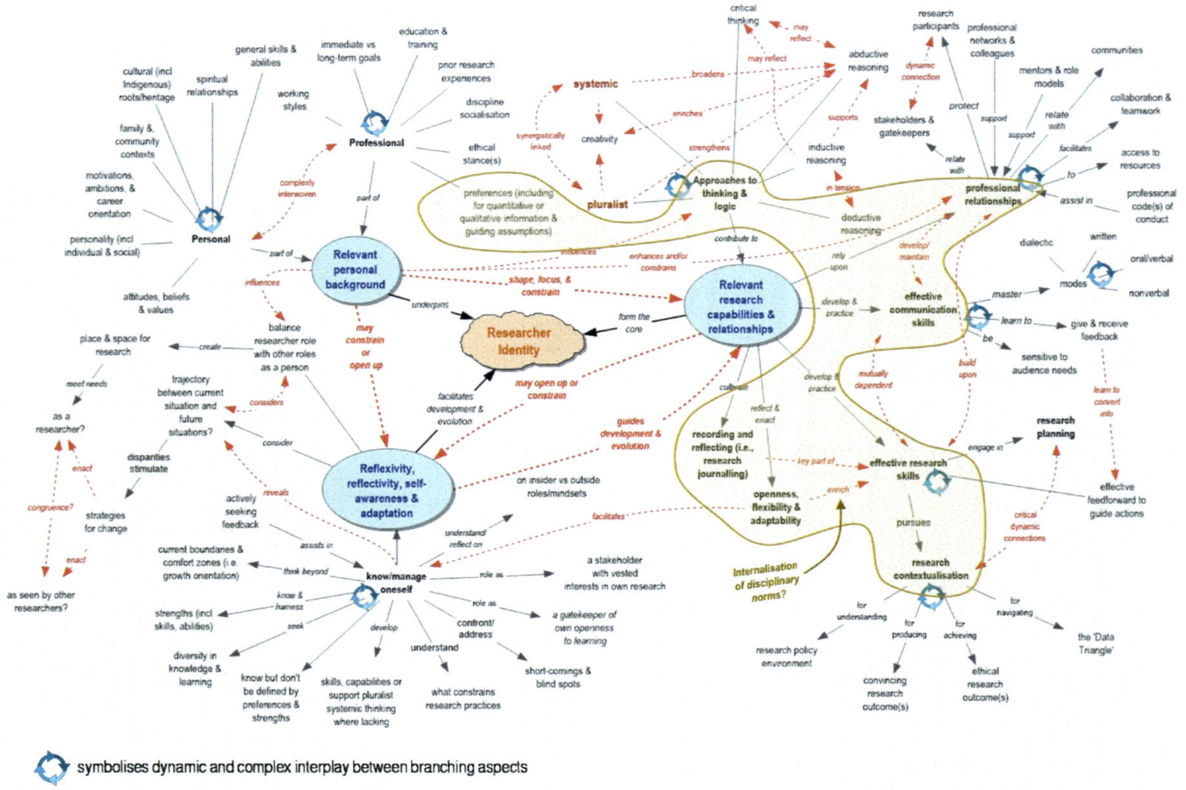

Fig. 1.2 Concept map of issues/considerations associated with researcher identity (yellow area encloses aspects where discipline norms can have strong impact)

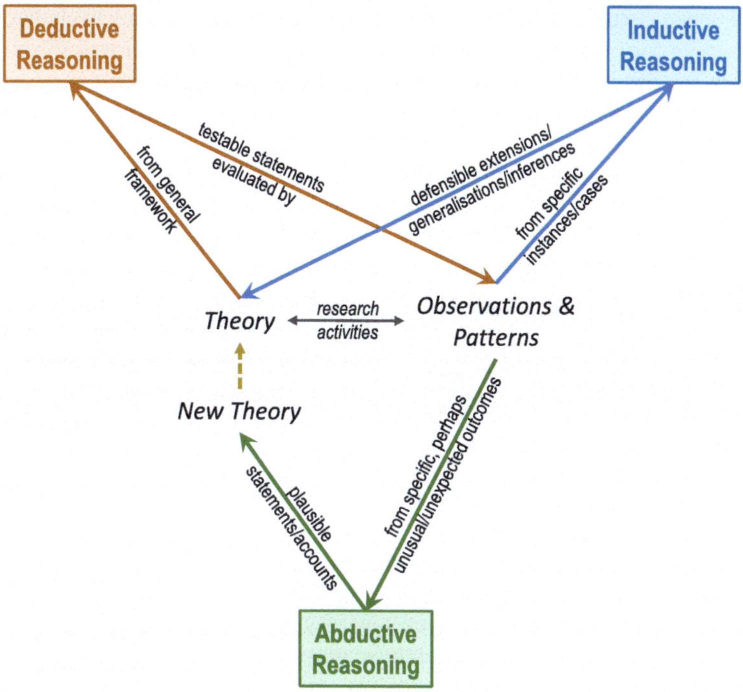

Fig. 1.3 Modes of logical reasoning in research

for the three modes. To these three modes of logical reasoning, we add two meta-levels of logical reasoning: *pluralist* (making specific choices drawing upon the least constrained set of possibilities) and *systemic* (thinking complexly about choices in anticipation of/response to relevant and interconnected contextual considerations). We argue that these two meta-levels of logical reasoning should be employed synergistically in order to appropriate shape the research journey. The researcher must therefore work to develop and incorporate these two meta-reasoning modes as a central part of their research identity. Deductive, inductive, and abductive reasoning help the researcher to make choices within a research journey to deal with the interplay between observations and patterns (i.e., data) and theory (however construed). Pluralist and systemic reasoning play their roles in helping the researcher to make choices about how the research journey itself should unfold.

In order for a researcher's identity to effectively evolve, the researcher must cultivate capacities for reflexive and reflective learning as pathways for self-questioning and enhancing self-awareness, self-learning, flexibility, and adaptation. Reflective learning focuses on learning from experiences and is oriented toward understanding and accounting for what has happened in the past, looking at learning from an outsider-perspective (see, e.g., Malthouse et al., 2014; Marshall, 2019; Niblock, 2007). Reflexive learning is ongoing, dynamic, critical, and contextually sensitive; it is a more holistic/systemic way of moving from what has happened before to what should happen next. Reflexivity is explicitly action-oriented, seeking to change what is done going forward based on what was done and learned under specific circumstances in the past. Reflexivity embodies an insider-perspective that considers the role/impact of the researcher in all that has happened (i.e., researchers "interrogate their own [research] practice over time or at a given moment", Niblock, 2007, p. 23) and what changes to self/identity might be needed to make current or new research practices work differently/more effectively in the future (see, e.g., Alvesson & Sandberg, 2021, Chap. 3; Bleakley, 1999; Harvey, 2013; Le Gallais, 2008). By effectively harnessing reflective and reflexive practices, the researcher can enhance their awareness of where they are situated with respect to the research they conduct and alter aspects of their identity in the face of such learning. As well, flexibility and adaptability can be enhanced as self-awareness and learning are converted into feedforward and action. Unless the researcher is aware of how their researcher identity impacts on/is impacted by the research they carry out, learning to adapt and change for the better will be very difficult to accomplish. When disciplinary norms are strongly internalised, such learning is constrained and typically refiltered and reinterpreted in light of those norms rather than facilitating an authentic response to feedback, anomalous outcomes, and signals about researcher impacts, both beneficial and adverse.

It is important to recognise that whenever a researcher's identity intersects the possibilities for a new research journey, tensions may be created in the stances that the researcher is willing to adopt and those tensions may not just represent binary 'this or that' considerations but may invoke a continuum of considerations. We reflect these tensions in a series of questions a researcher may ask of themselves:

- **Worldview stance**: what is the balance I want to achieve between objectivist and subjectivist orientations with respect to the world, i.e., do I value or privilege my own perspective as a person/researcher more or less than I value or privilege the perspectives of others?
- **Role**: is my role one of being an insider, an outsider, or somewhere in between with respect to the context(s) being researched?
- **Motivational stance**: what are my main ambitions, motivations, and needs for undertaking this research project and how do these mesh with the ambitions, motivations, and needs of others?
- **Comfort zone**: do I work with my current preferences and skill sets and maintain my personal boundaries or do I break down/go beyond/extend my personal boundaries and seek new relationships, challenges, skills, self-improvements, and opportunities?
- **Knowledge building stance**: Do I want to do this research to contribute to/advance learning in my discipline (e.g., testing/evaluating/learning/discovering; that is, knowledge building for its own sake) and/or do I aspire to build knowledge that has impact/meaning for others or to innovate for impact beyond the boundaries of my research (e.g., critiquing the status quo, finding better ways forward, innovating/applying/changing things; i.e., knowledge building for the sake of others)?
- **Collaborative stance**: Do I want to work on my own, with a research team, or in partnership with communities and lay researchers (as in participatory research)?

- **Temporal stance**: Do I need to consider past/historical, present, and/or future perspectives in positioning my research?
- **Psycho-socio-techno-contextual stance**: What is most relevant for me to draw upon or acknowledge when I consider how individual/personal, group, community, institutional/ organisational, ethical, legal, societal, cultural, spiritual, technological, and environmental perspectives may be relevant to my research?

Resolving these tensions (and the above list is certainly not exhaustive) essentially involves the researcher in serious introspection about where they are situated with respect to the research journey they are about to undertake versus where significant others are situated. Such introspection is a major step toward enabling pluralist systemic thinking.

1.5 Pluralist Systemic Thinking

The previous discussion brings us to the point where we must discuss pluralist systemic thinking in some detail and explore its implications for conceptualising, implementing, and communicating learning and impact associated with a specific instance of social and behavioural research. Pluralist systemic thinking involves two synergistically interconnected processes: pluralist thinking and systemic thinking. Pluralist thinking is breadth-oriented, considering the widest/most diverse range of available choices, irrespective of ideological or philosophical positioning. Systemic thinking is depth- as well as breadth-oriented, probing the nature and connections between various systems of human existence as they are or become relevant to the researcher's research journey and the choices they must make to successfully and convincingly complete that journey.

1.5.1 Systemic Thinking Highlights Potential Sources of Influence on Research

Systemic thinking demands that we *complexify before we simplify* and involves the researcher attempting to envisage, anticipate, and develop awareness of the human systems that their intended research will relate to, tap into, draw upon, be influenced by, or attempt to influence. Such systems build linkages between the research contexts themselves (including stakeholders and gatekeepers interested in or implicated by the research), the personal and professional lives of the researcher, the personal and professional lives of potential and actual participants as well as the systems within which other types of data sources (typically non-human) reside, and the systems in which potential and actual research users, sponsors, and communities (which may also include stakeholders and gatekeepers) reside, all *as they relate to the evolving context of the research.* Furthermore, all these more localised considerations are embedded within larger contextual systems relevant to human activities in space and time (Indigenous researchers refer to this as 'holistic' thinking, see, for example, Kovach, 2021, and it is clear that such thinking is both more widely applicable and valuable). What we attempt to do with systemic thinking is to metaphorically see the forest, the trees, and their interconnections. Figure 1.4 provides a visual representation of the sorts of influences and relevance considerations systemic thinking may invoke. The researcher needs to think carefully about which influences from which sources will be most important to address. The diagram shows that a researcher's specific journey emerges from the confluence of systemic considerations (represented by the simultaneous overlapping of the four localised research-focused systems. Systemic thinking also focuses on the nature and development of dynamic relationships between various relevant systems and people within those systems as they evolve throughout the research process. The cyclical arrows in the 4-way intersection region of systems of Fig. 1.4 signal that the systemic thinking, research shaping, scoping, and refinement process will be ongoing (learning from feedback being transformed into feedforward changes) for as long as the research journey lasts. Systemic thinking acknowledges that relevant systems vary in terms of the level of analysis they imply (e.g., cosmic, planetary, cultural, societal, community, technological, biological), may be hierarchically related (one system at least partially embedded within another (e.g., a community system embedded within a cultural system and a legal system) and that their influences on a specific research journey may be direct (e.g., a supervisor influencing the aspects of their postgraduate student's research journey) or indirect (e.g., a government agency influencing aspects of the journey of a researcher through an intermediate agent such as a university ethics committee or funding body).

For a researcher to engage in systemic thinking means devoting time, resources, and energy to figuring out which systems and which influences potentially emanating from such systems are most relevant to their evolving research (its character and its quality) and developing/implementing strategies for addressing those influences. Furthermore, the researcher must recognise that they will not be able to anticipate all possibilities and must therefore be constantly alert to ensuing feedback and emergent influences that may require flexible and adaptive responses (i.e., converting feedback learning into feedforward actions) if research character and quality is to be maintained. Systemic thinking is not just engaged at the start of the research process; it is continuously engaged and revisited throughout the researcher's journey. Some systemic considerations may not exist at the start of the researcher's journey but may emerge as relevant or strongly influential partway through that journey.

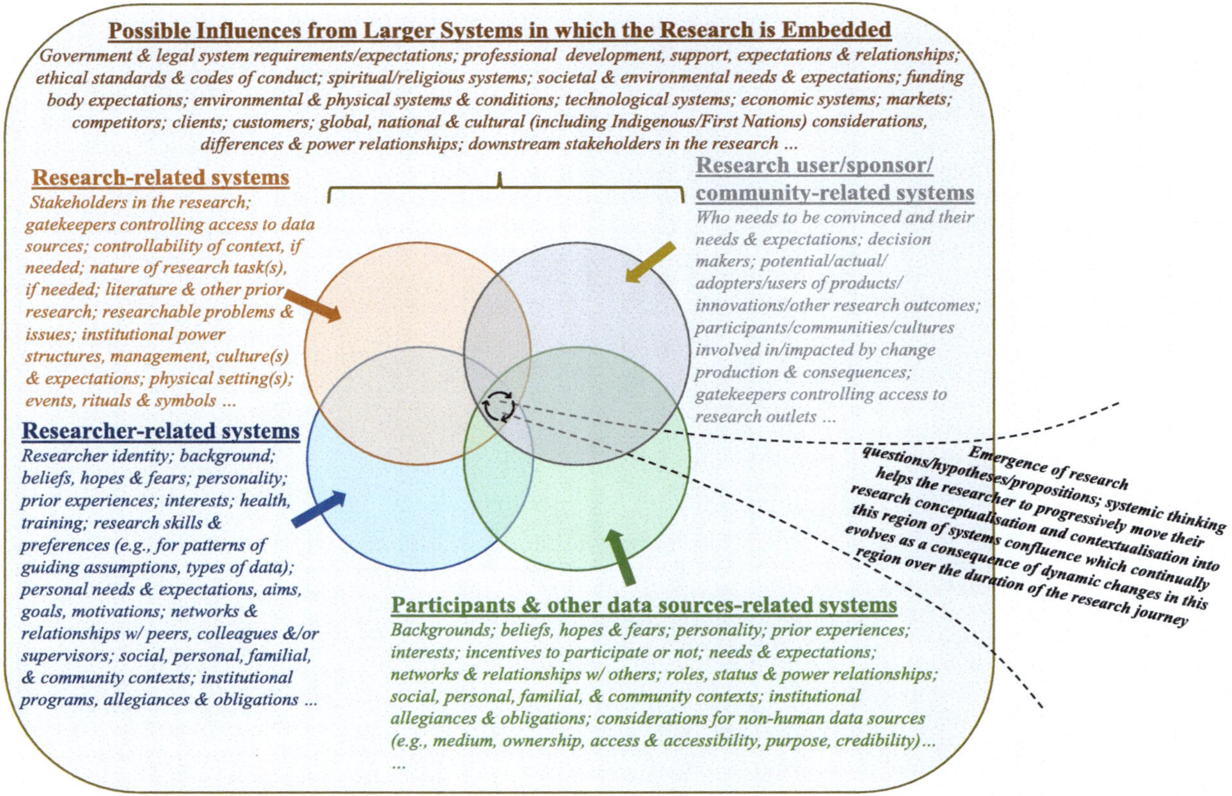

Fig. 1.4 Considerations for systemic thinking about social and behavioural research (adapted from Cooksey & McDonald, 2019, p. 306)

1.5.2 Pluralist Thinking Considers Potential Choices of Research Pathways and Strategies

Pluralist thinking involves continuous acknowledgement of the diversity of potential pathways, approaches, and strategies that could be adopted and employed in pursuit of a particular research journey and drawing upon that acknowledgement when making and enacting specific choices. The pluralist thinker keeps all options on the table, regardless of their ideological or philosophical position/stance, until reasoned choices have been made to take them off the table and does so while keeping their own preferences and predilections from prematurely adversely influencing or constraining the choice process (something that discipline norms tacitly reinforce, unfortunately). In short, the pluralist thinker remains open to possibilities while continually evaluating feedback that may be asking them to negate some possibilities and promote others; their toolkit remains as full as possible until contextual forces render some tools to be unviable or infeasible. Pluralist thinking is congruent with the open systems theory concept of 'equifinality' which refers to many potentially viable paths to the same goal (see, for example, Brunswik, 1952; von Bertalanffy, 1950).

Pluralist thinking can be much more difficult than systemic thinking to consciously engage in, simply because previous educational and research experiences and training, disciplinary norms, publication practices, and other contextual forces may have deliberately and/or historically removed some possibilities from consideration. Such forces generally tend to operate antithetically with respect to methodological diversity creating an imbalance in favour of homogeneity in thinking (e.g., consistency with disciplinary norms) in preference to heterogeneity in thinking (e.g., breaking boundaries or innovating new methodologies). This is especially the case when we consider patterns of guiding assumptions or 'paradigms' for research, such as positivism, interpretivism, postmodernism, and critical social science, each of which offers little room for alternative ways of thinking. Recent developments in mixed methods and transdisciplinary research approaches (see, for example, Plano Clark & Ivankova, 2016; Plowright, 2011; Poth, 2018; Tashakkori & Teddlie, 2010) have gone a long way toward rectifying this imbalance by reinforcing the need for and utility of diversity thinking in making methodological choices and for making those choices in light of the research goals and objectives as well as other contextual considerations. However, even mixed methods approaches impose certain limitations upon their practitioners (see the recent critiques offered by Flick, 2017). For example, mixed methods research must include both quantitative and qualitative forms of data, a stance that downplays the possibilities that do not require mixtures of data types in order to produce convincing research (although Bazeley, 2018; Gobo et al., 2022, have recently identified merged, hybrid or inherently mixed methods, which blur the distinction between quantitative and qualitative data).

What we are after, with the process of pluralist thinking, is an inclusionary logic, which consciously mitigates against premature or predetermined exclusion of certain choices. This is not easy, as it requires the researcher to withhold active commitment to a methodological pathway until much later in the research contextualisation process than they would typically prefer. Pluralist thinking, in pursuit of truly convincing research, should not be respectful of or beholding to researcher comfort zones. Rather, it should work like a funnel where all possible choices, including potential patterns of guiding assumptions and frames, are available at the start and methodological choices are progressively funnelled, shaped, pruned, and winnowed until a pathway forward for the research emerges. Premature choices, influenced by systemic forces such as one's own prior training, journal publishing practices, funding organisation preferences, and disciplinary norms, will likely force the thinking process to progress too far down the funnel too quickly, which means that viable and even creative alternatives are likely to be unrecognised or, worse, ignored/dismissed. Furthermore, a pluralist thinking attitude needs to be maintained throughout the research journey, so that if unanticipated obstacles, constraints, or opportunities present themselves, the researcher can take advantage of the widest array of potential trade-off and adaptive responses.

It is important to acknowledge that pluralist thinking is not without its challenges. For example, Midgley et al. (2017) argued that engaging in pluralist methodological thinking confronts three important obstacles: paradigm problems, psychological resistance, and cultural barriers. Kuhn (2012) used the term 'paradigm' to characterise what 'normal science' is and does within a particular discipline, to convey a sense of consensus professional commitment by a substantive group of researchers (what Kuhn referred to as "practitioners", p. 10) within that discipline, and to place boundaries around what counts as a worthy problem for investigation and what counts as worthy methods for addressing that problem within that discipline. [Note that, in Chap. 4, I will argue that continued use of the term 'paradigm' is no longer productive if we wish to fully understand the implications of pluralist thinking; instead, I advocate referring to patterns of guiding assumptions.] Taken as a whole, this means that a 'paradigm' has the power to influence a researcher's thoughts and actions. The paradigm problem thus refers to the issue that many researchers already have preferences or commitments to a 'paradigm' (due, perhaps, to prior training, prior research experiences, the influence of a mentor, disciplinary expectations) that they need to be aware of and acknowledge. Midgley et al. argued that [they believe] "it is not possible to be 'meta-paradigmatic'" (2017, p. 157), instead they suggested learning from other paradigms but reinterpreting this learning in terms of one's own preferred paradigm. This argument defeats the purpose of genuine pluralism and, as we will argue later in Chap. 4, with respect to meta-criteria, the learning they propose strongly risks misapplying concepts from their own 'paradigm' to make judgments about what can be learned from other 'paradigms'. To engage pluralist thinking, one's 'paradigm' preferences and commitments need to be consciously held at bay or relaxed while considering a new research journey. To do this effectively requires serious introspective cognition as well as emotional and motivational effort and may well be easier to accomplish through discussions with researchers working within other 'paradigms'. The second challenge to pluralist thinking is psychological resistance. Researchers may be aware of their 'paradigm' preferences and commitments but simply be resistant to changing or relaxing their perspective. One useful way of overcoming such resistance is to seek opportunities to actively expose oneself to or participate in research guided by other 'paradigms'. Nothing opens one's eyes and counteracts preconceptions so much as lived experience. Cultural barriers constitute the third challenge to pluralist thinking according to Midgley et al. who did not offer a way forward for meeting this challenge. The challenge is this: academic disciplines create and impose cultural expectations about the 'right' ways to do research (which reflect preferred 'paradigm' commitments) and that culture is itself resistant to change. These cultural expectations can exert influence over academic promotion expectations, research funding practices, journal reviewing practices, and postgraduate supervisory practices, thus, more broadly insinuating themselves into the professional lives/careers of academics and researchers who are part of that culture. Changing such cultural practices is a long-term cumulative process, one that we have seen unfold over time at different rates in many disciplines (e.g., psychology, nursing and health, marketing, management, education, sociology, information sciences). It requires inspiration and courage (and likely some personal career risk) to overcome inertia and initiate/stimulate changes and then the amassing of a groundswell of support and emergent evidence. The mixed methods movement (showcased in contributions such as Bazeley, 2018, 2019; Tashakkori & Teddlie, 2010; Creswell & Plano Clark, 2017; Plano Clark & Ivankova, 2016; Plowright, 2011; as well as the *Journal of Mixed Methods Research*) must be acknowledged to have strongly contributed to this groundswell over the years in the social and behavioural sciences. Important texts (e.g., Alvesson & Sandberg, 2021; Crotty, 1998; Layder, 1993; Mingers, 2014; Poth, 2018; Ragin & Amoroso, 2019; Sayer, 2000), offering more systemic or alternative perspectives, also contribute to the expanding groundswell as do emerging developments in Indigenous/First Nations research approaches (e.g., Chilisa, 2020; Kovach, 2021; Smith, 2021). Finally, this book aspires to contribute as well through its synergistic combining of pluralist thinking and systemic thinking.

1.5.3 Pluralist Systemic Thinking Engages a Synergistic Interdependence

When systemic thinking and pluralist thinking are synergistically combined, pluralist systemic thinking emerges. Systemic thinking identifies what influences might need addressing; pluralist thinking helps the researcher make choices that will help them to address those influences in relevant, sensible, defensible, and feasible ways. When harnessed in such an interdependent manner, pluralist systemic thinking helps the researcher to identify what needs to be done to produce convincing research and plan a pathway forward for achieving that goal. Down the track, pluralist systemic thinking will help the researcher to produce convincing research outcomes, shape how they communicate what they have learned to others, and facilitate potential impacts the research might have. Pluralist systemic thinking is more about identifying what choices are available/viable/feasible at which point in time (call this the 'choice set') than about which specific choices to make. It ensures that the choice set remains as populated as it can be at any one point in time, so that the researcher can make reasoned and logical decisions secure in the knowledge that particular choices have not been prematurely removed from consideration. For example, the researcher could logically choose amongst different modes of survey questionnaire administration (e.g., mail, internet, phone interview, face-to-face interview) in order to best meet their research needs, but not if a journal they aspired to publish the research in had a history of not publishing interview-based surveys.

In many cases, once the researcher has made a particular methodological choice, the choice itself will stimulate subsequent choices (i.e., every choice has downstream implications) which will include some possibilities and exclude others. Thus, some early choices may create the opportunity or need for certain other choices to be made down the line, generating a cascading effect, like travelling along the branches of a tree and encountering forks (i.e., choice points) along the way. Some generic examples: given choice X → choose to implement Y and Z together because they necessarily must be paired, because they create emergent synergy, or because the strengths of Y counteract the weaknesses of Z (simultaneous downstream choices); given choice X → choose to implement X first which then implies Y is required as a necessarily follow-on implementation or choose to implement X which rules out the possibility of implementing Y, and so on. The difference here is that diversity is being deliberately constrained by necessity, conditional upon a previous well-reasoned and fully-considered choice, rather than the vagaries of some other constraining force or influence. The hard part is not letting personal preferences, constraints, and background unduly influence these choice-narrowing processes. For example, choosing to use an internet survey questionnaire implies certain subsequent choices are necessary to make the internet questionnaire viable; choices that would not be required if a mail or interview questionnaire was being used; one would not make such a choice simply because their department just invested in a new internet survey software package. Specific choices may open up questions about necessary or desired technological support and mediation, resourcing, research-related relationships to pursue (e.g., working with gatekeepers to gain access to desired data sources or important contextually-relevant information), or personal development to be undertaken for certain aspect of the research process.

Pluralist systemic thinking would not be fully systemic and pluralist unless we devoted serious attention not only to western ways of thinking but also Indigenous/First Nations ways of thinking about research and research ethics. A major reason for this is that Indigenous/First Nations research is inherently systemic, linking people and their ancestors in relationships with/to environmental, ecological, social/community, cultural, spiritual, and cosmic worlds, which creates a complex yet unified well-spring for knowledge development, sharing, and safe-keeping. This is a strong form of relational reasoning and knowledge building that does not easily resonate with traditionally and academically privileged western ways of thinking (see, for example, arguments offered by Chilisa, 2020, Chap. 5; Kovach, 2021, Chap. 1; Smith, 2021, pp. 228–231). However, there is much that western researchers (which characterises the author of this book) can/should learn from Indigenous/First

Nations approaches to research, especially with respect to the attitudes we hold regarding knowledge creation and relationships with people and systems that help us to develop that knowledge. This learning emerges from the concepts of respect, reciprocity, responsibility, holism, interrelationships, and synergy (see, for example, Archibald, 2008; Archibald et al., 2022; Chilisa, 2020; Kirkness & Barnhardt, 1991; Kovach, 2021; Smith, 2021). While not claiming here to be using these concepts with anywhere close to the full senses of meaning they have for Indigenous/First Nations research, they can be seen to offer important and relevant lessons and implications for non-Indigenous research and researchers.

- **Respect** means that researchers should value/honour others and their culture, history, heritage, traditions, views, and perspectives, without imposing their judgments or preconceptions.
- **Reciprocity** means that researchers should seek to return knowledge value/benefit from their research to their relevant communities, however defined (this is an important reinterpretation and recontextualisation of the notion of research impact).
- **Responsibility** means that researchers should endeavour to keep their stories authentic as well as to convey authentic stories from/about/with participants in their research. The knowledge they share must be ethically created, co-created, or discovered (where ethically is taken to mean consistent with cultural, community, and societal traditions and expectations and in a manner that facilitates and maintains respect); and it should be appropriately shared and safeguarded.
- **Holism** means that researchers should acknowledge that the world they live in as well as do research in is comprised of complex systems and that their knowledge building activities should consider the interconnections between/within those systems when characterising the implications of that knowledge.
- **Interrelationships** means that researchers should build and value links between worlds and systems, from their own perspective as well as the perspectives of others, realising that social and behavioural research cannot be conducted or knowledge created without reference to and involvement of others at many levels.
- **Synergy** means that researchers should examine dynamics between the systems implicated in their research in order to harness synergistic connections that can facilitate their knowledge building and knowledge sharing work; thus, synergy is a well-spring for methodological innovation (for example, Blair, 2022, p. 203, created "lilyology as a transformative framework and storytelling practice to decolonize critical ethical spaces for research within academia" by "metaphorically engag[ing] waterlilys, sweet potatoes, spiders, and brick walls, creating a space to reflect on and privilege Indigenous Knowings in the academic context").

These concepts provide principles that are very useful in supporting the researcher through their pluralist systemic thinking processes. Respect, reciprocity, and responsibility speak to the development and management of relationships between knowledge building/discovery and those people and communities that facilitate and support such work throughout the research journey. Holism, interrelationships, and synergy speak to the worlds and systems that the researcher engages with during their research journey. For example, without respect, researchers and participants in their research will not truly connect in meaningful ways. Without reciprocity, research has diminished impact beyond its own boundaries. Without responsibility, there is no motivation to ethically create, discover, and share knowledge in ways that protect those involved in its development while providing a safe space for genuine views to be expressed. Without holism, researchers resort to reductionist linear thinking patterns that wall off or ignore the importance of the wider systems that people inhabit. Without a focus on interrelationships, researchers operate in a vacuum, creating meaning only for themselves. Without synergy, creative energies devoted to defining and conducting research in new and more meaningful ways dissipate leaving the researcher with little opportunity or capacity to move beyond the strictures of the status quo and dominant assumptions.

Figure 1.5 provides a concept map of the implications of pluralist systemic thinking for social and behavioural research to be more fully explored in this book. The left side of the map shows the different concepts that have been used to signal diversity in method choices in the research literature. For example, *triangulation* is "the combination of methodologies in the study of the same phenomenon" (Denzin, 1978, as cited in Jick, 1979, p. 602; see also Denzin, 2012; Jick, 1979; Moran-Ellis et al., 2006; Olson, 2004); *bricolage* is "the process of employing these [multidisciplinary] methodological strategies as they are needed in the unfolding context of the research situation" (Kincheloe, 2005, p. 324; see also Kincheloe & Berry, 2004; Patton, 2011, Chap. 9); *multimethodology* involves using "more than one methodology, or part thereof, possibly from different paradigms, within a single intervention" (Mingers & Brocklesby, 1997, p. 491; see also Bowers, 2011; Brocklesby, 1997; Flood & Romm, 1997; Midgley et al., 2017; Mingers & Gill, 1997; Taket, 1998); *mixed methods* (or *multimethod* or *mixed methodology* research) is "research in which the investigator collects, analyses, mixes, and draws inferences from both quantitative and qualitative data in a single study or a program of inquiry", *Journal of Mixed Methods*, 2006, cited in Cameron, 2011, p. 96; see also Morse & Niehaus, 2009; Plowright, 2011; Tashakkori & Teddlie, 2010); and cross-disciplinary approaches (e.g., *multidisciplinary, interdisciplinary, transdisciplinary*) inherently focus on multimethod approaches (see, for example, Fedorov, 2020; Leavy, 2011; Lury et al., 2018).

Some of these ways of signalling methodological diversity have emerged from specific disciplines. For example, 'multimethodology' emerged in the information systems literature (e.g., Mingers & Brocklesby, 1997); 'bricolage' emerged in the complexity and innovation literature (e.g., Kincheloe & Berry, 2004); 'triangulation 1.0' emerged from the psychometric literature, in association with the 'multi-trait-multi-method matrix' (e.g., Campbell & Fiske, 1959; Flick, 2017). Other ways are more generic in origin. Irrespective of origin, the emergence of these labels has signalled that diversity in methodological approaches to social and behavioural is something to be valued. We capture this intent by simply and more generally referring to pluralist thinking (the central cloud in Fig. 1.5). However, it is one thing to engage in pluralist thinking and quite another to harness it in service of conceptualising and undertaking convincing research. This is the role that systemic thinking

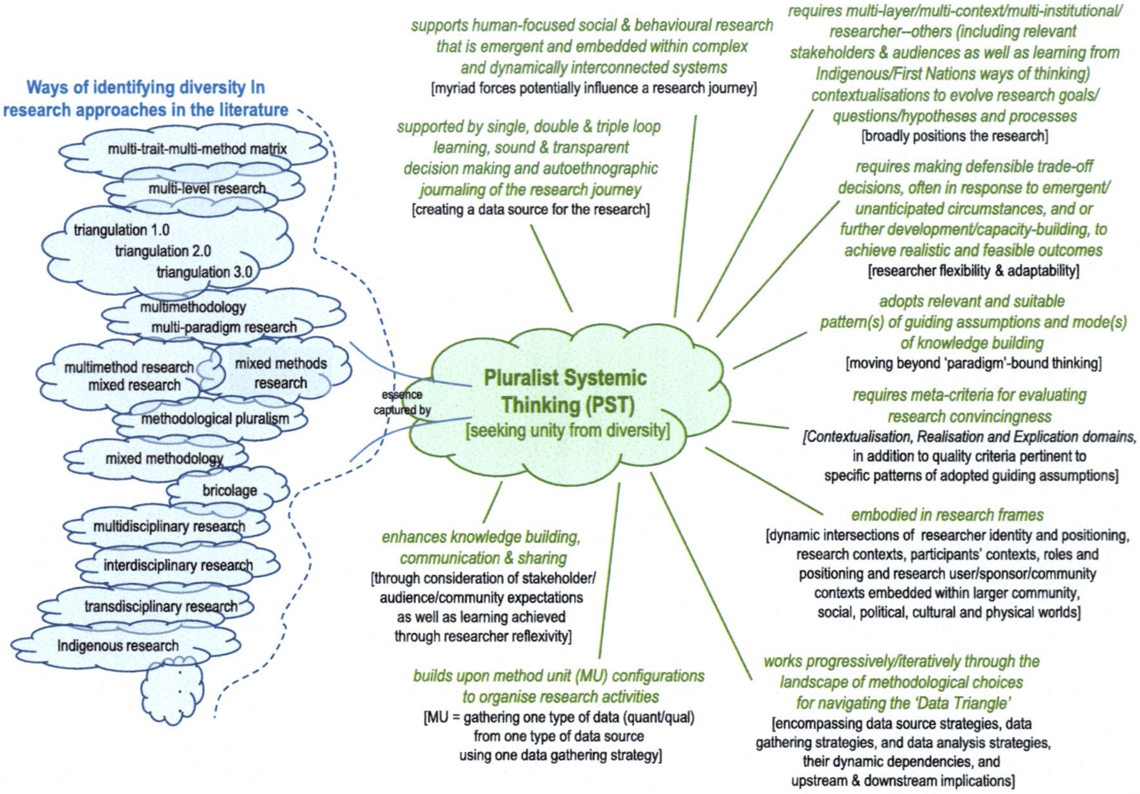

Fig. 1.5 Some implications of pluralist systemic thinking, to be further explored in this book

plays, working through the contextual forces, influences, relationships, and interconnectivity between and within systems, to help the researcher achieve feasible, viable, yet convincing research outcomes. In short, pluralist systemic thinking pursues 'unity from diversity', extending a concept offered by Ragin and Amoroso (2019), where 'diversity' exists in the multiplicities of choices a researcher could make to achieve their research goals for relevant audiences and 'unity', as knowledge, emerges from a diverse set of research considerations and implementations contextualised, scoped, and shaped into a convincing research outcome.

The right side of the concept map in Fig. 1.5 focuses on key features and implications of pluralist systemic thinking as it impacts on social and behavioural research, which provides the focus for the remainder of this book. The key message is that pluralist systemic thinking permeates every aspect of the research journey, not just the initial stages. It starts from the position of humans and the systems they inhabit then moves progressively (but not necessarily linearly) to consider where the researcher (and associated researcher identity) is situated, how the research itself is contextualised, what assumptions the researcher is willing to make to underpin the research (and the quality criteria that accompany such assumptions), how the researcher can meet the demands of the research quality meta-criteria, how to best frame the research, what choices are to be made to navigate the 'Data Triangle' to best address the researcher's goals, how best to configure and implement those research activities, and how best to shape, communicate and share what the researcher learns and with whom. During this journey, trade-off decisions will usually be required as circumstances change, and new challenges emerge. Such trade-offs will attempt to address two goals that work in tension with each other: keeping the research as convincing as it can be while making the research as realistic, feasible and viable as it can be. Pluralist systemic thinking involves trying to anticipate what might happen and why to help guide our actions, avoid blind alleys, and pursue productive paths for research as well as being flexible and adaptive enough to change course when the researcher's anticipations fail them. It also constantly reminds researchers that they don't control all they do and all they experience in research; contextual forces can influence their choices and their actions, for good or ill.

1.5.4 Critical Leverage Points

One important aspect of pluralist systemic thinking involves making productive and adaptive choices at *critical leverage points* during the research journey. Meadows (2008, p. 145) defined leverage points as "places in the system where a small change could lead to a large shift in behavior". If this logic is applied to the discussion of pluralist systemic thinking in social and behavioural research, we are talking about places that emerge during the research journey where a researcher could make a choice (e.g., methodological, conceptual, theoretical, contextual, goal-oriented) that would change how the research unfolds from that point in the journey onward so that a convincing outcome might still be achieved. The descriptor *critical* is attached to the concept of leverage point to signal that (1) choices are made by the researcher only after critical reflection on feedback and emergent events and may either move the researcher to make a trade-off that mitigates or avoids entirely the potential loss of convincingness at a critical juncture of the journey or move the researcher to capitalise on an emergent opportunity in such a way that convincingness is enhanced, and (2) that such choice points can influence the entire nature and trajectory of the journey (i.e., producing an evolutionary or transformative change in the journey rather than an incremental change).

Consider the following scenario.

> An Associate Professor at a major university, with 15 years research experience, undertakes an academic research project in organisational management. The project considered constructs such as job satisfaction, organisational commitment, satisfaction with leadership, and perceptions of the 'health' of the manufacturing sector in Australian society, theorised about their predictive relationships, and proposed a survey research process that would permit testing of those theorised relationships. The research was undertaken in a large manufacturing company, whose CEO was an acquaintance of the researcher. The CEO agreed to provide her with access to the company for the purposes of conducting survey questionnaire research. The researcher promised that both the company and all its employees and questionnaire participants would remain anonymous.

Figure 1.6 focuses on this scenario and demonstrates how dealing with critical leverage points can influence the trajectory of a research journey. The flow diagram pretty much tells the story, but we will unpack that story a bit further here. The ideal journey as initially planned by the researcher is depicted as a vertical dashed line. All subsequent trade-off choices cause this ideal trajectory to be jettisoned, commencing with emergent constraint/obstacle # 1, in favour of a more viable, feasible, and realistic trajectory (the realised journey, symbolised by the solid blue line), still with an eye on achieving a convincing research outcome. The scenario shows five emergent constraints/obstacles, two with important ethical implications, and an emergent opportunity. None of these emergent events could have been foreseen by the researcher and, at each critical leverage point, the researcher had to consider other options to pursue. Choice of an option reflects a trade-off, an adaptive response to recoup potential losses created by what the constraint/obstacle took away. While the scenario flow diagram contextualises each constraint/obstacle and opportunity, what is not shown are the cascading implications associated with each trade-off. For example, with emergent constraint/obstacle # 1, desired documents for helping the researcher to learn more about the organisation could not be sourced so the researcher opted for a small purposive sample of managerial interviews. This meant that ensuing choices needed to be made regarding how the interviews were to be conducted, what information would be sought, where, with whom, how the data would be recorded and analysed and so on (all implications that would not have existed had the desired documents been available to the researcher). Information coming out of these interviews created emergent constraint/obstacle #2 where the modality of the initially planned data gathering strategy had to be altered from online to mail, in order to prevent the possible occurrence of a very low response rate to the questionnaire. Note that the researcher was attracted to the option of conducting off-site interviews, she just didn't have the resources to do a large sample of them. However, this attraction created an emergent opportunity where the researcher modified her questionnaire to facilitate obtaining a small sample of post-questionnaire interviews. Analysing the content of these interviews created an ethical dilemma for the researcher, since it was clear that the sample of volunteer interviewees comprised mostly people who had been harassed in some way by a specific supervisor. Two of the three options would have required the researcher to either interfere with operations of the company (reporting the named manager; something she was afraid her CEO acquaintance would not approve of, since there were formal policies for reporting complaints about supervisory behaviour in place and none of the nine interviewees had filed a formal complaint) or expend resources she did not have to pursue a new sample of interviews (such a sample would have been ethically difficult to organise in any case because she could not identify countering perspectives if the alleged harassments had not been proven). Instead, she felt the most ethical and socially responsible choice was to sacrifice the data and live with the consequences to convincingness that such a sacrifice might bring. The CEO heard through the grapevine that there were problems arising from the researcher's post-questionnaire interviews and that a specific manager was implicated. While there had been no formal complaints, she worried about possible reputational damage from the research. The CEO created an ethical dilemma for the researcher by asking to preview her research manuscript and suggest changes prior to submission for publication. By complying, the researcher would have sacrificed both research quality and the academic independence of the research (which could hurt her publication chances). Instead, she offers the embargo solution (thereby foregoing short-term career advantage but safeguarding her trust relationship with the CEO) which the CEO accepts. A year later, she submits the manuscript, and it is considered sufficiently convincing to warrant publication.

While this scenario was not based upon any specific research journey, it is sufficiently like journeys I and other colleagues have experienced during their academic careers. There are several important things being illustrated here:

- researchers cannot anticipate everything that might occur during their research journey;
- researchers cannot control everything that happens during their research journey;
- critical leverage points create junctures in the research journey where a trade-off choice is required before further progress can be made;
- trade-offs inevitably create deviations from a researcher's ideal research journey trajectory;
- trade-offs can and do create cascading implications for downstream choices; and
- trade-offs need to be made with ethical/social responsibility concerns, practical concerns, and maintaining research quality/convincingness all in mind.

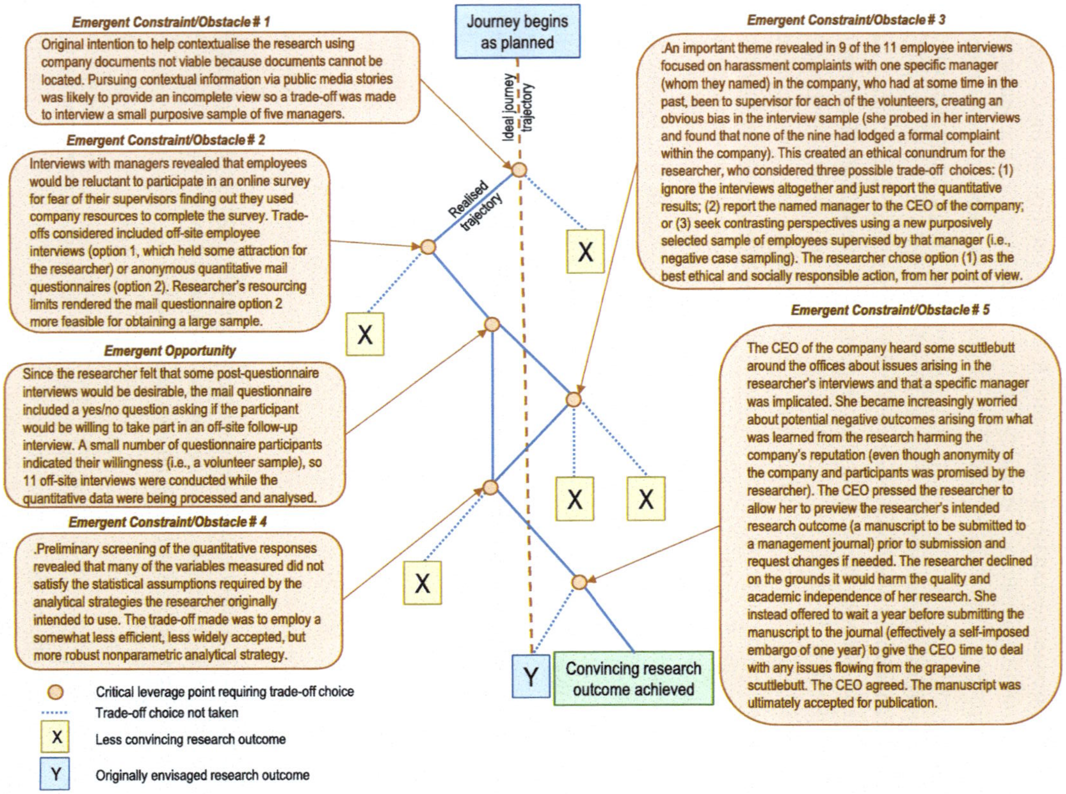

Fig. 1.6 Scenario flow diagram demonstrating the implications of critical leverage points during the research journey (timeline runs from top to bottom)

1.6 Implications of Pluralist Systemic Thinking

When thinking pluralistically, systemically, and complexly about social and behavioural research, the researcher needs to think about their research and strategic methodological choices in many ways. For example, there are:

- possible stakeholders to acknowledge and, perhaps, connect with to assemble contextual knowledge and understand political dynamics and possible agendas;
- possible patterns of guiding assumptions (or, more traditionally, 'paradigm' assumptions) to adopt;
- possible choices of research quality criteria to apply (cascading choices following from choice of pattern of guiding assumptions);
- possible choices of prior research/knowledge building outcomes to draw upon;
- possible choices with respect to the positioning of the research from both the researcher's perspective and their participants' perspectives;
- possible choices of research frames and MU configurations to implement;
- possible choices of data gathering strategies to implement;
- possible choices of form(s) of data (quantitative, qualitative, hybrid/merged) to gather;
- possible choices of data sources (humans, documents, artefacts, performances, multimedia recordings and so on) to connect with, associated with gatekeepers who control access to them; and
- considerations about how the choices they make during their research journey fit in with/relate to their other work and life roles and commitments.

The above list is certainly not exhaustive, but it is representative of the kinds of choices that confront the researcher. The range of choices available in the above list is what reflects diversity. Pluralist thinking ensures that, at least at the outset, all possible patterns of choices are on the table for consideration. Furthermore, each set of choices does not exist in isolation but is interconnected with other sets of choices and systemic thinking is needed to explore those interconnections in terms of potential synergies and implications. In many cases, exploring such interconnections and synergies may lead to the creative development of hybridised combinations. Pluralism does not imply that multiple choices at any level are mandatory, only that they are considered. Making research choices is not a matter of either/or but of which/when. The choices made depend upon the contextual circumstances in which they are being made and the purposes they can fulfil. In the end, the choices the researcher makes while pursuing their research journey will be those that simultaneously best suit their research problem *and* make their journey realistically feasible to complete. Along the way, it will become clear that some researcher choices will help them to make, or at least narrow down, other choices downstream (i.e., 'cascading'). For example, choosing a specific pattern of guiding assumptions to follow will make some data gathering strategies and quality criteria relevant while rendering others useless or meaningless. Choosing to include certain data sources in the research may narrow the range of possible data gathering strategies the researcher can use. Knowing that the researcher must meet certain stakeholder expectations may influence the type of research frame and data analytic framework they adopt.

Thus, the researcher must make all their research-focused choices in context; there is no single correct view, playbook, blueprint, schema, or formula that can claim to tell the researcher which choices will best help them to orchestrate their research journey. This is the tough lesson about social and behavioural research where it is all about making contextualised choices in pursuit of goal(s) (some of which may be the researcher's; some may come from others) in circumstances that are unlikely to remain static and predictable. It is also the beauty of social and behavioural research where there is ample room for creative thinking, for inclusionary processes, for discovering new things, for stimulating change. What can be stated with some certainty is that the more pluralist systemic thinking guides a researcher's choices, the broader their skill set must be, the more resources (including time and cognitive effort) will be demanded, the more diversity in stakeholder connections they will need to accommodate, the longer the research timeline will need to be, and the more challenging it will be for the researcher to bring everything together to convey a convincing story. Finally, there is no guarantee that research where choices have been made using pluralist systemic thinking will converge on a single unified research story, meaning that diverse, perhaps even contradictory, stories may have to be reconciled or, indeed, may be irreconcilable. However, if the researcher's choices are well-reasoned and viable/feasible to implement within the contextual constraints they confront and the opportunities that present themselves, a stronger and more convincing story will more likely emerge. The real strength of pluralist systemic thinking is that it demands the researcher not prematurely discount specific choices without good reasons. Working one's way through research choices is an actively conscious, open, and critical process that should not be derailed by status quo expectations, stymied by uncritical conformity with stakeholder demands, or constrained by the researcher's comfort zone.

1.6.1 Learning Loops and Modes of Thinking

In social and behavioural research, learning has traditionally focused on building knowledge from data that addresses one or more research questions/propositions/hypotheses using deductive, inductive, and/or abductive reasoning (the outward lens of research, focusing on others). However, researchers must also learn from their own research experiences (the inward lens of research, focusing on oneself), with respect to what worked and what didn't work and why, what could work, what should and shouldn't be done, shocks they encountered to their beliefs, assumptions, and preconceptions, how they felt throughout their journey, how well or how poorly their researcher role, either as insider or outsider, unfolded during their journey, what surprised them, what their journey means for their research identity, what their journey means for relationships with others, what they should change for future research endeavours, and so on. In short, researchers must learn about/from their situatedness in the research processes themselves, so that changes, adaptations, and creative leaps might be made for their future research journeys; this is, in part, what it means to be reflexive (Le Gallais, 2008; Malthouse et al., 2014). Such learning is deeply personal, contextually-focused, and oriented toward growth, development, and change. It demands some level of self-awareness, critical thought, flexibility, and willingness to change/adapt. It must also be both backward-looking (learning from feedback to make adjustments and changes) as well as forward-looking (converting feedback into feedforward learning to guide future actions). However, there are depths, differences, and nuances to understand about this researcher learning process, which pluralist systemic thinking can illuminate. We can achieve some of this understanding by applying and building upon a scaffold of 'learning loops' that complexity theorists and systems thinkers have provided in the organisational learning and management literature. This scaffold differentiates three distinct levels or depths of learning process which we can directly apply to contextualised researcher learning:

- *single* loop learning where changes are made to research processes when they do not or are anticipated to not achieve the researcher's aims, but those choices remain constrained by status quo expectations (including pattern of guiding assumptions) about what one should and could do in research (e.g., Argyris, 1976, 1990; Argyris & Schön, 1978; Gates & Cooksey, 1998);
- *double* loop learning where changes are made to research processes and/or assumptions when they do not or are anticipated to not achieve the researcher's aims, but those choices are made by questioning the status quo (including whether the researcher's aims and choice of pattern of guiding assumptions/'paradigms' are appropriate) and widening the set of possibilities, albeit falling short of fully systemic considerations (e.g., Argyris, 1976, 1990; Argyris & Schön, 1978; Gates & Cooksey, 1998);
- *triple* loop learning where the values orientation, researcher's positioning, environment, focus, and contextual circumstances of choices are fundamentally re-examined creating opportunities for new ways of building knowledge to emerge (e.g., Flood & Romm, 1997, 2018; Hummelbrunner & Reynolds, 2013; Peschl, 2007; Tosey et al., 2012)

If one looks at these learning loops as more than just pathways to error-correction (as was the case for single and double loop learning, especially), we can see that each mode of learning invokes a different mode of thinking (means-ends thinking—'doing things right'; critical thinking—'doing the right things', and meta-thinking—'learning what the right things are') about research, one's role in that research, and what one does or does not do in that research. Figure 1.7 presents a descriptive mapping of those three modes of thinking and learning. In methodological terms, means-ends thinking reflects mono-method/mono-'paradigm'/linear logic; critical thinking reflects multi-method/alternative 'paradigms'/nonlinear logic; meta-thinking reflects pluralist systemic/nonlinear/ complex logic. Each mode of thinking can be depicted as the researcher asking a particular type of introspective question (shown as coloured-coded text in Fig. 1.7). Note that each mode of thinking also invokes considerations of how to enhance research quality. The paragraphs underneath each question further unpack and clarify the implications of each mode of thinking. Figure 1.7 also shows how the modes of thinking vary along various dimensions relevant to research. For example, means-ends thinking is the least reflective, generally contextually-insensitive, and most change-resistant; meta-thinking is the most reflexive and contextually-sensitive, and may lead to new and more creative research practices. Meta-thinking is also more likely to enlist the aid of others (including participants as partners in their research) in knowledge building activities. Developments and advances in Indigenous/First Nations research approaches (e.g., Chilisa, 2020; Kovach, 2021; Smith, 2021) provide one clear example where meta-thinking processes play an important role.

1.6.2 Support Frameworks/Tools for Pluralist Systemic Thinking

Pluralist systemic thinking is a cognitive activity that requires support. Such support can take many forms and we offer a few frameworks here as examples. Some frameworks focus on prompting questions to engage pluralist systemic thinking in specific ways. Others offer ways of recording thinking that will help to capture interconnections between complex contextual considerations and research choices. Some frameworks such as mind mapping or concept mapping may be facilitated using a software support system (such as *Inspiration*, see https://www.inspiration-at.com/ or *MindManager*, see https://www.mindmanager.com/en/). In terms of envisaging/planning a research journey, these support frameworks all operate in the same way as a thought experiment, where the researcher imagines or anticipates undertaking a specific research journey, then uses one or more support frameworks to explore how that journey might unfold. Researchers need to anticipate not only the main pathways and decisions they need to make but also what they might have to do to make certain things happen or what they might have do to offset potential criticisms if they don't do certain things. Thus, researchers are basically trying to see not only the 'trees' that comprise the elements of their research, but also the 'forest' within which those trees are growing. Probing further, researchers need to think ahead about accessing research contexts and data sources as well as about data preparation, storage, and analysis processes. They need to try to anticipate at least some of the internal and/or external forces that might emerge to influence their research and modify their strategic or contextual choices accordingly. They need to plan for their approach(es) to data analysis, at least to some extent, as certain data gathering strategies may lead to certain analytical choices. Engaging in pluralist systemic thinking can help researchers to identify their own skill deficits and training/development needs (internal forces) as well as political, ethical, and research user expectations (external forces) that they will need to address if their research is to have a good chance of success. However, researchers should expect false starts, blind alleys, complexities, reversals in fortune,

Fig. 1.7 Learning loops/levels when pursuing new knowledge, learning, innovation, and/or change

unanticipated events, and even emergent opportunities throughout their journey. In most cases, their journey planned will not look exactly like their journey completed, partly because they cannot anticipate everything (researchers, after all, are not omniscient, they are human!) and partly because they do not control every aspect of their project. In this light, a good practice for effective research planning is to plan as well and as thoroughly as possible, but also plan to be flexible and adapt to changes and unanticipated events, which may eventually mean letting go of aspects of or the entirety of the original plan if circumstances require.

Systemic Implications Mapping

Recall that Table 1.1 described 15 different systems of human existence that could be potential relevant to/implicated by any researcher's journey. In terms of initial high-level thinking about a research journey, a researcher could find it useful to have a way to capture/record anticipated implications for their current or intended journey. Figure 1.8 (produced using Microsoft *PowerPoint*) presents a conceptual schema that could be useful for mapping such implications. The 15 systems of human existence have been grouped into four higher-level domains and each domain has two triangular sections: one for capturing/recording potential/actual influences on the researcher's journey (light blue triangles labelled with gold arrows) and the other for capturing/recording aspects of those systems that the researcher's journey could or would draw upon (light grey triangles labelled with brown arrows). In each triangular section, the researcher could write short phrases (perhaps connected to the relevant human system via an arrow) that capture the essence of each influence or contribution (perhaps recording on an A3-sized version of Fig. 1.8). If there are some aspects that could serve both roles, those phrases could be written across the boundary between the two triangles. The resulting schema would provide a material record of the researcher's thoughts about the connections between their research journey and *relevant* aspects of the systems of human existence.

Perspective Switching with Self-Questioning

One useful way to engage pluralist systemic thinking is perspective switching. Perspective switching is a self-questioning technique designed to prompt the researcher to consider not only their own perspectives on the developing research journey, but also the perspectives of relevant others as a pathway to facilitating and contextualising research choices (in the organisational change discipline, 'perspective swaps' offers a somewhat similar technique, see Hinds, 2023). Figure 1.9 offers a diagrammatic representation in which the researcher explores their own background as it is relevant to the research journey and considers a set of prompting questions intended to focus on the substance of the research journey from their perspective and as well as from relevant others (e.g., stakeholders, gatekeepers, potential participants, communities, and so on, where 'others' could be individuals or groups). At the intersections of these perspectives, the researcher begins to shape their journey thinking through to the outcomes they want to achieve and the learning and impacts that both themselves as researcher and relevant others might or could realise from the successful completion of the journey. Through perspective switching, the researcher

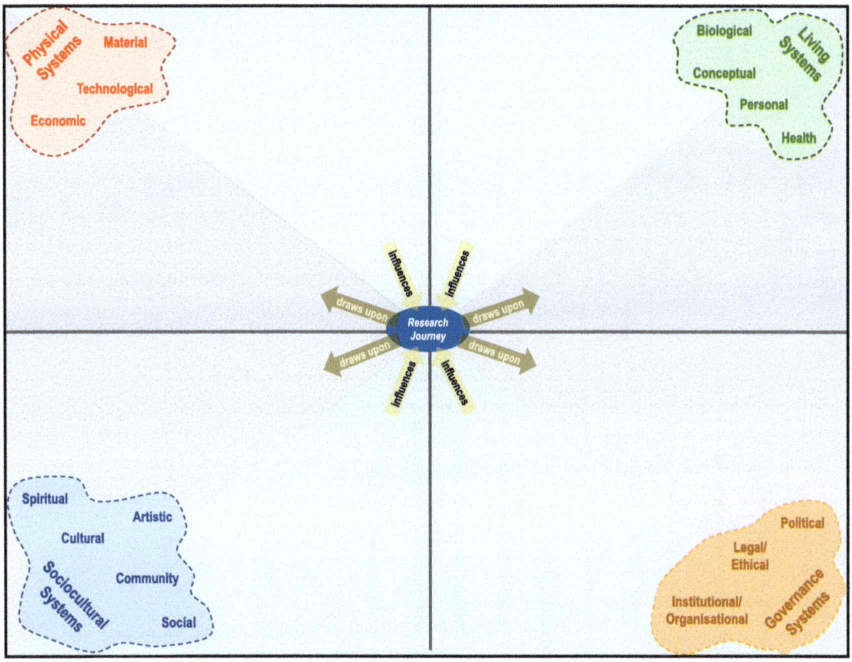

Fig. 1.8 Conceptual schema for mapping anticipated or actual systemic implications of various human existence systems

essentially views their journey both from the inside out as well as from the outside in. Additional value can be realised if the researcher talks through the prompting questions with relevant others which can help to substantiate as well as supplement their own thinking.

Scenario Thinking with Drill-Down Questioning

Scenarios are cognitive devices for constructing and exploring mental models (Healey & Hodgkinson, 2008; Kleiner, 1994). They are meant to stimulate thinking about how future events might unfold if certain choices are made. Scenario planning originated in the management literature as a way to facilitate/support strategic decision making in organisations (see, for example, Coates, 2000; Lindgren & Bandhold 2009; Schoemaker, 1995, 2004) and, in such contexts, multiple scenarios are typically generated and evaluated before a strategic decision is made. However, we can utilise a variant of the scenario planning process to facilitate/support pluralist systemic thinking in social and behavioural research. Think of a scenario as an anticipatory story which asks, 'what if I do my research in this way, how might that unfold?' It is a way of concretely implementing a thought experiment as a storytelling or narrative exercise (Beach, 2021; Kleiner, 1994). Once a research scenario has been created, it can be explored and critically evaluated for potential, caveats, critical leverage points, constraints, and opportunities using 'drill-down' questioning. Instead of generating

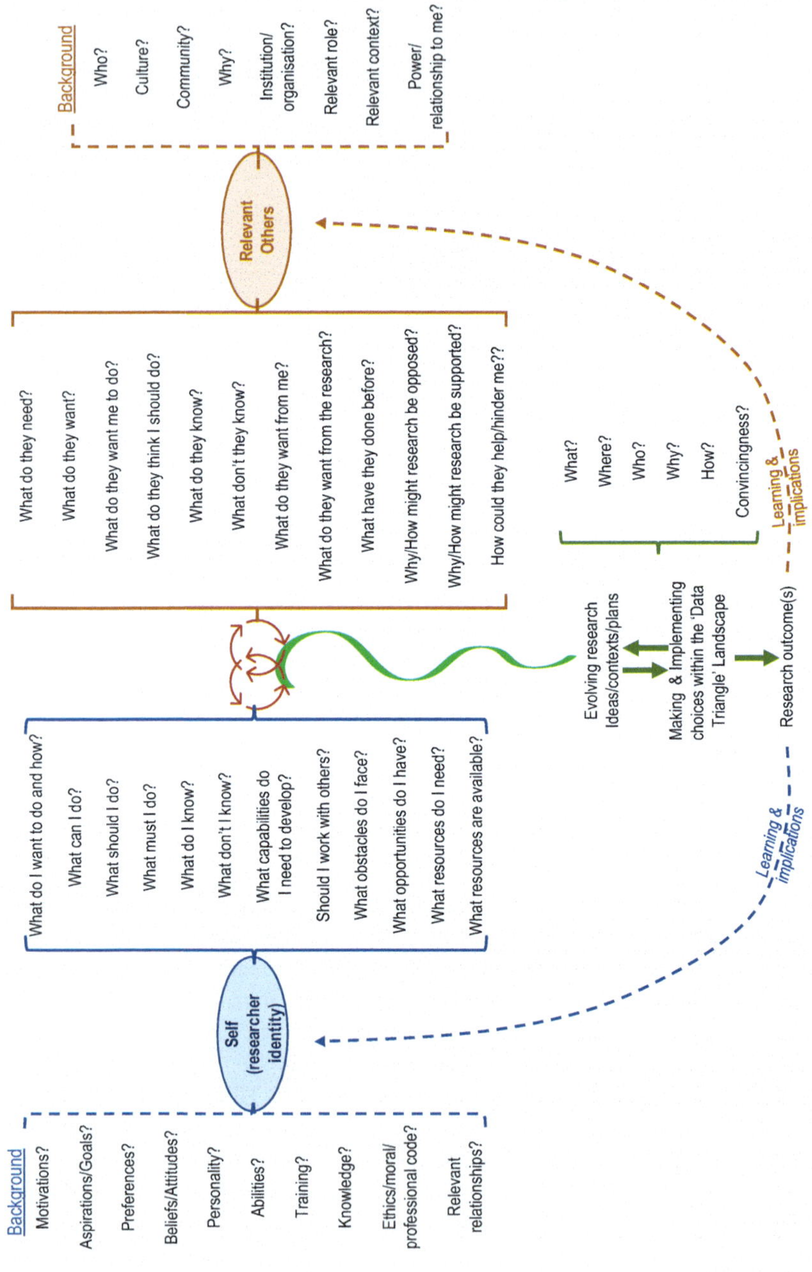

Fig. 1.9 Perspective switching using self-questioning prompts for engaging pluralist systemic thinking

multiple scenarios (although this could certainly be done for more complex types of research or for research involving teams), the researcher generates one main scenario regarding how they envisage their research journey/project would unfold, then, depending upon the outcomes of drill-down questioning, modifies aspects of the scenario and re-evaluates. Others may or may not be involved in either the scenario creation process or the scenario evaluation process. Figure 1.10 offers a possible schema for scenario thinking and drill-down questioning. The researcher generates the scenario by focusing on the research journey they anticipate taking, reflecting on how their researcher identity plays a role, the goals they wish to achieve, and how various contextual considerations and influences may present/exert themselves. The scenario is less of a complete story and more of a sketch of what the researcher thinks might happen. Through the drill-down questioning process, the sketch can be interrogated, evaluated, and potential modifications flagged. The drill-down questions shown in Fig. 1.9 are grouped into three domains (represented by the ellipses; other domains may certainly be possible, depending upon the particular research journey anticipated) and what is learned through one avenue of drill-down questioning may impact the directions and emphases for other avenues of drill-down questioning (the dashed arrows connecting the drill-down domains). When creating a scenario, Lindgren and Bandhold (2009, Chap. 4) offer some useful advice, which reflects pluralist systemic thinking:

- draw upon a diverse toolbox to generate the scenario (we offer such a toolbox in Chap. 2 in the form of the methodological landscape for navigating the 'Data Triangle');
- utilise critical thinking through clarifying and challenging assumptions and preconceptions (e.g., don't let personal comfort zones or preferences dictate how the scenario unfolds; one could think of this as being reflexive at the outset);
- think of the scenario as a story or a drama, fleshed out by actors (the 'who'), events (the 'what'), settings (the 'where'), timelines (the 'when'), resources and methodologies (the 'how'), and motivations (the 'why');
- try to anticipate what could happen in the future, in terms of emergent constraints or opportunities (drill-down questioning could then probe for possible pathways for dealing with these);
- go into the exercise with a healthy respect for uncertainty and the role it will play in how events might unfold and how circumstances might change over time;
- think systemically about the contexts in which the research scenario will unfold and about the costs and benefits associated with how the research journey might impact those contexts;
- think about who would potentially be involved or implicated in the research and what actions they could potentially take to either support or thwart the researcher's efforts; and
- try to visualise the potential costs/losses and the benefits/gains of the anticipated journey and if the costs/losses out-weigh/out-number the benefits/gains (i.e., if all else fails); be willing to let that scenario version of the journey go and conceptualise a new journey scenario.

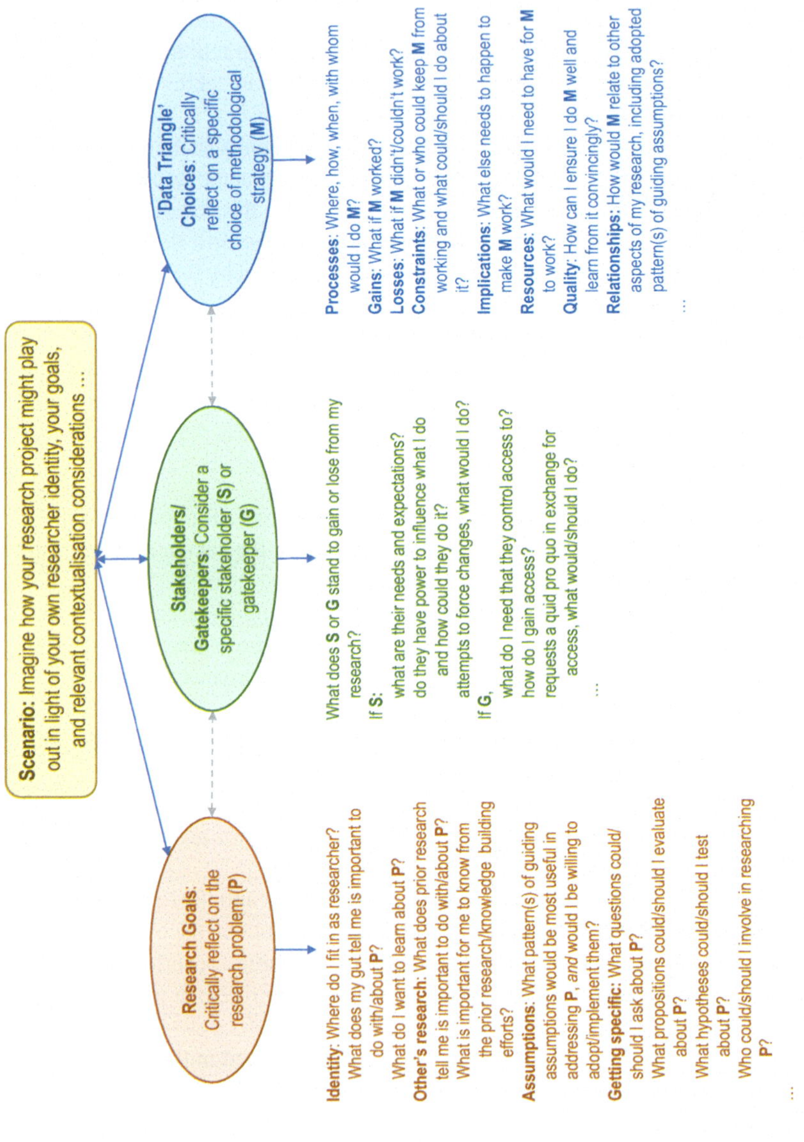

Fig. 1.10 Scenario thinking with drill-down questioning

The above advice applies equally to the drill-down questioning process which forces the researcher to critical examine the choices they have made in envisaging the scenario. Drill-down questioning has the effect of stimulating double-loop learning processes for the researcher, querying assumptions and goals as well as processes and choices. It is important to understand that scenario thinking with drill-down questioning requires commitment and cognitive effort. It is not an easy thing to do, and it is easy to fool oneself into thinking that one has all the answers when done (a danger called the 'illusion of control', which is a risk in all planning processes, creating an overconfidence cognitive bias). Healey and Hodgkinson (2008) warned that scenario planning and evaluation can have an unintended side effect of anchoring thinking too rigidly to a scenario, which then restricts just how critically the drill-down questioning will proceed (adherence to one's comfort zones would reinforce such rigidity, which would reduce the researcher's ability to adapt to any demands for change created by drill-down questions). They also warned about the potential for emotion to cloud or interfere with judgments made about the scenario as it may inhibit rational exploration of costs and benefits. For example, if one's research journey is closely tied to one's livelihood or employment (as when doing research for a postgraduate degree or within one's own organisation), then emotional entanglements need to be acknowledged and dealt with as part of the scenario creation and evaluation processes. For these reasons, it may be helpful to involve others (e.g., supervisors, mentors, team members) in the creation and evaluation of the research scenario. Note that, in a way, a research proposal is a more formalised and detailed version of a research scenario, so, by inverting the order, a research scenario, evaluated and modified through drill-down questioning, could eventually evolve into a coherent research proposal.

Anticipatory Imagining Using Reasoning Chains

Another support tool for pluralist systemic thinking involves the use of reasoning chains (see, for example, Sniderman et al., 1986). In his theory of narrative thought, Beach (2016, 2021) characterised narratives as connected chains of causal thinking and that, with imagination, we can move from the past (prior experiences) through the present (current experiences) to the future (anticipated experiences) when we make choices. Reasoning chains are useful for mapping these moves between past, present, and future by exploring the possible/anticipated causal links between choices and consequences. They operate at a smaller scale than more systemic frameworks such as scenario thinking and complexity mapping (to be discussed later in this chapter), allowing the researcher to probe issues more specifically. Here, we are not talking about causal reasoning that researchers use to create or evaluate theoretical propositions; instead, we focus on the researcher anticipating what could happen if they make certain choices or realise a certain consequence, rather like a miniature highly focused (and necessarily partial) version of scenario thinking. In this regard, we refer to the use of reasoning chains as anticipatory imagining. We can identify three distinct types of reasoning chains: (1) *forward implications* involving if–then anticipations from a specific choice to its potential consequences; (2) *backward linkages* where a specific anticipated consequence is traced back to potential choices;

and (3) *cascading chains* where an initial choice (or a series of connected choices) produces one or more consequences which then leads to another choice leading to other consequences and so on. Figure 1.11 provides a generic representation of each type of reasoning chain (note there are multiple variants of cascading chains; two possibilities are shown). Choices could concern contextual options, data sources, data gathering strategies, data analysis strategies, research outcome dissemination/knowledge sharing strategies, or any other aspect of the research process. Reasoning chains could be particularly useful in helping the researcher to evaluate responses to stakeholder/gatekeeper pressures, resource limitations, ethical dilemmas, and, importantly, responses to critical leverage points. Reasoning chains could also be usefully combined with scenario thinking as a way of engaging drill-down questioning. When reasoning chains are used, it is essential for the researcher to initially clarify/specific the contextual circumstances in which the choice or consequence is being considered prior to evaluating the links between choices and consequences in that context (represented by the yellow thought clouds in Fig. 1.11).

Cognitive/Concept Mapping and Mind Mapping

Cognitive maps (also known as cognitive networks) or concept maps visually display a "person's mental representation of concepts or processes about a particular domain, showing the relationships, flows, and dynamics among them" (Miles et al., 2019, p. 185; see also Davies, 2011; McDonald et al., 2004; Novak & Cañas, 2006; Pidd, 2009). Concepts in a cognitive map are usually defined by brief textual descriptions. Linkages between concepts may encode (using, for example, solid lines, dashed lines, varying colours or line thicknesses, lines with and without arrowheads) a wide range of possible meanings, such as causal implication, time flow or progression, association, 'is part of' and 'is related to'. Cognitive maps can thus be used to capture patterns and flows (e.g., passage of time, cause and effect) of linkages between concepts. In concept maps, the linkages are often defined as well by brief textual descriptions signalling causation, hierarchical category membership, implication, consequence, characterisation, qualification, or action (see Novak & Cañas, 2006). If causation is the predominant focus of the linkages, the map would be referred to as a causal or cause map (see, e.g., Bryson et al., 2004). Figure 1.12a illustrates a generic cognitive map framework that could be useful in harnessing pluralist systemic thinking for displaying, linking, and evaluating potential research choices. Each concept rectangle has an associated memo area for recording notes, elaborations, considerations, and issues.

Mind mapping (Buzan, 2018; Davies, 2011) is a visualisation technique that provides a nonlinear full-colour method for organising concepts, ideas and issues associated with a central concept, idea, or process. Mind maps are useful for more general conceptual mapping of research journey considerations, evolution of research problem/goals/questions/propositions/hypotheses, and choices of methodological strategies. Figure 1.12b displays a generic mind map structure that could be used to explore and evaluate the implications of choosing a particular methodological strategy. The mind map not only helps to surface important issues but also anticipates associations and potential connections that the researcher could further explore.

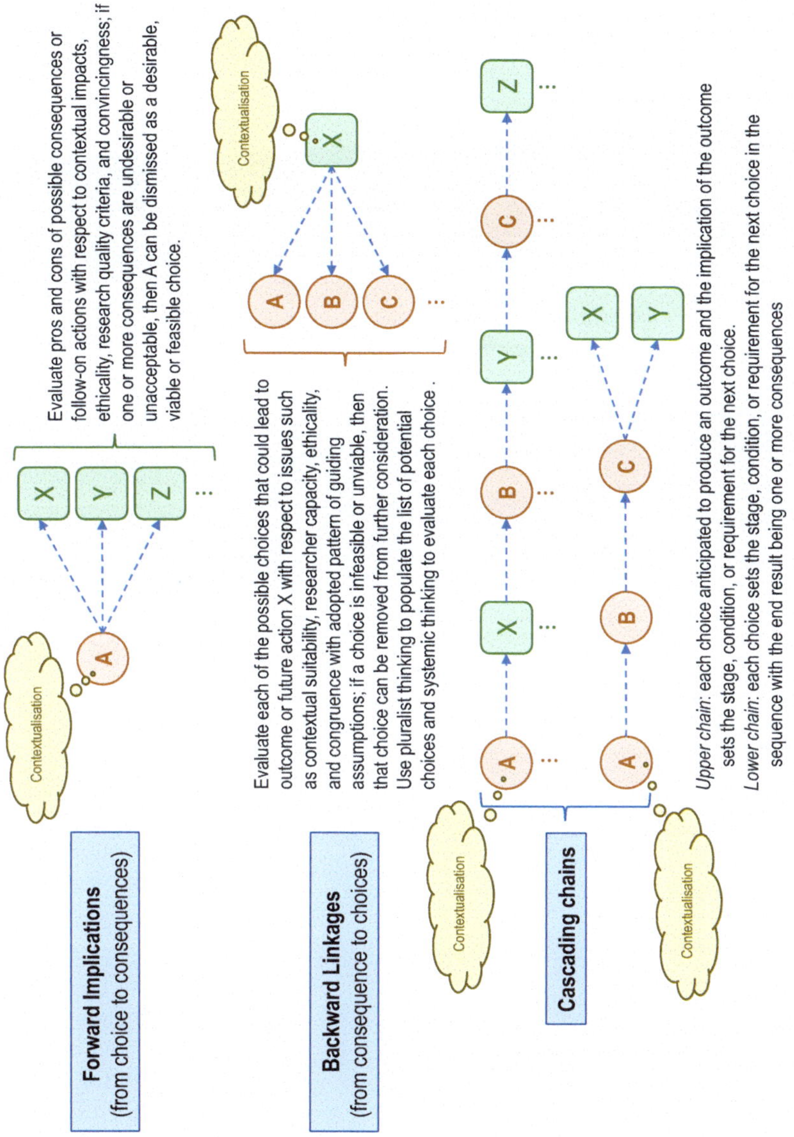

Fig. 1.11 Anticipatory imagining using choice – consequence reasoning chains

(a)

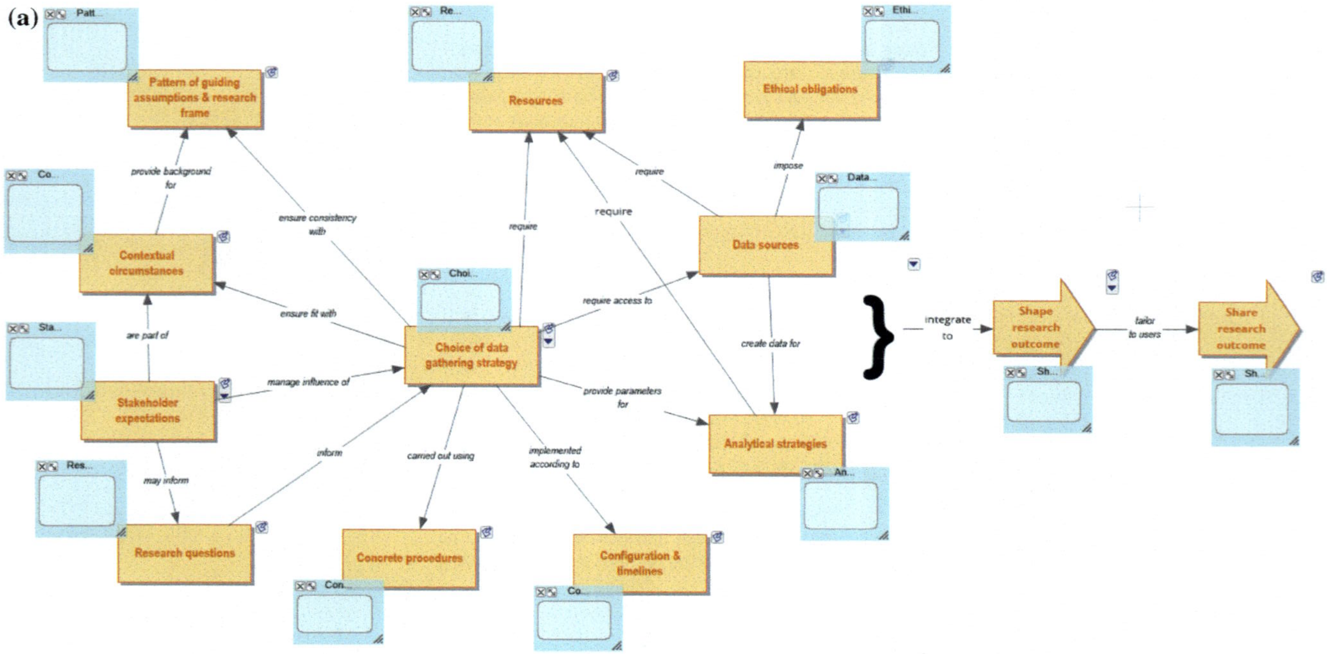

Fig. 1.12 **a** Generic cognitive mapping framework to support pluralist systemic thinking. **b** Generic mind mapping framework to support pluralist systemic thinking

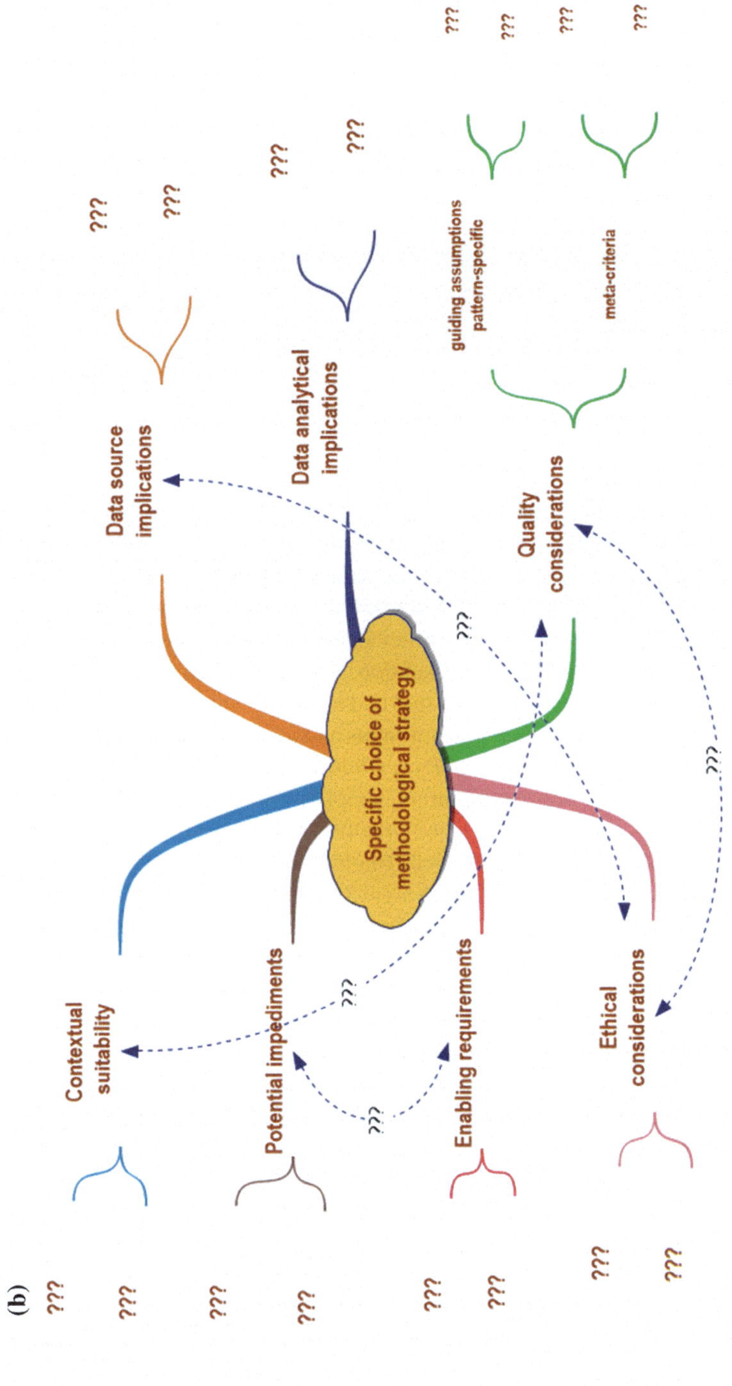

Fig. 1.12 (continued)

Furthermore, dynamic and conceptual connections between concept branches can be established to enrich the nonlinear structure and connectedness of meanings in the map. The mapping frameworks depicted in Fig. 1.12a, b are not fixed in nature but malleable and adaptable. This means they can be elaborated, reconfigured, and relabelled to help researchers explore a range of potential research-oriented methodological choices. Cognitive or mind mapping can generally be supported using a software package such as *Inspiration* (https://www.inspiration-at.com/; both Fig. 1.12a, b were produced using version 10 of *Inspiration*), *FreePlane* (https://docs.freeplane.org/home.html), *Decision Explorer* (https://banxia.com/dexplore/), *MindManager* (https://www.mindmanager.com/en/), or *PowerPoint* (https://www.microsoft.com/en-au/microsoft-365/powerpoint).

Contextualised Journey Mapping

Contextualised journey mapping is a visualisation process for capturing systemic contextualised thinking about the evolution of the researcher's journey. Contextualised journey mapping is similar to complexity mapping (see, for example, Cooksey (2000, 2001). It works by offering a prompting framework for recording thinking about contextual constraining and facilitating considerations and influences, and identifying critical cross-contextual influences that may impinge upon the researcher's journey; considerations and influences may be anticipated, actual, or a mixture. Figure 1.13 (produced using Microsoft *PowerPoint*) shows a visual framework for structuring a contextualised journey map that can be used to unpack, elaborate, and record contextualisation considerations and anticipations of critical leverage points tied to pivotal cross-contextual issues and tensions. Note that the framework in Fig. 1.13 contains a range of mnemonic prompts and guides to thinking. Researchers could sketch (or use software such as *PowerPoint* to produce) a skeletal version of such a framework in their research journal, simply by stripping out all the prompts (an A3-sized printout of such a sketched or software-produced skeletal framework would work well for the researcher to record their thinking). The four blue rectangles in Fig. 1.13 prompt thinking in different areas of contextual concern for a research journey (building on the representation offered in Fig. 1.4). The rectangles overlay the amorphous grey area dealing with larger contextual considerations, forming the background within which other contextualising considerations are embedded within or emerge. Each rectangle is dynamically linked to every other rectangle (via the red dashed arrows) to capture cross-contextual implications (some of these may reveal or create a critical leverage point for the researcher to consider). The red box is there simply as a reminder to the researcher to record their thinking about these cross-contextual influences. Contextualising thinking then feeds into the gold box, where mapping of the researcher's journey occurs; the prompts signalling choice points for the researcher to consider. An icon representing the research journey (extracted from Fig. 1.1) is included as an additional visual prompt in the gold box for mapping the journey. While mapping the journey, it may be prudent to revisit one's contextualising thinking in certain areas in light of choices made while mapping/undertaking the journey. The one dimension that cannot be easily captured in this 2-dimensional contextualising journey map is the time dimension (see the discussion and illustration in Cooksey, 2001, pp. 90–91), that is, the researcher should not assume that their map provides a static representation, rather it should be thought of as a snapshot in time and that, as time passes, there may be new dynamic changes that emerge that will impact their research journey. Thus, it will be important to regularly revisit this map as the research journey unfolds.

Quality Mapping

When making choices with respect to aspects of the research journey, the researcher should never lose sight of how their choices influence the quality of the research they are undertaking. There are various quality criteria the researchers need to address, some associated with a specific pattern of guiding assumptions they have adopted and some that operate above the level of patterns of guiding assumptions (differentiated as *pattern-specific criteria* and *meta-criteria*, respectively; to be explored much more fully in Chap. 4). Convincingness is the primary driver of quality considerations, from the perspective of the intended audience(s), and inferences about convincingness are informed by how the research journey maps out against the meta-criteria, which are, as part of their considerations, informed by relevant pattern-specific criteria. Quality mapping (see Fig. 1.14) offers a visual framework for recording one's thinking about how their choices may impact on research quality. The researcher should describe the pattern of guiding assumptions they are building upon and should try to elaborate how convincingness might be perceived by relevant audience(s) (which may, in turn, highlight critical leverage points for meta-criteria). The left (blue) side of the diagram focuses on pattern-specific quality criteria; the right (green) side on meta-criteria (mostly the contextualisation and realisation domains of meta criteria but anticipations with respect to the explication domain can also be captured; see discussions in Chap. 4). How each criterion or meta-criterion is relevant can be unpacked and elaborated using Notes and Steps to address. Note that this framework produces a high-level holistic perspective, so it would be more productive to employ this framework with another support system that facilitates drilling-down.

1.6.3 Caveats Associated with Research Planning

In Sect. 1.6.2, we introduced a range of potential support frameworks to help researchers contextualise and understand the choices they make during their research journey. This range of frameworks is more illustrative of possibilities rather than exhaustive and each framework is malleable and adaptable, able to be shaped to meet specific researcher needs. We have argued that the research journey can be influenced by a diverse range of forces, such as demands, expectations, constraints, and opportunities that emerge from the researcher's own context(s) ('internal forces') as well as from sources other than the researcher ('external forces'). Referring back to Fig. 1.1, external forces are depicted by the arrows with the question marks,

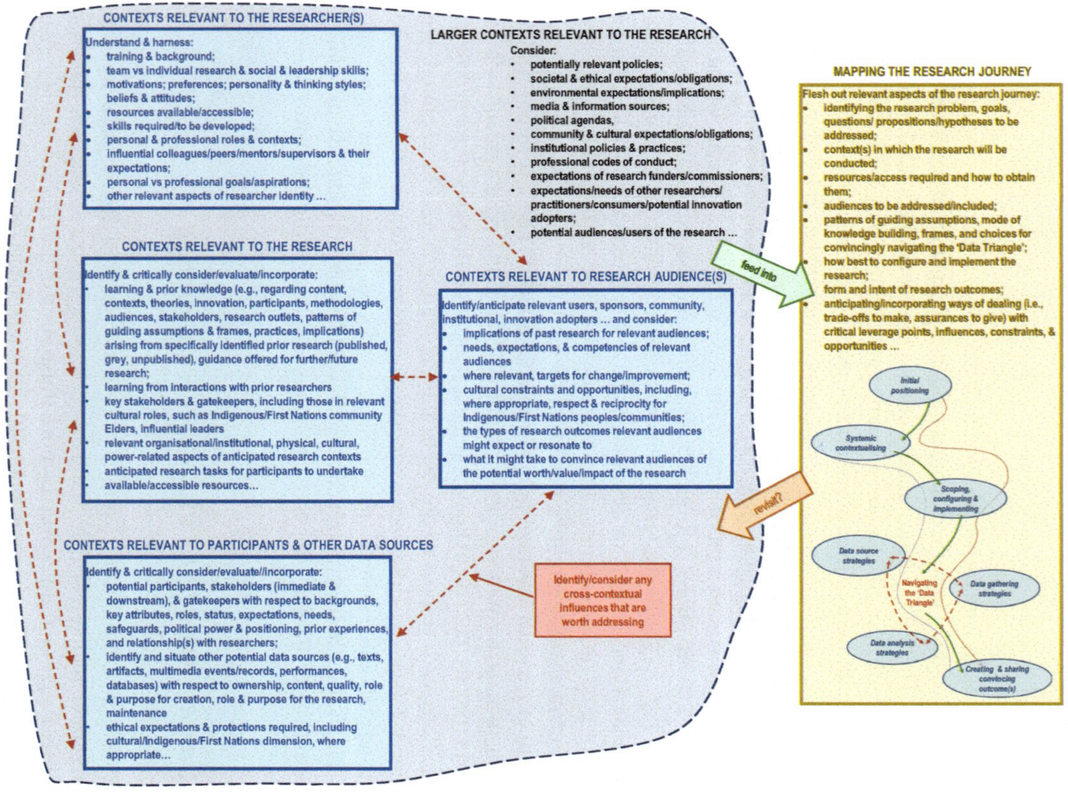

Fig. 1.13 Stylised contextualised journey map with mnemonic prompts and guides

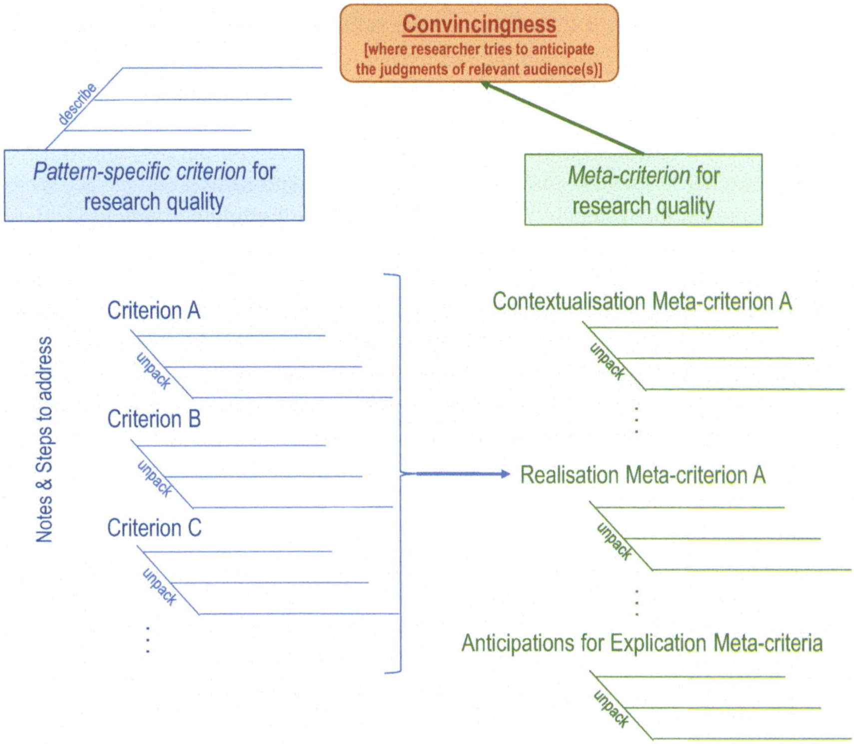

Fig. 1.14 Mapping holistic journey considerations against pattern-specific research quality criteria and meta-criteria

and some may have the power to help shape a research journey, usually by stimulating trade-offs and/or adaptations. The researcher only controls some of these forces (mostly internal forces) whilst being controlled, directed, or constrained by other forces (mostly external forces). Some of these forces can be anticipated and managed from the start (the green ? arrows in Fig. 1.1) whereas others may emerge during the research journey (the red ? arrows in Fig. 1.1), requiring the researcher to adapt. The existence of external forces creates the need to consider the political and ethical implications and potential applications of research and may strongly influence the researcher's choice of research frame (to be discussed in more detail in Chap. 4). The final goal of the research journey, of course, is to feasibly achieve an outcome that speaks convincingly to one or more target audiences. Any outcome from research may have impact on, or implications for, not only the researcher's immediate research context(s) but also for certain context(s) outside the boundaries of the immediate research journey. How well and how transparently the researcher traverses this network of connections and addresses the forces that create pressures on those connections, ultimately influences how convincing the final research outcome(s) will be. Different researchers may follow very different pathways between the various phases of research, even if they focus on the same issue, problem, or topic (highlighting the phenomenon of equifinality referred to earlier). Thus, there is no single best way to plot a pathway through the various phases depicted in Fig. 1.1; that is, there is no one right pathway for carrying out social and behavioural research.

Planning for a research journey is intimately connected to, but also constrained by, the researcher's research goals and questions/propositions/hypotheses, access to needed resources, the research frame(s) and pattern(s) of guiding assumptions they adopt, pattern-specific research quality criteria and meta-criteria, and any decisions and sacrifices they make along the way. Effective research planning cannot occur in a vacuum, nor can effective planning anticipate everything that might happen along the way. As we mentioned earlier, it is important to realise that the research journey is different for every researcher. A research journey is just one episode, unfolding in parallel with other journeys in a researcher's life (a very absorbing, time-intensive, tiring and potentially rewarding episode, to be sure) and the journey they take will be an intensely personal one. A high degree of involvement and personal investment in their research is essential for (but not determinant of) their success. However, the downside of this level of commitment is that it may make letting go of aspects of the research plan, if circumstances demand, more difficult. Cultivating flexibility and adaptability as well as an appreciation for diversity and pluralism in patterns of guiding assumptions, research frames, and strategy choices is one important way to combat the sometimes-paralysing effects of the researcher feeling like they must give away an important part of their research plan because it simply isn't working or is no longer feasible to implement. The adaptive researcher is one who can (1) see other pathways along the journey to their end goal besides the one originally planned and (2) is willing to switch paths when required, even if it takes their journey in a different direction than they originally intended.

Feasibility constraints, such as constrained resources, lack of access, attrition amongst participants, and ethical obligations, will strongly influence the researcher's choice of strategies for navigating the 'Data Triangle' in a convincing manner. Many such constraints will be imposed by others and the researcher will simply need to adapt to them by making trade-offs. This is where planning hits its limits. Researchers cannot anticipate how their journey will unfold to the last detail, simply because many constraints and opportunities unexpectedly emerge downstream. All that researchers can do is try to anticipate those emergences, where possible, but be ready to adapt and change as circumstances require. This does not mean that planning is a waste of time and resources; it means acknowledging that planning only goes so far. It is important not to be so wedded to one's plan that flexibility and adaptability are inhibited. Planning creates a sense of certainty and security, and it is a natural human reaction to allow that sense of certainty and security to create an illusion of control (Mintzberg, 1993 initially flagged the illusion of control as a problem associated with strategic planning in organisations, but the problem arguably holds for research planning as well). This natural human proclivity to reduce uncertainty must be resisted because if a research plan becomes set in concrete, the researcher will not be motivated to deviate from it, even if circumstances are sending clear signals that deviations are indeed required. A rigidly adhered-to plan becomes an anchoring point for downstream choices as well as a filter for interpreting feedback and generating feedforward. It also stifles pluralist thinking and circumscribes/short-circuits systemic thinking. The best way forward with respect to the role that research planning should play in a research journey is to engage in planning using pluralist systemic thinking to envisage as much of the research journey as possible but treat it as a dynamically evolving process that needs to be continually revisited, revised, reconsidered, and/or reshaped. To do this effectively requires serious, open-minded, and non-defensive attentiveness to feedback and re-engagement of pluralist systemic thinking when generating feedforward for downstream research choices and activities. One strategy to consider, where a downstream constraint or an opportunity is anticipated, to have a back-up plan (i.e., a Plan B) in place, if they do, in fact, emerge. For example, if the researcher anticipates that they might be denied access to certain participants, they could have a Plan B which sets out an alternative pathway to sourcing participants. Another strategy to employ could be for the researcher to build into their plan a pilot test or trialling process for evaluating their initial choices for navigating the 'Data Triangle', study the feedback information that those events produce, and then make adaptive changes (i.e., feedforward) where needed.

1.7 Where to from Here?

Chapter 2 will sketch out the continuously evolving landscape of potential data-focused methodological choices available to the researcher. A critical aspect of pluralist systemic thinking involves making choices about what to include and what to exclude when navigating this landscape for localised research purposes, while recognising that some choices may change in response to emergent feedback and learning. Some choices are made by the researcher in advance, other choices may emerge downstream as the research unfolds, and still other choices may be thrust upon the researcher by outside contextual forces. The researcher's localised landscape for a specific research journey can be conceptually represented as the 'Data Triangle'. The 'Data Triangle' has three vertices: (1) data source strategies; (2) data gathering strategies; and (3) data analysis strategies. The researcher's goal is to effectively navigate the 'Data Triangle' to construct a clear and convincing story about what has been learned from the data. Often, researchers must iteratively cycle between the 'Data Triangle' vertices to achieve a complete picture of what they are learning, and emerging feedback as well as adaptive feedforward may cause them to alter their navigational trajectory. Feasibility/viability concerns may cause the researcher to rethink their pathways through the 'Data Triangle'.

This landscape represents the toolkit for the pluralist systemic thinker. However, it is a lifeless and colourless landscape of choices devoid of guidance about how to navigate through it for a specific research journey. One cannot do social and behavioural research simply by knowing about the landscape itself. Motivation, context, knowledge, assumptions, skills, goals and intentions, and reasoning are all needed to steer the research journey through the landscape. These are the considerations that will bring the landscape to life, giving it shape, colour, texture, sensitivity, meaning, and purpose and the remaining chapters are devoted to exploring those considerations.

Chapter 3 will explore the systemic contextualisation of research in order to situate and position the research with respect to relevant and potentially influential considerations. These contextual considerations include: where the researcher is coming from (e.g., their identity, systems they are embedded within) and what they wish to learn (problems, goals, research questions/propositions/ hypotheses); who/what potential data sources might be and why; what has been learned before (published and unpublished); where and under what conditions the researcher might wish to conduct their research; who relevant stakeholders might be and their expectations; who relevant gatekeepers might be that control access to desired data sources; as well as larger contextual influences which may be relevant to the research (e.g., legal, ethical and regulatory, resource-based, environmental, community-based, market-based, institutional/organisational, governmental, sociocultural, technological, political, economic). Contextualisation provides colour, shape, and texture to the methodological landscape, bringing some potential choices into the foreground and allowing other potential choices to recede into the background. Contextualisation is not just a one-off 'upstream' activity for the researcher, it is continually revisited throughout their research journey. Nothing remains static and the researcher must be actively vigilant for contextual changes that may exert pressure to influence how their journey is progressing.

Chapter 4 will concentrate on the focusing and framing aspects of systemic research contextualisation. To provide focus, the researcher chooses one or more *pattern(s) of guiding assumptions* to underpin their intended research project/journey. To establish one's guiding assumptions, we downplay the more restrictive notion of 'paradigm' (Kuhn, 2012) and rely instead on the patterning of answers to pivotal questions about the specific orientation and intentions of the research. A pattern typically rules some methodological choices in and other choices out. We will see that every pattern of guiding assumptions invokes specific criteria for evaluating research quality. We also discuss meta-criteria for evaluating research quality, centred around the concept of *convincingness*, which transcend specific guiding assumption patterns (yet without ignoring or dispensing with pattern-specific considerations) and facilitate comparisons of research processes and outcomes arising under different patterns. Finally, we consider choices of *research frame* as a holistic and systemic way of capturing the essence of the research journey, encompassing situating, positioning, and guiding assumption contextualisations, choices of mode of knowledge building/production, and learning focus for intended audiences. Within the ambit of a research frame, downstream methodological choices are scoped, configured, and implemented and approaches to communicating and sharing research learning are enacted.

Chapter 5 will explore the choices a researcher makes to convincingly scope, configure, and implement their activities around the 'Data Triangle'. These enacting choices follow from and are more focused than choices related to contextualisation, patterns of guiding assumptions, and framing, and provide concrete pathways for the researcher to traverse in order to gather and analyse the data they need, from the sources they intend to or need to connect with. Here is where the landscape of methodological choices gets shaped, configured, pruned, and detailed (in short, 'brought to life') to produce the 'Data Triangle' the researcher navigates. We examine the issue of trade-offs, namely the adaptive decisions a researcher must sometimes make during their journey in anticipation of future situations or events (feedforward) or in response to emergent, perhaps unexpected, situations or events (feedback). With respect to configuring research activities, we utilise the concept of a 'Method Unit' (MU, which synergistically combines one data gathering strategy, one form of data, and one type of data source) as the fundamental building block. MUs provide the most flexible way of implementing pluralist systemic thinking. Following on from research configuration, we discuss implementation considerations underpinning each component of each MU that must be addressed.

Chapter 6 will explore the 'downstream' end of the research journey, where research learning, knowledge, and outcomes are shared with/communicated to relevant audiences and interested stakeholders. The researcher shapes the stories they wish to convey about their research, translating their choices made throughout their journey into forms that will resonate with intended audiences. Communication may be largely unidirectional (researcher to audience). In other cases, research sharing may be multi-directional, participative, and reciprocal, involving the researcher and interested and/or potentially impacted people/communities, and may also encompass direct or indirect actions or impacts. The researcher's stories/actions should be influenced by the choices made throughout their journey, critically reflect on relevant contextualisations, acknowledge the influences/impacts of relevant stakeholders/gatekeepers, where appropriate, and be guided by meta-criteria for research quality. A well-maintained research journal will prove immensely helpful in shaping those stories/actions. If the research produces a concrete innovation as an outcome, then at least some of the research sub-stories will need to address efforts to evaluate, share, facilitate, and spread adoption of that innovation to interested potential users and communities. If the research recommends specific actions to be undertaken, the arguments, contexts, and ultimate goals for such actions need to be clearly spelled out.

Figure 1.15 displays a cognitive mapping of the nonlinear relationships between the six chapters in this book. This nonlinearity is reflected in the reference to/use of concepts and processes in an earlier chapter which are discussed in more detail in a later chapter (especially true of Chaps. 3 and 4). The green solid arrows depict the general feedforward progressive linkages between the chapters. Note, though, that the connections between Chaps. 2 and 5 are strongly influenced by contextualisation considerations in Chaps. 3 and 4 such that landscape choices reflected in Chap. 2 create cascading implications for choices made in Chap. 5 (which are also shaped by choices made in Chaps. 3 and 4). The red dashed arrows depict feedback loops where choices made in a later chapter are revisited by thinking about/reflecting on earlier considerations and, where required, moving the researcher to make one or more trade-off choices or reshape one or more of the stories they are conveying in a research outcome. Such linkages are stimulated by the emergence of new contextual information, constraints, or opportunities. The shaded diamond

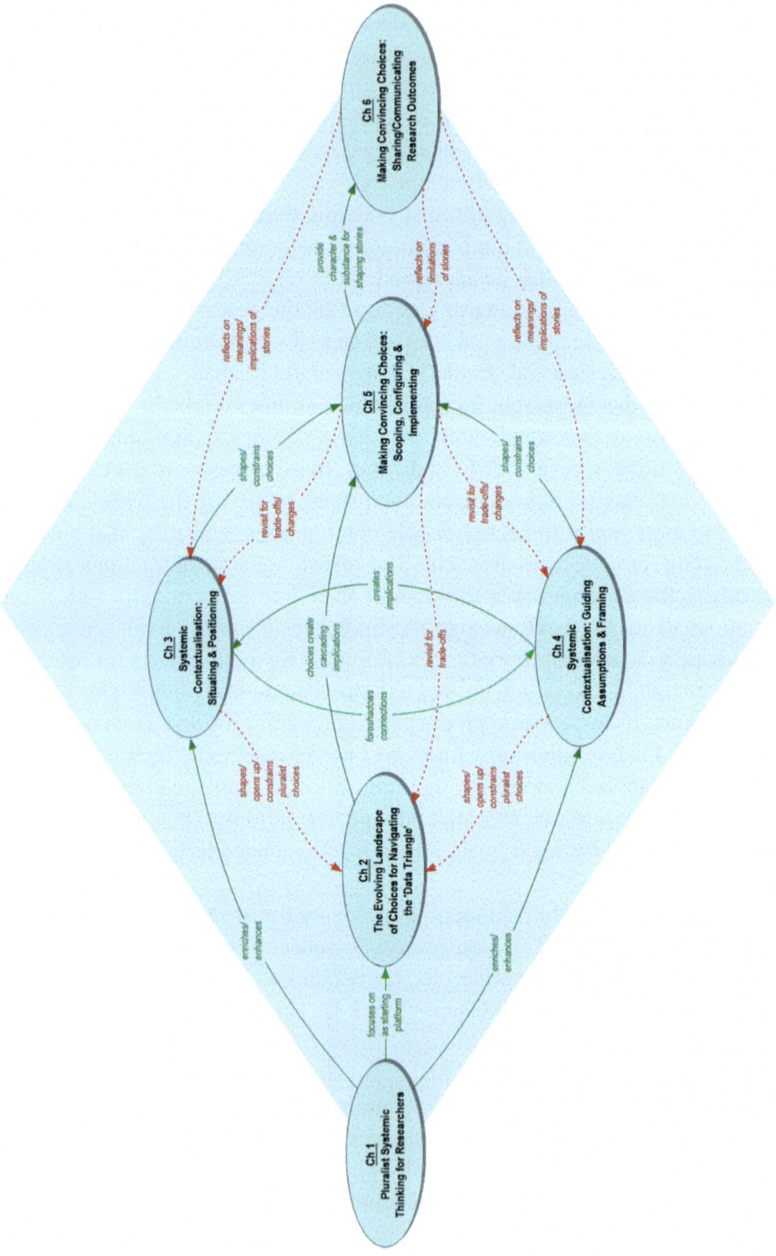

Fig. 1.15 The nonlinear flow between chapters showing cross-linkages, feedforward and feedback loops

represents the continuing influence of pluralist thinking in all aspects of the research journey, initially through shaping/opening up/possibly constraining thinking about contextualisation choices, then through narrowing/constraining choices and helping to shape the stories in research outcomes, based on the emergent contextualisations and subsequent trade-offs. This means that pluralist systemic thinking permeates the entire research journey.

References

Aiello, J., & Nero, S. (2019). Discursive dances: Narratives of insider-outsider research tensions. *Journal of Language, Identity, & Education, 18*(4), 251–265.

Alvesson, M., & Sandberg, J. (2021). *Re-imagining the research process: Conventional and alternative metaphors.* Sage Publications.

Archibald, J. (2008). *Indigenous storywork: Educating the heart, mind, body, and spirit.* UBC Press.

Archibald, J., Xiiem, Q., Lee-Morgan, J., & Santolo, J. (Eds.). (2022). *Decolonizing research: Indigenous storywork as methodology.* Bloomsbury.

Argyris, C., & Schön, D. (1978). *Organizational Learning.* Addison-Wesley.

Argyris, C. (1976). Single-loop and double-loop models in research on decision making. *Administrative Science Quarterly, 21*(3), 363–375.

Argyris, C. (1990). *Overcoming organizational defenses: Facilitating organizational learning.* Allyn & Bacon.

Åsvoll, H. (2014). Abduction, deduction and induction: Can these concepts be used for an understanding of methodological processes in interpretative case studies? *International Journal of Qualitative Studies in Education, 27*(3), 289–307.

Ballenger, B. (2004). *The curious researcher: A guide to writing research papers.* Pearson Longman.

Banks-Wallace, J. (2008). Eureka! I finally get IT: Journaling as a tool for promoting praxis in research. *Association of Black Nursing Faculty Journal, 19*(1), 24–27.

Bazeley, P. (2018). *Integrating analyses in mixed methods research.* Sage Publications.

Bazeley, P. (2019). *A practical introduction to mixed methods research for business and management.* Sage Publications.

Beach, L. (2016). *A new theory of mind: The theory of narrative thought.* Cambridge Scholars Publishing.

Beach, L. (2021). Scenarios as narratives. *Futures & Foresight Science, 3*:e58. https://doi.org/10.1002/ffo2.58.

Blair, N. (2022). Lilyology as a transformative framework for decolonizing ethical spaces within the academy. In J. Archibald, J. Lee-Morgan & J. De Santolo (Eds.), *Decolonizing research: Indigenous storywork as methodology.* Bloomsbury Academic, pp. 203–223.

Bleakley, A. (1999). From reflective practice to holistic reflexivity. *Studies in Higher Education, 24*(3), 315–330.

Bowers, T. (2011). Towards a framework for multiparadigm multimethodologies. *Systems Research and Behavioral Science, 28*(5), 537–552.

Boyd, P., & Smith, C. (2016). The contemporary academic: Orientation towards research work and researcher identity of higher education lecturers in the health professions. *Studies in Higher Education, 41*(1), 678–695.

Brocklesby, J. (1997). Becoming multimethodology literate: An assessment of the cognitive difficulties of working across paradigms. In J. Mingers & A. Gill (Eds.), *Multimethodology: The theory and practice of combining management science methodologies* (pp. 189–216). John Wiley & Sons.

Brunswik, E. (1952). *The conceptual framework of psychology* (International Encyclopedia of Unified Science, vol. 1, no. 10). University of Chicago Press.

Bryson, J., Ackermann, F., Eden, C., & Finn, C. (2004). *Visible thinking: Unlocking causal mapping for practical business results*. John Wiley & Sons.

Buzan, T. (2018). *Mind map mastery*. Watkins Media Limited.

Cameron, R. (2011). Mixed methods research: The five Ps framework. *Electronic Journal of Business Research Methods, 9*(2), 96–108.

Campbell, D., & Fiske, D. (1959). Convergent and discriminant validation by the multitrait-multimethod matrix. *Psychological Bulletin, 56*(2), 81–105.

Carcary, M. (2009). The research audit trail: Enhancing the trustworthiness in qualitative inquiry. *Electronic Journal of Business Research Methods, 7*(1), 11–24.

Castelló, M., McAlpine, L., Sala-Bubaré, A., Inouye, K., & Skakne, I. (2021). What perspectives underlie 'researcher identity'? A review of two decades of empirical studies. *Higher Education, 81*, 567–590.

Charmaz, K. (2014). *Constructing grounded theory: A practical guide through qualitative analysis*. Sage Publications.

Chilisa, B. (2020). *Indigenous research methodologies* (2nd ed.). Sage Publications.

Clarke, K. (2009). Uses of a research diary: Learning reflectively, developing understanding and establishing transparency. *Nurse Researcher, 17*(1), 68–76.

Coates, J. (2000). Scenario planning. *Technological Forecasting and Social Change, 65*(1), 115–123.

Cooksey, R. (2000). Mapping the texture of managerial decision making: A complex dynamic decision perspective. *Emergence: A Journal of Complexity Issues in Organizations and Management, 2*(2), 102–122.

Cooksey, R. (2001). What is complexity science? A contextually-grounded tapestry of systemic dynamism, paradigm diversity, theoretical eclecticism, and organizational learning. *Emergence: A Journal of Complexity Issues in Organizations and Management, 3*(1), 77–103.

Cooksey, R., & McDonald, G. (2019). *Surviving and thriving in postgraduate research* (2nd ed.). Springer Nature.

Creswell, J., & Plano Clark, V. (2017). *Designing and conducting mixed methods research* (3rd ed.). Sage Publications.

Crotty, M. (1998). *The foundations of social research: Meaning and perspective in the research process*. Allen & Unwin.

Davies, M. (2011). Concept mapping, mind mapping and argument mapping: What are the differences and do they matter? *Higher Education, 62*, 279–301.

Denzin, N. (2012). Triangulation 2.0. *Journal of Mixed Methods Research, 6*(2), 80–88.

Engin, M. (2011). Research diary: A tool for scaffolding. *International Journal of Qualitative Methods, 10*(3), 296–306.

Fear, F., Rosaen, C.L. Bawden, R., & Foster-Fuishman, P. (2006). *Coming to critical engagement: An autoethnographic exploration*. University Press of America.

Fedorov, S. (ed.). (2020). *Research trends in multidisciplinary research and development* (Volume 6). Weser Books.

Flick, U. (2017). Mantras and myths: The disenchantment of mixed-methods research and revisiting triangulation as a perspective. *Qualitative Inquiry, 23*(1), 46–57.

Flick, U. (2023). *An introduction to qualitative research* (7th ed.). Sage Publications.

Flood, R., & Romm, N. (1997). *Diversity management: Triple loop learning*. John Wiley & Sons.

Flood, R., & Romm, N. (2018). A systemic approach to processes of power in learning organizations: Part I–literature, theory, and methodology of triple loop learning. *The Learning Organization, 25*(4), 260–272.

Gates, G., & Cooksey, R. (1998). Learning to manage and managing to learn. *Journal of Workplace Learning, 10*(1), 5–14.

Giampapa, F. (2011). The politics of "being and becoming" a researcher: Identity, power, and negotiating the field. *Journal of Language, Identity, & Education, 10*(3), 132–144.

Glaze, J. (2002). PhD study and the use of a reflective diary: A dialogue with self. *Reflective Practice: International and Multidisciplinary Perspectives, 3*(2), 153–166.
Gobo, G., Fielding, N., La Rocca, G, & van der Vaart, W. (2022). *Merged methods: A rationale for full integration.* Sage Publications.
Hamilton, K., Dunnett, S., & Downey, H. (2012). Researcher identity: Exploring the transformatory power of the research experience. *Journal of Consumer Behaviour., 11*(4), 275–282.
Harvey, J. (2013). Footprints in the field: Researcher identity in social research. *Methodological Innovations Online, 8*(1), 86–98.
Hayfield, N., & Huxley, C. (2015). Insider and outsider perspectives: Reflections on researcher identities in research with lesbian and bisexual women. *Qualitative Research in Psychology, 12*(2), 91–106.
Hayman, B., Wilkes, L., & Jackson, D. (2012). Journaling: Identification of challenges and reflection on strategies. *Nurse Researcher, 19*(3), 27–31.
Healey, M., & Hodgkinson, G. (2008). Troubling futures: Scenario planning for organizational decision making. In G. Hodgkinson & W. Starbuck (Eds.), *The Oxford handbook or organizational decision making* (pp. 565–585). Oxford University Press.
Hinds, R. (2023, May 16). How "perspective swaps" can unlock organizational change, *Harvard Business Review Change Management*, https://hbr.org/2023/05/how-perspective-swaps-can-unlock-organizational-change.
Hughes, S., & Pennington, J. (2016). *Autoethnography: Process, product, and possibility for critical social research.* Sage Publications.
Hummelbrunner, R., & Reynolds, M. (2013). Systems thinking, learning and values in evaluation (pp. 9–10). *The European Evaluation Society Newsletter*, June.
Jick, T. (1979). Mixing qualitative and quantitative methods: Triangulation in action. *Administrative Science Quarterly, 24*(4), 602–611.
Kennedy, B., & Thornberg, R. (2018). Deduction, induction, and abduction. In U. Flick (Ed.), *The Sage handbook of qualitative data collection* (pp. 49–64). Sage Publications.
Kerstetter, K. (2012). Insider, outsider, or somewhere between: The impact of researchers' identities on the community-based research process. *Journal of Rural Social Sciences, 27*(2), 99-117.
Kincheloe, J., & Berry, K. (2004). *Rigour & complexity in educational research: Conceptualizing the bricolage.* Open University Press.
Kincheloe, J. L. (2005). On to the next level: Continuing the conceptualization of the bricolage. *Qualitative Inquiry, 11*(3), 323–350.
Kirkness, V., & Barnhardt, R. (1991). First Nations and higher education: The four R's—respect, relevance, reciprocity, responsibility. *Journal of American Indian Education, 30*(3), 1–15.
Kleiner, A. (1994). Creating scenarios. In P. Senge, A. Kleiner, C. Roberts, R. Ross, & B. Smith (eds.), *The fifth discipline fieldbook: Strategies and tools for building a learning organization.* Currency Doubleday, pp. 275–278.
Kovach, M. (2021). *Indigenous methodologies: Characteristics, conversations, and contexts* (2nd ed.). University of Toronto Press.
Kuhn, T. (2012). *The structure of scientific revolutions* (Rev. ed.). The Folio Society.
Lamb, D. (2013). Promoting the case for using a research journal to document and reflect on the research experience. *Electronic Journal of Business Research Methods, 11*(2), 84–92.
Layder, D. (1993). *New strategies in social research.* Polity Press.
Le Gallais, T. (2008). Wherever I go there I am: Reflections on reflexivity and the research stance. *Reflective Practice, 9*(2), 145–155.
Leavy, P. (2011). *Essentials of transdisciplinary research: Using problem-centred methodologies.* Left Coast Press.
Lindgren, M., & Bandhold, H. 2009). *Scenario planning: The link between future and strategy.* Palgrave Macmillan.
Lury, C., Fensham, R., Heller-Nicholas, A., Lammes, S., Last, A., Michael, M., & Uprichard, E. (eds.). (2018). *Routledge handbook of interdisciplinary research methods.* Routledge.

Malthouse, R., Roffey-Barentsen, J., & Watts, M. (2014). Reflectivity, reflexivity and situated reflective practice. *Professional Development in Education, 40*(4), 597–609.
Marshall, A. (2019). The concept of reflection: A systematic review and thematic synthesis across professional contexts. *Reflective Practice, 20*(3), 396–415.
McDonald, S., Daniels, K., & Harris, C. (2004). Cognitive mapping in organizational research. In C. Cassell, & G. Symon (Eds.), *Essential guide to qualitative methods in organizational research* (pp. 73–85). Sage Publications.
Meadows, D. (2008). *Thinking in systems: A primer.* Chelsea Green Publishing.
Mertkan, S., & Bayrakli, H. (2018). Re-inventing researcher identity: When the individual interacts with the contextual power dynamics. *Higher Education Research & Development, 37*(2), 316–327.
Meyer, K., & Willis, R. (2019). Looking back to move forward: The value of reflexive journaling for novice researchers. *Journal of Gerontological Social Work, 62*(5), 578–585.
Midgley, G., Nicholson, J., & Brennan, R. (2017). Dealing with challenges to methodological pluralism: The paradigm problem, psychological resistance, and cultural barriers. *Industrial Marketing Management, 62*, 150–159.
Miles, M., Huberman, A., & Saldaña, J. (2019). *Qualitative data analysis: A methods sourcebook* (4th ed.). Sage Publications.
Mingers, J. (2014). *Systems thinking, critical realism and Philosophy: A confluence of ideas.* Routledge.
Mingers, J., & Brocklesby, J. (1997). Multimethodology: Towards a framework for mixing methodologies. *Omega, 25*(5), 489–509.
Mingers, J. & Gill, A. (eds.). (1997). *Multimethodology: Towards theory and practice and mixing and matching methodologies.* John Wiley & Sons.
Mintzberg, H. (1993). The pitfalls of strategic planning. *California Management Review, 36*, 32–32.
Moran-Ellis, J., Alexander, V. D., Cronin, A., Dickinson, M., Fielding, J., Sleney, J., & Thomas, H. (2006). Triangulation and integration: Processes, claims and implications. *Qualitative Research, 6*(1), 45–59.
Morse, J. M., & Niehaus, L. (2009). *Mixed methods design: Principles and procedures.* Left Coast Press.
Muhammad, M., Wallerstein, N., Sussman, A., Avila, M., Belone, L., & Duran, B. (2015). Reflections on researcher identity and power: The impact of positionality on community based participatory research (CBPR) processes and outcomes. *Critical Sociology, 41*(7–8), 1045–1063.
Nadin, S., & Cassell, C. (2006). The use of a research diary as a tool for reflexive practice: Some reflections from management research. *Qualitative Research in Accounting & Management, 3*, 208–217.
Niblock, S. (2007). From "knowing how" to "being able": Negotiating the meanings of reflective practice and reflexive research in journalism studies. *Journalism Practice, 1*(1), 20–32.
Novak, J., & Cañas, A. (2006). The origins of the concept mapping tool and the continuing evolution of the tool. *Information Visualization, 5*(3), 175–184.
Olson, W. (2004). Triangulation in social research: Qualitative and quantitative can really be mixed. In M. Holborn, & M. Haralambos, (Eds.), *Developments in Sociology.* Causeway Press Ltd., pp. 1–30.
Ortlipp, M. (2008). Keeping and using reflective journals in the qualitative research process. *The Qualitative Report, 13*(4), 695–705.
Patton, M. (2011). *Developmental evaluation: Applying complexity concepts to enhance innovation and use.* The Guilford Press.
Peschl, M. (2007). Triple-loop learning as foundation for profound change, individual cultivation, and radical innovation: Construction processes beyond scientific and rational knowledge. *Constructivist Foundations, 2*(2–3), 136–145.
Pidd, M. (2009). *Tools for thinking: Modelling in management science* (3rd ed.). John Wiley & Sons.

Plano Clark, V. & Ivankova, N. (2016). *Mixed methods research: A guide to the field.* Sage Publications.

Plowright, D. (2011). *Using mixed methods: Frameworks for an integrated methodology.* Sage.

Ponterotto, J., & Grieger, I. (1999). Merging qualitative and quantitative perspectives in a research identity. In M. Kopala, & L. A. Suzuki (Eds.), *Using qualitative methods in psychology* (pp. 49–62). Sage Publications.

Poth, C. (2018). *Innovation in mixed methods research: A practical guide to integrative thinking with complexity.* Sage Publications.

Ragin, C., & Amoroso, L. (2019). *Constructing social research: The unity and diversity of method* (3rd ed.). Sage Publications.

Reichertz, J. (2004). Abduction, deduction and induction in qualitative research. In U. Flick, E. von Kardorff, & I. Steinke (Eds.), *A companion to qualitative research* (pp. 159–164). Sage Publications.

Richards, L. (2020). *Handling qualitative data: A practical guide* (4th ed.). Sage Publications.

Salmons, J. (2020). *Little quick fix: What kind of researcher are you?* Sage Publications.

Sanbonmatsu, D., Cooley, E., & Butner, J. (2021). The impact of complexity on methods and findings in psychological science. *Frontiers in Psychology, 11*, 1–16.

Sayer, A. (2000). *Realism and social science.* Sage Publications.

Schoemaker, P. (1995). Scenario planning: A tool for strategic thinking. *MIT Sloan Management Review, 36*(2), 25–33.

Schoemaker, P. (2004). Forecasting and scenario planning: The challenges of uncertainty complexity. In D. Koehler & N. Harvey (eds.), *Blackwell handbook of judgment and decision making.* Blackwell Publishing, pp. 274–296.

Smith, L. (2021). *Decolonizing methodologies: Research and Indigenous peoples* (3rd ed.). Zed Books.

Sniderman, P., Hagen, M., Tetlock, P., & Brady, H. (1986). Reasoning chains: Causal models of policy reasoning in mass publics. *British Journal of Political Science, 16*(4), 405–430.

Taket, A. (1998). Book review: Diversity management: Triple loop learning. *Journal of the Operational Research Society, 49*(3), 293–296.

Tashakkori, A., & Teddlie, C. (eds.). (2010). *Sage handbook of mixed methods in social and behavioral research* (2nd ed.), Sage Publications.

Thomson, P., & Gunter, H. (2011). Inside, outside, upside down: The fluidity of academic researcher 'identity' in working with/in school. *International Journal of Research & Method in Education, 34*(1), 17–30.

Tosey, P., Visser, M., & Saunders, M. (2012). The origins and conceptualizations of 'triple-loop' learning: A critical review. *Management Learning, 43*(3), 291–307.

Uluğ, Ö. M., Acar, Y. G., & Kanık, B. (2021). Reflecting on research: Researcher identity in conflict studies from the perspectives of participants. *European Journal of Social Psychology, 51*(6), 847–861.

von Bertalanffy, L. (1950). The theory of open systems in physics and biology. *Science, 111*(2872), 23–29.

Yu, C. (1994). Abduction? Deduction? Induction? Is there a logic of exploratory data analysis? Paper presented at the Annual Meeting of the American Educational Research Association (New Orleans, LA).

Chapter 2
The Evolving Landscape of Choices for Navigating the 'Data Triangle'

Abstract This chapter sketches out the continuously evolving landscape of potential data-focused methodological choices available to the researcher. A critical aspect of pluralist systemic thinking involves making choices about what to include and what to exclude when navigating this landscape for localised research purposes, while recognising that some choices may change in response to emergent feedback and learning. Some choices are made by the researcher in advance, other choices may emerge downstream as the research unfolds, and still other choices may be thrust upon the researcher by outside contextual forces. The researcher's localised landscape for a specific research journey can be conceptually represented as the 'Data Triangle'. The 'Data Triangle' has three vertices: (1) data source strategies; (2) data gathering strategies; and (3) data analysis strategies. The researcher's goal is to effectively navigate the 'Data Triangle' to construct a clear and convincing story about what has been learned from the data. Often, researchers must iteratively cycle between the 'Data Triangle' vertices to achieve a complete picture of what they are learning, and emerging feedback as well as adaptive feedforward may cause them to alter their navigational trajectory. Feasibility/viability concerns may cause the researcher to rethink their pathways through the 'Data Triangle'.

In the social and behavioural sciences, there is a continuously evolving landscape of potential data-focused methodological choices available to the researcher. This landscape dynamically evolves as methodological innovations continue to emerge. A critical aspect of pluralist systemic thinking involves making choices about what to include and what to exclude in this landscape for localised research purposes, while recognising that those choices are being made within a dynamically changing environment in response to emergent feedback and learning. The researcher may make some choices in the early stages of their journey whereas other choices may emerge downstream as the research unfolds and the researcher needs to adapt to changing circumstances. Still other choices may be thrust upon the researcher by outside contextual forces. In this chapter, we summarise key facets of this landscape. However, it is important to realise that this landscape has no real substance, texture, or meaning for a specific research endeavour until and unless the researcher

R. W. Cooksey, *Unity from Diversity*, https://doi.org/10.1007/978-981-97-3462-7_2

makes choices based on relevant contextualisations, choices about patterns of guiding assumptions and frames, and choices about research activity scoping, configuration, and implementation (to be discussed in Chaps. 3, 4 and 5) to 'terraform' this landscape into a workable shape for their project.

The landscape of potential methodological choices (who or what to gather data from, how to gather the desired data and how to make sense of data that are gathered) in social and behavioural research can be displayed in a conceptual representation called the 'Data Triangle' (see Fig. 2.1; but also recall where the 'Data Triangle' sits in the visualisation of the research journey in Fig. 1.1). The 'Data Triangle' is a dynamic and systemic representation which depicts the connective pathways and feedback/feedforward links that operate between the three categories of strategies that anchor the vertices of the triangle as well as 'upstream' and 'downstream' from the triangle (these feedback/feedforward links will be discussed in more detail in Sect. 2.4). Navigating the 'Data Triangle' is an essential part of the research journey, comprising the data-oriented activities a researcher needs to undertake to address their research goals/ objectives/questions/propositions/hypotheses. A broad view of what constitutes information or data for research purposes is adopted: numerical/ measurement-based information and non-numerical information of various kinds (e.g., textual, symbolic, linguistic, para-linguistic, visual, multi-media, social media, performances, and creative works). This yields two broad forms of data for the researcher to draw upon: quantitative and qualitative, with choices ranging from one type alone to various blended/hybrid/merged mixtures. Technological supports of various kinds (recording devices, storage devices, hardware, software, etc.) may be drawn upon to assist in navigating the 'Data Triangle' at any stage, but each support system must be appropriately, ethically, sensitively, and contextually situated.

2.1 Data Source Strategies

One vertex of the 'Data Triangle' shown in Fig. 2.1 concerns strategies associated with choosing data sources. Figure 2.2 provides a mind map overview of data source strategies and their potential synergies. For our purposes, a data source will be considered as someone or something that provides data specifically relevant to the researcher's research questions/propositions/hypotheses or as someone or something that provides contextual or background information relevant to the research context or other data sources within that context. The latter instance may be associated with a gatekeeper role within that context. The traditional way of referring to how a researcher chooses data sources for their research is sampling (see, e.g., Eichhorn, 2021). However, I find the term 'sampling' to be overly simplistic, mechanistic in feel, and linear in execution; it also tends, unfortunately, to have statistical connotations which are not always applicable. I find it more appropriate to speak of data source choice strategies as pathways for focusing data gathering strategies.

Data sources may be human or non-human (e.g., texts/documents, YouTube segments, social media posts, research articles, movies, policies, paintings/drawings,

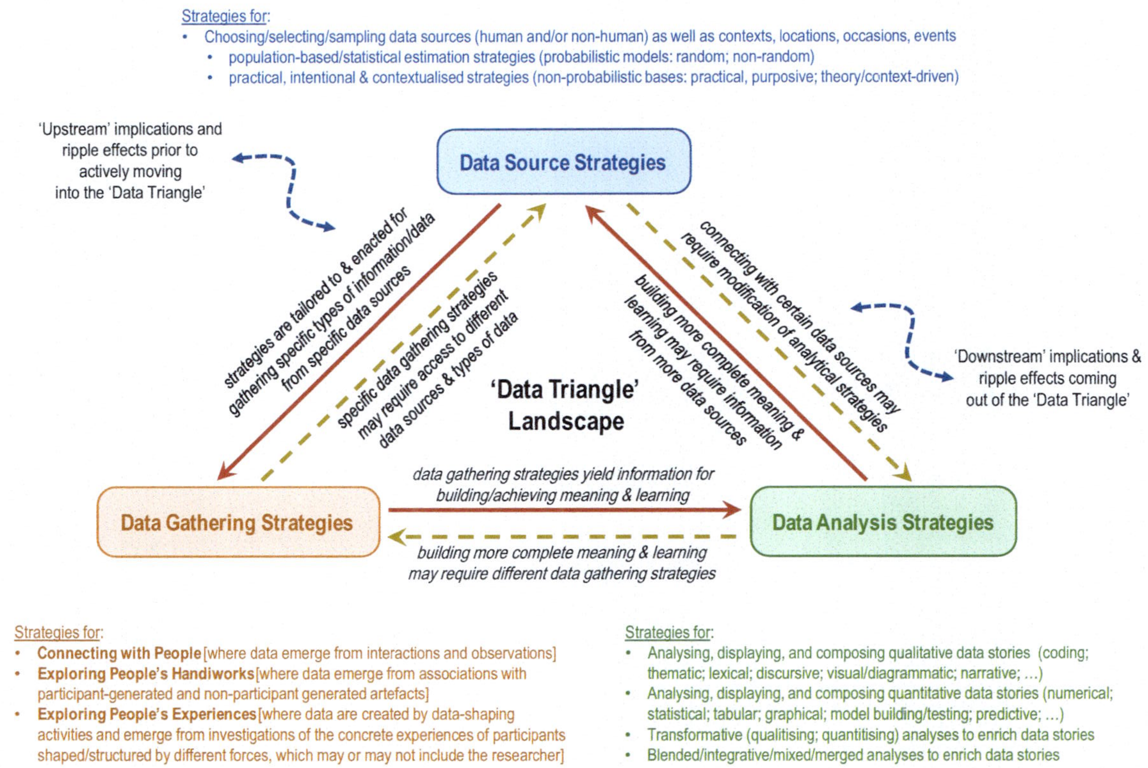

Fig. 2.1 Conceptual mapping of the 'Data Triangle' landscape of potential choices with dynamic interconnections between three sets of data-focused research activities

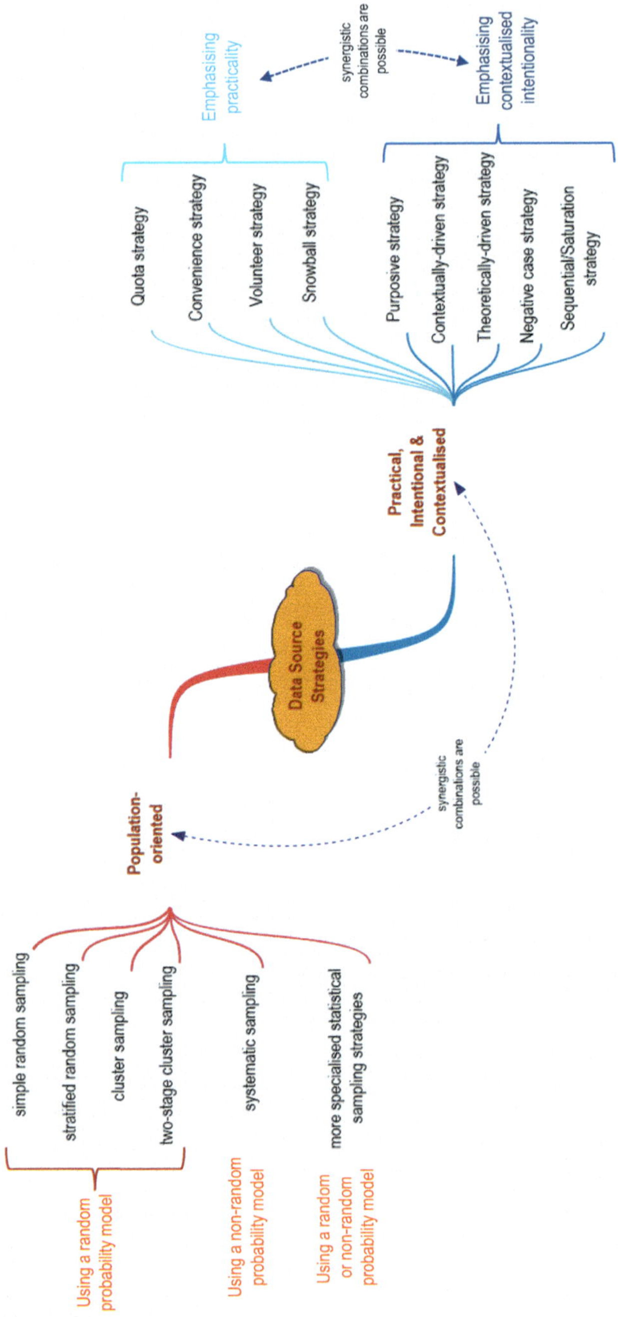

Fig. 2.2 Strategies and synergies for choosing data sources

cultural artefacts, secondary databases, performances, animals) in nature. Data source choice strategies may involve the use of numerical identifiers within a larger list of possible choices (e.g., *population-based choice strategies*) or specific researcher-identified, self-identified, or other-identified data sources (e.g., *practical, intentional, and contextualised strategies*). If the data sources that the researcher wants to choose are human (i.e., research participants), then their ethical right to give informed consent to participate in the research overrides the researcher's desire for them to participate, which means that participation is their choice, not the researcher's. This then becomes one of the aspects of social and behavioural research that researchers must acknowledge and cope with but cannot completely control.

Data source choice strategies also extend beyond simply choosing who or what one wants to gather research information from (consistent with the traditional view of sampling). They may also involve choices of the circumstances, contexts, and occasions in which data sources are encountered including, for example:

- how the researcher chooses organisations and/or groups to approach for participation in case studies and other types of field-based or naturalistic research;
- how the researcher chooses specific tasks or task design features for participants to experience or undertake;
- how the researcher chooses which handiworks, documents (e.g. text, digital, creative arts, multi-media) or other artefacts to collect/read/view/experience/analyse in their research;
- how the researcher chooses which contexts, situations, times, places, and/or events to observe/participate in and which conversations to engage in and/or record; and
- how the researcher chooses which participants to use as examples to illustrate/support their interpretations as well as how the researcher chooses which quotations or paraphrases to report.

Population-oriented sampling strategies are statistical in nature and require quantitative data for their implementation. The researcher implements the sampling strategy, but the actual choices made are identified through the mechanisms of different statistical probability models. Practical, intentional, and contextualised choice strategies do not rely upon probability models and may be relevant for research that emphasises qualitative and/or quantitative data. The researcher implements such strategies, but the mechanism for making choices may reside with the researcher themselves, with potential human data sources, or with other individuals—the common thread being that a person makes the choice, not a statistical algorithm.

It is important to note that in social and behavioural research, sampling human participants adds much more complexity to the research process, compared to sampling non-human data sources and artefacts. This complexity arises because human participants, acting as data sources, are motivated, perceiving, thinking, feeling beings who have their own lives, work, roles, and experiences and they don't cease being so when they participate in research. Rather it is their motivations, perceptions, cognitions, feelings, life and work roles and experiences that the researcher must, in some way acknowledge and/or account for in their data source choice strategies and, downstream, their data gathering and analysis activities.

Data sources choice strategies should maximise the chances of accessing the data the researcher needs to address their research questions, propositions, and/or hypotheses. A sound strategy can strongly influence how convincing the researcher can be in addressing research questions propositions, or hypotheses and how far the researcher can extend their findings beyond the specific boundaries of their study. Data source choice strategies have one of two main interests (which may or may not overlap) when actual choices are being made to assemble a sample: (1) sample size and (2) sample composition. In research that uses a strategy for gathering quantitative data, both sample size and sample composition are important to consider; sample size because it influences statistical efficiency, effectiveness, and stability (larger samples are better for these purposes); composition because it influences the extent to which defensible generalisations to a target population can be made. Sample size is much less of a concern in research that uses a strategy for gathering qualitative data. In this case, it is not how large the sample is that wins the day but how, who, when, where and what has been chosen, i.e., composition. In the discussions that follow, we will use the term 'data source' as a collective label to refer to the focal unit for sampling, be it human data sources such as a person, a group, an organisation or non-human data sources and handiworks such as documents, journal articles, YouTube clips, articles in newspapers or other media outlets, social media postings, performances, photographs, drawings, or other artworks. We will explore more fully potential criteria a researcher might employ to choose a specific data source strategy to implement in Chap. 5.

2.1.1 *Population-Oriented Sampling Strategies*

In research that uses statistical analysis to draw generalising inferences from a sample to a population using quantitative measurements, population-oriented sampling strategies, also known as probabilistic sampling strategies, provide a range of statistically appropriate methods for assembling a sample (see additional discussions and illustrations in Cooksey & McDonald, 2019, pp. 843–857). In any population-oriented sampling strategy, the researcher must: (1) describe precisely what the intended population for generalisations is and (2) describe who (or what) in that population they can feasibly gain access to, thus establishing the *sampling frame*. A sampling frame defines the relevant and accessible population for the purposes of the research. It may differ from the actual population to which one would like to generalise findings because the sampling frame may not contain all the members of the population. The sampling frame may also differ from the actual population to which one would like to generalise findings because of lags in its updating and maintenance. Thus, the sampling frame may contain data sources who are no longer members of the population or may not contain data sources who have most recently become members of the population. In a population-oriented sampling strategy, the probability that any specific data source in the sampling frame of the population would be selected for inclusion in the researcher's sample is known. There are two basic types of probability sampling model available: (1) *random*, where every potential data source in a specified sampling frame within the population has an equal chance of being selected in the sample (and every sample of a given size has an equal chance of being assembled); or (2) *non-random* or *systematic*, where some data sources in a specified sampling frame within the population have a greater chance of being selected in the sample than other data sources. Any data source strategy, where the probability of data sources being selected into the sample remains unknown or is irrelevant, is non-probabilistic and therefore non-random in nature; these will be discussed later as practical, intentional, and contextualised choice strategies.

For the purposes of making strong statistical inferences and generalisations with known precision and/or power, a probabilistic population-oriented sampling strategy is preferred over any non-probabilistic sampling strategy, a random probability model is preferred over a non-random probability model, and an adequate sample size is a necessary precondition for sound inferences. Research investigations that use a sample of inadequate size, even if that sample has been obtained using a random probability model, typically yield conclusions that should be carefully qualified with respect to how far the authors try to generalise their findings. The goal of a random probability model is to try to achieve a representative sample (with respect to a specified population of interest) with a given or desired level of precision in statistical estimation and/or with a given level of power to detect relationships of a specified size.

Simple Random Sampling

Simple random sampling is the most basic scheme that relies on a random probability model and can be accomplished via a very simple set of steps. To draw a random sample, the researcher assembles or procures the sampling frame in the form of a list showing all potential data sources uniquely identified using some key index such as an ID number, tax file number, or phone number. Random sampling then occurs by choosing specific values of this key index, identified by a random number generating algorithm. If this list (perhaps from a census or a complete listing of all current employees in a specific organisation) can be imported into a spreadsheet statistical program like Excel or SPSS, it is a relatively straightforward matter to draw a simple random sample.

Stratified Random Sampling

The stratified random sampling strategy is useful in situations where the simple random sampling strategy scheme, with its whole-of-population emphasis, runs the risk of under- or over-sampling data sources in specific mutually exclusive categories (i.e., non-overlapping strata) within the population. If membership in those categories is relevant to the propositions or hypotheses being tested, it becomes critical that those categories are adequately sampled. The process involves segmenting the population into relevant strata (e.g., categories defined, for example, by gender, geographical location, ethnic background, income band, tax bracket, type of organisation, management level) that the researcher suspects might influence patterns of relationships and therefore need to be controlled in some way. Stratified random sampling provides one such control mechanism through generating separate simple random samples within each stratum, each of a size that the researcher determines.

Depending upon needs, there are two distinct approaches to determining how large the sample within each stratum needs to be: (1) proportionate stratified random sampling, where sample sizes are proportional to the size of the stratum in the population/sampling frame (pursues proportional representation, based on stratum size, and if the strata are relatively homogeneous, results in more precise statistical estimates) or (2) disproportionate stratified random sampling, where, for example, there are several larger strata and one very small stratum (e.g., as with the Indigenous/First Nations sector of the population of Australia, Canada, or the US) and the very small stratum is over-sampled to ensure adequate representation and capacity to generate sound estimates at the level of that stratum whereas all other strata may be sampled proportionally.

Cluster Sampling

Cluster sampling is a somewhat more complicated scheme designed to enhance the feasibility of accessing a sample. For example, if the intended population is widely geographically dispersed, one may not have the resources needed to reach all areas for data gathering purposes. In this case, a logical sampling strategy would be to divide the population by geographic areas and randomly select a subset of areas to focus on. Each selected area would constitute a cluster, and one would then gather data from every individual entity within each cluster. Thus, cluster sampling is basically simple random sampling applied to clusters as the sampled entities. In general, clusters may be defined in various ways, but most often are defined by some type of geographical, physical, political or organisational boundary, such as states or territories in a country/nation, local councils, counties or shires in a state, suburbs in a city, departments or divisions in an organisation, police or fire brigade precincts, wilderness, recreation or national park areas, school districts, electoral districts, water catchment areas, and so on.

Two-Stage Cluster Sampling

Two-stage cluster sampling is a synergistic combination of cluster sampling and simple random sampling that is useful in situations where it is not feasible for a researcher to sample every individual within each cluster. The process is straightforward: (Stage 1) randomly choose clusters and (Stage 2) within in each chosen cluster, take a random sample of potential data sources. This sampling strategy largely negates one of the main disadvantages of cluster sampling, namely that every data source within a sampled cluster must be accessed.

Systematic Sampling

Systematic sampling relies on a non-random probability model since the basis of the scheme is a sequential progression through a listing of the sampling frame. To assemble a systematic sample, one randomly chooses a data source from the first k data sources in the sampling frame list, then selects every kth successive data source from the list. Such a sample is called a 1 in k systematic sample. It is considered a probability sample because the first selected data source has a probability of 1 in k of being chosen and, after that, every kth data source has a 1.0 probability of being selected (all data sources in between have a probability of 0.0 of being selected). The researcher must establish the value for k on the basis of the proportion of the population desired in the sample; k thus depends upon desired sample size.

2.1.2 Practical, Intentional, and Contextualised Strategies

The use of population-oriented sampling strategies is often (but not always) driven by statistical requirements for estimation, relationship testing, and generalisation using quantitative data. However, population-oriented sampling strategies form a smaller part of a much larger set of data source choice strategies that can be used to facilitate achieving a much wider range of research purposes, under more diverse patterns of guiding assumptions, using qualitative and/or quantitative data. The larger portion of this set comprises what can be termed practical, intentional, and contextualised sampling strategies (see also discussions and illustrations in Cooksey & McDonald, 2019, pp. 857–887). These are strategies that do not depend upon knowledge of the probability of data sources being chosen for a sample and do not utilise the principle of randomness as a sample selection criterion. They depend upon a human decision maker rather than a statistical algorithm to make the actual data source choices for and in specific contexts. Practical, intentional, and contextualised sampling strategies have evolved to offer the researcher new avenues for choosing data sources under more diverse sets of contextual circumstances and under different patterns of guiding assumptions. The quota, convenience, volunteer and snowball strategies all emphasise practicality for the researcher; the remaining five strategies all emphasise contextualised intentionality, choosing data sources to achieve specific goals.

Quota Strategy

The quota strategy is the non-probabilistic analogue to the stratified random sampling strategy and involves three discrete steps: (1) identify key dimensions, which should be relevant to establishing sample representativeness in light of the research questions, and associated categories of participants to be chosen, in very much the same way as for stratified random sampling; (2) establish the desired number of participants in each category of each dimension to establish the sampling quotas, judged against both the total desired sample size and the desired pattern of representation; (3) selection of participants proceeds until the requisite quota for each category of each dimension is filled (various sampling strategies could be used at this stage). A quota sample can help the researcher ensure that they access a desirable range and volume of perspectives from specific types of data sources.

Convenience Strategy

The convenience strategy, as the name suggests, involves choosing data sources that are easy to access, usually at much lower cost, relative to a population-oriented sampling strategy. Convenient access can be achieved when data sources are found in close geographical proximity or work in organisational proximity to the researcher, when the researcher has a personal connection to specific gatekeepers who control organisational or institutional access or attendance at specific events for data gathering purposes, or when the researcher is part of a professional network or knows a colleague that can facilitate access to participants. Postgraduate research students and early career researchers, for example, frequently must rely on the convenience strategy because of the resource and access constraints they face. University academic researchers frequently use convenience sampling (either of undergraduate students, who may be promised course credit as an incentive, or postgraduate students or students undergoing professional development, such as MBA students or student teachers) because it is an efficient way to produce research outcomes quickly. For university academics, the driving motivation behind using a convenience sample may be less about adapting to contextual constraints and more about taking advantage of contextual opportunities, perhaps as a pathway toward furthering their career.

Volunteer Strategy

The volunteer strategy is relevant only to choosing human participants in research. The researcher publicises the nature of their research project (including any preconditions participants must meet in order to qualify for participation) and provides their contact details for any people interested in volunteering to participate. In many cases, an offer of remuneration or recompense for time spent participating accompanies the request for volunteers. The people who respond and agree to participate then comprise the sample. With this strategy, sample size is determined by the number of people who agree to participate, making it very difficult to anticipate the final sample size that will be achieved. The key distinguishing feature is that potential participants approach the researcher to participate on their own volition, based on information they have acquired or been given about the research, rather than the researcher approaching the participants. There are now online resources, constituting very large sometimes multi-national participant pools or online panels assembled, maintained and regularly updated by various companies, where researchers can access volunteers willing to participate in many types of research projects, often with an expectation of remuneration for their efforts (see, for example, Qualtrics Online Sample (https://www.qualtrics.com/online-sample/; see also Chandler et al., 2019).

Snowball Strategy

The snowball strategy, also known as network sampling, generally commences with selection of one or more key participants or other data sources and, as part of the data gathering process from those sources, the researcher seeks information on who else might be useful to approach for participation or what other data sources (e.g., documents, reports, media stories) might be useful to seek out. The snowball strategy works well in contexts that the researcher is not very familiar with. The researcher relies, instead, on the knowledge of 'insiders' and social/collegial/professional networks to expand, extend, and enrich their sample. This cascading effect of data source identification can yield reasonably large samples with relatively little effort and expenditure of resources. The trick to making the snowball sampling strategy work most effectively is knowing how to solicit the information about other data sources to connect with. It is good practice for the researcher to be very clear about the reasons they are seeking such information and to be very open about what they hope those other data sources might be able to provide. The researcher's goal is to invite each participant to suggest the most appropriate data sources, i.e., to leverage their contextual and personal knowledge in ways that will best benefit the research aims. Using the snowball strategy exposes the researcher to an ethical problem in that relying on referrals for potential participants means that they are asking for private contact details to be revealed without the knowledge of those being referred and some institutional ethics oversight committees or panels may not approve of such a strategy.

Purposive Strategy

The purposive strategy is a very important and commonly used strategy, particularly for research guided by certain patterns of guiding assumptions. The essence of this sampling strategy is quite straightforward: the researcher specifically and deliberately chooses the precise data sources they want to connect/engage with because they meet one or more of the selection criteria relevant to the research questions (we say data sources here because while human participants are most commonly purposively sampled, it is also possible to purposively sample events, locations, documents, media stories, websites, and so on). The researcher might choose data sources because of their role or purpose in a specific group or organisation, because of the perspective they think they might offer, because of information they might possess (and be willing to share in the case of human data sources), and/or because of a specific characteristic or background the data source possesses or reflects (such as their historical or strategic value, gender, ethnic/Indigenous/First Nations background or management level or leadership position). With this strategy, what is important is not how many people (or other data sources) are accessed, but who or what is accessed. For this strategy to work well, the researcher should be familiar with the context in which they are collecting data so that the appropriate data sources can be identified and approached. Different modes of the purposive strategy for choosing research participants have been identified by Cohen, Manion and Morrison (2018, pp. 156–158):

- *Typical mode* involves the researcher identifying those data sources they think offer the most typical or representative of perspectives available in the research context.
- *Atypical mode* involves the researcher identifying those data sources they think are atypical (e.g., unusual, extreme, or deviant, in the minority, silenced or hidden) in

the research context in order to expand and enrich the perspectives and information they gain access to.

- *Diversity mode* is a weaker and more affirmational form of the atypical mode where the researcher purposively chooses data sources that will likely provide access to a more diverse range of perspectives.
- *Criterion-based mode* involves choosing data sources that meet a specific criterion for the researcher, such as newly-hired employees hired or retirees, or current policy documents.
- *Critical/Pivotal mode* focuses on choosing data sources that the researcher identifies as critical or pivotal in the research context, such as being associated with or involved in a critical incident or event in the research context, holding pivotal knowledge about an issue, or providing good test cases for emerging theoretical propositions.
- *Opportunistic mode* cannot be planned or implemented from the commencement of the research; rather it is an emergent sampling mode focusing on connecting with specific data sources that were initially unknown or unavailable at the start of the research, but later become known/available to the researcher.
- *Intensity mode* pursues data sources that provide the researcher with a deeper level of exposure to the perspectives of a specific group, event, or situation.
- *Homogeneous mode* focuses on choosing data sources based on their similarity to each other, where the researcher explicitly defines the dimensions along which similarity will be gauged (e.g., high or low performers at work or in school).
- *Reputational mode* is a variant of the snowball strategy, where key data sources are identified based on the reputations of those data sources in some relevant context. Reputation may be based on specialist knowledge, insights, or experiences, on political or cultural importance or power, or because it would be politically prudent for the researcher to seek out that data source.
- *Revelatory mode* involves the researcher seeking access to the first members/founders (or documents they have produced) of a particular group or organisation or developer(s) of an innovation in order to gain preliminary developmental insights with respect to that group or innovation, including learning about the historical context of the group or innovation as a backdrop against which to gauge new learning.

Contextually-Driven Strategy

The contextually-driven strategy involves (a) identifying and choosing data sources, both inside and outside the research context, that will facilitate acquiring a wider range of perspectives as well as a greater depth of understanding of that context and (b) selecting times, occasions, events, and/or situations for data gathering purposes within that context. With (a), the researcher seeks information from specific data sources; with (b), the researcher chooses when and where they should seek information from data sources. Thus, the general goal of this strategy is to enrich contextual knowledge from both inside and outside viewpoints. The contextually-driven strategy is often successfully realised when used in tandem with the purposive strategy, especially when it comes to selecting events and situations in which to gather data and data sources from which information will be sought about those events and situations.

Theoretically-Driven Strategy

The theoretically-driven strategy is a type of purposive sampling where emerging theoretical ideas, concepts, and relationships drive the sampling logic. It is an iterative form of intentional sampling that provides the basis for methodological approaches such as grounded theory (see, for example, Bryant & Charmaz, 2007, Chap. 11). The theoretically-driven strategy offers a mindset for sampling whereby the researcher commences data gathering from an initial sample of participants, documents, events and so on, and undertakes data analysis activities at the same time. As the researcher begins to form theoretical interpretations and ideas about meanings, relevant concepts, and relationships from this initial sample, these are then used to guide further purposive sampling in pursuit of additional evidence from events, situations, and data sources that will help to refine, shape, and explicitly evaluate and extend the initial theoretical ideas and interpretations that have emerged. The theoretically-driven strategy is inherently adaptive in that early gaps in sampling and/or gaps in knowledge from the first stage of sampling can be easily rectified, qualified, extended and/or expanded upon by subsequent stages of sampling.

Negative Case Strategy

Unlike the theoretically-driven strategy, the negative case strategy (also known as deviant case sampling; see, for example, Neuman, 2020, Chap. 8) employs disconfirmatory logic in the theory-building/evaluation process. It involves the deliberate yet challenging search for extreme, unusual, or disconfirming evidence relevant to specific interpretations and theoretical explanations, ideas, concepts and/or relationships. Using this strategy, evidence that contradicts emerging theoretical ideas is explicitly sought out to provide pathways for the researcher to enhance/enrich/ enlarge/prune/refine/correct their interpretations and theoretical understandings. Instead of deleting, discounting, or transforming outlier perspectives (a logic employed, for example, in statistical analysis), the negative case strategy seeks to identify, understand, and incorporate them. However, the strategy only works if the researcher commences data analysis simultaneously with data gathering, thus providing preliminary evidence to form the initial impressions and interpretations needed to identify and pursue negative or disconfirming cases. Researchers also need to be aware that there are several different ways to define a negative case. A negative case might be a data source that offers/provides/reveals a minority or marginalised viewpoint, a disaffected viewpoint, a viewpoint that challenges the dominant stories that are emerging, or a completely different perspective (perhaps one considered deviant or abhorrent by the dominant culture, group, or organisation). The choice of negative cases is always purposive in intent and can work most effectively if employed in tandem with the theoretically-driven strategy.

Sequential/Saturation Strategy

The sequential/saturation strategy represents the logical iterative extension of the theoretically-driven strategy taken to a concrete endpoint. The sequential/saturation strategy depends upon the continuous iterative interplay between information arising from various data sources and analysis of that information to yield interpretations and theoretical explanations, concepts, and relationships that are progressively enhanced, extended, revised, and improved while, at the same time, stimulating the gathering of additional data. Eventually, a point will be reached in the iterative process where the researcher learns nothing new that requires them to make further changes to their interpretations and theoretical explanations. This constitutes the point of saturation and serves as a signal that the researcher can cease gathering data. Saturation is more likely to be achieved with longer-term contextual immersion where the researcher cycles into and out of the research context over a long period of time. It may not always be clear when or even if the point of saturation will or can be reached. This uncertainty may emerge because the research context is simply too complex and multi-faceted, offering many potential data sources, events, situations to consider. The researcher will often need to use their best judgment to make a practical trade-off, weighing the benefits of the learning achieved against the resources required to gather additional data. In such instances, the principle of diminishing returns for benefits and costs can be employed to establish a quasi-saturation point where new learning is considered minimal or inconsequential rather than zero (i.e., full saturation).

Synergistic Combining of Data Source Choice Strategies

Throughout our discussion of the various data source choice strategies, we have highlighted the potential benefits to be achieved by using synergistic combinations of more than one choice strategy, either simultaneously (as when the purposive strategy is used with the quota strategy or the contextually-driven strategy) or sequentially (as when the snowball strategy or purposive strategy is used at different stages of a theoretically-driven or negative case strategy). Onwuegbuzie and Collins (2007) and Teddlie and Yu (2007) discuss a range of potential synergistic combinations of sampling strategies, focusing on sampling to yield quantitative and qualitative data (i.e., sampling for traditional mixed methods research). These strategies are usually implemented as multi-stage sampling strategies, which means they can be very useful when different patterns of guiding assumptions are adopted for different stages of the research. Note though that the synergistic use of data source choice strategies need not be tied to obtaining mixtures of different forms of data (quantitative and qualitative); they may be equally valuable for obtaining data of a single form (quantitative or qualitative) from different types of sources.

Every synergistic strategy combination will be accompanied by advantages and disadvantages, which means that when planning for their use, the researcher must consider how best to balance costs and benefits. For example, the theoretically-driven strategy is most effective if combined with the negative case strategy as this helps the researcher to achieve a more well-rounded story, but this synergy becomes much more resource-intensive if saturation is being pursued. The quota strategy paired with either the convenience or volunteer strategy can be a very resource-efficient way to build up a sample but may create deficits in sample composition. The purposive strategy can be combined with simple random sampling to target specific organisations/institutions and randomly sample their employees or clients. However, generalisations are very difficult to defensibly extend beyond the boundaries of those purposively-sampling organisations/ institutions. Whatever synergistic combination is used, the researcher will need to ensure that they convey a transparent story about the logic, benefits, and limitations of the specific data source choice strategies and processes they employed.

2.2 Data Gathering Strategies

Data gathering strategies form the second vertex of the 'Data Triangle' landscape (refer to Fig. 2.1). Many data gathering strategies can work synergistically together to enrich the learning the researcher can achieve and, where appropriate, enhance participant involvement. Some strategies are oriented toward gathering quantitative data and others toward qualitative data. While mixed method practitioners argue that gathering both forms of data is needed to convey a convincing story, this thinking is too limiting. Gathering the same form of data using different data gathering strategies focusing on different types of data sources can be equally convincing. This reinforces the value of pluralist systemic thinking rather than just mixed methods thinking. Figure 2.3 presents a three-level mind map showing that 20 specific types of data gathering strategies can be organised into six primary categories as well as three superordinate categories (denoted by the superimposed ovals):

- Primary categories:
  - *Interaction-based strategies*—the researcher (or one or more trained interviewers) interacts directly with people;
  - *Observation-based strategies*—the researcher (or one or more trained observers) directly or indirectly observes the behaviour of people in isolation, with each other, and/or in the context of a particular environment;
  - *Participant-centred strategies*—the researcher invites participants to produce their own data in some form and context;
  - *Artefact-based strategies*—the researcher chooses, extracts data from, or analyses artefacts that serve as sources of data (a very wide definition of 'artefact' is applied here, following the arguments set forward by Plowright, 2011, pp. 92–97);
  - *Data-shaping strategies*—the researcher constructs specific instruments, questionnaires, devices, coding systems, or other data-generating or data-transformation processes to produce quantitative, qualitative, and/or hybrid/ merged forms of data; and

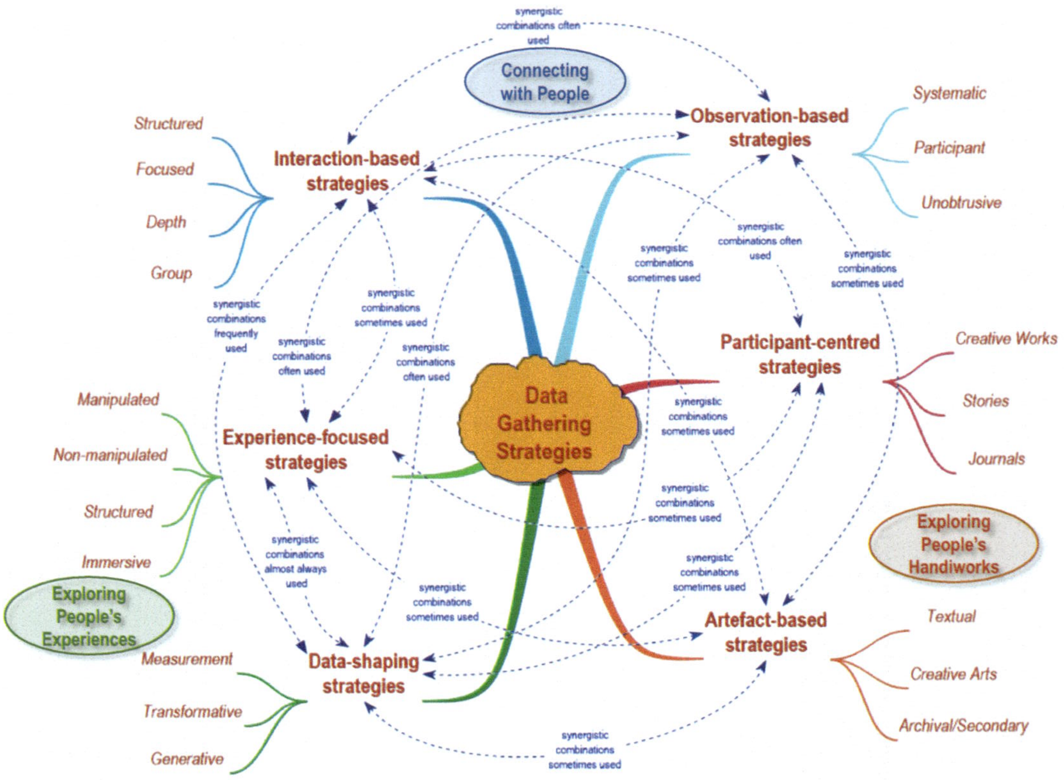

Fig. 2.3 Three-level mind map showing 20 specific types of data gathering strategies, grouped into primary and superordinate categories, as well as dynamic synergistic pairings (adapted/revised from Cooksey & McDonald, 2019, Fig. 14.1, p. 556)

  - *Experience-focused strategies*—the researcher gathers data in the context of specific researcher-produced structured experiences that participants undergo, in the context of experiences structured and provided by others, or in the context of natural events that participants experience and deal with.

- Superordinate categories:
  - *Connecting with People*—strategies that facilitate connecting with people through direct or indirect interaction or observation;
  - *Exploring People's Handiworks*—strategies that facilitate exploration and understanding of the handiworks people create, produce, or perform in specific contexts; and
  - *Exploring People's Experiences*—strategies that facilitate construction, provision, and/or evaluation of designed, structured, or naturally occurring experiences of people.

In light of the pluralist systemic thinking perspective, we consider each data gathering strategy, not only as a distinct and self-contained strategy (which may encompass a range of more specific data gathering activities), but also as part of a larger more interconnected and dynamic toolkit for social and behavioural researchers. Some strategies form natural obligatory pairings such as experience-focused strategies and data-shaping strategies for quantitative experimentation, and where this happens, we would not consider that pairing to constitute a pluralist approach. However, many other pairings of strategies can provide synergistic gains, notably in pluralist investigations where different data gathering strategies, different forms of data (quantitative and qualitative) and/or different types of data sources are to be included. Figure 2.3 signals some of the potential synergistic combinations between the primary strategy categories. [Note that the limitations of two-dimensional space prevent the visualisation of viable synergistic combinations between more than two data gathering strategies in the mind map. It is important to realise that three or more data gathering strategies can be used synergistically.] My intention in this chapter is to provide a helicopter overview of each data gathering strategy (adapted and revised from more extensive discussions in Cooksey and McDonald (2019, Chap. 14). More details may be found in any of the many research methods reference books and textbooks that are available (see, for example, Banyard, et.al. 2022; Bell, Bryman & Harley, 2022; Cohen, Manion, & Morrison, 2018; Gray, 2021; Liamputtong, 2019; Neuman, 2020).

2.2.1 Strategies for Connecting with People

One important way a researcher can gather data for research purposes is to connect directly or indirectly with the people whose behaviours, social interactions, relationships, perspectives, cultures, beliefs, values, attitudes and/or thoughts are of focal interest. With interaction-based strategies, the researcher or some other interviewer creates and maintains a direct social connection with participants as a fundamental part of the data gathering process. Just like any other social interaction, particularly between strangers, things can go well, things can go badly, or things can start well but then head south. The key is developing and maintaining trust. This means that effective social interactions need to be planned for and practiced including how to get things going (e.g., 'breaking the ice' and building rapport/trust) and how to wind things up. With observation-based strategies, the researcher employs perception rather than social interaction for data gathering. Here, the tricky part is ensuring that those perceptions are not biased or misconstrued and, for the researcher or some other observer, this means learning how and where to look, what to look for, and what to take note of and record.

Interaction-Based Strategies

Interaction-based data gathering strategies involve some mode of social connection and interaction, either face-to-face or virtually through technological mediation and support (e.g., telephone or mobile/smart phone, videoconference, Skype, FaceTime, Zoom, or chat room). Interviews are essentially conversations with the explicit purpose of providing the researcher with data. They are often conducted one-on-one with a participant, but can be done at a group-level, as with a group interview or focus group. The data gathered in an interview may be quantitative, qualitative or both, depending upon the type of interview conducted. Typically, the less structured the interview, the more in-depth the researcher can go with respect to exploring participants' perspectives and the less the researcher will be interested in condensing or transforming the data into quantitative indicators. Researchers may conduct the interviews themselves or may enlist and train others to conduct them. Where possible, it is preferable that the researcher conducts at least a portion of the interviews to make sure they stay connected to and familiar with the data being gathered. Interviews, particularly those involving less researcher-determined structure, combine well with other data gathering strategies, making them a popular choice for pluralist research because data are directly sourced from participants. While such direct sourcing of data is a strength of interaction-based strategies, unless the researcher is very careful in attending to how the interactions are managed from start to finish, that strength can turn into a weakness, resulting in very poor quality data as well as potential adverse cascading effects for other researchers (e.g., one bad interview can put a person off participating in any future interviews).

Interviews are necessarily a more personal form of data collection, especially as their structure decreases and expectations of participant involvement increase. This can enhance response rates, especially for face-to-face interviews, because once the interview has commenced, it is more difficult, both psychologically and socially, for a participant to pull the plug unless they become truly uncomfortable. Equally, however, some people find interviews, especially face-to-face, difficult to participate in precisely because they involve social interaction, typically with a stranger, and this may not suit their personality or social preferences. This may then turn them off participating in a face-to-face interview. Having another modality for interview participation available as a Plan B to suit such instances could be a viable data gathering tactic for the researcher to adopt.

Face-to-face interviews allow the researcher access to nonverbal as well as verbal cues, which can add richness to the data gathered. Interviews mediated by some sort of technological interface will typically sacrifice access to at least some types of nonverbal and other contextual information; non-visual media more so than visual media. Thus, in terms of richness of access to verbal and nonverbal information as well as contextual information, the ranking of interview modalities, in decreasing order of data richness, would be: face-to-face (visual, auditory, olfactory, tactile and 360-degree contextual information all available); Skype, smartphone, FaceTime, Zoom, or other video-conferencing technology (narrowly-focused visual and auditory information plus limited and narrowly-focused contextual information available); landline phone and voice-only mobile phone (auditory and very limited and narrowly focused contextual information available); and web-based (computer-interface only with no camera, typed text without contextual information). Thus, a critical consideration for choosing the most useful modality for an interview is how important it is for the researcher to be able to access nonverbal and contextual information along with interview content.

Interviews are resource-intensive, especially if conducted one-on-one. The less structured the interview, the more effort the researcher must invest to carry them off successfully and many researchers can find this exhausting. The risk of errors created by negative interviewer attitudes and inadvertent/inappropriate behaviours is relatively high, especially as the degree of structure increases. For interviews, even highly structured ones, to work well, the researcher needs to devote some initial attention to building at least a minimal level of rapport with the participant. The less structured the interview where proportionally greater depth of participant contribution is expected, the more attention the researcher or other interviewers need to devote to building rapport. Establishing rapport helps to build trust and trust facilitates genuine access to and revelations of what participants think and feel. Participant responses to questions or the information/insights they offer or are willing to offer during an interview may be intentionally or unintentionally influenced by their attitudes, beliefs, fears, worries about what others might think of them, suspicions about the interviewer or the research itself, feeling threatened or unsafe, feeling resentment or frustration that they can't voice their own views (a risk especially for structured interviews), and political or other agendas (real or imagined). Success in an interview depends on how well the researcher plans for and consistently works at building and then maintaining participant rapport and trust.

The setting for interviews is often a critical consideration. For example, interviewing participants at their workplace may create worries about what their bosses might think, so finding a neutral and comfortable place to conduct interviews away from their workplace could help to build both trust and comfort levels. The problem would be magnified if topics being discussed in their interviews are workplace-related. It is therefore important to create a comfortable safe context for interviews, free from distractions and possible interruptions. Another important consideration is congruence between interviewer and interviewee. Here, the issue is whether differing backgrounds of the interviewer and interviewee could affect trust building and perspective sharing in an interview. In cross-cultural contexts, matching the ethnicity and/or cultural/religious background or even the gender of the interviewer to that of the participant could be important. Interviewing Indigenous/First Nations participants can be a difficult proposition unless the researcher gives proper consideration to where the interviews occur, whether the interviewer is a member of the colonising culture or from an Indigenous/First Nations background (even the nature of their Indigenous/First Nations background vis-à-vis the participant could be relevant). To get to a position where interviews with Indigenous/First Nations people can be properly conducted, the researcher will have to show proper respect for Elders, community, and traditions by ensuring they have agreed to the interviews and the contexts in which they occur.

The four interviewing strategies discussed below each unfold with differing degrees of structure. They should not be seen as distinct non-overlapping strategies but as strategies whose characteristics may be blended to some extent to help the researcher achieve their purposes. For example, a focused interview may have attributes of a semi-structured interview and focus group interviews may be highly structured, semi-structured, or unstructured. Cooksey and McDonald (2019, Sect. 14.1.1), Flick (2021), Gubrium et al. (2012) and Brinkmann and Kvale (2018, especially Chaps. 5, 6, and 7) provide very useful information about a range of interviewing methods.

Structured interaction-based strategy

This strategy employs structured interviews to gather data, which typically involves the oral delivery of and responses to a structured questionnaire (thus pairing this strategy with the Measurement data-shaping strategy). A structured interview creates two interacting roles, interviewer and participant, where the power dynamic strongly favours the interviewer, and the nature of the interaction is generally not adaptable. Instead, the interview follows a strict script, even down to the probes requesting further information, and any deviations from that script can create problems with the data. To work effectively, the researcher needs to develop a minimal level of rapport to establish a baseline level of trust. Interviewers must be highly trained, ideally to an identical level of proficiency, and the design, ordering, complexity, response scaling, and flow of interview questions (i.e., the interview schedule) needs to be carefully thought through and pretested. Throughout a structured interview, the participant's role is passive—only responding to questions. Any information participants volunteer or questions they raise (with respect to their own context or to research issues they are being interviewed about) on their own initiative are actively discouraged and, if offered, are typically ignored.

Structured interviews may be conducted face-to-face or via some technological medium. Face-to-face interviews are the most personal and tend to yield higher response rates/lower withdrawal rates relative to technologically-mediated interviews. Importantly though, the potential impact of nonverbal behaviours tends to be much higher in face-to-face or visual technologically-mediated interviews. Interviews mediated by non-visual technology run a higher risk of participant withdrawal in that it is far easier for a participant to hang up or disconnect a phone call or switch away from a webpage, because they don't risk seeing disappointment or disapproval on the interviewer's face. Non-visual technologically-mediated interviews may be more efficient from a data gathering and recording perspective and the impact of nonverbal behaviours (except those evident in speech) will tend to be lower, but generally at the cost of the interview seeming less personal and engaging to the participant.

Structured interviews are especially useful where research purposes include obtaining comparative responses to a standard set of questions, where literacy or other constraints may preclude obtaining written responses to questions from participants, and/or where the researcher feels that face-to-face communication may yield better data. Structured interviews may also be used in conjunction with experience-focused strategies where questions focus on participants' experiences.

Focused Interaction-Based Strategy

The focused interaction-based strategy employs interviews that target one or more specific issues that have been identified, either by the researcher, or by the participant, as important or useful to explore. Focused interviews can be conducted in a structured manner, often in combination with data-shaping strategies and/or Textual artefact-based strategy (where documents associated with the focal events, incidents or experiences are content-analysed), where the target focus is normally chosen by the researcher. They may also be conducted in a semi-structured or unstructured manner (thus combining it with the Depth interaction-based strategy), especially when the participant is invited to nominate the focal event or issue. The focused interaction-based strategy may also be combined with other approaches that yield qualitative data such as the Creative Works participant-centred strategy, the Participant observation-based strategy, or the Textual artefact-based strategy, and can be particularly useful in conjunction with the Non-manipulated experience-focused strategy.

The critical incident technique (e.g., Chell, 2004, 2014; Woolsey, 1986) is an important type of focused interview. A critical incident or event is one that may have occurred in the past (giving a retrospective focus), may be currently unfolding (giving a present-day focus), or may be planned or anticipated in the future (giving a prospective focus) and should have clear boundaries, purposes, and outcomes. The interview questions and discussion are then used to explore this incident or event with the participant in some depth including issues, feelings, attitudes, causes, consequences, and reactions associated with each incident. If participants are asked to identify or describe critical incidents, then it is important that the researcher concentrate at least part of the interview on why they identified those incidents as critical. If the researcher identifies the critical incidents, then they must tell a clear story about how and why they selected specific incidents as being critical.

Focused interviews need to have time allocated early on for the building of rapport and trust followed by a period for contextualising, setting up, or negotiating the focal events/incidents with the participant. Trust is important to build as, in many cases, there may be sensitivities attached to the incidents/events being discussed (e.g., feelings of guilt, shame, or blame). The focal incidents/events must be relevant to or meaningful to the participants, something that can be ensured if the participant helps to identify them. Focused interviews may be less useful if participants have not directly experienced the incidents/events/issue being discussed. However, in cases where one or more future scenarios are being explored, prospective focused interviews may provide insights into potential reactions and consequences. The focused interview may be constrained by its focal incidents and events, which may mean that the researcher does not gain access to the larger contextual picture. This may be countered by tapping into other data sources, by devoting a portion of the interview to discussing the larger context surrounding the critical incidents/events, or by pairing focused interviews with a different data gathering strategy.

Depth Interaction-Based Strategy

The Depth interaction-based strategy can be enacted using either semi-structured or unstructured interviews (see, for example, Flick, 2021; Galletta, 2013). Semi-structured interviews (sometimes called 'in-depth' interviews) are largely open-ended but guided by an overarching framework of conversation topics, called a 'topical landscape'. Unstructured interviews (sometimes called 'non-directive' interviews) are the most open-ended qualitative data gathering strategy, closest to what might be considered a natural conversation between two people.

Semi-structured Interviews

Semi-structured interviews are generally conducted face-to-face, one-on-one and yield qualitative data. Unlike structured interviews where the conversation is intentionally highly scripted, semi-structured interviews involve a more natural conversation, but with an interest in ensuring that certain specific topics or issues are touched upon or discussed. The researcher maps out the general range of issues to be covered in the interview by creating a 'topical landscape'. The issues informing the topical landscape should emerge from the research questions and should resonate with the participant's context. The topical landscape is typically not traversed in any fixed order and the researcher should look for early discussions to trigger natural shifts to other topics. The flow of the interview is guided more by the participant than by the researcher. The researcher's role is largely one of redirecting the conversation where necessary or possibly introducing a topic for discussion. Questioning during the interview should be open and neutral so as not to suggest or presuppose specific meanings or interpretations. If there is a need to redirect the focus of discussion or get the interview back on track, the researcher should use conversational cues to subtly nudge the conversation in a desirable direction. A semi-structured interview is highly flexible and adaptable in that the interview can cover issues in any order while remaining open to following up any emergent issues and associated leads, which may result in the topical landscape evolving through the course of the interviews. Semi-structured interviews can be effectively combined with participant observation or with almost any Participant-centred or experience-focused data gathering strategy (particularly the Non-manipulated experience-focused strategy).

Semi-structured interviews take effort and practice to become proficient at conducting. The flexible and adaptable nature of the semi-structured interview imposes a greater demand on the researcher to keep clear notes on how and why the topical structure of the interviews changed over time as well as from one interview to the next and to consider this evolution in all subsequent analyses. In this regard, a research journal becomes an indispensable resource. Contextualisation of every interview is essential (including researcher reflections on the interview process itself and how it unfolded) so that the context of the interview can be considered as part of the analytical process. Equally, where possible, the researcher should look for and note any contextualising commentary offered by the participant. In some cases, the participant can be invited to further clarify their discussion by providing some context for it. The biggest danger associated with semi-structured interviews is the potential for the researcher's preconceptions to inadvertently influence or dominate the course of the conversation. If this occurs, trust will be eroded and the interviewer-interviewee relationship as well as data quality/quantity may be irreparably damaged. Accordingly, the researcher must be constantly vigilant in letting the participant steer the flow of conversation as far as it is possible to do so. Shifts between topics should be as natural as possible, with only minimal 'nudges' from the researcher.

Unstructured Interviews

Of all the interviewing strategies, unstructured interviews are the most difficult to carry out successfully, being dependent upon developing a high level of rapport and trust. They require the most delicate balancing act between the researcher wanting to learn things in the context of their research and needing to listen to the unprompted perspectives of the participants. Unstructured interviews are very well-suited for constructing oral or life/biographical histories. Unstructured interviews are frequently combined with the Stories or Creative Works participant-centred strategy. The unstructured interview is not scripted, rather it unfolds as a natural conversation. Accordingly, the building of strong rapport and deep trust is critical and, in some cases, may not be fully developed until several interviews with the same person have been conducted. The researcher must be non-directive and non-judgmental throughout the interview process in order to strengthen and maintain the bond of trust. The unstructured interview is entirely driven by the participant in that the researcher goes where the participant leads. The researcher's role is as an active listener or mirror for the participant's views, following the participant down the paths they are exploring. Essentially, the participant is in the foreground of the interview, the researcher/interviewer is in the background. An unstructured interview is completely open in that the participant can discuss any issues that come to mind. The interviewer's contributions to the conversation will consist mostly of requests for clarification or for more detail on some issue the participant has raised. Throughout the interview process, the researcher needs to be sensitive to cues from the participant that signal just how far they are willing to go in discussing any issue; the more sensitive the issues discussed, the more trust needs to be built up and maintained. If social cues suggest it is time to move on or steer away from an issue, then the researcher should do so immediately.

Unstructured interviews take a great deal of effort and practice to become proficient at conducting. The participant-driven nature of the unstructured interview imposes a much greater demand on the researcher to maintain clear and detailed notes, observations, and reflections on the entire interview process. In this regard, the split page research journal format becomes an indispensable resource. Contextualisation of every interview is essential (including reflections on the interview process itself and how it unfolded) so that the researcher can consider the context of the interview as part of the analytical process. Equally, where possible, the researcher should look for and note any contextualising commentary offered by the participant. In some cases, the participant might be invited to further clarify their discussion by providing some context for it.

Like semi-structured interviews, the biggest drawback associated with unstructured interviews is the potential for the researcher's own preconceptions to inadvertently influence or bias the course of the conversation. Any hesitancy or judgmental signals on the researcher's part with respect to whatever issues the participant is discussing runs the risk of eroding trust, harming the interactive relationship, and degrading the quality/quantity of data provided by the participant. This includes the researcher exhibiting (even inadvertently) insensitive/negative reactions to what the participant is trying to communicate or insensitive or inappropriate language/ nonverbal behaviours during the interview. To mitigate against these potential problems in depth interviews, the researcher may find it beneficial to adopt active listening practices, including mirroring back what someone has said to ensure correct interpretation, showing empathy rather than judgment, and using gentle prompts to show interest and request further information (see, for example, Louw et al., 2011). The goal is to help participants feel safe and know that their voices/viewpoints are actually being heard.

Group Interaction-Based Strategy

The Group interaction-based strategy is implemented when an interviewer or facilitator interviews a group of individuals simultaneously. One of the more popular forms of group interview is the focus group (see, for example, Barbour, 2018). A focus group is essentially a planned group-based interview process where one or more researchers/interviewers/facilitators engage in a conversation on specific topics with a group of people collected together according to some common interest, role, or shared characteristic or experience. The nature of the group interview can range from fully structured (using scripted questions) to unstructured (open group discussion) as researcher needs dictate. The twist is that it is conducted in a group setting, often to take advantage of group dynamics and synergies to enhance and enrich the conversation. Focus groups can be combined effectively with a range of Participant-centred, Artefact-based or Manipulated experience-focused data gathering strategies and may often precede the design and implementation of the Measurement-data shaping strategy.

Focus groups or group interviews may involve an intact pre-existing group or a group of strangers gathered for research purposes. If the group is pre-existing, then trust will likely be present, but viewpoints might also be more homogeneous, which may be problematic if the researcher is seeking a diversity of views. If the group comprises strangers, the interview will need a much longer lead-in time for the group to get used to discussing things with and revealing things to each other (interpersonal rapport and trust and at least a minimal sense of being a 'group' needs to be built up). In all cases, the researcher must allow time for the group to get used to their presence and role in the group (especially in a pre-existing group, where the researcher might be seen as an interloper or viewed with some suspicion as to their agenda). In a focus group interview, the researcher may determine the focal topic for discussion but will need to ensure that the topic is relevant to or meaningful to the participants. The use of tangible objects or experiences (such as a product, innovation, multi-media presentation, policy, or political event such as an election), where appropriate, can assist in achieving this goal. Using a highly-structured interview protocol in a focus group will mean that every group member is asked the same questions, which means that group interaction may not be the main point of the exercise. A focus group requires a facilitator (which may also be the researcher) who adopts a directive role and must be carefully trained/practiced in how to communicate the guidelines for the session, how the group will be run and what it will focus on, how to manage the group, how to manage social interactions without dominating them, how to multi-task (manage the conversations, observe, record and ensure group remains on task), and how to move the conversation on between topics. A focus group is generally more productive if run as an interactive group discussion, rather than as a question-and-answer exercise between the facilitator and individual group members. Full group interaction works better for building trust and for generating ideas and sharing viewpoints. If a pre-existing group is serving as a focus group, the researcher needs to understand the history and nature of the relationships between all members of the group and whether power dynamics (e.g., manager/subordinate, acquaintances/friends vs non-acquaintances, different departments) are likely to influence group interaction and willingness to discuss topics. This becomes even more important if the researcher is conducting a more general group interview in a pre-existing group.

There are other types of group interview which do not constitute focus groups. In such groups, the interview will tend to be of a semi-structured or unstructured nature, and, in that context, group interaction will always be an important aspect of the process. If a semi-structured or unstructured interview format is used, the researcher's role will be more active listener than directive. In an unstructured group interview, the topics for a group interview will tend to be free ranging, following the natural flow of group interaction. In a semi-structured group interview, the researcher may follow a topical landscape, exerting only subtle influences to redirect group discussion as the conversation must still be the group's own. The researcher will also tend to be take on a participant observer role, at least in terms of observing how the group dynamics play out during the session. Trust is very important to build and maintain in more general group interviews, where members don't know each other, as, in some cases, there may be sensitivities attached to the topics being discussed. In a pre-existing group, the researcher's observations should be alive to the possibility that some group members might be marginalised or silenced, either because of the topic under discussion or because of their relationships or status with other group members. Contextualisation of a group interview is essential (including researcher reflections on the interview process itself, the people involved in the group and how the interview unfolded) so that the context of the interview can be considered in the analytical process. Equally, where possible, the researcher should look for and note any contextualising commentary offered by participants.

Observation-Based Strategies

Observation-based strategies for data gathering involve purposeful watching and recording of activities, behaviours, and interactions in a specific context. Observations may be partially or completely recorded using a specifically-designed coding protocol (pairing with the Transformative data-shaping strategy) or they may be entirely open-ended. In many cases, observations occur in the participant's natural environment (i.e., in situ), but it is also possible to gather observations in contrived settings (such as a laboratory or simulated environment, which means pairing with the Manipulated or Immersive experience-focused strategies). Observation-based strategies can be differentiated along several dimensions:

- **Depth of participation**: Depth of participation varies along a continuum ranging from complete observer to complete participant according to the extent of researcher immersion and contact with participants in the observation context (e.g., see Cohen, Manion & Morrison, 2018, Chap. 26).
  - *Complete observer*: The researcher observes and records or assesses participants' naturally-occurring activities, behaviours, and/or interactions without any contact with participants. Contextual immersion is minimal, and a highly structured researcher-designed observational protocol is typically used.
  - *Observer as participant*: The researcher is partially immersed in the research context as an 'outsider' participant but with a role defined as 'being there to observe'. One variant of the role of observer-as-participant is 'shadowing', where the researcher follows a participant around as they carry out activities. Another variant is the 'fly on the wall' where the researcher is in the room with the group or individuals being observed but keeps their interactions with those participants to a minimum. Observational protocols tend to be semi-structured at best, where the researcher looks for certain things but remains open to new and unexpected occurrences.
  - *Participant as observer*: The researcher is more deeply embedded in the research context with a role of an 'insider' participant member of the group who is also observing. For example, this could characterise some types of action research in a group that the researcher joins temporarily within their own organisation. It could also characterise ethnographic research or the researcher's stance in a group interview. Observational protocols also tend to be semi-structured at best, where the researcher looks for certain things but remains open to new and unexpected occurrences.
  - *Complete participant*: The researcher is completely immersed in the research context as a full group or community member where observation is simply a by-product of their participation for their 'researcher role' to reflect upon; it is the ultimate form of insider research. This characterises some forms of action and ethnographic research. Observations tend to be very unstructured,

where what the researcher observes, and its meaning/significance, emerges rather than being pre-determined.

- **Observer mode**: Observer mode refers to whether, in the context under observation, the researcher's role as observer/researcher is overt, covert, or absent from the context (e.g., see Gray, 2021, Chap. 16).
  - *Overt mode*: In this mode, participants know that the researcher is observing and recording their activities, behaviours, and interactions. This mode of observation is ethically more defensible because participants must give their informed consent to be observed. However, the fact that people know they are being observed may alter the very phenomena the researcher is observing, creating what are called reactive effects. The more sensitive the activities, behaviours, or interactions the researcher observes, the greater the risk that those phenomena may change or even disappear under observation. Overt observation may be conducted in a way where the researcher's presence is not visible to those observed (e.g., the researcher observes from behind a one-way mirror) as long as those observed have given their prior consent to being observed. This can help to reduce reactive effects, as long as the technology or structures used to hide the researcher's presence do not themselves create reactive effects because they serve as a reminder to participants that that they are being watched.
  - *Covert mode*: In this mode, participants do not know that the researcher is observing and recording their behaviours, i.e., the researcher is operating 'undercover'. The people being observed cannot be considered as 'participants' per se. Thus, this mode of observation is much less ethically defensible because participants are not asked for their informed consent to be observed and recorded. For this reason, university ethics committees, for example, may not approve research proposing to employ covert observation. Covert observation does mean that the researcher can observe genuine behaviours unaffected by their researcher role, but great risks potentially confront them if they are discovered, i.e., if the researcher's 'cover is blown' and their role as researcher is revealed. In addition, depending upon what the researcher observes and in what context (e.g., criminal, ethically, or morally suspect behaviours), covert observation may create a situation that could constitute entrapment of the participant, which may have legal ramifications, or the researcher could become aware of behaviours that should be reported to authorities, thereby creating an ethical dilemma.
  - *Absent mode*: Unobtrusive observation is a third observer mode which occurs when the behaviours of and interactions between people are not directly observed at all; in fact, the cooperation of participants is not required (Webb et al., 2000). Instead, the researcher observes the contextual/environmental impacts that result from people's activities, behaviours, and interactions. With this mode, the concept of 'participant' effectively disappears; what is of interest is the evidence/traces of their past presence. This mode is non-reactive because the researcher cannot influence the patterns of behaviour and interaction they are observing. In many instances, it can be an ethically safe mode (i.e., informed

consent not required) since no individuals or groups are directly observed. However, informed consent may be required in cases where the traces left by people's activities that the researcher analyses could potentially be linked back to individual identities (e.g., receipts, shopper dockets, security camera recordings in stores or community environments).

- **Focus of learning**: Focus of learning considers the extent to which the researcher is interested in or able to make sense of/ascribe meaning to the activities, behaviours, and interactions being observed (see, e.g., Cash & Snider, 2014).
  - *Manifest/surface level*: For observations to have meaning at the manifest or surface level, the researcher must be able to perceive them using their six main human senses. Aural and visual observational data are typically the most important sensory sources of data; however, tactile, proprioceptive, taste, or olfactory data may also be relevant in certain types of research (e.g., in ergonomic research, product development and testing research). This is a much more descriptive level of observation, finding meaning through counting, recording, summarising, and/or displaying what the researcher observed and in what context.
  - *Latent/deep meaning*: The researcher is looking to understand the latent and deeper, perhaps even hidden or taken-for-granted, meanings of activities, behaviours, and interactions, going beyond what they observed and in what context to make interpretations about why those activities, behaviours, and interactions were observed and what they might mean in that context. Finding/learning about deep/latent meaning requires the researcher to convincingly demonstrate the links between the inferences they make and the data from which they arose.

Systematic Observation-Based Strategy

The systematic observation-based strategy typically involves the use of objective observers (either the researcher or others trained in the observation task) to record activities, behavioural observations, and interactions, using a paper-and-pencil or electronic recording form/instrument or technological support such as a video camera and/or hand-held data entry device (see, for example, discussions in Cohen et al., 2018, Chap. 26; Gray, 2021, Chap. 16). Systematic observation involves the researcher or other observers adopting the 'complete observer' stance where observations are typically obtained overtly in that participants are aware of the researcher's presence and intentions to observe their behaviours. They may be obtained overtly but unseen, where the researcher remains out of sight, perhaps behind a one-way mirror or by using a video recording camera in some context of interest but are known to be observing by participants. Finally, they may be obtained covertly, where the fact that the researcher is making observations of their behaviours is not known to participants (e.g., using GPS tracking or a hidden observation post) or if the researcher records behavioural observations of people from surveillance recordings obtained for other purposes, such as maintaining store, community, or airport security. It is far easier to access the manifest/surface levels of behaviour (what is happening) using systematic observation than to access the latent/deep meanings behind those behaviours (why are they happening and what they signify).

The systematic observation strategy is always paired with either the Measurement or Transformative data-shaping strategy to focus the observational process and facilitate recording observational categorisations and/or ratings appropriately and consistently. This creates the most challenging task in the systematic observational strategy: developing a defensible scheme for recording the observations made by observers. It is important that the researcher constructs (or adopts) a recording scheme that minimises the intrusion of irrelevant and erroneous judgments on the part of the observers. This becomes even more critical if the researcher intends to gather inferential data where the internal state of an observed person (i.e., latent/deep meaning) will be inferred from their overt (i.e., manifest/surface) actions. For this reason, systematic observation studies frequently use multiple observers observing a sample of the same people at the same time to assess inter-observer/rater reliability. The systematic observation strategy is also frequently employed in conjunction with the Manipulated, Non-manipulated or the Immersive experience-focused strategy. Sometimes, in the spirit of implementing a pluralist approach, the researcher might employ both the Systematic observation and self-report Measurement strategies and look for convergence and/or divergence between the conclusions each type of approach supports. This tends to strengthen the research, but at the cost of greater research complexity.

Systematic observation with observer present versus observer hidden creates different issues that the researcher must deal with. The researcher's physical presence as an observer can create potential reactive effects where behaviours change because of their presence (especially if they are seen to be writing) and the researcher needs to allow time for such effects to dissipate. The researcher also needs to take care not to interfere with behaviours (even inadvertently). If the researcher is a hidden observer, but participants are aware they are being observed, this may reduce some reactive effects but create others, especially if a camera or one-way mirror (devices that signal the hidden presence of the observer) is visible. In such cases, the researcher needs to allow time for such effects to dissipate or effectively hide/screen the technology. If the researcher's role as observer is covert where people don't know they are being observed, this creates ethical concerns about lack of informed consent from the people being observed. If the researcher is observing by shadowing (e.g., following a shopper around a store as they shop, or following a manager around their workplace), then recording observations, using pre-constructed forms or a hand-held recording device, can make observing on-the-run in real-time much easier. If activities, behaviours, and interactions are video recorded, observations can be made while watching the recording rather than watching them live.

Participant Observation-Based Strategy

Participant observation is a general data gathering strategy that describes observational research occurring at various depths of researcher involvement in context. It typically involves greater depth of participation (participant as observer or complete participant). As a data gathering strategy, participant observation is clearly distinguishable from systematic observation which distances the researcher/observers, as far as possible, from potential involvement in the research context. Participant observation is often combined with other data gathering strategies, such as the Textual artefact-based and/or Depth interaction-based strategy. It can also be effectively combined with the Non-manipulated experience-focused strategy, in cases where a group experiences change/upheaval due to an outside event or with the Manipulated experience-focused strategy in cases where a specific program is being experienced by a group.

The central feature of participant observation is the researcher's partial or complete immersion in the research context for a period of time—a characteristic of ethnographic research (see, for example, Angrosino, 2007; Gray, 2021, Chap. 17; Jorgensen, 1989; Waddington, 2004). The deeper the level of participation, the greater the involvement in co-creating any data gathered and the greater the chance that the researcher could inadvertently influence contextual changes. Also, the deeper the level of participation, the longer the research process will generally take, especially if the researcher has adopted a grounded theory approach. Taking a grounded theory approach (see, e.g., Bryant & Charmaz, 2007; Charmaz, 2014) involves a process whereby repeated immersions (partial or complete) in the research context are punctuated by phases of data analytic/interpretive activity and what is learned during earlier contextual immersions at least partly shapes what the researcher looks for in later immersions. Maintaining extensive field notes in a research journal is an indispensable tool for managing the data recording and preliminary analysis processes in participant observation.

Where the researcher's participation is toward the 'shallow' end of the spectrum (observer as participant or being a 'fly on the wall'), they remain somewhat at a distance from central involvement in the day-to-day activities of the groups/people being observed (which limits the risk of inadvertently influencing the context). Researcher participation may be short-term and limited to a single event (such as shadowing a consumer while they shop or observing a small sample of days or events like meetings), which restricts the depth of learning. Even though researcher participation is minimised in the observer as participant role, the group must still get used to their presence. This suggests that very early observation sessions should be treated as adaptation/habituation sessions, rather than as full data gathering sessions. Where the researcher's participation is toward the 'deeper' end of the spectrum (participant as observer or complete participant), they deliberately seek a closer connection to the research context and its people. Here, researcher participation would require a greater length of time in the research context in order to achieve an understanding of participants' perspectives through observation of interactions, rites and rituals, symbols, and so on.

If researcher participation is both deep and covert, care must be taken to avoid role discovery ('getting one's cover blown') or else researcher access may be destroyed, and the researcher themselves may be put at grave risk, depending upon the context. Covert participant observation can be very dangerous and is very difficult to justify ethically. Furthermore, the researcher needs to plan very carefully how they will record their observations. Many participant observers record their day's observations in their field notes at night, after hours, before going to sleep, because recording data during the day as they observe is not feasible without risking their cover. The nightly recording strategy may help to reduce memory problems, but it is likely that the researcher will remember very early and very late observations more clearly and more often (because of primacy and recency effects in human memory) as well as any surprising events (novelty effect). What the researcher may not remember well are the more mundane daily occurrences, events, and behaviours and this can create important gaps in their observations. Some memory effects can be offset if the researcher has access to other data sources in the context.

The risk of going native is very real in participant observation research. It involves the researcher losing their 'researcher attitude' and, in some instances, risking distortion of their values. This risk is especially potent with complete participant contextual immersion or with the covert observer mode. Mitigating this risk requires the researcher to take active steps such as planning for periods of time out of context and re-connecting with colleagues and peers. There is also a risk of inadvertently influencing context and changing behaviours/relationships as a consequence, which comes about if the researcher is not effectively managing their researcher role. Managing this risk demands that the researcher constantly monitor and transparently record reflections on their own behaviours in context.

Unobtrusive Observation-Based Strategy

Unobtrusive observation (sometimes referred to as non-reactive research; see Gray, 2021, Chap. 19; Kellehear, 1993; Webb, et al., 2000) emerged as a data gathering strategy intended to circumvent the ethical problems associated with covert observation. Data are generally gathered in forms not directly traceable to, or directly observed as being produced by, specific people. It is thus a very indirect way of connecting with people. The focus of observations is on the physical and/or virtual traces that people leave behind as a consequence of their behaviours and in ignorance of their identity (e.g., litter, walking pathways, products or services purchased, websites visited (only if the user's IP address is anonymised), wastage, or recycling of materials). The unobtrusive observation strategy can work well in combination with the Participant observation-based strategy, Creative Works, or Journals participant-centred strategy.

Some forms of unobtrusive observation can be seen as truly anonymous (such as analysing patterns of littering at various city sites, patterns of wear and tear on surfaces, pathways, and roads, patterns of waste materials around machines, patterns of stationery use in an office) in that the people leaving the traces are never observed nor can their identity ever be recovered. In other forms of unobtrusive observation, the activity traces that the researcher accesses (e.g., Facebook postings, internet browser histories, graffiti (where tags could be recognised), scanner data/shopping dockets; security video recordings in stores or in the community) may inadvertently reveal clues to, or provide evidence of, a person's identity, whereupon ethical concerns may be raised regarding lack of informed consent from those people to use their traces as data sources. When gathering such data, the better ethical practice would be for the researcher to seek individual informed consent to access and use trace data such as scanner purchase patterns and always deal with the resulting data in aggregate form rather than at the level of the individual. Alternatively, the researcher could give a guarantee to people in a specific context, such as a shop or meeting room, about exactly which data are and are not being recorded when they enter that context.

Inferences about human behaviour and relationships are always indirect with unobtrusive observations. Patterns may be detected, but the reasons why patterns have developed may be very hard to discern or convincingly argue for, unless other data gathering strategies are also used. Thus, manifest/surface content is easy to access; latent/deep meaning may not be. Seeking latent/deep meaning creates a burden of argument for the researcher to convincingly show how the chain of logic for their inferences connects to what they have observed.

2.2.2 *Strategies for Exploring People's Handiworks*

During their lives, as individuals, as members of groups, as workers in organisations and institutions, as members of communities and societies, as members of cultures or nations, as inhabitants of environments, people produce an amazing array of handiworks. Such handiworks can have personal and social meaning, may be socially, culturally, environmentally and/or spiritually relevant, may offer insights into and records of the past, may carry implications for the future, and so on. We are using the more general term 'handiworks' so that we can distinguish between tangible works produced by research participants and artefacts in the more general sense intended by Plowright (2011, p. 92) who defined them as "those objects or events that are produced by people", which we would modify slightly to 'produced and used or enjoyed by people'.

Exploring handiworks produced by people anchors the second important region in the system of data gathering strategies. We broadly classify/differentiate strategies available for exploring people's handiworks into participant-centred strategies and artefact-based strategies (recall the 3-level mind map in Fig. 2.3). In participant-centred strategies, research participants actively create handiworks during the research process, at the invitation of the researcher, or the researcher may have created handiworks themselves as a 'participant' in their own research process (e.g., diagrams, maps, research journal, or field notes), based upon their prior interactions with people. In artefact-based strategies, artefacts may be completely disconnected from their creators/users or may have come into existence or use well before the researcher becomes interested in them. For all but a couple of strategies, the researcher has little or no opportunity to influence the original shape, meaning, content, or use of artefacts.

Participant-Centred Strategies

Participant-centred strategies rely on research participants to generate representations, creative works, stories/narratives, texts, diaries/journals, and/or data records of experiences and events that provide the information needed for the researcher to achieve an understanding of their modes/patterns of thinking, experiences, and perspectives or to undertake an evaluative test for a specific research question, proposition, or hypothesis. A key characteristic of almost all participant-centred strategies is that the researcher gives control over the data recording/generation/creation process to the research participants. This represents an important shift in the researcher-researched power dynamic where data quality and quantity are placed squarely in the hands of the participants. The researcher may provide a loose framework or some guidance on or technological support for the data generation/creation process, but the final say over form, content, consistency, quality, and quantity rests with the participants. For participant-centred strategies to work, the researcher needs to invest time and effort in building up a level of trust with participants sufficient to interest and motivate them to take responsibility for their data generation/creation process and to maintain that motivation until the data recording/generation/creation process is completed. The researcher must be prepared to accept that not all participants will do equally well in managing their data recording/generation/creation process. For example, some people are not artistic or creative thinkers, which could make the Creative Works strategy very difficult for them to implement. The researcher should have a fall-back data gathering strategy to implement for such participants, such as an interaction-based strategy, the Stories strategy, or the Structuring experience-focused strategy.

The strengths of participant-centred strategies are two-fold: (1) an enhanced feeling of respect and involvement for participants engendered by the trust the researcher places in them to manage their own data recording/generation/creation, and (2) minimal opportunity for the researcher's preconceptions and perspective to influence that process. This in turn enhances participant commitment to stay the course until data gathering is completed. The weakness of participant-centred strategies is that they rely on participant commitment and diligence to manage the data recording/generation/creation well enough that sufficient data of appropriate quality are captured. Commitment means little if poor quality data are produced. This can be especially problematic for the Journals strategy, which may require days or weeks of consistent data recording. The Creative Works and Stories strategies are much more episodic in nature where the commitment is more related to depth and clarity of thinking and meaning than length of time. The researcher needs to acknowledge that using a participant-centred data gathering strategy may result in sacrificing some elements of data quality (because participants have control over what they produce and at least partial control over how they produce it) to gain important aspects of data content (i.e., deeper and more open revelations of participants' perspectives and thinking).

Creative Works Participant-Centred Strategy

The Creative Works participant-centred strategy encompasses any approach where the researcher asks participants to produce a specific type of creative representation which can provide insights into their thinking and understanding as well as into their perspectives on an issue or event. Such representations, which are typically but not exclusively non-textual, could include visual art (e.g., paintings, sketches, drawings, cartoons, paintings, rich pictures, photographs, systems thinking diagrams, mind maps, geographical maps, cause maps, concept maps, cognitive maps, flow charts, graphs, Venn diagrams; see Eppler, 2006 for some comparisons), physical art (e.g., sculptures, clay models, building block models, paper forms) as well as performance art (e.g., poetry, dance, music, pantomime, drama or theatre). Creative works may be produced according to a set of rules that the researcher provides (creating a pairing with the Structured experience-focused strategy), or free-form as determined by the participant, or some hybrid combination thereof. This strategy may be associated with arts-based research (see, for example, Saldaña, 2024; Siegesmund, 2024). Creative works as visual art may be hand-drawn, drawn using some type of technological support system (such as a software program, tablet and stylus, camera) or created using materials such as Post-It Notes®, a whiteboard, or other physical materials/tools. Creative works as physical art may require materials such as clay, modelling tools, Lego® bricks. Creative works as performance art may require musical instruments, camera or smart phone, costumes, stage, or other performance space. The goal is for the participant to produce a creative work that provides a meaningful representation of how they think/feel about some object/problem/process/ event of interest and to have access to the tools they need to produce the work. Creative works may be requested with some specific target object/issue/ process/event in mind, usually at researcher's invitation, in order to provide insights into a participant's mental model or perspective. The Creative Works strategy can be combined with a Structured experience-focused strategy such as process tracing via think-aloud protocol (participant verbalises what they are doing during the process of producing the creative work), focused interviews (participant discusses their creative work) or storytelling to accompany/explain the creative work. It could also be combined with the Group interaction-based strategy to produce group-level creative works.

For many research participants, producing a creative work as part of data gathering will likely make their participation more meaningful as well as more interesting and involving. The researcher, where appropriate, should define any specific rules, characteristics, and materials for participants to use in their creative process beforehand, as well as level of detail or quantified features expected in any labels, objects, and links. If a software support system (e.g., *Inspiration* for diagramming or mind mapping, or a sketching app) or other technology is used to create visual art, the researcher may need to provide training and/or ensure prior experience on the part of participants and this would need to be factored into the timeline of their research. Some participants may struggle with or prefer not to work with technology, so a back-up process for producing the creative work should be put in place for these people. The researcher also needs to plan an image capturing method for any visual artworks that are produced (unless they are automatically captured in electronic form by a software package) and ensure that the capture method works to produce a permanent image that can be worked with. It is important to realise that many people are verbal thinkers, rather than visual or spatial thinkers, and may therefore resist the researcher's request to produce a specific type of creative work. People who lack creative skills in the desired medium or who are self-conscious about their skills may actively refuse to produce a creative work. For performance art, this can be a particular problem. Accordingly, the researcher should formulate a fall-back data gathering strategy (e.g., a think-aloud process-tracing protocol, a focused interview, a storytelling session) or an alternative medium to work with to accommodate such participants.

Stories Participant-Centred Strategy

Storytelling and the creation of narratives are natural modes of communication between people and form frequently informal ways of sharing culture, impacts of people and events, history, beliefs and values, information, emotions, fears, hopes and so on. Furthermore, storytelling and narratives may follow different genres, such as autoethnography, metaphors, myths, poetry, fables, and may accompany performances, such as dance, theatre, and film. In many cultures, including Indigenous/First Nations cultures, oral histories are handed down from generation to generation, often in the form of stories. Such stories embody important cultural/spiritual/historical knowledge and can serve as "socialization instruments" (Chilisa, 2020, p. 194) between generations (e.g., see Archibald et al., 2022; Windchief & San-Pedro, 2019). We can see the same things occur in organisational and institutional stories. Stories are typically conveyed in a rich, extended, coherent, and densely interconnected form. Stories and narratives can provide vital data for achieving a deeper understanding of specific contexts and people, hence their intrinsic value as data for social and behavioural research. Stories and narratives may be generated explicitly for purposes of the research (e.g., see Czarniawska, 2004; Elliott, 2005; Gabriel & Griffiths, 2004) or specifically sought in data that participants have already provided for the researcher or in contexts where story sharing can be observed (thus combining the Stories strategy with the Participant observation-based strategy, for example). Narratives typically have a sequential flow, connecting events, thoughts, feelings, memories, and the like into a coherent discourse that explains or accounts for those experiences (i.e., facilitating sense-making in a specific context). Storytelling is a more informal type of narrative that can harness images, employ metaphors, involve characters and plot trajectories and use other linguistic devices to help add texture and meaning to what is being conveyed. Stories may have specific target audiences in mind as they are being conveyed and may be intended to create a specific learning or emotional outcome for the listener such as a moral, a social norm/expectation, or empathy/enmity for a situation or character (see, e.g., Archibald et al., 2022; Boje, 1991; Gabriel & Griffiths, 2004).

Soliciting stories and narratives is not a question–answer conversational type of data gathering approach, at least in western contexts. It is a much more of a one-sided affair, where the participant is given the power to tell their story in whatever fashion they wish. Typically, the researcher's role is to get things kick-started by inviting the participant to share a story about their life, workplace, a person or persons (including themselves) or a specific event. If a specific event, series of events or person(s) is to be the focus of the story, the researcher could suggest this focus in their invitation. The researcher must limit their interference with or prompting of the unfolding of a narrative or telling of a story once it has commenced. Asking questions during storytelling or narrative creation can disrupt the participant's flow of thinking, which can lead to distortions of originally-intended meanings. In Indigenous/First Nations research, however, storytelling as a research methodology has evolved in ways consistent with cultural traditions, taking forms such as Indigenous storywork (e.g., Archibald et al., 2022), sharing or talking circles (e.g., Chilisa, 2020, pp. 256–257; Kutz, 2013; Tachine et al., 2016), and yarning (e.g., Byrne et al., 2021; Fredericks et al., 2011; Geia et al., 2013; Kovach, 2021, pp. 162–164), to encourage open story sharing and equal opportunity for speaking. It is important to realise that stories may represent or contain privileged or culturally sensitive knowledge and that this will need to be explicitly and sensitively negotiated and handled. Gathering retrospective life histories or event histories may suffer from recall bias and other memory distortions, with the risk increasing the further back in time the participant reaches. Having access to other corroborating sources of data can assist in managing this problem and, where appropriate and available, a physical reminder of a specific event or point in time (e.g., a photograph, report card, newspaper article) may also stimulate recall.

Narratives or stories may be invited or sought in written or oral form or as a performance (thereby combining with the Creative Arts artefact-based strategy) and may or may not be accompanied by tangible objects and symbols. Many stories/narratives capture elements of life in the form of metaphors that create an anchoring image for the person's story, such as a machine, a river, a sports team, an island, a brain, unity, an organism, or animal. Inviting participants to create and elaborate on an appropriate metaphor for an aspect of their life or work (perhaps illustrating their metaphor with a sketch or drawing) may provide a safer less threatening way for participants to convey meaning for difficult, perhaps, sensitive subjects. If the researcher seeks a story in the form of a metaphor, then this request can form part of their initial invitation to participate. Participants may need some clarification as to what the researcher is seeking, as they may not have thought of their work or life in this manner before (doing this effectively pairs the Stories strategy with the Structuring experience-focused strategy). In instances where the researcher has not invited the creation of a narrative or story but instead looks for them in the context of other data being gathered from participants or seeks them from other social contexts entirely (pairing with the Creative Arts artefact-based strategy), they will need to depend heavily upon field notes in their research journal (and, where available, a digital/video recording) to capture the essentials of the narrative or story for later analysis (thus effectively pairing the Stories strategy with the Journals strategy). The researcher's goal would be to achieve as close to a verbatim record as possible, so writing the notes as soon as possible after encountering the narrative or story would be paramount to maximise data quality.

Autoethnography is a recently developed, very specific, type of self-reflective/reflexive and highly analytical storytelling, where the researcher becomes their own source of data in their own context (see, e.g., Anderson, 2006; Chang, 2016; Duncan, 2004; Jones, 2005). It is a reflexive narrative of the self and has emerged as a specific way of presenting one's inner knowledge accumulated through the researcher's research journey. The autoethnographic approach requires a very high degree of openness, honesty (i.e., uninfluenced by the need to create a false or self-serving image of oneself) and critical self-reflection. The (very difficult) trick is to sensitively maintain a balance between researcher attitude (involving some degree of contextual distance) and the researcher's subjective experiences (on which they are reflecting). The interaction of memory with length of retrospective timeline also needs to be managed through assiduous and consistent attention to recording in a journal (thereby combining the Stories strategy with the Journals strategy).

Journals Participant-Centred Strategy

The Journals participant-centred strategy encompasses diaries, journals, and other self-monitoring and self-recording techniques where data are generated by research participants acting, often out of the researcher's sight and hearing, as quasi-researchers (what Charmaz, 2014, p. 47, referred to as "elicited texts"; see also Symon, 2004). The strategy may rely on the support of technology such as a laptop or tablet computer (for building blogs on the internet, managing a Facebook timeline, recording events against a calendar page or timetable or keeping an electronic version of a diary), voice recorder, smartphone or video camera, or may simply rely on written records kept by the participant according to some agreed upon schedule. Such strategies typically generate qualitative data and are entirely dependent upon the diligence of the participant in adding to and keeping the record. It is possible, however, to obtain quantitative data (e.g., time spent doing specific tasks, number of tasks undertaken, various types of health measurements) using a diary or journal that can take handwritten or electronic form. The unique feature of the Journals strategy is that it is time-anchored; entries are placed in chronological order, which can help to build an evolutionary or developmental perspective and can relate entries to other events that occurred at the same time.

The Journals strategy works synergistically with the Stories and/or Creative Works participant-centred strategies. If there is an event, program, innovation, or treatment implicated in a development/change process, then this strategy could be used in conjunction with experience-focused strategies, such as Manipulated, Non-manipulated or Immersive strategies. Researcher insights and understanding of participants' diaries or journals can be enhanced by combining their use with post-diary/journal collection interviews. It can be an effective strategy in combination with the Measurements and/or Transformative data-shaping strategies (e.g., for recording tasks and activity characteristics, physiological data, such as heart rate, steps and distances walked and/or logging answers to standardised questions about activities, emotional states, feelings of stress, physiological measurements). Typically, this yields a logbook that may be recorded manually by the participants at scheduled times or may be recorded automatically, if appropriate technological support is available.

When a researcher maintains a research journal or set of field notes, they are taking the role of 'self' in 'self-recorded', as a participant in their research journey. A research journal forms a generally useful source of information providing contextual background for different aspects of the research journey. Field notes are more commonly assembled in conjunction with other strategies, such as participant observation. When writing a research journal or field notes, the researcher is actually creating data; data that can analysed and interpreted in their own right as well as in the context of other forms of data, perhaps gathered using other strategies. Where journal or field notes are in handwritten form, they may need to be transcribed into an electronic form for analysis (thus creating a synergy with the Transformative data-shaping strategy and the Textual artefact-based strategy). The research journal can also be critical for the researcher to call upon if they are writing an autoethnography, providing contextual, reflective, and personal details against a timeline. Using the research journal for this purpose shifts the focus from outward (understanding and making sense of what others do and experience) to inward (understanding self, as the researcher experiences and makes sense of their own research journey).

The researcher should realise that participants will vary in their capabilities and motivations for maintaining a diary or journal. For some, this will not be a natural activity and, if so, participants should be given space for learning and developing their capabilities, which means the researcher should downplay focusing on their earliest recorded entries. This data gathering strategy is highly dependent upon participants' motivation and commitment to cooperate (including their willingness to reveal diary contents to the researcher), their diligence in attention to the task, and their care and reliability in the diary recording activity. The researcher needs to take care to ensure that no incentives to falsify or impression manage their diary/journal entries are created. Effective safeguards and assurances, design of the recording medium, instructions/expectations can help in this regard as can the provision of external rewards once their diary has been completed and submitted to the researcher. Maintaining participant commitment without forcing (which may lead to inauthentic or incomplete entries) is critical and can be enhanced through periodic contact between the researcher and participant. Participants need to know how much time they need to commit to maintaining their diary/journal before they return it to the researcher and they need to be clear as to the schedule for completing diary entries (e.g., daily, weekly, time of day). The researcher also needs to make any expectations they have with respect to depth and level of detail of recorded information, including contextualisation of that information, clear. Missing data (e.g., missing entries) is always a risk, due to forgetfulness or lack of commitment on the participant's part or lack of clarity of instructions/expectations on the researcher's part.

If the research involves participants keeping a quantitative logbook, accuracy will be a necessary requirement, as the contents of the record will eventually become the source of measurements for statistical comparisons. The researcher may need to provide participants with specific equipment or instruments—such as smart watches, stop watches, pulse rate meters or oximeters, or rating scales—to facilitate their logbook entry activities. The quantitative logbook variant of the Journals strategy is not a strong stand-alone method but can be a useful initial step in exploratory investigations. Some training of participants may be required where quantitative measurements are to be gathered, especially in relation to which specific events are to be recorded, how to record them, what contextual details, if any, they should record, and what units of measurement and levels of precision are required in entries. If wearable technology or diary-keeping software is to be used, participants will need to be trained in their use (and perhaps basic maintenance, such as changing or charging batteries, calibrating, or resetting the device). The researcher is ethically responsible for safeguarding who has access to the contents of a participant's diary/journal while it is in their possession and what uses the information will be put to. This needs to be clarified up front before commencing the research. If the researcher plans to access pre-existing diaries/journals (i.e., that pre-date the research, thus creating a pairing with the Textual artefact-based strategy), they need to carefully negotiate with the writer about who will access the diary/journal, how its contents/information will be protected (as well as the identity of the writer) and how the researcher will access its contents.

Artefact-Based Strategies

Artefact-based strategies focus on handiworks of potentially any form that may have been produced for other purposes at other times and places and by other people. Here, the widest possible view of what counts as an artefact is taken, consistent with Plowright's (2011) view. Plowright (2011, p. 92) indicated that artefacts may be text-based (e.g., books, academic journal articles, essays, editorials, theses, dissertations and portfolios, emails, letters, historical documents, newspapers, diaries and journals, web pages, magazines, brochures, leaflets, reports, blogs and other kinds of technologically-transmitted text, such as SMS and Twitter messages and Facebook comments, transcripts of interviews) or may reflect other types of sensory experiences and combinations of sensory experiences:

- visual (e.g., photos, videos, YouTube clips, maps, illustrations and drawings, artworks, paintings, pottery, clothes, buildings and other architectural manifestations);
- sound (e.g., stories, oral histories, music, television, films, radio, recorded or live streamed content);
- kinesthetic/spatial (e.g., theatre, dance, plays, operas, concerts, gigs & community-based busking, treatments for improving health and well-being)
- smell (e.g., food, drink, herbs and spices, medicines, places, plants, animals, other people); and
- taste (e.g., food, drink, herbs and spices, medicines).

Any artefact can potentially serve as a (non-human) data source, without necessarily needing to connect with its creators or users (who may, in fact, no longer be amongst the living). Plowright (2011) stated that artefacts serve different purposes: carry/store information, present something, represent something, and/or interpret something. The researcher's task is to decide which artefacts, if any, may be relevant to pursue to advance their learning with respect to their research questions, access those artefacts, and then work to understand their purpose(s), their meanings, and/or their significance relative to the research context(s).

People typically produce and use or enjoy the artefacts in other times, places, and contexts. The researcher has no opportunity to influence their content, production process, or use. Artefact creation thus predates the commencement of the research journey, serving purposes established by others; their existence, form, and substance must be taken as given. Artefacts provide a 'clean' source of data, but one where the meanings, purposes and intentions of the author/producer/user can only be indirectly inferred from the artefact itself, unless direct access to the author/creator/producer/user becomes possible. What the researcher must decide is which artefacts will be chosen, how they will be accessed, and how they will be used for research purposes. Choice of artefacts to focus on as data sources thus becomes a very important consideration. The researcher's task is to discover the intersections between each artefact and the research stories they are relevant to. This may require reading between the lines and drawing inferences about relevance and meaning that go beyond the original purposes of the artefact while, at the same time, revealing something about the context(s) and circumstances in which the artefact emerged.

Textual Artefact-Based Strategy

The Textual artefact-based strategy involves choosing and analysing various forms of pre-existing written texts (e.g., presentations, commissioned research and other types of reports, journals, newspaper and magazine articles, annual reports, minutes of meetings, emails, media stories, transcripts of interviews, speeches) that the researcher deems relevant to their research questions. The researcher can assemble a collection of hardcopy texts (e.g., papers, reports, manuscripts, books) and/or texts in electronic form (e.g., emails, pdf documents, rich text format (rtf), Word or PowerPoint documents, text images, social media posts), with the latter being far more common nowadays. Even in situations where 'texts' are accessed in hardcopy or non-textual format, the researcher may convert them into an electronic text format (which may involve pairing with the Transformative data-shaping strategy). Every text the researcher gathers should be accompanied by contextual notes regarding its purpose, source, date, context of production, authorship, if relevant, why it is relevant for the research at hand, and so on. In some cases, texts (e.g., company or school annual reports or policy documents, historical records, minutes of meetings, media stories) may form background and contextual material for other data gathering strategies as well as for the research context itself. Thus, for some research purposes, texts may facilitate contextualisation (an elaborative or supplemental role) and, for other purposes, texts may constitute the main evidence the researcher will use to address their research questions (a primary data role). The Textual artefact-based strategy is effectively combined with the Transformative data-shaping strategy and possibly the Creative Arts artefact-based strategy when transcripts of interviews and other spoken word events are produced from different recording media.

It is important to avoid potential biases in selecting texts to gather. Normally, this would require some type of data source choice strategy to ensure that the appropriate types of texts are gathered for research purposes. Such a strategy may be random, purposive, systematic, opportunistic, or guided by what participants may offer up as documents to which they can provide access (a kind of snowball strategy). Access to some documents and texts can be problematic, especially if the documents and texts are not on the Internet (as might be the case with certain company documents, minutes of meetings, older documents that predate electronic document storage, historical documents of various kinds, and so on). The researcher may have to invest effort and legwork to chase up texts that exist only in hardcopy format and, if they cannot be located, this rules them out of their sample. Certain texts and documents the researcher might want access to might only be available on an organisational or institutional intranet, in which case, they will need to negotiate access (this may be especially important if the researcher is seeking texts and documents to help them contextualise a group or a case study, for example). Government documents may or may not be publicly accessible and, if the latter, the researcher may need to negotiate access or forgo access (if such texts are classified, for example). The Internet makes accessing many kinds of publicly available texts quite easy, but the researcher needs to establish a transparent basis on which to identify and select them for inclusion in their research. Such documents may be more difficult to contextualise with respect to their original purpose, authorship, and authenticity. Any constraints on researcher access to the range and types of documents and texts needed for research purposes will, of necessity, alter the researcher's sampling intentions, thereby adversely impacting on their capacity to appropriately contextualise what has or has not been learned.

Meta-analysis and meta-synthesis are important variants of the Textual strategy. Meta-analysis and meta-synthesis are approaches for choosing, assessing, theorising, and generalising the learning amassed in a series of, most commonly, published research studies (however, books, unpublished research articles, theses and reports can also be considered). Thus, the texts chosen are the text-based research outcomes conveyed and displayed by researchers, primarily in the academic literature. In short, meta-analysis and meta-synthesis provide ways to conduct and display systematic literature reviews. Meta-analysis focuses on research that gathers, analyses, summarises, and interprets quantitative data and meta-synthesis focuses on research that gathers, analyses, summarises, and interprets qualitative data. Both approaches provide avenues for systematically building up learning/cumulating knowledge in a specific topic area which transcends individual research investigations, hence the term 'meta'. Meta-analysis and meta-synthesis are both useful strategies for making sense of patterns and guiding learning from many research investigations of the same topic or issue. They may be implemented as a stand-alone research endeavour where the meta-analysis or meta-synthesis is the intended, typically published, outcome of the research project. Alternatively, a meta-analysis or meta-synthesis may comprise a preliminary phase of a larger investigation conducted in order to set the stage and inform the logic and focus of the research to follow.

Meta-Analysis

Meta-analysis focuses on published (most frequently, articles published in refereed academic journals) and, occasionally unpublished research articles and reports (see, for example, Cooksey, 2020, Procedure 8.8; Durlak, 1995; Glass, McGaw & Smith, 1981; Konstantopoulos & Hedges, 2004; Rosenthal, 1984; Schmidt & Hunter, 2014). Meta-analysis employs samples of researcher-authored stories that report quantified or quantifiable data and statistically analyses those stories in the search for more aggregated and generalisable patterns. Avoiding sampling biases (e.g., ignoring unpublished or grey literature or only sampling published outcomes form specific journals) is a key concern in meta-analysis. A central quantitative concept associated with meta-analysis is effect size. Different systems for meta-analysis have differing statistical approaches to defining/measuring effect size (e.g., a standardised mean difference measure, an odds-ratio measure, or a correlational measure), but all approaches have roughly the same goal: to quantify, in some standardised/comparable fashion, the size of group differences or the relationships between two or more variables, taking account of the size of the sample (N) as well as quantitatively-coded aspects of the contexts in which those differences or relationships were observed. The onus is on the researcher to clearly argue for the measure(s) of effect size they plan to adopt and to indicate how and why they will code certain aspects of each investigation for comparison purposes. Once the researcher has converted the statistical outcomes from individual research investigations into the selected standardised measure of effect size and coded aspects of interest about each research story being considered, they are able to make various types of statistical comparisons between investigations to inform higher-level knowledge generalisations.

Meta-Synthesis

Meta-synthesis (or meta-ethnography, as it was first known) was initially proposed by Noblit and Hare (1988) as a way of making sense of and perhaps theorising (at a 'meta' level) about what has been learned in a sample of qualitative investigations. It can be very loosely considered as the qualitative analogue of meta-analysis (Paterson et al., 2001, for example, refer to it as qualitative meta-analysis). It has proven to be particularly useful in the medical, nursing, and health sciences, but is much more widely applicable. Synthesising is a comparative translational process where the goal is to see if meaningful patterns and explanations for outcomes can be identified across the studies. Synthesis is not an aggregative generalisation process in the same way as a meta-analysis. Rather, through the comparative translation process, meta-synthesis attempts to construct an integrative perspective; one where both commonalities and uniquenesses among the studies are considered together in order to achieve a higher-order level of learning (one form of which may be a meta-theoretical account). The most challenging aspect of meta-synthesis is adopting a sufficiently nuanced and balanced approach to the individual studies being considered, in full awareness of any contextual differences that may exist between the different investigations (Noblit & Hare, 1988, offer one such approach).

Creative Arts Artefact-Based Strategy

The Creative Arts artefact-based strategy involves choosing and analysing various forms of artefacts that are represented in a single medium or multi-media form other than simple written text (e.g., photographs, artworks, poetry, theatre performances, recitals, recordings, cartoons, videos, music, dance, films, plays, media releases, websites, postings on Flipboard, Google News and other multi-media news apps; image and interactive postings on Facebook, Snapchat, Instagram, Twitter, Pinterest, YouTube and other social media platforms). If we take a very broad view of the term 'artefact', we can extend the creative arts focus to encompass cultural symbols, objects, icons, rituals, and any other artefact that has meaning for/is significant to people in specific contexts. The artefacts gathered using the Creative Arts artefact-based strategy may have emerged from previous arts-based research efforts.

In general, the artefacts gathered using the Creative Arts artefact-based strategy will tend to emphasise a range of audio-visual and other sensory inputs (Banks, 2018; Plowright, 2011; Ross, 2001). Some creative arts artefacts expressed in multi-media forms, such as films, websites, media releases, YouTube clips, social media postings, news postings, and speech recordings, may incorporate written text or spoken words as one component of the mix. However, if the researcher only is interested in the embedded written text or a transcript of the embedded spoken words in that artefact, this strategy effectively morphs into the Textual strategy, which means that the researcher is deliberately sacrificing the multi-media nature of the artefact. Other creative arts artefacts (e.g., cartoons, photographs, artworks, emojis and other communication devices, performances, symbols, icons, rituals) can only be dealt with in their original form. Furthermore, some creative arts artefacts based on performances may be ephemeral (perhaps even one-off or unique) time-specific events (e.g., a play, dance, lecture, or recital) that the researcher must directly experience (pairing with the Participant observation-based strategy). In such cases, the researcher will need to acknowledge this in their contextualisations of the artefacts and put procedures in places for recording/reflecting on their experiences.

Creative arts evidence typically focuses on sampling artefacts that were originally produced for other purposes, but which the researcher believes are relevant to their research questions. The researcher must transparently describe their processes for identifying and choosing creative arts artefacts, facilitated by detailed annotations in their research journal about when, how, where, from whom and why specific artefacts were brought into their research journey. Sampling in the Creative Arts strategy may be random, purposive, systematic, opportunistic, or guided by what research participants or other data sources may offer up as artefacts to which they can provide access (a kind of snowball strategy). It is possible to conduct research that synergistically combines the Creative Arts strategy with the Journals participant-centred strategy in instances where the researcher invites participants to record their journal or diary in a multi-media format (by taking photos or shooting video clips, perhaps using a smart phone) or the Stories participant-centred strategy in instances where the researcher invites participants to record or perform their story in a multi-media context. It is also possible to conduct research that synergistically combines the Creative Arts strategy with the Participant or Systematic observation-based strategy where the researcher records their observations, typically in a multi-media format, for later analysis. The Creative Arts strategy presents the researcher with some unique challenges. For example, one challenge is inherent in the very nature of the artefacts, namely the fact that they rely on multiple sensory channels to convey meaning. The researcher must therefore make strategic decisions about how those various channels should be handled, holistically or in isolation. Another challenge may be variable quality in various creative arts artefacts, which may degrade or obscure meaning.

Archival/Secondary Artefact-Based Strategy

The Archival/Secondary artefact-based strategy focuses on choosing, retrieving, and analysing data sources such as organisational and institutional records, historical and other text-based records and archives, financial and other types of government, publicly or privately-maintained online or spreadsheet databases, and censuses and company annual reports (see, for example, Bell et al., 2022, Chap. 14; Chandola & Booker, 2021; Cohen et al., 2018, Chap. 28; Cowton, 1998). Such data are gathered/maintained by people other than the researcher for purposes other than those of the researcher (data gathered directly by the researcher constitute primary data, see Veal, 2005). Archival data usually refers to collections of records and texts that are more historical and qualitative in orientation, which effectively pairs this strategy with the Textual artefact-based strategy. Such archives can be useful for historical, biographical, ethnographic, and life history research. Note that the Textual artefact-based strategies of meta-analysis and meta-synthesis, technically, depend upon archival data sources that store published and/or unpublished research texts. Secondary data typically refers to maintained databases containing quantitative data (e.g., accounting, finance, education, health/medicine, governmental, management, political and policy sciences, geographical, operations research, and economics databases).

Two major advantages of using archival or secondary data methods are: (1) substantial cost savings often accrue from not having to collect primary data and (2) the extent or reach of the data will often permit a wider scoping of the research project than would otherwise be feasible if the researcher were left to their own data gathering resources (e.g., accessing more representative samples, capturing longitudinal trends across years, access to more obscure records). The second advantage is especially important if the units of analysis are in a highly-aggregated form (e.g., at company, industry, or nation level). A major difficulty, particularly with quantitative secondary databases, is that the nature/definitions of variables, accuracy of the data (unless they are independently audited), the motivations of the data source creators, ethical protections or problems (e.g., data harvesting issues) associated with the data, and maintenance activities (e.g., frequency of updating, how data are vetted for input) remain largely hidden from the researcher, unless the researcher can access background information (i.e., 'meta-data') relevant to the database itself. Searchability of the database is also important to understand as different data bases may have different search functionalities.

Archival and secondary data methods require the researcher to accept a very important assumption, namely, that a data source they intend to use provides the data necessary for evaluating their research questions. The issue is often one of completeness here. The researcher will have no control over which data are gathered and how they are represented in the data source. If there are important content, conceptual, or measurement gaps in the data, there may be no way the researcher can independently fill those gaps. This may force the researcher to revise the scope of their study and perhaps even their conceptual framework. Thus, the researcher must, of necessity, accept the contents of the secondary or archival data source as a given. In the case of a quantitative secondary database, this means accepting the operational definitions of the variables it contains and the unit(s) of analysis (e.g., company, geographic region, group, industry, country) represented in the data. There will normally be insufficient information accompanying such a secondary data source to permit the researcher to verify the measurement quality of the data, unless they have access to other data sources that can be used for triangulation purposes. In the case of a qualitative archival database, this means accepting their processes for acquiring the documents and ensuring the stored quality of any documents (even if marginally readable in the case of handwritten records).

Recently, there have been many advances in large-scale secondary database development, management, storage, and use, giving rise to a new discipline called 'data science' (Provost & Fawcett, 2013). The resulting databases are often colloquially referred to as 'Big Data' (see, for example, Boyd & Crawford, 2012; Gandomi & Haider, 2015; Walker, 2014). Such databases are typically assembled by governments or large companies (e.g., census bureaus, health departments, labour departments, education departments, large organisations, such as advertising agencies and marketing organisations), typically quite massive, and often managed through a process called data warehousing. Their use in research has created a new field called knowledge discovery (see, for example, Fayyad et al., 1996), where warehoused databases are made publicly available to researchers in order to 'mine' the databases for previously unknown patterns to build new knowledge. Tools for knowledge discovery in Big Data include machine learning and data mining, text mining (for qualitative databases), and predictive analytics (building up predictive statistical patterns and relationships using large-scale databases; Cohen et al., 2014, discuss the legal and ethical issues associated with the use of predictive analytics in the health care industry). There are also programming languages such as Python, R and Wolfram that are designed to support learning from large databases (see, for example, EMC Education Services, 2015; VandePlas, 2022; Wolfram, 2023).

2.2.3 *Strategies for Exploring People's Experiences*

The domain, exploring people's experiences, encompasses strategies that researchers can use to shape the data that they gather from participants into a quantitative or qualitative form suitable for analyses (i.e., data-shaping strategies) as well as several strategies for focusing directly on contextualised concrete experiences participants are having, or have had, about which researchers will be gathering data (i.e., experience-focused strategies). There are three broad data-shaping strategies, two of which focus on shaping data gathered from or about research participants (Measurement and Transformative) and one (Generative) which focuses on the creation of quantitative data, without the direct need for participants. There are four different experience-focused strategies, three of which generally focus on scenarios where the researcher exerts partial or complete control over context to structure participants' experiences (Manipulated, Structuring, and Immersive) and one (Non-manipulated) where the researcher must take the contextual events experienced by participants as a given (i.e., shaped by other people or other forces) rather than as something the researcher has structured. In many cases, one or more data-shaping strategies are paired with an experience-focused strategy.

Data-Shaping Strategies

Data-shaping strategies generally focus on ways of ensuring that the researcher gathers or produces the appropriate kind of quantitative or qualitative data for their research needs. The researcher would implement the Measurement strategy when they need to design processes for obtaining quantitative data in the forms and with the meanings that they require, thus shaping data at the time the researcher gathers them. The researcher would implement the Transformative strategy if they are gathering or have already gathered quantitative or qualitative data but need to change their form for analysis and interpretation purposes, thus reshaping data either at the point they are gathered or after the researcher has gathered them. The researcher would implement the Generative strategy if they needed to create data from scratch, without having access to or needing actual human participants (in effect, using computer-generated 'participants'). The Measurement and Transformative strategies tend not to be used on their own. Instead, they are typically used in conjunction with a wide range of other strategies to support data gathering activities. The Generative strategy can be used in a stand-alone capacity but will often be combined with the Manipulated experience-focused strategy through implementing a specific experimental design.

Measurements Data-Shaping Strategy

The Measurements strategy (also known as *psychometrics*, see, for example, Furr, 2021; Price, 2017), in the social and behavioural sciences, focuses on constructing and implementing procedures for quantifying the characteristics and behaviours of people and the systems they inhabit. The goal of the Measurements strategy is to provide the quantitative variables that a researcher needs to describe samples, test hypotheses, build models, and/or search for patterns and relationships and, in this sense, measurement is most often used as a strategy to support one or more other data gathering strategies. We consider the Measurement strategy to be data-shaping because measurement processes produce quantitative data at the outset, i.e., a systematic process for using numbers to represent some phenomenon of interest with respect to a specific characteristic or property (Athanasou, 1997). Quantitative data are only as good as the measurement processes that give rise to the numbers. Good measures will facilitate analyses and their meaningful interpretation; poor measures will have an adverse effect on analyses, leading to indefensible interpretations. It is important to realise that no amount of sophisticated data analysis will save an investigation that uses poor measurement processes to obtain the quantitative data being analysed.

Measurement is typically focused on constructs or theoretical concepts/ideas that emerge in relation to research questions/propositions/hypotheses. Those constructs may focus on the internal mental (e.g., what/how people or groups think, believe, value, and feel as conveyed by how they behave) and physical (e.g., metabolic rates, blood flow, nutrients, biochemical, and neurochemical processes) states of people or on objects, people, and other entities in the environments that people inhabit and with respect to behaviours they exhibit and actions they take (which encompasses demographic, social, organisational, employment, educational, creative, cultural, political, geographical, economic, natural, and ecological facets) (see, for example, Hand, 2004). Such measurements require some sort of instrumentation that converts a phenomenon into meaningful numbers and that instrument may be observational (e.g., reflecting what can be seen, heard, felt, tasted, smelled, counted, sorted, and so on), technological (e.g., reflecting what can be read on a ruler, dial, clock, display, gauge, scale, meter or other measuring device) or self-reported (e.g., reflecting patterns of answers to questions on a questionnaire or of responses to some stimulus). Measurement may also focus on counting observed phenomena where, in many cases, the researcher does the counting. In such cases, the researcher must be very careful to describe exactly what constitutes an instance of a phenomenon that will be counted. Some measures or indices may be formed as mathematical or conceptual composites of other measures (as with economic indicators, financial ratios, personality or attitudinal factors) and here the researcher needs to ensure that the process for constructing the composite or index is credible, defensible, and theoretically meaningful (effectively combining the Measurement strategy with the Transformative strategy). When the Measurement strategy is implemented, the researcher must devote adequate attention to ensuring that appropriate measurement scales are used and that the measurement systems demonstrate validity, reliability, and sensitivity with respect to the aspects of phenomena that the researcher intends for them to be reflecting (we will have more to say about these matters in Chap. 5).

Transformative Data-Shaping Strategy

The Transformative strategy focuses on helping the researcher to transform one form of data into another form. This data-shaping strategy often operates as a bridging strategy between the data gathering and data analytic vertices of the 'Data Triangle' but can also play a role in more directly recording observations. The Transformative strategy encompasses the use of formalised recording protocols, templates, schedules, mathematical formulas, or devices to record or code observations or segments of text or other multi-media content. Often, when data are gathered using another strategy, they may not immediately be in the most appropriate form or type for analysis and interpretation. This strategy is almost always combined with the Structured observation strategy and/or Immersive experience-focused strategy (to develop/implement observational protocols/recording forms). It is also used in combination with the meta-analysis or meta-synthesis approach to the Textual artefact-based strategy to systematise, at least partially, the coding of journal articles and other documents. Finally, it is used in combination with the Textual and/or Creative Arts artefact-based strategies to systematise, at least partially, the coding of texts, videos, and other types of multi-media evidence. As well, the Transformative strategy is used to manage the process of producing transcripts or translations from recorded interviews or other multi-media recordings (including from other cultures or in other languages) or researcher-produced journal or field notes.

There are four potential transformational pathways the researcher can follow: transforming quantitative data into another quantitative form; transforming quantitative data into a qualitative form; transforming qualitative data into another qualitative form; and transforming qualitative data into a quantitative form. More than one pathway may be followed in any specific research endeavour, depending upon the researcher's needs.

Quantitative → Quantitative

This pathway for the Transformative strategy is implemented when one or more quantitative measures, gathered in one form need to be transformed or recoded into other quantitative forms; the data remain in a quantitative form, but the nature of measurements is changed. This pathway for transformation can be used to alter the distributional shape of a quantitative measure when its distribution does not match what is assumed to be the case for a specific statistical analysis procedure (e.g., non-normal data can be transformed using mathematical operations, such as log10, reciprocal, arcsin, or square root, to produce a more normally-distributed variable). If the researcher wants to use nonparametric statistical procedures, such procedures typically transform the scale of measurement for the dependent variables being analysed by reducing them to ordinal ranks. Another common version of this pathway is to mathematically combine individual quantitative measures in specific ways to make new composite measures or indicators (as when a factor or scale score, a financial ratio, or economic index is produced). In all cases, this pathway produces new quantitative measures that the researcher can then use for building statistical models and testing hypotheses. If the researcher follows this pathway, they need to be very clear as to their reasoning, which can be recorded in their research journal for later reference.

Quantitative → Qualitative (Qualitising)

In this pathway for the Transformative strategy, the researcher changes data form and nature. Here. one or more quantitative measures are transformed into a meaningful or higher-order qualitative category or thematic system or are used to define meaningful groups. This is often done when simplified descriptions/interpretations for more complicated measures are desired (e.g., low, medium, and high categories). The new qualitative categories may reflect inferences made about participants who are positioned contiguously along the quantitative scale of the original measure (e.g., on a series of questions about voting patterns and policy preferences, a researcher may classify a participant as being liberal/left- or conservative/right-leaning). For knowledge discovery purposes, there are certain statistical procedures, such as cluster analysis, neural network analysis, or data mining (pairing with the Archival/Secondary artefact-based strategy), that researchers can use to discover hitherto unknown categories of participants using a set of quantitative measures and, once discovered, they can create qualitative descriptions of each cluster or category using information contained in the patterning of their quantitative measurements. In general, qualitising is a tactic for enhancing interpretability and meaningfulness from the point of view of the research reader/user. It may also pave the way for clearer policy development and decision making in instances of applied or transdisciplinary research, because social policy can be targeted more readily to specific subgroups rather than to a larger heterogenous group of individuals.

Qualitative → Qualitative

This pathway for the Transformative strategy is implemented when qualitative data are transformed or coded into a meaningful or higher-order qualitative category or thematic system (e.g., pairing with interaction-based strategies, or Textual or Creative Arts artefact-based strategies). A qualitative coding system may be data-driven, inductive, and emergent (e.g., as in some forms of grounded theory) where codes are built up from the data and where subsequently gathered qualitative data are coded into the emergent categories as well as into any new categories that make sense. Additionally, data may be categorised into new higher order categories (what Saldaña, 2021, describes as a form of second-cycle coding). In other forms of qualitative research, coding may be concept-driven, where the researcher uses a pre-designed category system (often called a structural codebook or coding template, see Guest et al., 2012, pp. 55–60) to code important contextual aspects/features of texts to be analysed or to ensure that specific ways of looking at the data are not overlooked (e.g., see Schreier, 2012, pp. 85–87, see also template-guided coding, King, 2004). These are types of deductive approaches to qualitative data management (distinguishable from inductive or data-driven approaches, see Schreier's, 2012, Chap. 5 discussion of coding 'frames') where coding categories are at least partly formulated a priori by the researcher and may be theoretical, experiential, literature-based, or logical in origin. In content analysis or meta-analysis (pairing with the Textual artefact-based strategy), the researcher may classify written responses or segments of research articles into pre-designed sets of categories. Typically, in content analysis or meta-analysis, the researcher would have more than one person code the same texts using the same coding protocol/schedule/ template/codebook to facilitate checking for inter-rater (or inter-coder) reliability.

The qualitative → qualitative transformation pathway also relates to the preparation of interview transcripts from digital or multi-media recordings (i.e., aural or multi-media data are transformed into textual data). When the researcher creates new texts via some transcription or translation (if the data are in another language) process, they must first decide what kinds of information they wish to have represented in the transcripts (e.g., verbatim speech, stutters, ums, pauses, laughs, over-speaks, emotional emphasis, inaudible or unclear utterances, background noises/intrusions/ interruptions) and how that information should be represented. Transcript preparation needs to be consistent, and the researcher needs to ensure that all of what they intend to be recorded in the transcript is represented. If texts are being translated from another language, the researcher must be very clear about how the translation process was implemented and managed.

Qualitative → Quantitative (Quantitising)

This pathway for the Transformative strategy can be implemented when qualitative data are transformed into counts, numerical ratings or rankings (e.g., rating the content of utterances according to the level of hostility evident in them; counting the number of phrases in transcripts that have been coded against a specific theme or category, thus pairing with the Textual or Creative Arts artefact-based strategies). The interview schedule used in a structured interview constitutes one type of quantitising data transformation (vocalised answers to recorded quantitative answers) protocol, one which offers specific prompts for the data to be requested and recorded. For content analysis or analysis of open-ended questionnaire items, this transformation often follows initial qualitative → qualitative transformations to go one step further and produce quantitative measures that are appropriate to use in statistical tests and models. In the Manipulated and Non-manipulated experience-focused data gathering strategies, events, groups, treatments, or conditions are often simply qualitatively represented as group membership. However, for modelling purposes, these qualitative classifications need to be transformed into quantitative variables having the most appropriate and statistically interpretable form in the context of the model the researcher is constructing and testing. In such cases, the Transformative strategy works by recoding the group membership using specific numerical values (e.g., using dummy coding, effect coding, orthogonal contrast, or orthogonal polynomial coding—see, for example, the discussions in Cohen et al., 2003; Cooksey, 2020, Fundamental Concept VI).

The qualitative → quantitative transformation is also implemented in combination with the Systematic observation-based strategy. Through the design and use of an observation recording protocol, qualitative, even holistic, observations can be numerically rated or counted for purposes of statistical analysis. When designing such a protocol, ease of use (minimising the risk of recording errors) for the observer is essential to ensure, especially if they are required to make judgments, as when evaluating latent/deep meaning in observations. Efficiency of use also is essential to ensure for observational coding protocols in situations where observations are obtained under tight schedules or in rapidly-changing situations. Efficiency (and in some cases accuracy) can be improved if the structured observational protocol can be supported or automated using technology. For observational protocols, inter-rater (or inter-coder) reliability is important to establish.

Generative Data-Shaping Strategy

The Generative strategy encompasses a set of approaches for creating quantitative data, using specific parameters and processes, to test social and behavioural models. Such models are evaluated using computer simulation rather than gathering data from human participants. While it might appear that creating data is ethically problematic, the difference here is that the researcher is not 'faking' research data to support a research hypothesis, which is ethically reprehensible; rather, they are openly and transparently creating artificial data so as to explicitly test some model or hypothesis of interest, using computer simulation processes (see, for example, Axelrod, 2007, and Gilbert & Troitzsch, 2005; note that computer simulations differ from the types of live-action simulations used in the Immersive experience-focused strategy, which do involve human participants). Use of the Generative strategy can follow two different pathways. The first pathway is artificial data generation, such as Monte Carlo research, which involves the production of data, using random sampling algorithms, which exhibit specific statistical properties in order to test hypotheses about their behaviour. The second pathway is computational modelling, such as agent-based modelling, which involves the creation of a computer-simulated environment, based on theory, where assumptions, variables, and parameters can be explicitly controlled and their effects on simulated entities precisely tracked (for illustrations, see Cooksey, 2020, Procedure 9.10).

Monte Carlo Pathway

The Monte Carlo pathway (named after the famous Monte Carlo Casino, Monaco) involves sampling via random data generation (see, for example, Carsey & Harden, 2013; Cooksey, 2020, pp. 670–673; Mooney, 1997). With the Monte Carlo pathway, the researcher's goal is to decide which data distribution(s) and associated parameters are to be used to define the population(s) to be randomly sampled and the nature of the samples to be generated (e.g., how many observations to generate, how many groups to create, which data generation parameters to systematically vary). All Monte Carlo methods depend upon random number generators, which most statistical packages make available. Categorical as well as continuous data can be created using the Monte Carlo approach and any number of distributions can be simulated (e.g., uniform, normal, binomial). Once the researcher has created the artificial dataset(s), they can then analyse them as if they were empirically gathered data. One major purpose for generating random data involves conducting experimental research explicitly to study the behaviour of data, generated with different characteristics, assumptions, and/or generating algorithms, and seeing how those differences impact on the conclusions one might draw, using different analytical approaches (pairing with the Manipulated experience-focused strategy). The advantage of Monte Carlo research is that the researcher can control virtually every aspect of the research from start to finish. What they need to be careful to do, though, is to be thoroughly transparent about all the decisions they made through the data generation process and anchor their choices to theory and/or the literature.

Computational Modelling Pathway

Computational modelling (see Cooksey, 2020, pp. 673–681; Taber & Timpone, 1996, for overviews and illustrations; see Wolfram, 2002, who describes a new kind of computational science) allows the researcher to set up and test theoretical models about the social, political, physical, and/or economic world without needing to gather empirical observations. In some instances, the theoretical models are so complicated that it is not feasible or too costly for the researcher to gather the necessary empirical data. However, this limitation can be partially circumvented by creating a simulated computational environment. Computational modelling may also allow the researcher to test certain types of hypotheses that are infeasible/unethical to test in the real world. Computational modelling generally involves structuring a theoretical model, generating the data needed to monitor the behaviour of the model, and providing information useful for validating both the model processes and its outcomes. We will examine two broad approaches here: dynamic systems modelling and agent-based modelling. [Note that, technically speaking, *statistical bootstrapping* (see, e.g., Cooksey, 2020, Procedure 8.4; Ditka & Scheer, 2021; Mooney & Duval, 1993) is a kind of hybrid Monte Carlo/computational modelling generative data-shaping strategy, where an original empirical sample of data (data in this sample are real, not artificially generated) is repeatedly randomly resampled to create a very large number of synthetic samples, whose statistical characteristics can be calculated and used to make justifiable inferences about the original sample, without the restrictive assumptions required by parametric statistical methods. Statistical bootstrapping is thus a kind of nonparametric statistical technique.]

Dynamic Systems Modelling. Dynamic systems modelling evolved as a way of developing and testing complex simulation models and has been integrated with systems thinking principles (see Anderson, 1997: Meadows, 2008) as well as complexity theory and new mathematical approaches for modelling complex nonlinear systems (see discussions in Maani & Cavana, 2007; Sterman, 2000). Such models are typically so complex that they cannot be constructed and implemented without the support of computer hardware and software (such as *Insight Maker* (web-based platform), *Stella, iThink, Vensim,* or *PowerSim*). Systems dynamics models may be determinist in emphasis (where a system of equations calculates exact quantities for each step of the simulation process) or stochastic (where simulated random data and probability models are used as part of the modelling process; this is where Monte Carlo methods can play a role). Dynamic systems modelling tries to capture and simulate the behaviour of a defined system over time. The models are dynamic because they depend upon feedback loops over time just as real human systems do, using the concepts of stocks (supplies of necessary resources) and flows (movements of resources over time).

What is important to understand about building these types of models is that they are critically dependent upon the nature and quality of the theory that gives rise to the model. The difficult part is identifying all the control parameters the researcher will need to make the model workable as well as all the key variables and their functional relationships and feedback loops. Unless this is done well, the simulation will simply produce nonsense. What this often means is that the researcher needs to couple a deep exploration of the relevant literature with observations of the live system in action and/or interviews with experts. The researcher will need training not only in dynamic systems modelling itself, but also in using the computer environment in which they will program the simulation (note that modern-day software platforms offer graphical interfaces so that models can be constructed without writing a computer program or wrestling with differential equations). Along the way, the researcher needs to document all their thinking in a research journal, including the reasoning behind all the choices and assumptions they have made, how each of the variables in the model were defined and quantified, and how and why specific functional relationships and feedback loops linking the variables over time were set up. The researcher's story must be convincing but also must contain sufficient openness and extent of detail so that another researcher would be able to replicate what they had done.

Agent-Based Modelling. Agent-based modelling is another form of computational modelling (see, for example, Gilbert, 2008; Gilbert & Troitzsch, 2005, Chaps. 8 and 9; Miller & Page, 2007; Railsback & Grimm, 2019; Wilensky & Rand, 2015).

Agent-based modelling focuses on models involving simulated agents (e.g., individual people, teams, departments, organisations, nations, cars, planes, animals, trees, molecules) interacting with each other dynamically through time in a virtual world. This virtual world is controlled by model parameters and simulation conditions that the researcher identifies. Interactions between agents may constitute information flows (verbal, nonverbal, observational, decisions and choices, wealth), relational connections and qualities (formal, informal, cross-cultural, conflict, trust, prior history, flocking, disease, culture), and action-reaction linkages, all occurring in discrete steps along a timeline. Agent-based modelling requires very precise theorising/conceptualisation and mathematical modelling within a computer support system, the most popular model development interface for the social sciences is probably *NetLogo* (Railsback & Grimm, 2019; Wilensky & Rand, 2015). An agent-based model is conceptually and technically challenging to construct. It will likely require extensive background work in the literature, perhaps incorporating information from experts and potential users to verify the model's conceptualisation, relevant variables and control parameters and its embedded processes. The researcher will likely have to make simplifying assumptions about the 'world' they are about to construct, which concepts and relationships to incorporate and which they cannot include, what parameters are needed, what relationships need to be modelled, how variables should be represented/measured, and what kinds of outputs they want the model to produce.

The researcher's assumptions shape and delimit the specifications and characterisation of their model; any computational model is only as good as its underlying assumptions and model specifications. This imposes an extra burden of transparency on the researcher to be very clear about the entire process for building and testing their model from conceptualisation to validation. Furthermore, the performance of the researcher's model will be critically dependent upon their anchoring assumptions and the choices for controlling parameters. This means that thorough experimentation (thus pairing with the Manipulative experience-structuring strategy, but using virtual rather than real participants), sensitivity testing and validation (wherever possible) of their computational models is essential, if the model is to be anything other than an academic exercise. In some cases, external validation of model outcomes may be difficult or impossible to obtain because the researcher is unable to access the data they would need to do so. Docking (i.e., testing one computational model against another, see, for example, Olaru, Purchase & Denize, 2009) has emerged as one way of trying to cope with this potential limitation, but to be successful, docking requires an appropriate and viable alternative model to compare outcomes with—something that may not exist.

Experience-Focused Strategies

Experience-focused strategies create and implement or examine/evaluate a specific structuring or type of experience or activity for research participants. In essence, they establish the context within which or about which data gathering will focus. If making causal inferences is a goal of the research, then the stage can be set by manipulating, where possible, the contextual environment or tasks in which participants are providing/generating information and/or data. Manipulation of contextual conditions allows the researcher to ensure that putative causes change before putative effects are observed. If contextual manipulation is not possible, then causal variables can only be measured, not controlled, at the cost of lessened capacity to infer causality. In some cases, the researcher may not be able to create or manipulate the conditions, events, or situations that participants experience; another agency may play this role. Other agencies encompass social or organisational/institutional agencies, government entities, or natural environmental forces. It may also be the case that a social, cultural, or physical event forms the basis of participants' experience. In these latter instances, researcher control over context is minimal, consequently weakening the strength of causal inferences even further. Experience-focused strategies thus range from non-manipulated experiences (imposed by agencies or forces other than the researcher) to completely manipulated experiences (where the researcher controls and manipulates all contextual conditions, usually in some kind of laboratory) to structured experiences (where the researcher provides a framework for participants to work within to produce data) to immersive simulation experiences (where the researcher or another agency creates an artificial but holistic and realistic task environment within which participants have their experiences).

Manipulated Experience-Focused Strategy

The Manipulated experience-focused strategy is the prototypical strategy in the social and behavioural sciences for controlling people's experiences in a research context, i.e., the experiment (see, e.g., Ariel, Bland & Sutherland, 2021). Generally, the goal is to gather quantitative or quantifiable data to justify defensible inferences about the existence of causal relationships, what has been termed the 'causal comparative' approach (see, for example, Johnson, 2001). In a true experiment, the researcher manipulates and randomly determines all the conditions/events that participants experience (the hypothesised causal factors), controlling for as many alternative plausible yet extraneous (i.e., unwanted or confounding) explanations as possible, then measures the effects of those conditions/events on aspects of participants' thinking and/or behaviour. The true experiment is the strongest data gathering strategy for making defensible inferences about external causation and is most likely to be achievable in a laboratory setting. The researcher's ability to randomly assign participants to conditions/events is what differentiates a true experiment (where random assignment to groups or conditions is used) from a quasi-experiment (where random assignment to groups or conditions is not possible or not used). The strongest true experimental designs incorporate not only one or more treatment groups (e.g., where different levels of a causal factor have been manipulated, such as when members of each group experience a different specifically defined dosage level for a drug in a clinical trial) but also at least one control group (e.g., a placebo condition, where members of the group experience a 'non-treatment' such as a saline injection for a clinical trial) to establish a comparative standard/baseline against which to gauge experimental treatment or condition effects and, importantly, the researcher randomly determines which participants experience which conditions. In health and medical research, such designs are termed 'randomised control trials' or RCTs (considered the gold standard of medical research and clinical trials; see, for example, Kaptchuk, 2001).

In theory, random assignment of participants to groups is intended to statistically equate the groups on all extraneous causal factors and the use of a control group provides a comparative standard for judging the magnitude and directionality of causal effects or changes. Tracking influences/changes over time leads to repeated measures experimental designs. The simplest version of such a design is the Pre-test-Post-test experiment where the causal factor or intervention occurs between the gathering of pre-test measurements and the gathering of post-test measurements. Cooksey (2020, Fundamental Concept VII) illustrates and discusses a range of experimental and quasi-experimental designs (see also discussions of experimental and quasi-experimental designs in Ariel, Bland & Sutherland, 2021; Campbell & Stanley, 1966, Cook & Campbell, 1979, and Kirk, 2013).

If the researcher is not able to fully determine which condition(s)/event(s) participants experience, the experiment becomes a quasi-experiment. Thus, with a quasi-experiment, contextual manipulation can only be partially achieved, making it a weaker data gathering approach for drawing defensible causal inferences, relative to a true experiment. A quasi-experiment is more likely to be used in field settings or naturalistic conditions, where the researcher cannot control contextual situations (e.g., which class a student is in or what company or department an employee works for). A quasi-experiment also emerges in situations where the causal factors are defined by demographic or social characteristics of participants (such as their sex, age, socioeconomic status, income, years of experience or education level, ethnic background, religious background) since participants cannot be randomly assigned to a category of such factors. For some quasi-experimental designs, control groups can be used, but the researcher does not have the ability to randomly assign participants to such groups (which means the groups must be considered to be non-equivalent; e.g., classrooms in an educational evaluation of a new curriculum program in schools, where different classes experience different versions of the curriculum).

To defensibly draw conclusions about external causation, the researcher needs to meet three necessary and sufficient conditions in their research:

- *Temporal priority of causes*—where causal factors precede effect measurements in time, that is, cause factors must change before effects can be measured (e.g., when the researcher arranges circumstances where causal factors are manipulated or measured before effect variables are measured).
- *Covariation of causes and effects*—where if/when a causal factor changes values, the theorised effect measurement is observed to change (thus, the cause and effect must co-vary, i.e., when a causal factor increases in value, the effect measurement tends to increase or decrease in value, a pattern that statistical models are well-suited for substantiating).
- *Ruling out alternative plausible explanations*—where other potential causal influences are removed as possible confounding explanations for the covariations that have been observed in order to facilitate unambiguous external causal inferences (e.g., the researcher acts, using one or more control procedures, to prevent a potential causal factor that they are not interested in from changing at the same time as the causal factor(s) that are of interest).

The concept of control is central to the Manipulated experience-focused strategy as it is through the exercise of control that the researcher minimises the most critical of alternative plausible or extraneous influences. 'Most critical' is a judgment the researcher makes in trying to anticipate potential alternative plausible explanations which could interfere with achieving convincing outcomes. Sound control practices are all about choosing sensible and defensible approaches, in full recognition of what the researcher stands to gain or lose by making each choice. From a statistical model building and testing perspective, control creates opportunities to remove systematic influences from estimates of model errors, ideally reducing model errors to only unpredictable and uncontrollable random influences.

The Manipulated experience-focused strategy is always combined with one or more data-shaping strategies and may frequently be combined with the Systematic observation-based strategy and/or the Structured interview interaction-based strategy. The meta-analysis approach within the Textual artefact-based strategy can be used in conjunction with an experimental design as can the Generative data-shaping strategy (especially for Monte Carlo simulation research).

Non-manipulated Experience-Focused Strategy

The Non-manipulated experience-focused strategy is implemented when the researcher undertakes research that either: (1) employs a correlational, instead of causal-comparative, approach where all factors relevant to causal hypotheses are explicitly measured or (2) follows (or, in some cases, anticipates) the occurrence of an event in people's life, community, cultural, or work context (e.g., new social or environmental policy, regulatory scheme, social change program, conflict or war, natural or man-made disaster, the occurrence of which the researcher would have no control over) in order to examine the impacts, side effects, and longer-term consequences/implications of that event.

Correlational Non-experiment

A correlational non-experiment (often referred to as the 'correlational approach' or a 'correlational design'; see, for example, the discussion in Johnson, 2001) is achieved when the researcher has no capacity or intention to manipulate conditions or experiences for participants. All causal factors and effects of interest are measured rather than manipulated. This means that control occurs through effective measurement processes (a data-shaping strategy), combined with appropriate statistical model specifications and analyses. Causal inferences are often sought but are generally weakest in a non-experiment. However, convincing causal inferences can sometimes be supported using more sophisticated multivariate modelling techniques (e.g., in marketing, psychological, educational, and organisational research where structural equation modelling, partial least squares modelling, or multi-level modelling approaches are used) and/or time series designs where causal factors and effects can be tracked through time (an intervention at some point in the time series may or may not play a role in such designs). For a correlational non-experiment (i.e., a correlational design), the causal factors may, in fact, be latent constructs, reflected by a set of observed measurements (e.g., ratings on questionnaire items).

Both structural equation modelling and partial least squares modelling permit testing of theoretically-specified causal models. A multi-level model is used when the researcher has observations at different levels of analysis (e.g., individual, group, institution).

In the correlational approach, there are several things the researcher must understand when they are configuring their research. Generally speaking, correlational designs that use sophisticated multivariate statistical techniques require very large samples. This is so the statistical estimation procedures have a chance of providing good estimates of all parameters being tested in the models. Sample size is somewhat less of an issue for partial least squares modelling than it is for structural equation modelling or multi-level modelling (see discussions in Cooksey, 2020, Procedures 8.7 and 9.5). Stable results when testing a complicated multivariate structural or multi-level model may not be achieved (usually because of anomalies in the data, such as non-normality), in which case the researcher will need a back-up analytical strategy, and this may incur costs in terms of capacity to make the kind of causal inferences the researcher was hoping to make. Sometimes, data anomalies can be fixed using a quantitative to quantitative Transformative data-shaping strategy. For time-series designs based on secondary data sources (e.g., economic, financial, education, marketing/consumer, or pharmaceutical databases), missing or incomplete data for some time periods (e.g., data for one year might be missing) can be a problem, which means the researcher will have to decide how best to handle the issue. There are approaches for estimating missing data, none of which has universal acceptance, so the researcher would have to argue persuasively for the approach they plan to take.

Life Events Non-experiment

The life events non-experiment strategy focuses on natural contextual events that are beyond the researcher's control or will to influence. The research, in such situations, is more likely to be opportunistic or solicited by parties interested in the impacts of the natural contextual event. Natural contextual events may be created or caused by human activities (e.g., social or economic policy interventions, new laws/regulations, organisational merging/restructuring, social programs, social events, elections, war/conflict, advertising programs). Natural contextual events may also be created or caused by natural forces (e.g., earthquake, flood, fire, severe weather events, seasonal changes, climate change, etc.) that are often unpredictable. If the research focuses on a change or event that has already occurred, it would be a type of ex post facto non-experiment that seeks retrospective feedback and perspectives and the researcher would need to be sensitive to possible memory issues, the further back in time the event occurred. Irrespective of the type of change or event being focused on, one of the researcher's responsibilities would be to ensure that they appropriately contextualise and describe the event, which will likely require them to implement a pluralist approach. In that light, the life events non-experiment will likely be synergistically combined with one or more data-shaping strategies and/or one or more observation-based, artefact-based, or interaction-based strategies. Many changes and events that people must live through may be very stressful and emotional and all stages of the research need to be sensitive to this possibility when the researcher examines one of those occurrences. It is important to realise that the researcher may not get access to anyone who had experienced the change or event they are studying but who has since moved on or passed on because of it. In such cases, sampling will invariably be somewhat biased.

Structuring Experience-Focused Strategy

The Structuring experience-focused strategy encompasses frameworks that employ structured thought- or experience-organising stimuli, tools and/or templates (e.g., decision support procedures and tools, scenarios, repertory grids, card or photo sorting, projective techniques using various kinds of stimuli) to guide or focus the experiences of research participants. The researcher's goal in using a structuring framework is to focus participants' attention on a specific problem/issue or set of stimuli in a specific way in order to more fully investigate how they think about or deal with that problem/issue (which may range from very simple to highly strategic or impactful in nature) or understand participants' reactions to/interpretations of those stimuli.

Structuring Templates

A structuring template is a task-configuring tool that the researcher provides to participants or asks them to work within. Some examples of structuring templates for data gathering include: decision support procedures and tools such as decision trees and multi-attribute matrices (Brown, 2005; Pidd, 2009); judgment analysis tasks (Cooksey, 1996); scenarios (Maani & Cavana, 2007, Chap. 5); the Complex Dynamic Decision Perspective (CDDP see Cooksey, 2000); metaphor/storytelling development frameworks (Kaye, 1996); repertory grids (Cassell & Walsh, 2004), creative thinking and problem solving tools (e.g., Proctor, 2014) and the Five Whys procedure (Senge et al., 1994). Structuring templates may focus on real or hypothetical issues, events, situations, or stimuli and may be used for individuals or in group settings. With planning-oriented templates like scenarios or decision support tools, data gathering may focus on hypothetical events or choices that have yet to occur.

Structuring templates are a very useful strategy for investigating problem solving and decision making and can also be useful in helping to understand participant perceptions and thinking patterns. As such, a structuring template is a distinctly cognition-oriented applied research strategy. One important consideration attached to the use of a structuring template is that participants will generally need to be trained or at least have some experience with using or working with the template. This training requirement will add to the timeline of the research and there may be variations in how well individual participants cope with the template that training would not necessarily level out.

Sorting Tasks

Another type of structuring framework is the object sorting task (where the objects may be pictures, drawings, words or phrases, concepts, items, or photographs; see, for example, Hackett et al., 2023; Kane & Trochim, 2007). To implement this strategy, the researcher would design or sample a set of stimuli to provide the objects for sorting and ask participants to sort them into piles/categories, either according to some criterion that the researcher provides (e.g. along a scale indicating degree of preference or liking) or by simply grouping them together in some way that makes sense to the participant. One variant of this sorting strategy is to have participants verbalise their logic as they sort (see process tracing methods below). If the researcher asks participants to verbalise as they sort things into piles/categories, they can record their verbalisations and analyse the resulting qualitative data to uncover their perceptions, interpretations, and logic behind the piles/categories they create.

Projective Techniques

Projective techniques constitute a different type of structuring framework. Such techniques employ physical stimuli, such as words, statements, pictures, photographs or drawings, for the express purpose of eliciting reactions in the form of interpretations, stories, comments, emotional reactions, or evaluative ratings from participants (see, for example, Bond & Ramsey, 2007; Chrzanowska, 2002, Chap. 8; Hackett et al., 2023; Mesias & Escribano, 2018). As a set of data gathering strategies, they tend to be used more frequently in marketing and psychological research, as well as in some cross-cultural research. A variation of projective technique is photo-elicitation, where photos are used to stimulate storytelling and narratives (pairing with the Stories participant centred strategy; see Banks, 2018). A key feature of projective techniques is that they implement an indirect mode of structured data gathering where the researcher draws inferences about participant motivations, attitudes, beliefs, and/or emotions based on their responses to intentionally ambiguous stimuli. The theory is that participants will project evidence of these internal states into their responses without the researcher directly asking them about those states. Thus, projective techniques are dependent upon a minor deception of participants in order to work effectively.

Mesias and Escribano (2018) indicated that there are four broad classes of projective techniques: association (e.g. word association tasks); completion (e.g. sentence or story completion tasks); construction (e.g. picture response tests or cartoon tests where the participants build a story around the stimulus being presented); and expressive (e.g. role playing where the participants try to put themselves in the 'shoes' of someone else). A major advantage of projective techniques is that, because of their indirect nature, they may provide insights into a participant's state of mind which they otherwise might be unwilling to provide. Thus, these techniques are robust against social desirability and other impression management strategies that participants might use to cover up uncomfortable thoughts or feelings. On the flip side, projective techniques are notoriously difficult to validate, and suffer from relatively low levels of reliability. They may require the services of skilled interviewers and interpreters (e.g., a trained psychologist). In general, the researcher would be advised not to depend too heavily on projective techniques as a single source of data and, in most cases, to only employ them when other avenues are not viable. A projective technique would generally not be used by itself. Instead, projective techniques are almost always paired with the Stories participant-centred strategy or the focused interview interaction-based strategy.

Process Tracing

Yet another type of structuring framework is process tracing. Process tracing takes several forms including techniques such as monitoring information acquisition and use and the think-aloud (verbal) protocol (see, for example, Ericsson, 2003; Ericsson & Simon, 1993; Payne et al., 1993). These methods are most useful for investigating problem-solving and decision-making, often at the individual level. Process-tracing strategies are frequently used in combination with other data gathering strategies. For example, certain types of decision-making research are conducted by gathering verbal think-aloud protocols in conjunction with the Manipulated strategy (where specific task conditions are manipulated). In the think-aloud verbal protocol process tracing approach, the researcher invites participants to consider a specific problem (which may be real or hypothetical) and asking them to think out loud as they solve the problem, verbalising whatever they are thinking. Their verbal think-aloud talk is usually recorded and transcribed for later analysis, yielding a 'think-aloud protocol'. The protocol can then be analysed either using a pre-designed coding scheme or more openly and thematically (thus pairing this approach with the Transformative data-shaping strategy and the Textual artefact-based strategy). Think-aloud protocols can be produced with or without on-going input or prompting from the researcher, as appropriate to their research goals. The technique is limited by the extent to which participants have conscious access to they are thinking about a problem and by their potential susceptibility to social desirability or normative pressures (in which case, what participants say they are thinking may not agree with what they are actually thinking; see discussions in Nisbett & Ross, 1980, for example). To circumvent this limitation, the researcher might pair think-aloud protocols with another data gathering strategy (such as the Creative Works participant-centred strategy, any of the interaction-based strategies or observation-based strategies) to provide triangulating data.

Monitoring information acquisition and use is a software/technology-mediated process tracing approach. It involves the use of a software system (such as *MouseLab* or *MouselabWEB* (see https://www.mouselabweb.org/), see also Payne et al., 1993) to configure a manipulable experimental environment in such a way that all movements of a mouse or other control device, which a participant makes to uncover and examine information on a computer display relevant to solving a problem or making a (typically hypothetical) decision, can be tracked and stored for analysis (thus pairing with the Manipulated strategy). It is also possible to monitor information acquisition (as well as other visual learning tasks) using a wearable device that tracks eye movements as participants read a document or look at computer screen (pairing with the Measurements data-shaping strategy). Another type of process tracing approach is used in the social and behavioural sciences, especially the political sciences, as a way of identifying, unpacking, and clarifying the causal mechanisms underpinning social and political processes (see, e.g., Beach & Pedersen, 2013; Bennett & Checkel, 2014). It focuses on causal processes operating at the more macro-levels of society, e.g., institutions, governments, nations. What is interesting about this variant of process tracing is that it employs qualitative data gathering strategies such as 'elite interviewing' (see Tansy, 2007) to connect with key players in the political process being researched and almost invariably is used in a pluralist manner in conjunction with other data gathering strategies, such as the Non-manipulated experience-focused strategy and the Archival/Secondary artefact-based strategy. For other kinds of problem-solving research, the process tracing strategy may be used as a stand-alone approach (as when expert chess players are asked to think-aloud as they plan their next move, see Ericsson & Simon, 1993, for example). The think-aloud protocol approach can be useful in the development and pilot-testing of a complex questionnaire, where participants are asked to think-aloud as they work through the questionnaire (e.g., Trenor, Miller & Gipson, 2011).

A major drawback with using structuring frameworks of any type is that they may bound the problem or issue so tightly that issues outside the framework get overlooked, or in a way that is not congruent with participant values or their preferred mode of cognition, meaning that they will tend to struggle with using the framework and produce data that may be very suspect in quality. A pluralist investigation, combining the structuring strategy with other data gathering strategies, may be the most effective way to combat this drawback. Another potential drawback is that structuring templates and projective techniques are usually created based on a specific theory and if that theory is incomplete or makes problematic assumptions, the usefulness of the framework could be greatly diminished.

Immersive Experience-Focused Strategy

The Immersive experience-focused strategy provides goal-oriented participant experiences that are sometimes structured within a high-fidelity replica of some environment of interest, the parameters of which the researcher can control. For example, researchers into aviation ergonomics and flight crew behaviour may use extremely high fidelity (and quite expensive) flight simulators for jet aircraft or spacecraft (Lee, 2005). Training and research on medical teams can now be conducted in the context of high-fidelity operating room or hospital ward simulations, complete with simulated patients (Small et al., 1999). There are also system modelling 'flight' simulators, which have less fidelity than complete environment simulators, but which comprise formal integrated dynamic systems for tracking and providing feedback on decisions and choices (see, for example, Pidd, 2009, Chap. 9; Maani & Cavana, 2007, especially Chaps. 3–6). With today's computing technologies, it has become feasible to build highly realistic ('immersive') virtual worlds for the study of organisational and managerial behaviour (Pierce & Aguinis, 1997). Web-based virtual worlds/games such as Second Life have also become potential environments in which research can be conducted (for example, see Mennecke et al., 2008; Rymaszewski et al., 2007). Games are typically of much lower fidelity, relative to simulated environments but are distinguished by the requirement that participants follow a set of rules that all players share. Simulations are much more open-ended in terms of permitted behaviours. The Immersive strategy is seldom used on its own, instead it is usually combined other data gathering strategies, such as the Manipulated strategy, the Measurement data-shaping strategy, and/or the Systematic observation-based strategy. Thus, the Immersive strategy works best in a pluralist environment.

Immersion can vary in depth from shallow, where participation is more surface level and the lines between the research experience and real-world experience are minimally blurred, to almost complete, where participants may behave as if their research experience is a real-world experience and lines between the research experience and real-world experience are maximally blurred. Immersion can also vary from very low fidelity (i.e., low verisimilitude, where it is obvious that participants' research experiences are not much like their real-world experiences, except with respect to certain features) to extremely high fidelity (i.e., high verisimilitude, where participants' research experiences are almost indistinguishable from the real-world experience with respect to nearly all features). For some research goals, all that the researcher may require is the capacity to control some formal aspects and key variables of the tasks they are asking participants to perform, in which case, lower fidelity, shallower depth games may be best choice, where the larger environment within which the game is played is of little importance. For other research goals, embedding participants in an environment that matches, as closely as possible, one they would or could confront in the world outside of the research but one over which the researcher still maintains control, would be the only ways to study the behaviours the researcher is interested in.

Participation in a simulation or game may create emotional stress for some participants, particularly if there is a power dynamic involved (as in prisoner dilemma games where participants play prisoner or guard roles). This risk increases where engagement/immersion in a role may create experiences that have emotional consequences. For this reason, participants should normally be debriefed after their participation so that any adverse consequences can be discussed and dealt with. If the researcher is employing the Immersive strategy in conjunction with the Manipulated strategy, they must attend very closely to choosing which aspects of the simulation or gaming environment they want to manipulate and at what levels. The researcher must be alive to what is realistic for people to experience. The researcher does not want to create conditions that will inordinately stress people or place them in unrealistic scenarios. The higher the fidelity and deeper the immersion, the greater the risk of placing participants in situations they cannot adapt to or work within. Normally, the researcher would not want to expose their participants to a no-win scenario in a high-fidelity simulation (e.g., the Kobayashi Maru no-win scenario from Star Trek®; a no-win scenario for a patient in a medical simulation, an irrecoverable spin in an aircraft simulator; but see Lee & Skog, 2014 for an alternative viewpoint).

2.3 Data Analysis Strategies

Data analysis strategies comprise the third vertex of the 'Data Triangle' (refer to Fig. 2.1). Quantitative, qualitative and/or hybrid/merged data that have been gathered (or created) need to be analysed. Quantitative analysis deals with data in the form of numbers, measurements, and indices whereas qualitative analysis deals with data that are non-numerical in form, which can include recordings, texts, documents and transcripts, images, websites, music, objects, and films/videos. The researcher's goal is to build meaning from data, in whatever form, so that they can convey the stories the data reveal. Research questions, propositions, and/or hypotheses give direction and intent to analyses and connections between the original data and the analytical outcomes arising from those data should always be preserved and, where appropriate, displayed to relevant audiences. While analyses of quantitative and qualitative data draw upon very different procedures, the end goal is the same, namely, to discover what the data are telling the researcher (i.e., building meaning) and to create displays and narratives that convey those stories to others (i.e., communicating/sharing meaning). Hybrid/merged analytical strategies simultaneously harness quantitative and qualitative data (where the two data forms are so blurred together, they become indistinguishable) in order to generate a coherent story (see, for example, Gobo et al., 2022). If the stories that arise from quantitative and/or qualitative data or their hybridisation can be integrated to highlight points of convergence, divergence, qualification, or deeper understanding, this can add to the convincingness of the research.

In general, when we talk about data analysis strategies, we are talking about activities that the researcher undertakes to finding meaning in the data as well as activities for displaying such meaning. Such activities are simply a means to an end, helping the researcher to build up the stories they want or need to convey; they are not the end in themselves. The analytical approaches the researcher takes should be appropriate for the data they have (or will have) to hand. Too often, the choice is made the other way around, especially with respect to quantitative data, where data are forced through a specific analytical approach, perhaps because the researcher is familiar with the approach, good at it, has no experience with other approaches, or because it is an approach valued by a specific discipline or audience they want to reach (e.g., research funders, readership of a journal). However, we would argue that such forcing makes the research less convincing and inhibits making clear and logical arguments about why the researcher adopted a specific analytical approach. It may also be the case that there is no suitable analytical approach available for building meaning from the researcher's data which may stimulate innovation and creative development of new approaches and new hybridisations or require the researcher to implement a less optimal approach. Data analytic strategies therefore do not constitute a static domain of choices, but dynamically evolve as needs and new technologies emerge.

No matter what type of data the researcher has gathered, analyses will almost always transform, condense, filter, aggregate, re-represent, or categorise the raw data to build meaning and such meaning will always need to be contextualised in specific ways, including showing/arguing for how the original data connect to analytical outcomes and interpretations. While it is certainly important for the researcher to plan their strategies for data analysis, they should also remember that when they are working through the analysis process, there will be many choices and decisions they will have to make on the fly, especially when surprises, opportunities, constraints, and anomalies arise. Also, the researcher needs to realise that awareness of their audience(s) should influence their choices of data analysis strategies and displays, at least to some extent. Who the researcher is trying to convince, therefore, becomes an important consideration for analytical and data display decisions. Thus, as part of the analysis process, the researcher will need to think carefully about how they want to display their results, convey their stories, and draw their conclusions, keeping in mind the context(s) in which the data were gathered or produced, their research goals/questions/propositions/hypotheses, as well as their intended audience(s). The researcher may not be able to anticipate all the decisions they may need to make with respect to strategies for analysing their data, but the more they can anticipate and plan for, the smoother their analytic journey will be.

2.3.1 Data Analysis Strategies Inform Multi-layered Stories

Figure 2.4a (adapted and revised from Cooksey & McDonald, 2019, Fig. 21.1a, p. 926) provides a high-level visualisation of different analysis strategy pathways the researcher could follow, depending upon their research questions, propositions, or hypotheses. The left (blue) branch of Fig. 2.4a focuses on quantitative data analysis strategies, which tend to progress in a relatively linear fashion. The right (orange) branch focuses on qualitative data analysis strategies, which tend to yield their stories in a more dynamic and iterative fashion. The centre (grey) branch focuses on hybrid/merged data forms which integrate aspects of both the quantitative and qualitative pathways. For certain purposes, via the Transformative data-shaping strategy, qualitative data may be transformed into a quantitative form, i.e., 'quantitised', prior to analysis (e.g., participants, words, codes, themes, and documents can be categorised, counted, ranked, or rated). As well, quantitative measurements and patterns can be 'qualitised' through categorisation and labelling, such that richer interpretive meaning is attached to the numbers. Such cross-data form transformations can help to support stories emerging from the original form, where the need arises, in mixed methods research, for example. Equally, however, within data form transformations are also possible (i.e., one quantitative form to another quantitative form or one qualitative form to another qualitative form), as needed, to better focus or strengthen the analytical strategy being used.

There are several important lessons in Fig. 2.4a: (1) there are layers of stories (identified down the left side, splitting the diagram up into horizontal regions) that can be assembled and conveyed using each form of data; (2) analysis strategies and data displays influence each other in dynamic ways which will help to shape the layered stories the researcher tells, and (3) under a pluralist approach, these stories emerge from a dynamic mixture of within data form analyses and/or between data form or hybrid/merged analyses, as appropriate (symbolised by the large colour-coded rectangle of story-elaboration strategies). When focusing between data forms, the researcher is working in the space carved out by mixed methods researchers (conveyed by the **green** arrows) but saying this does not demote or devalue the possibilities associated with working entirely within one data form, which, of course, can still be done in a very pluralist and convincing manner as shown by the **blue**

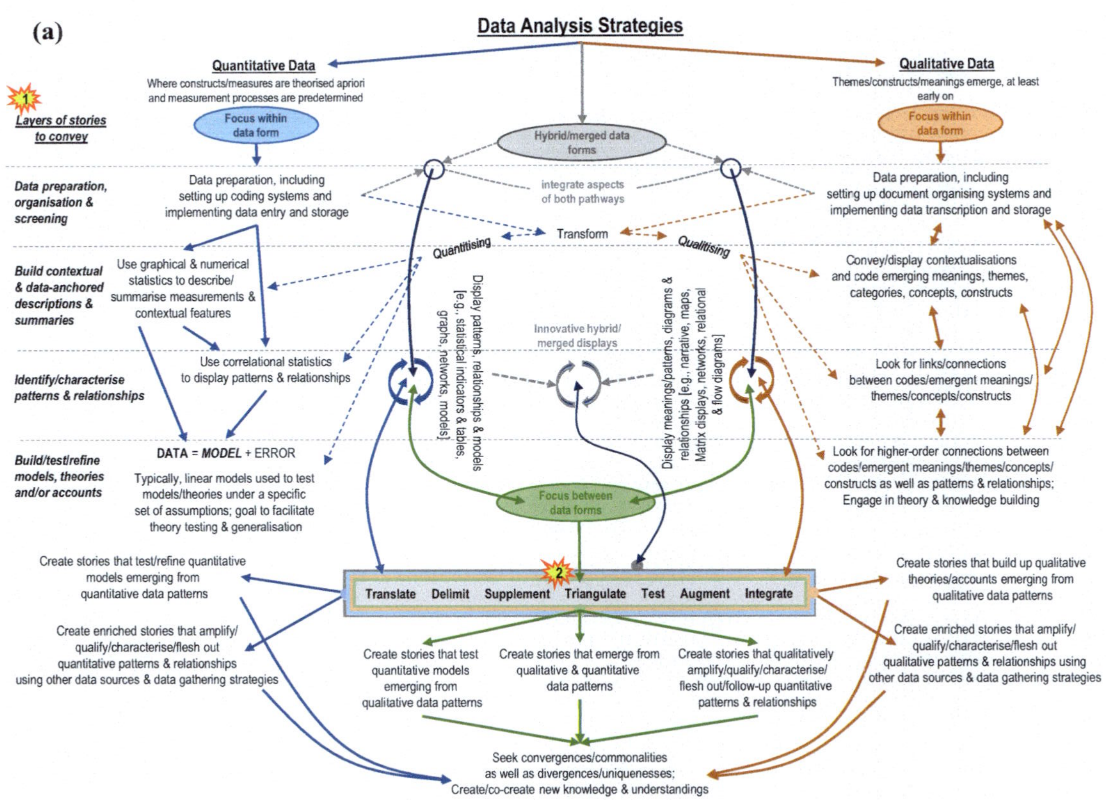

Fig. 2.4 **a** Analysis strategy pathways for pursuing stories at different layers of analysis, within and between quantitative and qualitative data forms, as well as for hybrid/merged data forms, as required. **b** Exploded view within the layers of stories following on from Fig. 2.4a and **c** Exploded view of the story elaboration rectangle following on from Fig. 2.4a

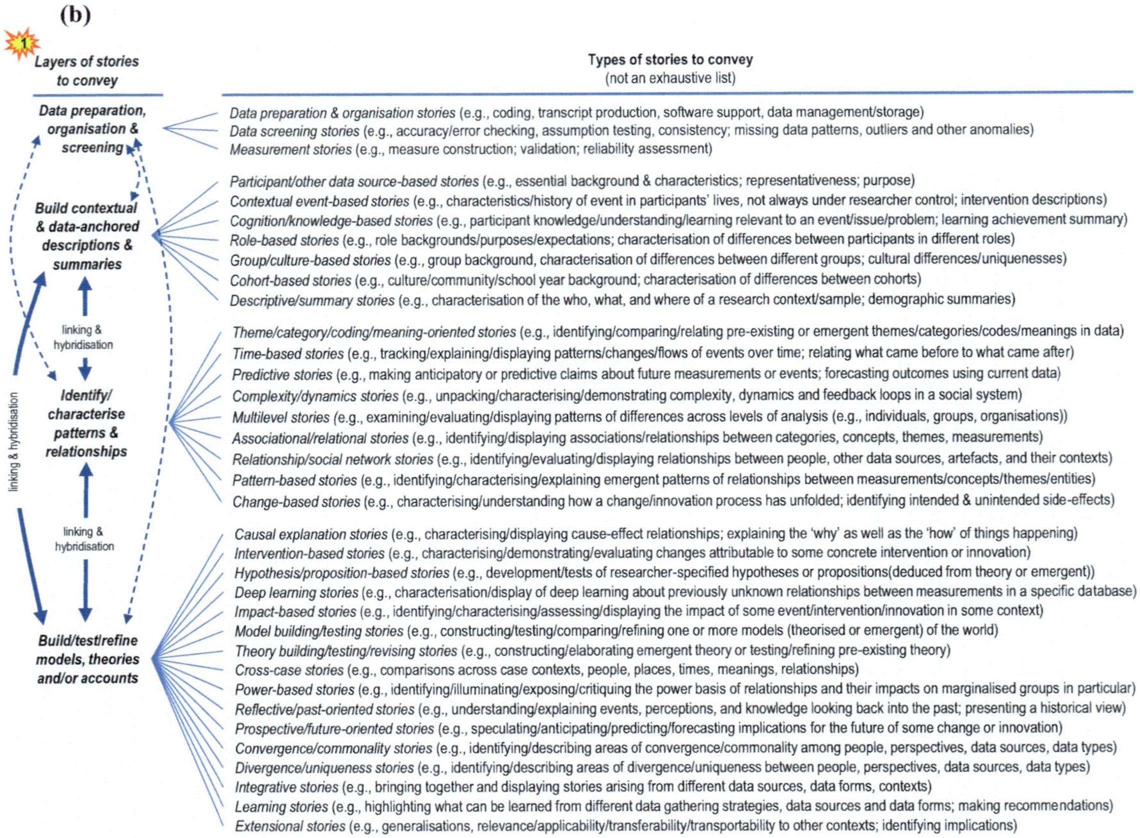

Fig. 2.4 (continued)

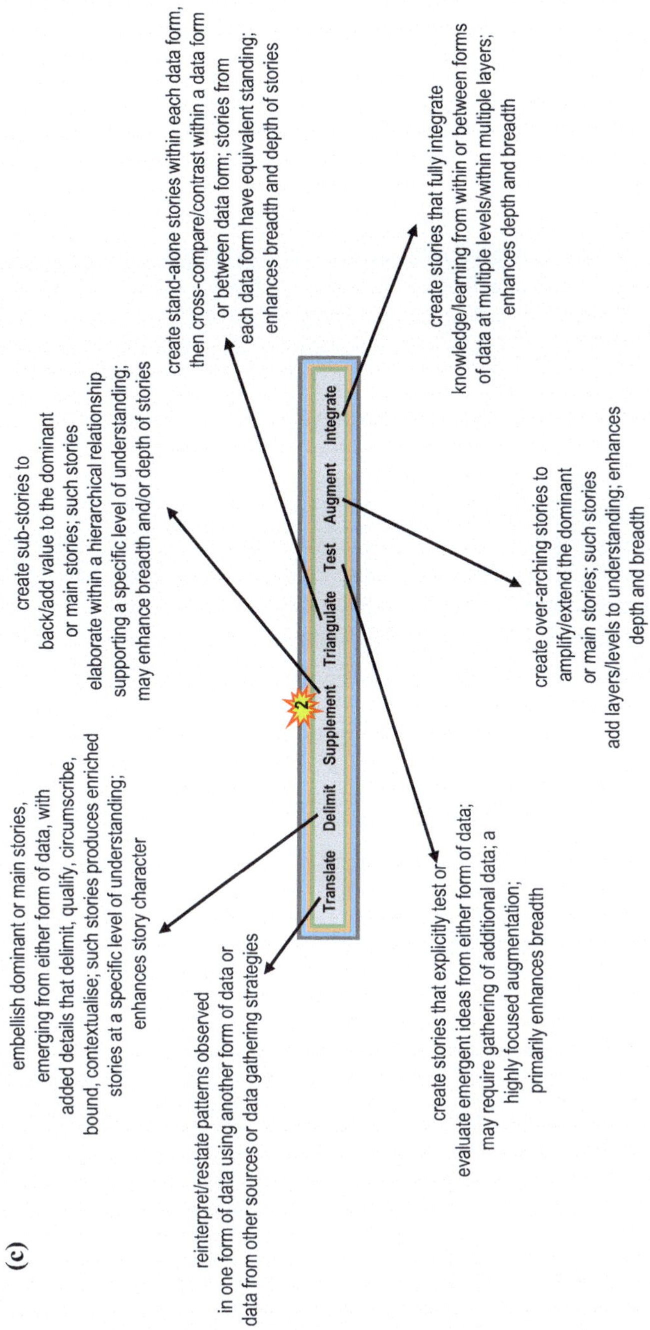

Fig. 2.4 (continued)

and **orange** solid arrow pathways to and from the story elaboration rectangle. The **grey** solid arrows depict hybrid/merged analyses simultaneously feeding into considerations about data displays, which may lead to creating innovative hybrid/merged displays, informed by cycling to and from the story elaboration rectangle. Figure 2.4b provides an exploded view of the types of stories that researcher might convey within each story layer. This list of potential stories is certainly not exhaustive, and no story should be considered in isolation from other stories. Stories can and should cross-feed from and to each other. In many cases, story hybrids may be developed in order to convey more complex learning. Some stories might emerge as important to convey depending upon the stakeholders who have an interest in the research. With a pluralist approach, the researcher can achieve a more richly diverse set of goals in terms of the kinds of stories they might wish to support through their analyses, and these are further unpacked in the exploded view of the story elaboration rectangle offered in Fig. 2.4c (adapted and revised from Cooksey & McDonald, 2019, Fig. 21.1b, p. 927).

Figure 2.5 drills down to identify more specific data analytic strategies useful for dealing with quantitative data, with qualitative data, and with hybrid merging of both forms of data. Note that connections with the logic and colour coding in Fig. 2.4a, b and c are maintained in Fig. 2.5. The possibilities for purposefully and creatively building links between specific analytical and data display strategies and the stories they might inform are far too numerous to convey in any single graphic. Our intention here is simply to signal that diversity, without implying any one-to-one connection between any particular analysis strategy and any particular type of story. In the end, what will influence and shape the researcher's choices are the various contextualisations associated with the research, including choices regarding patterns of guiding assumptions, the research goals/questions/ propositions/hypotheses, the requirements of any software tool adopted to assist in the analytical process, as well as the ebb and flow of data source choices and data gathering activities as they unfold during the research journey.

2.3.2 *Quantitative Analysis Strategies*

The analysis of quantitative data presents a diverse array of choices and considerations, as indicated by the diversity of strategies arrayed down the left side of Fig. 2.5. To make defensible choices, the researcher needs to possess a minimum level of statistical knowledge. Such knowledge should be sufficient to (1) understand which choices are appropriate given the research questions/propositions/hypotheses being pursued, (2) understand which choices are appropriate given the nature and quality of the data, (3) understand conceptually (as distinct from computationally) what each analysis procedure is designed to do and what it produces, and (4) know how to interpret and display the analysis outcomes and how to competently write them up. Statistical analysis of quantitative data can provide evidence for each of the layers of stories highlighted in the top centre section of Fig. 2.5. In addition, many statistical procedures can inform multiple types of stories, and, in many cases, a sequence

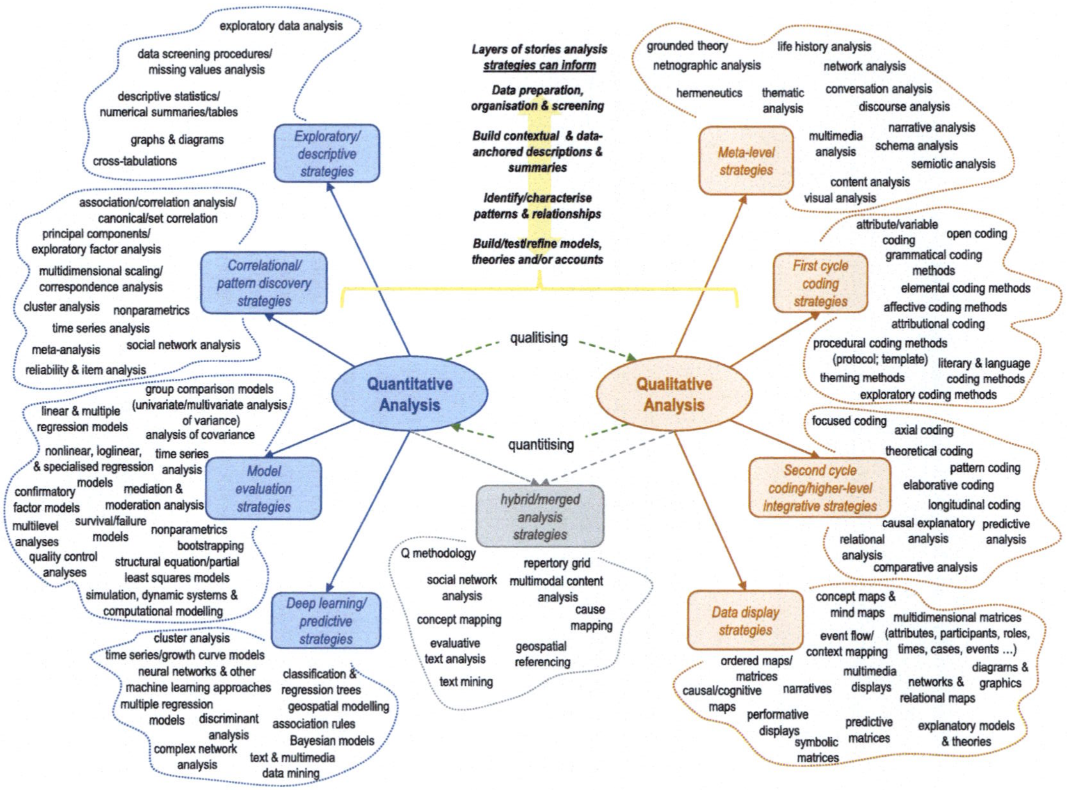

Fig. 2.5 Drilling down to specific data analysis strategies that can inform research stories

of connected analyses may need to be undertaken before a particular story can be fully fleshed out. In what follows, we provide a very brief overview of potentially useful statistical analysis strategies for extracting meaning from quantitative data (for more complete details on these strategies, consult texts such as Cohen et al., 2003; Cooksey, 2020; Field, 2018; Judd et al., 2017; Little, 2013; Kaplan, 2004; Sprent, 2011; Tabachnick & Fidell, 2018; as well as the various volumes in the *Sage Quantitative Research Kit*, Williams et al., 2021). In the discussions to follow, we will use the term 'variable' to refer to a set of measurements (e.g., an instrument reading, a test score, an item rating, category membership) obtained from a sample. The brief previews below are based largely upon the Procedures and Fundamental Concepts discussed in Cooksey (2020).

Exploratory/Descriptive Analysis Strategies

Exploratory/descriptive strategies utilise numerical, tabular, and graphical techniques to summarise and display important characteristics of samples and variables (see, e.g., Cooksey, 2020, Chap. 5; Jones & Goldring, 2021). These strategies also provide very useful screening processes for checking data for anomalies and errors, extreme values or outliers, patterns/issues associated with missing data, and checking to see if the variables to be used in specific analyses meet the parametric assumptions (e.g., normality, homogeneity of variance, linearity, independence of errors) required by such analyses. Based upon what is learned, the researcher may decide to transform certain variables to better meet assumption requirements, manage outlier problems, or to better facilitate meaningful interpretations of analyses. There is a range of numerical and visual analysis strategies for exploring/describing the 'behaviour' of variables in a sample.

- *Exploratory data analysis procedures* comprise strategies that use innovative visualisation and numerical summary techniques, such as stem-and-leaf plots and boxplots, to examine the distributional characteristics of variables, whilst keeping the connections between data and display more transparent.
- *Data screening and missing values analyses* employ strategies that examine data for anomalies and data entry errors, extreme values (i.e., outliers), patterns/relationships associated with missing data, and, where possible, correct for emergent problems with specific variables (e.g., using data transformations, procedures for handling outliers, missing data estimation/imputation).
- *Descriptive statistics & numerical summary/tables* reflect strategies that provide statistical summaries of the distributional characteristics of variables, using indices such as mean, median, variance, standard deviation, skewness, and kurtosis, often displayed in tabular form for more efficient presentation.
- *Cross-tabulations* are produced by strategies that examine joint membership distributions between two or more sets of categorical variables (useful for more efficiently summarising demographic characteristics of a sample and their interrelationships, for example).
- *Graphs & diagrams* emerge from strategies that visually display distributions/characteristics of variables in a variety of meaningful ways (e.g., bar charts,

histograms, line graphs, density plots, Venn diagrams, scatterplots, multivariate plots, with various embellishments as needed to highlight the story being conveyed).

Correlational/Pattern Discovery Strategies

Correlational/pattern discovery strategies employ numerical and graphical techniques to summarise and display relationships between variables and/or between entities (e.g., participants, companies, schools) in a data set (see, for example, Cooksey, 2020, Chaps. 6 and 8; Cohen et al., 2003). Such analyses cannot, in and of themselves, support claims of external causality, only of covariation (i.e., the tendency for values on two variables to move together in the same, opposite, or random directions across entities in a sample). [Correlation is a normalised index of covariation that characterises both the strength (closeness to 1.0 or −1.0) and direction (sign attached to the index) of the relationship.] In some cases, correlational relationships reveal previously unknown grouping patterns in the data (e.g., variables grouping together to form factors; entities grouping together to form clusters). Correlations may be found between two individual variables (bivariate relationship), between a variable and a composite (i.e., linear combination) of other variables or between two composites of different variables, giving rise to a very broad range of quantitative analysis strategies for undertaking correlational and pattern discovery in a set of data.

- *Association/correlation analysis/canonical & set correlation* emerge from strategies that examine relational patterns between variables (e.g., categorical, continuous, and various mixtures thereof) using various types of statistical indicators (e.g., Pearson correlation, partial correlation; chi-squared), ranging from very simple bivariate assessments to multivariate assessments of the relationships between composites of many variables simultaneously.
- *Principal components/exploratory factor analysis* encompass strategies that search for patterns of relationships amongst a set of variables with a view to reducing their dimensionality by forming interpretable composites of variables (e.g., analysing 40 attitudinal questionnaire items to see if the information they contain can be captured/represented by a much smaller number of composites, generically called 'components' or 'factors'). Alternatively, if the researcher had certain constructs in mind when they created the variables, exploratory factor analysis can be used to hopefully reveal patterns that match the constructs (i.e., the researcher creates an instrument to measure general political orientation and writes items designed to reflect liberal or conservative leanings, hoping the items will group themselves into composites accordingly).
- *Multidimensional scaling/correspondence analysis* reflect strategies that look for interpretable metric or nonmetric spatial patterns in a set of either continuous or categorical variables or entities. Relating other variables to the resulting spatial patterns yields a 'property fitting' analysis.
- *Cluster analysis* comprises strategies that search for previously unknown groups (i.e., clusters) of entities (e.g., research participants) in a sample using profiles of

a set of variables; such groups can be said to represent a typology. Either a hierarchical (e.g., minimum variance, complete, or average linkage) or non-hierarchical partitioning (e.g., K-means, mixture) algorithm can be used to construct the typology. Cluster analysis can also be used to cluster variables/items as an alternative strategy to exploratory factor analysis (see, for example, Cooksey & Soutar, 2006).

- *Nonparametrics* encompasses strategies that relax the more restrictive assumptions about the behaviour of variables associated with parametric correlation analyses. These strategies typically work by transforming variables to a lower scale of measurement.
- *Time series analysis* reflects strategies that look for trends, patterns, and cycles in variables measured at many time intervals (e.g., longitudinal data measured at daily, monthly, or yearly intervals).
- *Social network analysis* invokes strategies that examine and summarise (numerically and graphically) relational connections between entities (e.g., research participants; members of a group, organisations). The strengths and directionality of the relationships can also be explored.
- *Meta-analysis* is associated with strategies that numerically and/or graphically assess emergent patterns of findings in a sample of research literature using a specified measure of effect size. The use of these strategies typically follows implementation of the meta-analysis Textual artefact-based data gathering strategy.
- *Reliability & item analysis* encompass strategies that assess the reliability of composites of variables/items as well as drilling down to examine the behaviour of individual variables or items (and, in the case of Rasch model analysis, simultaneously scaling the behaviour of variables/items and participant performance/abilities/attitudes). Reliability analysis is typically used to evaluate the character of components/factors identified by principal components/exploratory factor analysis.

Model Evaluation Strategies

Model evaluation strategies utilise numerical techniques to construct, test, and refine statistical models and numerical, tabular, and graphical techniques to display aspects of the story about what has been learned from the model development/evaluation process (see, e.g., Byrne, 2016; Cohen et al., 2003; Cooksey, 2020, Chaps. 7, 8 and 9; Hayes, 2018; Hair et al., 2022; Martin, 2021a, 2021b; McBee, 2021; McCoach & Cintron, 2021). Model evaluation strategies will permit external causal inferences to be drawn under certain conditions. To effectively use model evaluation strategies, the researcher must be able to identify the role that each variable to be included in the analysis is to play: variables used to construct the model are called *independent variables* or predictors (the putative causes in the researcher's hypothesis); variables that the model is to explain or predict are called *dependent variables* or criterion variables (the putative effects in the researcher's hypothesis); and variables that are not really of interest to the researcher but need to be statistically controlled for in order to strengthen causal inferences are called *extraneous variables* or covariates (alternative plausible causes that could interfere with the researcher's hypothesised causal inferences). Variables that are hypothesised to modify the relationship between an independent and a dependent variable are called *moderators* and variables whose influences are hypothesised to operate in between an independent and a dependent variable are called *mediators.*

The most common type of model that social/behavioural researchers construct is what is called a general linear model (see discussions in Cooksey, 2020, pp. 256–264; Judd et al., 2017). A linear model can be conceptually depicted as **DATA** = ***MODEL*** + ERROR (shown initially in Fig. 2.4a). The **DATA** component contains one or dependent variables (more than one dependent variable gives rise to multivariate analyses). The ***MODEL*** component contains all of the independent/predictor variables the researcher wishes to explicitly evaluate plus, where appropriate for the model being tested, any hypothesised moderating, and/or mediating variables. The ERROR component captures all other unspecified/unknown influences on the **DATA** (lumping together *random* unpredictable/unknowable/uncontrollable influences as well as *systematic* yet uncontrolled influences). Thus, in a linear model, the observed **DATA** are completely explained/predicted by ***MODEL*** + ERROR**.** Tests of hypotheses emerge as ratio comparisons between the behaviour of the ***MODEL*** and the behaviour of ERROR**.** The smaller the researcher can make the ERROR component (through the exercise of various forms of control to shrink the *systematic* component), the easier it will be to show the ***MODEL*** is significant. Measures of effect size emerge from ratio comparisons between the behaviour of the ***MODEL*** and the overall behaviour of the **DATA**. The larger the effect size, the better quality, more substantive/meaningful, and practically useful the model will be.

- *Group comparison models (univariate/multivariate analysis of variance)* use strategies that facilitate making comparisons amongst groups of research participants, classified according to some method of treatment by the researcher and/or some pre-existing characteristic (i.e., one or more categorical independent variables), using one (i.e., univariate) or more (i.e., multivariate) dependent variables. The intent is to show that the independent variable used to create or define the groups is what causes any observed changes in the dependent variables. Such models are a primary tool used to analyse experimental and quasi-experimental data resulting from implementing Manipulated or Non-manipulated experience-focused data gathering strategies.
- *Analysis of covariance* also reflects a strategy that facilitates making comparisons amongst groups of research, classified according to some method of treatment by the researcher and/or some pre-existing characteristic (i.e., one or more categorical independent variables), using one (i.e., univariate) or more (i.e., multivariate) dependent variables, but with the added capability of statistically controlling for the effects of one or more potentially confounding extraneous variables or 'covariates' (e.g., comparing student academic performance under three different curriculum programs controlling for student's reading ability).

- *Linear & multiple regression models* comprise strategies that construct and evaluate a linear model (where components of the model are added together) using one or more sets of one or more independent variables (i.e., 'predictors') to predict the behaviour of a focal dependent variable of interest (i.e., the 'criterion').
- *Nonlinear, loglinear & specialised regression models* comprise strategies that construct and evaluate a more complex model, which is not necessarily linear or is linear only after some transformation process, using one or more sets of one or more independent variables (i.e., 'predictors') to predict the behaviour of a focal dependent variable of interest (i.e., the 'criterion').
- *Time series analysis* applies strategies that evaluate the parameters of specific models that capture trends, patterns, and cycles in variables measured at many time intervals. The effect of an intervention or event of some description (e.g., policy change, treatment program, assassination or death of a key national/ international leader) may or may not be present, when evaluating a time series model.
- *Confirmatory factor models* use strategies that seek to confirm a researcher-specified model representing patterns of relationships amongst a set of variables with a view to showing how the variables relate to constructs in ways the researcher has hypothesised (i.e., deduced from a theory). This differs from exploratory factor analysis in that variable-factor relationship patterns are specifically hypothesised and tested rather than being left simply to emerge from correlational data.
- *Mediation & moderation analysis* reflect strategies that use regression models to evaluate theorised mediating and/or moderating effects of specific variables on the causal/predictive relationships between predictor variables and a criterion variable.
- *Multilevel analyses* encompass strategies that use regression models (also called 'mixture models' or 'hierarchical linear models') to evaluate predictor/criterion relationships at different levels of analysis (e.g., individual, classroom, school, district or group, department, organisation) within a sample. Patterns and relationships are evaluated both within and between the different levels of analysis.
- *Quality control analyses* apply strategies that help researchers (and decision makers) track measurements that are relevant to the quality control of products, services, and processes over time to identify where, how and under what conditions quality is adversely affected (i.e., products, services, and processes fail to meet quality expectations).
- *Nonparametrics* refers to strategies that relax the more restrictive assumptions about the behaviour of variables associated with testing parametric group comparison or prediction models. These strategies typically work by transforming dependent variables to a lower scale of measurement.
- *Survival/failure models* use strategies that model 'time to event' outcomes (e.g., failure of a product; death of patient; drop out from school; workplace accident/ injury) using a set of predictor variables.
- *Bootstrapping* refers to strategies that employ an alternative approach to nonparametric analysis, whereby a very large number of samples, constructed by randomly resampling the members of the original sample, is used to empirically examine/

estimate statistical properties of variables involved in group comparisons and/or relationship tests, instead of making parametric assumptions.

- *Structural equation/partial least squares models* reflect strategies that help to develop, refine, and evaluate theorised causal models that interrelate a network of variables and latent constructs. Testing such models typically requires very large stable samples. Structural equation models are covariance-based (i.e., models estimate covariation/correlation patterns) whereas partial least squares models are variance-based (i.e., models partition and explain variance in a set of variables).
- *Simulation, dynamic systems & computational modelling* emerge from strategies that follow from implementing the Generative data-shaping strategy, permitting the creation, simulation, and evaluation of numerical, statistical, and dynamical behaviour models (e.g., Monte Carlo methods, dynamic systems modelling; agent-based modelling) using artificial data.

Deep Learning/Predictive Strategies

Deep learning/predictive strategies are primarily associated with the thriving interdisciplinary domain of data science which focuses on knowledge discovery and prediction typically within massive data stores (i.e., 'big data', stored in data warehouses and archives, spreadsheets, databases, servers, intranets, internet) as well as complex systems. There are two types of deep learning strategies: *unsupervised learning*, which seeks to discover initially unknown classes or categories of entities then find patterns and relationships amongst them, and *supervised learning*, which seeks patterns and relationships amongst known classes or categories of entities. Many of these strategies are heavily computational as well as statistical in nature, requiring sophisticated computer systems and algorithms. There are very strong connections to data mining, predictive analytics, machine learning, and artificial intelligence. Gandomi and Haider (2015) offered a useful typology for strategies to deal with 'big data': text analytics, audio analytics, video analytics, social media analytics, and predictive analytics. Within this typology, a range of useful strategies for analysing both structured (e.g., spreadsheets; databases) and unstructured (e.g., documents, live-captured text, audio, social media, and video files) data can be identified. For more details on many of these strategies, see, for example, Castellani and Rajaram (2021), Cooksey (2020, Chaps. 8 and 9), and EMC Education Services (2015). Some of the more prominent strategies are:

- *Cluster analysis* reflects unsupervised learning strategies that search for previously unknown groups or clusters of entities, thereby creating a typology, in a database using profiles described by a set of variables. Hierarchical or non-hierarchical partitioning algorithms are used to construct the typology. In some cases, the emergent cluster typology can be used as a schema for classifying new observations/profiles.
- *Time series/growth curve models* use strategies that evaluate time series or growth curve mixture models which capture trends, patterns, and cycles in variables measured at many time intervals. The resulting model is then used to forecast and evaluate future trends and outcomes.

- *Neural networks & other machine learning approaches* reflect supervised or unsupervised learning strategies (such as neural networks, support vector machines, genetic algorithms, latent class analysis, FP-growth algorithms) that employ computational models and algorithms to generate predictions and classifications from large sets of variable information. These strategies have a close association with the concept of artificial intelligence.
- *Classification & regression trees* are produced using supervised learning strategies that progressively split a large sample comprising many variables into predicted sub-groups (classification trees) or range of scores on a variable (regression trees) using a variable that maximises the information gained at each particular split point. The resulting data display typically reflects a tree-like predictive structure.
- *Discriminant analysis* is a multivariate supervised learning strategy that uses information from a set of independent variables to form one or more mathematical rules for predicting membership in two or more groups (the dependent variable). For example, financial and credit history data for loan applicants can be used to discriminate and therefore predict membership in either a 'likely to default' or 'not likely to default' group.
- *Multiple regression models* reflect supervised learning strategies that construct and evaluate a model using a set of variables to predict the behaviour of a focal criterion variable of interest. The resulting model is then used to predict outcomes from new observations.
- *Geospatial modelling* invokes the use of supervised learning strategies that seek to discover and evaluate how geospatial and geographic variables relate to other economic, biological, social, cultural, and political variables of interest (e.g., mapping how an epidemic disease progresses across the globe; mapping labour movements in between countries; mapping how key climate change indicators change across the world's oceans).
- *Association rules* reflect unsupervised learning strategies that seek to describe, not predict, relationships hidden in large datasets, using computational algorithms such as 'frequent itemsets' and the 'apriori algorithm' to construct association rules (e.g., rules such as 'X implies Y' or 'X is associated with Y' such that when X is observed, Y also tends to be observed).
- *Complex network analysis* encompasses strategies that extend the principles of social network analysis to large volumes of complexly interconnected and dynamically changing networks and webs.
- *Bayesian models* are created using predictive analytic strategies that generate and update probabilities, associated with forecasts, future decisions, or choices (which reflect hypotheses about what will happen next), based on new or additional data. The update predictions are conditional upon known or assumed prior probabilities, using a well-known probability relationship called Bayes theorem.
- *Text & multi-media data mining* comprise unsupervised learning strategies that particularly useful for building meaning from unstructured data. Such strategies rely upon algorithms to search and retrieve, classify, count (e.g., occurrences and co-occurrences of words), and analyse content in the hunt for interpretable

patterns and relationships, including concepts, themes, images, sentiments, topics, and issues.

2.3.3 Qualitative Analysis Strategies

Qualitative data analysis typically focuses on systematically assembled texts. These texts may have produced by the researcher recording field notes or observations, by other researchers (e.g., reports, journal articles, critiques), by organisations (e.g., company documents, web pages), by participants completing open-ended survey questions, diaries or journals, or by participants participating in conversations with an interviewer or with others, face-to-face or virtually. These texts, even if not originally in written form, will either be analysed in hardcopy form or in electronic form after appropriate transcription (via a Transformative data-shaping strategy). Some data sources may not be 'textual' in form (e.g., video clips, radio broadcasts, multi-media communications, cultural artefacts, video recordings, photos, performances, artworks, and other types of handiworks), and the researcher will need to decide how they want to treat these sources and how they will link them to other sources of data and to emerging interpretations.

Qualitative analysis strategies, arrayed down the right side of Fig. 2.5, offer far fewer set rules than for the analysis of quantitative data and a wide range of conceptual and procedural approaches exists, depending upon researcher needs (see, for example, Flick, 2014). In qualitative data analysis, the researcher seeks to construct contextualised meaning (possibly multiple layers of meaning) from the data. Qualitative analysis can adopt a top-down (conceptually-driven) orientation, a bottom-up (data-driven) orientation, or a blend of the two orientations. Top-down or conceptually-driven analysis means the researcher works downward into the data applying some type of preconceived framework (e.g., a theory-driven or practically-driven concept or coding system). Bottom-up or data-driven analysis means the researcher works upward from the data to create, modify, and elaborate a coding system and develop meaning which generally demands an iterative approach to the analysis. A blended orientation involves either simultaneous or sequential application of conceptually-driven and data-driven orientations (e.g., starting with an a priori coding framework, but allowing for on-the-fly evolution, revision, and elaboration or starting with a data-driven orientation, then using the emergent coding system as a quasi-standardised framework within which to do further coding and analysis).

The first stage in many qualitative analysis approaches is some form of coding process and there are many strategies that can guide this activity. Saldaña (2021, p. 5) defined a code as "a word or short phrase that symbolically assigns a summative, salient, essence-capturing, and/or evocative attribute for a portion of language-based or visual data" and argued that there are at least two cycles involved in the coding process: first cycle coding and second-cycle coding. In first cycle coding, the initial coding system is developed or applied, reflecting analysis in the form of breaking down the data. In second-cycle coding, the researcher looks for categories, concepts, patterns, and relationships relating to the first-cycle codes and strives for higher and more integrative levels of meaning, possibly even theory, reflecting synthesis in the form of building up new ways to understand the data. The role of the researcher in qualitative analysis is active, reflective, and reflexive. They do not stand apart from the data, they are immersed in the data actively seeking structure and meaning, reflecting on emerging interpretations, and reflexively considering how the interpretations they build relate to their own stances, worldviews, and research processes (May & Perry, 2018).

Qualitative analysis requires continual consideration of not only the data and their interpretations but also strategic choices about how best to display those interpretations to relevant audiences. An important part of any qualitative analysis process is the use of analytic memoing (Charmaz, 2014; Saldaña, 2021). Analytic memos are useful for recording thoughts, elaborations, qualifications, interpretations, and reflections on every aspect of the analysis process from initial data source/document preparations and contextualisation through the coding process and on into the interpretation and data display processes. In short, memos are extensions of the researcher's journalling process and keep the researcher from having to rely too heavily on memory, prompt the researcher to do/check/remember certain things, help the researcher to stay close to the data analysis process, and serve as an additional source of qualitative data that can be coded and analysed in their own right. Conveying a convincing story from qualitative analyses requires data, data coding, and data displays to be coherently and transparently connected and analytic memos serve invaluable supporting roles for making and interpreting these connections (Miles et al., 2019). In what follows, we provide a very brief overview of potentially useful analysis strategies for building meaning and understanding from qualitative data (for more complete details on these strategies, consult texts such as Bazeley (2021); Bernard et al. (2016); Grbich (2013), Flick (2014); Kuckartz (2014); Miles et al. (2019); Saldaña (2021); as well as the various volumes in the *Sage Qualitative Research Kit* (Flick, 2018a).

Meta-Level Strategies

Meta-level strategies for qualitative data analysis invoke different overarching orientations toward the kinds of understanding/meanings the researcher is seeking to achieve from qualitative data as well as informing choices of preferred analytical pathways. There is no necessary one-to-one correspondence between meta-level strategies and specific analytical procedures; rather, a meta-level strategy simply provides guidance with respect to what might work best. Some meta-level strategies focus on specific forms of qualitative data (e.g., social media for netnographic analysis; film and video for visual analysis, text for hermeneutics) whereas others can encompass a more diverse range of qualitative data forms.

- *Grounded theory* is a meta-level strategy that seeks to inductively build theory from the data using a cyclic pattern of conjoined data gathering, analysis, and interpretation activities, constantly comparing new data to what has already been learned, in order to refine and elaborate the emerging explanatory theory (see, e.g.,

Bernard et al., 2016, Chap. 10; Flick, 2018b; Grbich, 2013, Chap. 7; Bryant & Charmaz, 2007).

- *Netnographic analysis* is a meta-level strategy which employs a quasi-ethnographic orientation that seeks to produce a "cultural analysis of social media and online community data" (Kozinets et al., 2018, p. 262).
- *Life history analysis* is a meta-level strategy that seeks to understand "the ways in which individuals account for and theorize about their actions in the social world over time". Past, present, and/or future perspectives may be pursued (Musson, 2004, p. 34).
- *Network analysis* is a meta-level strategy that seeks to understand the connectedness of people, concepts, ideas, events, policies, and places from the perspective of the participants or actors in the network (see, e.g., Ahrens, 2018; Bernard et al., 2016, Chap. 19; Bliss et al., 1983; Kilduff & Tsai, 2003; Löblich & Pfaff-Rüdiger, 2011);
- *Hermeneutics* is a meta-level strategy that seeks to understand the layers of meaning and interpretation in texts, whether the authors of such texts are alive or not (e.g., see Kuckartz, 2014; Wernet, 2014). Note that the concept of 'text' in the modern era has been expanded to encompass other cultural products such as pictures, music and art.
- *Thematic analysis* is a meta-level strategy that seeks to identify, categorise/code, interrogate, and interrelate emergent themes in textual data (see, e.g., Bernard et al., 2016, Chap. 5; Gibbs, 2018, Chap. 4; Guest et al., 2012).
- *Conversation analysis* is a meta-level strategy that seeks to explore and interpret the dynamics of natural conversation, from both sociocultural mechanics (e.g., turn taking, components of speech and interaction) and communication (creation of shared meaning) standpoints (see, e.g., Grbich, 2013, Chap. 19; Peräkylä & Ruusuvuori, 2018; Rapley, 2018; Toerien, 2014). Note that the concept of 'conversation' has evolved to encompass nonverbal/visual, symbolic, and virtual/online interactions, such as chat rooms.
- *Discourse analysis* is a meta-level strategy that seeks to identify, explore, and interpret patterns of written or spoken discourse with a view to understanding their rules, cultural evolution, historical significance, and roles in maintaining/manipulating power in social and cultural relationships (see, e.g., Bernard et al., 2016, Chap. 14; Grbich, 2013, Chap. 20; Rapley, 2018; Tannen et al., 2015; Willig, 2014).
- *Multimedia analysis* is a meta-level strategy that seeks to understand the meaning and significance of sounds, images, virtual/computer-mediated interactions, and communications (such as SMS/MMS, chat rooms, bulletin boards, blogs, YouTube or TikTok postings), in both static and dynamic modes (see, e.g., Maeder, 2014; Marotzki et al., 2014).
- *Narrative analysis* is a meta-level strategy that seeks to understand and interpret, from a sociolinguistic (i.e., narrative structure) and/or sociocultural (i.e., interpretation, meaning, and sensemaking) perspective, elicited or pre-existing stories conveyed by people, such as research participants, community members, members

of a network or group (see, e.g., Bernard et al., 2016, Chap. 13; Grbich, 2013, Chap. 18; Esin et al., 2014).

- *Schema analysis* is a meta-level strategy that seeks to identify, understand, and elaborate individual and cultural schemas, mental models, and metaphors as cognitive sensemaking devices in social and cultural contexts (see, e.g., Bernard et al., 2016, Chap. 12; Kaye, 1996).
- *Semiotic analysis* is a meta-level strategy that seeks to identify, understand, and interpret the evolution and usage of signs and symbols in human socio-cultural communication and interaction, often through a process of deconstruction to reveal layers of meaning (see, e.g., Grbich, 2013, Chap. 15; Peräkylä & Ruusuvuori, 2018).
- *Content analysis* is a meta-level strategy that seeks to understand and display the concepts and themes evident in a body of texts (broadly defined as encompassing written, visual, and audio materials) using one or both enumerative (emphasising content categorisation and counting) and ethnographic (emphasising social and cultural meaning) modes (see, e.g., Bernard et al., 2016, Chaps. 11 and 17; Grbich, 2013, Chap. 16; Schreier, 2014).
- *Visual analysis* is a meta-level strategy that seeks to explore, understand, interpret, and historically/culturally situate pre-existing, researcher-created, and/or research participant-created, physical and/or virtual visual images (see, e.g., Bazeley, 2021, Chap. 11; Banks, 2014, 2018; Grbich, 2013, Chap. 17; Knoblauch et al., 2014; Margolis & Zunjarwad, 2018; Mikos, 2014).

First Cycle Coding Strategies

One of the hallmarks of effective qualitative data analysis is its iterative nature. The researcher often needs to make many passes through the data to build up the interpretations necessary to tell a convincing story. Saldaña (2021) nicely captures this iterative dynamic in his references to first cycle coding strategies (early passes through the data to build up initial codes), second cycle coding strategies (continuing passes through the data as well as through the first cycle codes to build an understanding of emerging patterns, relationships and connections) and beyond (building higher-level more holistic and integrative interpretations, including theory, where appropriate). Coding simply refers to the process of attaching conceptual labels to identifiable units of analysis (e.g., words, phrases, sentences, paragraphs, segments of audio or video recording, memos, documents) to capture something meaningful about each unit. Here, we briefly describe a number of first cycle coding strategies.

- *Attribute/variable coding* is a initial coding strategy used to log and categorise essential characteristics or attributes (e.g., background characteristics, dates, places, and times) of data sources, documents, and events. Such codes are especially useful in downstream analytical and data display activities where, for example, data sources with different attributes are compared with respect to themes and viewpoints.
- *Open coding* is an initial coding strategy, often used in grounded theory approaches, where the researcher attaches codes to qualitative data units, which

capture some analytical reflection on meaning, going beyond simple description, to answer questions such as who? what? where? when? why? how? (see, e.g., Gibbs, 2018).

- *Grammatical coding methods* comprise a collection of coding strategies intended to capture intensity, value, or magnitude, elaborate sub-code hierarchies, and/or apply multiple codes to the same qualitative data unit (see, e.g., Saldaña, 2021, Chap. 5).
- *Elemental coding methods* encompass coding strategies that seek to represent structural features, descriptive characteristics, processes or actions, verbatim content (in-vivo coding), or concepts evident in qualitative data units (see, e.g., Saldaña, 2021, Chap. 6).
- *Affective coding methods* reflect coding strategies that seek to represent the affective (e.g., emotional, judgmental, evaluative, conflict-oriented) meanings of experiences reflected in qualitative data units (see, e.g., Saldaña, 2021, Chap. 7).
- *Attributional coding* is a type of process-oriented coding that seeks to highlight explicit causal explanations or attributions (e.g., explaining the 'why' of things, such as 'Y happened because X happened' or someone did X and then Y happened') embedded in qualitative data units. Causal attributions may be directly reflected as when linking words such as 'because' or 'therefore' are used or indirectly reflected by implication as when one statement is used as an explanation for another (see, e.g., Silvester, 2004).
- *Procedural coding methods (protocol; template)* utilise prescriptive coding strategies that apply pre-existing frameworks, coding templates, or coding protocols to capture meaning in qualitative data units, with or without researcher flexibility to adapt and extend them as ill-fitting data arise (well-suited for top-down conceptually-driven analyses; see, e.g., Saldaña, 2021, Chap. 10).
- *Theming methods* comprise coding strategies that seek to capture emerging themes (e.g., consistencies in story, meaning and intent) across multiple qualitative data units (see, e.g., Bernard et al., 2016, Chap. 3; Saldaña, 2021, Chap. 11).
- *Literary & language coding methods* invoke coding strategies that seek to capture literary and language-oriented aspects of meaning (e.g., metaphors, motifs, dramaturgical concepts, genre, verbal exchanges and features of talk) in qualitative data units (see, e.g., Saldaña, 2021, Chap. 8).
- *Exploratory coding methods* are preliminary broad-brush coding strategies that capture first-pass explorations of meaning (e.g., holistic meaning, suggesting hypotheses, provisional meanings based on some a priori schema) in qualitative data units (see, e.g., Saldaña, 2021, Chap. 9).

Second-Cycle Coding and Higher-Level Integrative Strategies

First-cycle coding will typically yield a plethora of specific codes. However, more needs to be done before concrete and convincing interpretations and stories can be produced. First-cycle codes need to be organised (e.g., mapped into a coherent hierarchy, a landscape, or some other larger-scale organisational system such as coding the codes) so that both the fine grain of coding (the 'trees') as well as the emerging relationships between them (the 'forest') can be seen. Once the first cycle codes have been mapped or landscaped, second-cycle coding can commence. Here, we briefly describe a number of second cycle and higher-level/ more complex coding strategies (Saldaña, 2021).

- *Axial coding* is a coding strategy, often used in grounded theory analyses, that seeks to identify and codify how first-cycle codes and sub-codes relate (e.g., as causes or consequences, or as conditions and actions) to each other and which codes are most important to continue concentrating on with respect to the research questions. Axial coding typically occurs within a cyclical process of induction and deduction (e.g., see Flick, 2018b, 2023, Chap. 25; Saldaña, 2021, Chap. 13).
- *Focused coding* is a coding strategy (also called selective coding), often used in grounded theory analyses, that seeks to drill down deeper into the data as well as into the refined coding system in search of the emergent story or theory where the focus is guided, in many instances, by outcomes from axial coding (e.g., see Flick 2018b, 2023, Chap. 25; Saldaña, 2021, Chap. 13).
- *Theoretical coding* is a somewhat different take on higher-level coding in grounded theory in that it seeks to identify the central or core category/primary theme for the research and integrate all that has been learned through open, axial, and focused coding into a single coherent theoretical account or explanation of that central or core category/theme (e.g., see Flick, 2023, Chap. 25; Saldaña, 2021, Chap. 13).
- *Pattern coding* is a coding strategy that seeks identify, condense, refine, and, in some cases, augment a category system (what Saldaña calls 'meta-coding') so that patterns and connections between codes as well as cases become more apparent (see, e.g., Saldaña, 2021, Chap. 14).
- *Elaborative coding* is a top-down coding strategy, useful, for example, in meta-synthesis, which seeks to integrate the researcher's current analytical codes and learning with what has emerged in previous research on the same topic or issue (see, e.g., Saldaña, 2021, Chap. 14).
- *Longitudinal coding* is a coding strategy, very useful in life history analysis and when looking for patterns of change, which seeks to identify, refine, and compare the emergent stories from a coding system through time (see, e.g., Saldaña, 2021, Chap. 14).
- *Causal explanatory analysis* is a higher-level analysis strategy that seeks to understand, explain, and display the 'why' of coding patterns by identifying, theorising, and interpreting connections between antecedent conditions (i.e., 'causes'), mediating/mitigating factors, and outcomes (i.e., 'effects') within as well as between cases (see, e.g., Miles et al., 2019, Chap. 9).
- *Predictive analysis* is a higher-level strategy that goes beyond causal explanatory analysis, by seeking to generate and display predictions about what might, could, or should happen in the future (i.e., anticipated outcomes), given the causal explanatory relationships identified in the data (see, e.g., Miles et al., 2019, Chap. 10).

- *Relational analysis* is a higher-level analysis strategy that seeks to identify, interpret, and display connections and relational codes and patterns within and between entities, such as data sources, case studies, social networks, organisations, people, and documents/texts (see, e.g., Bazeley, 2021, Chap. 13).
- *Comparative analysis* is a higher-level analysis strategy that seeks to compare, contrast, and display codes, patterns and interpretations of qualitative data across different entities, such as data sources, case studies, organisations, people, and documents/texts, with a view toward identifying/revealing patterns of commonality and distinctiveness (see, e.g., Bazeley, 2021, Chap. 12; Bernard et al., 2016, Chap. 15; Gibbs, 2018, Chap. 6).

Data Display Strategies

Qualitative data analysis requires a continuous cyclical interplay between coding and interpretive sensemaking of the data and displays of such sensemaking. Thus, the researcher needs to be thinking about not only what the data are telling them but how best to display what they have learned. In order to maximise convincingness, data displays need to remain as close/connected to the data as possible (through contextualisations and links to specific participants, documents, codes, and quote materials), so that audiences are clear about how the researcher has achieved the interpretations. Here, we describe a range of potential data display strategies, many of which are discussed more fully in Miles et al., (2019, Part II). Note that this is not a comprehensive list of descriptions as there are always opportunities to create new, relevant, and meaningful data displays depending upon the contextual circumstances and data content of the individual research project.

- *Concept maps & mind maps* constitute a visual data display strategy that summarises/characterises and diagrammatically depicts emerging concepts and the linkages (often labelled or qualified) between those concepts (see, for example, Bazeley, 2021, Chap. 13; Bryson et al., 2004, Part II; Buzan, 2018; McDonald et al., 2004; Kane & Trochim, 2007). Mind maps are typically anchored by a central concept from which hierarchical relational branches emanate, whereas a concept map is more free-form in structure.
- *Multidimensional matrices (attributes, participants, roles, times, cases, events …)* reflect a data display strategy, which summarises content, patterns, comparisons, and meanings in a tabular row-column format (where rows and columns may be anchored by the categories of one or more variables, attributes, codes, cases, or concepts), designed to convey a specific story, such as changes over time, elaboration of concepts, comparisons of viewpoints between participants in different roles or between cases, summaries of demographic characteristics (see, for example, the extensive discussions and illustrations in Bazeley, 2021, Chaps. 4, 9 and 12; Miles et al., 2019, Chaps. 5, 6, and 7). By careful design, up to four different dimensions can be incorporated (typically, up to two dimensions anchoring the rows; up to two the columns). Trying to incorporate more than four dimensions in a single matrix will generally place an undue cognitive burden on the audience.

- *Predictive matrices* emerge from a data display strategy that seeks to depict relational links between 'predictor' codes/categories, 'effect' codes/categories, and anticipated or observed consequences (see, for example, Miles et al., 2019, Chap. 10).
- *Symbolic matrices* reflect a matrix-style data display strategy that seeks to summarise and depict relationships, strength of sentiment, and other orderable or rateable aspects of qualitative coding and learning using symbols, shadings, and colours (see, for example, Henderson & Segal, 2013).
- *Ordered maps/matrices* reflect a data display strategy that seeks convey a sequencing (time-, process-, strength-, or case-based) story evident in the data (see, for example, Miles, Huberman & Saldaña, 2019, Chap. 8).
- *Event flow/context mapping* constitutes a visual data display strategy that seeks to summarise essential characteristics of context, the sequencing/flow of events, influences through time and space, or an event-state network (see, for example, Bazeley, 2021, Chap. 4; Carriger et al., 2018; Miles et al., 2019, Chap. 8).
- *Networks & relational maps* emerge from a data display strategy that seeks to build visual representations of social and code networks and other relational structures (e,g., hierarchies, kinships, sets/collectivities; see, for example, Bazeley, 2021, Chap. 9; Bliss et al., 1983; Miles et al., 2019).
- *Causal/cognitive maps* comprise a data display strategy that seeks to build visual representations of cognitions and perceptions of causality, relationships, and influences (see, for example, Bryson et al., 2004; Eden & Ackermann, 2002; Pidd, 2009, Chap. 6).
- *Diagrams & graphics* emerge from a creative data display strategy that uses researcher-created visual representations of concepts, meanings, and relationships tailored for specific communication purposes (see, for example, Cooksey & McDonald, 2019, Chap. 21).
- *Narratives* represent an interpretive strategy that uses storytelling as a way of conveying what has been learned from qualitative data, where the story interlaces narrative commentary and interpretive flow from the researcher (while displaying reflexive awareness of the researcher's identity), and quotes/material from relevant data sources (see, for example, Bazeley, 2021, pp. 468–470; Elliott, 2005, Chap. 9).
- *Multimedia displays* emerge from a data display strategy that employs multimedia performative constructions (e.g., cyberethnography) and technologically-mediated outlets to convey research findings, (including YouTube sessions, podcasts, blogs, and other types of social media posting; see, for example, Banks, 2018, Chap. 5; Grbich, 2013, Chap. 13).
- *Performative displays* reflect a data display strategy where the researcher orchestrates one or more performances (e.g., ethnodrama, ethnotheatre, poetry, resistance politics) to convey meaning, learning, and arts-based critique arising from qualitative inquiry (see, for example, Finlay, 2018; Grbich, 2013, Chaps. 11 and 12; Saldaña, 2018).

- *Explanatory models & theories* emerge from a data display strategy that seeks to depict relational links between antecedent conditions/codes/categories, mediating/mitigating variables/codes/ categories, and outcome codes/categories, with a view to explaining or theorising why things happened as they did (see, for example, Bazeley, 2021, Chap. 14; Miles et al., 2019, Chap. 9). For example, this may be a way to visualise the ultimate form of a grounded theory.

2.3.4 Hybrid/merged Analysis Strategies

Thus far, we have treated quantitative and qualitative data analysis strategies as two distinct pathways for making sense of data (recall the major pathway fork in Fig. 2.4a). Hybrid/merged analysis strategies emerge in the central portion of Fig. 2.4a and are arrayed in the lower centre section of Fig. 2.5. One major feature distinguishes hybrid or merged analysis strategies from straight-down-the-line parallel or sequential qualitative or quantitative analysis strategies, namely, the boundary between qualitative data and quantitative data is so integratively blurred that it makes no sense to separate them (see, for example, Bazeley, 2018; Gobo et al., 2022; Onwuegbuzie & Combs, 2010, discussion of cross-over analyses, pp. 422–423). Here, we describe a range of potential hybrid/merged data analysis/display strategies but note that this list is gradually expanding as researchers become more innovative and the technological support available to them becomes more sophisticated.

- *Q methodology* offers a hybrid/merged analysis strategy that aims to provide an 'objective' analysis of 'subjective' points of view by qualitatively developing, sorting (technically called a Q-sort, individually produced by participants), and statistically analysing, using procedures such as factor analysis or multidimensional scaling, sets of items related to a particular issue or problem, (see, for example, Gobo et al., 2022, Chap. 7; McKeown & Thomas, 2013).
- *Social network analysis* can work as a hybrid/merged analysis strategy that seeks to characterise and analyse patterns of relationships and connectedness in virtual or face-to-face social networks by qualitatively characterising network actors as well as the nature of linkages and statistically analysing and summarising network nodes and linkages (see, for example, Bazeley, 2018, pp. 245–250; Borgatti et al., 2022; Crossley, 2010; Scott, 2017; Nooraie, et al., 2020).
- *Complementary analysis* is a hybrid/merged analysis strategy that builds up a "deep understanding of complex processes gained through intensive case studies by using a systematic approach to (associative) causal analysis based on cross-case patterning of critical variables drawn from the case studies", where the critical variables may have been qualitatively or quantitatively identified (see, for example, Bazeley, 2018, pp. 250–253; Ragin, 1987).
- *Text mining* offers a hybrid/merged analysis strategy that statistically searches for and analyses patterns, sentiments, opinions, and/or emotions in text (knowledge discovery from textual databases), usually with the support of sophisticated software systems such as *Leximancer*, *QDA Miner* or *R* (see, for example, Gobo

et al., 2022, Chap. 9; Lewis & Maas, 2007; Liu, 2012; Silge & Robinson, 2017; Smith & Humphreys, 2006).

- *Geospatial referencing* comprises a hybrid/merged analysis strategy which capitalises on the fact that data as well as data sources are situated in both time and space. By explicitly linking geospatial location information (geotagging) to emerging quantitative and qualitative data, patterns, relationships, distributions, and emphases across space and time can be highlighted (see, for example, Bazeley, 2018, pp. 243–245; Fielding & Fielding, 2015).
- *Repertory grid* is a hybrid/merged analysis strategy, based upon personal constructs theory (Kelly, 1955), that maps the cognitive understanding people have of a phenomenon (e.g., an event, idea, object, experience) in their lives by statistically analysing patterns in the relationships between constructs (the dimensions a person uses to conceptualise aspects of their world) and elements (the target objects that the person evaluates using their constructs) (see, for example, Bazeley, 2021, pp. 255–257; Bell, 2018).
- *Multimodal content analysis* is a hybrid/merged analysis strategy that combines content analysis (listing, counting, and categorising key words in context or coded themes, images, emojis) with multimodal discourse analysis (focusing on narratives and topics), often with the support of sophisticated software systems such as *Leximancer, QDA Miner, MAXQDA*, or *NVivo* (see, for example, Gobo et al. 2022, Chap. 8; Lewis & Maas, 2007; Oliviera et al., 2013; Smith & Humphreys, 2006).
- *Cause mapping* offers a hybrid analysis and visualisation strategy for capturing, displaying, and quantitatively assessing cognitive maps that link participants' perceived causes to perceived effects with respect to some issue of interest (see, for example, Eden & Ackermann, 2002; Hine et al., 2005; Laukkanen, 1994, 1998). They are often produced with the support of a software system such as *Decision Explorer* or *Inspiration.*
- *Concept mapping* is an integrated mixed method strategy for capturing, rating, sorting, clustering, and visually representing/displaying concepts and ideas that emerge from collaborative group (especially stakeholder) discussions (see, for example, Bazeley, 2018, pp. 253–255; Kane & Trochim, 2007, particularly Chaps. 3–6; Trochim & McLinden, 2017).
- *Evaluative text analysis* is a hybrid/merged analysis strategy that seeks to assess, classify, and evaluate qualitative content of texts, using a sequential process of coding, sorting, and ordering/rating text segments (see, for example, Kuckartz, 2014, pp. 88–101).

2.4 Dynamic Feedback/Feedforward Links Within, 'Upstream', and 'Downstream' from the 'Data Triangle'

Thus far, we have considered the components of the 'Data Triangle' in relative isolation from one another in the landscape of methodological choices. However, these three components do not stand in isolation, they are dynamically interconnected and choices made with respect to one component will often create implications for the other components (either as feedback in terms of modifying a previous choice based on a subsequent choice or as feedforward in terms of a current choice influencing subsequent choices) as well as creating potential 'upstream' and 'downstream' implications and ripple effects (recall Fig. 2.1). Pluralist systemic thinking requires researchers to be constantly aware of such dynamics when making choices at any one point in time and for any particular context.

2.4.1 Dynamics Within the 'Data Triangle'

Choice of data gathering strategy may influence data source choice strategies and vice versa. For example, choosing a participant observation strategy may, for feasibility reasons, demand the use of a convenience data source choice strategy or, if grounded theory is being pursued, demand theoretically-driven, negative case, and/or sequential/saturation data source choice strategies (feedforward). If a large random sample is desired, resource limitations may mitigate against using a face-to-face interview strategy (feedforward). Specific data source choices (i.e., participants) may, upon onset of data gathering, not be able to cope with a Creative Works data gathering strategy, necessitating the use of an alternative data gathering strategy, such as a Structured experienced-focused approach like a think-aloud protocol (feedback). A specific data gathering strategy (e.g., a Manipulated experience-focused strategy) might create such strong reactive effects that a large proportion of chosen data sources (i.e., participants) decline to continue their participation in the research, necessitating a rethink of both data choice strategy and data gathering strategy (feedback).

Choice of data gathering strategy may influence the choice of data analytic strategy and vice versa. For example, using an unstructured interview strategy will require the use of qualitative data analytic strategies (feedforward). If the researcher's goal is to conduct a structural equation model, multi-level model, or neural network analysis, this will demand a large, perhaps stratified, random sample (feedforward). Encountering anomalous or ambiguous data from a questionnaire developed using the Measurement data gathering strategy may require the additional use of a structured or semi-structured interview data gathering strategy (coupled, perhaps, with a purposive sampling strategy) in order to more fully explore the anomalies (feedback). Learning achieved using a particular data analysis strategy (e.g., meta-analytic Textual artefact-based strategy) may require the use of an additional data gathering strategy (as well as, perhaps, different data sources) in order to more fully explore the implications of that learning (feedback).

Data source choice strategies may influence choice of data analytic strategy and vice versa. For example, a two-stage cluster sampling strategy may require the use of parametric statistical estimation/model building techniques (feedforward). Using a contextually-driven sampling strategy will require a more diverse or sophisticated set of data analysis strategies, especially if time is one of the contextual dimensions driving the choices of data sources (feedforward). Learning from a theoretically-driven sampling strategy creates the need to develop a more elaborated and diversified second-cycle coding strategy (feedback). The specific nature and composition of the realised sample of data sources may require the researcher to reconsider the types of data analytic approaches they intend to apply (feedback).

Lack of contextual fit/suitability and/or lack of fit with adopted pattern of guiding assumptions stimulates methodological creativity, innovation, and evolution around the 'Data Triangle', yielding new data source selection and management strategies, new data gathering strategies and/or new data analytical strategies. Lack of suitability with respect to the emerging data may stimulate data analysis innovation or adoption of new approaches from other sciences. These evolutionary processes flow from feedback about what is and is not working well for the researcher during their research journey or from feedforward thinking about what might be needed to work best in a particular research context or with specific types of data sources.

2.4.2 'Upstream' Dynamics

When we speak of 'upstream' dynamics associated with activities around the 'Data Triangle', we are talking about feedforward influences that research planning creates for the researcher prior to commencing the journey around the triangle or about feedback influences that stimulate changes in contextual thinking about the research as a consequence of what is emerging from early activities within the triangle. This is a major reason why it is important to realise that the research journey is nonlinear in nature, where early thinking influences later choices (i.e., shaping of choices), but later choices may also stimulate revisiting of earlier choices (i.e., reconsideration of choices). Researchers need to be conscious of the possibilities of creating ripple effects with the choices they make, many of which will signal the need for flexibility and adaptability. In terms of 'upstream' dynamics, researchers may need to grapple with issues such as:

- ensuring the contextual fit of choices for navigating the 'Data Triangle' is and remains strong (which includes, for example, considerations about decolonisation with respect to research in Indigenous/First Nations contexts; considerations about cultural relevance and meaningfulness in cross-cultural research; and considerations about the evolution of the research questions/propositions/hypotheses that

'Data Triangle' activities are intended to pursue within the anticipated research context(s));

- ensuring that the choices made for navigating the 'Data Triangle' are and continue to remain ethically and culturally appropriate and well-managed (which includes, for example, considerations, where appropriate, about reciprocity, respect, and relevance in Indigenous/First Nations research; considerations about continuing/maintaining the ethical and cultural focus in cases where additional data source, data gathering, and data analysis choices are made; and considerations about congruence of the researcher's ethical and cultural stance with relevant regulatory, institutional, and cultural policies and practices);
- ensuring that stakeholder and gatekeeper relationships are and continue to be effectively managed (which includes, for example, deciding what, where, how, when, and, most importantly, why specific expectations of stakeholders need to be addressed in making choices within the 'Data Triangle'; deciding what, where, when, and, most importantly, why and how specific expectations of relevant gatekeepers will be met when seeking access to desired data sources and research contexts; and being alive to the possibilities that the pool of relevant stakeholders and gatekeepers or the mindsets of those within the pool may change over time, meaning that expectations to be addressed may also change);
- ensuring that choices for navigating the 'Data Triangle' are congruent with the pattern(s) of guiding assumptions adopted for the research (which includes, for example, considerations of how to best meet the expectations of the researcher's adopted pattern(s); considerations of the on-going relationship between the contextual, ethical, and cultural fit of the researcher's choices and their choices regarding patterns of guiding assumptions; and, in cases where multiple patterns of guiding assumptions are in play (either simultaneously with a team of researchers or sequentially through different phases of the research journey), considerations about how choices of activities around the 'Data Triangle' can help the researcher(s) to effectively manage the tensions between different patterns of guiding assumptions).

2.4.3 'Downstream' Dynamics

When we speak of 'downstream' dynamics associated with activities around the 'Data Triangle', we are talking about feedforward influences that research learning arising from the journey around the triangle has on how that learning is shared/communicated with relevant audiences/stakeholders or about feedback influences that stimulate changes in activities within the triangle because of anticipated or real adverse impacts on or potential benefits for relevant stakeholders, the wider public, the environment, or policy matters. 'Downstream' dynamics are another hallmark of the nonlinear research journey, only here, the researcher needs to be conscious of the possibilities of creating ripple effects with the learning they achieve via activities around the 'Data Triangle', again signalling the need for flexibility and adaptability.

In terms of 'downstream' dynamics, researchers may need to grapple with issues such as:

- ensuring that the research, and the 'Data Triangle' choices that produce learning, remain relevant and meaningful for the most appropriate audiences and stakeholders (which includes, for example, considerations about who the most relevant audiences and stakeholders are for research sharing. communication, and impact purposes; considerations about reciprocity, cultural relevance, and respect when sharing research learning and outcomes with Indigenous/First Nations people and communities; and considerations, where appropriate, about cultural relevance and meaningfulness for audiences and stakeholders in cross-cultural settings);
- ensuring that the choices made for navigating the 'Data Triangle' and the learning they yield are contextually relevant for potential downstream users/consumers of the research (which includes, for example, considerations about how best to share/communicate outcomes and learning; and considerations about the balance to be achieved between communicating/disseminating knowledge and listening to/addressing feedback and other informative signals about that knowledge, even if such signals are critical of what has been learned or how it has been learned);
- ensuring that the researcher remains alive to learning and responding to information about potential or actual adverse/unexpected outcomes/impacts or beneficial aspects of their research (which includes, for example, deciding how best to minimise the downstream impact of adverse or unexpected outcomes, perhaps through engaging in further activities around the 'Data Triangle' or in further communication/knowledge sharing activities; deciding how best to harness/expand beneficial outcomes and learning within as well as beyond the boundaries of the research (e.g., deciding how best to share a program or treatment, demonstrated, by the research, to be beneficial or effective, with people who did not get to experience it in the research); and deciding, where appropriate, how best to stimulate adoption of an innovation shown by the research to be useful, cost effective, and/or beneficial).

2.5 Integrative Foreshadowing

This chapter set out to provide a comprehensive helicopter overview of the evolving methodological landscape for social and behavioural research. This landscape is populated with potential choices of strategies for navigating the 'Data Triangle', i.e., strategies for choosing data sources, strategies for gathering data, and strategies for analysing data. The landscape is rich and dense, but devoid of principles and guidance for making choices appropriate to a specific research journey. We will see, in the chapters to come, that such principles and guidance emerge from pluralist systemic thinking about the contextualisation and positioning of the researcher's journey, including considerations about pattern(s) of guiding assumptions, choice of research frame(s) and mode(s) of knowledge building to be enacted, and research quality considerations. Furthermore, as the research journey unfolds, contextualisation influences will continue to emerge/evolve and, in doing so, may create circumstances that demand that the researcher make trade-off choices in order to keep their journey moving forward. Such trade-offs will often mean that something of value is lost in order to gain something else of value. This means that the methodological landscape, from the point of view of research contextualisation and implementation, remains a dynamic resource, to be drawn upon when needed. The researcher will likely be revisiting the landscape more than once to revise or alter the choices they have made in order to keep the research journey feasible as well as ensuring it is as convincing as possible, within the constraints imposed upon them and in light of the opportunities that they have taken advantage of.

References

Ahrens, P. (2018). Qualitative network analysis: A useful tool for investigating policy networks in transnational settings? *Methodological Innovations, 11*(1), 1–9.

Anderson, L. (2006). Analytic autoethnography. *Journal of Contemporary Ethnography, 35*(4), 373–395.

Anderson, V. (1997). *Systems thinking basics: From concepts to causal loops*. Pegasus Communications.

Angrosino, M. (2007). *Doing ethnographic and observational research.* Sage Publications.

Archibald, J., Xiiem, Q., Lee-Morgan, J., & Santolo, J. (Eds.). (2022). *Decolonizing research: Indigenous storywork as methodology.* Bloomsbury.

Ariel, B., Bland, M., & Sutherland, A. (2021). *Experimental designs.* Sage Publications.

Athanasou, J. (1997). *Introduction to educational testing.* Social Science Press.

Axelrod, R. (2007). Simulation in the social sciences. In J-P. Rennard (Ed.), *Handbook of research on nature inspired computing for economy and management* (pp. 90–100). Idea Group Reference.

Banks, M. (2014). Analyzing images. In U. Flick (Ed.), *The Sage handbook of qualitative data analysis* (pp. 394–408). Sage Publications.

Banks, M. (2018). *Using visual data in qualitative research.* Sage Publications.

Banyard, P., Winder, B., Norman, C., & Dillon, G. (eds.). (2022). *Essential research methods in psychology.* Sage Publications.

Barbour, R. (2018). *Doing focus groups.* Sage Publications.

Bazeley, P. (2018). *Integrating analyses in mixed methods research.* Sage Publications.

Bazeley, P. (2021). *Qualitative data analysis: Practical strategies* (2nd ed.). Sage Publications.

Beach, D., & Pedersen, R. (2013). *Process tracing methods: Foundations and guidelines.* University of Michigan Press.

Bell, E., Bryman, A., & Harley, W. (2022). *Business research methods* (6th ed.). Oxford University Press.

Bell, R. (2018). Personal constructs. In L. Cohen, L. Manion, & L. Morrison, *Research methods in education* (8th ed.) (pp. 496–509). Routledge.

Bennett, A., & Checkel, J. (eds.). (2014). *Process tracing: From metaphor to analytic tool.* Cambridge University Press.

Bernard, H., Wutich, A. & Ryan, G. (2016). *Analyzing qualitative data: Systematic approaches* (2nd ed.). Sage Publications.

Bliss, J., Monk, M. & Ogborn, J. (1983). *Qualitative data analysis for educational research: A guide to uses of systemic networks.* Croom-Helm.

Boje, D. (1991). The storytelling organization: A study of story performance in an office-supply firm. *Administrative Science Quarterly, 36*(1), 106–126.

Bond, D., & Ramsey, E. (2007, Sept.). Going beyond the fence: Using projective techniques as survey tools to meet the challenges of bounded rationality. In Proceedings of the 2007 Association for Survey Computing Meeting, *The challenges of a changing world: Developments in the survey process* (pp. 259–270). Southampton, UK.

Borgatti, S., Everett, M., & Johnson, J., & Agneessens, F. (2022). *Analyzing social networks using R*. Sage Publications.

Boyd, D., & Crawford, K. (2012). Critical questions for big data: Provocations for a cultural, technological, and scholarly phenomenon. *Information, Communication and Society, 15*(5), 662–679.

Brinkmann, S., & Kvale, S. (2018). *Doing interviews*. Sage Publications.

Brown, R. (2005). *Rational choice and judgment: Decision analysis for the decider*. John Wiley & Sons.

Bryant, A., & Charmaz, K. (eds.) (2007). *The Sage handbook of grounded theory*. Sage Publications.

Bryson, J., Ackermann, F., Eden, C., & Finn, C. (2004). *Visible thinking: Unlocking causal mapping for practical business results*. John Wiley & Sons.

Buzan, T. (2018). *Mind map mastery*. Watkins Media Limited.

Byrne, A., McLellan, S., Willis, E., Curnow, V., Harvey, C., Brown, J., & Hegney, D. (2021). Yarning as an interview method for non-Indigenous clinicians and health researchers. *Qualitative Health Research, 31*(7), 1345–1357.

Byrne, B. (2016). *Structural equation modelling with AMOS: Basic concepts, applications, and programming* (3rd ed.). Routledge.

Campbell, D., & Stanley, J. (1966). *Experimental and quasi-experimental designs for research*. Houghton Mifflin.

Carriger, J., Dyson, B., & Benson, W. (2018). Representing causal knowledge in environmental policy interventions: Advantages and opportunities for qualitative influence diagram applications. *Integrated Environmental Assessment and Management, 14*(3), 381–394.

Carsey, T., & Harden, J. (2013). *Monte Carlo simulation and resampling methods for the social sciences*. Sage Publications.

Cash, P., & Snider, C. (2014). Investigating design: A comparison of manifest and latent approaches. *Design Studies, 35*(5), 441–472.

Cassell, C., & Walsh, S. (2004). Repertory grids. In C. Cassell & G. Symon (Eds.), *Essential guide to qualitative methods in organizational research* (pp. 61–72). Sage Publications.

Castellani, B., & Rajaram, R. (2021). *Big data and complexity*. Sage Publications.

Chandler, J., Rosenzweig, C., Moss, A. J., Robinson, J., & Litman, L. (2019). Online panels in social science research: Expanding sampling methods beyond Mechanical Turk. *Behavior Research Methods, 51*(5), 2022–2038.

Chandola, T., & Booker, C. (2021). *Archival and secondary data*. Sage Publications.

Chang, H. (2016). *Autoethnography as method*. Routledge.

Charmaz, K. (2014). *Constructing grounded theory* (2nd ed.). Sage Publications.

Chell, E. (2004). The critical incident technique. In C. Cassell & G. Symon (Eds.), *Essential guide to qualitative methods in organizational research* (pp. 45–60). Sage Publications.

Chell, E. (2014). The critical incident technique: Philosophical underpinnings, method and application to a case of small business failure. In E. Chell & M. Karatas-Ozkan (eds.), *Handbook of research on small business and entrepreneurship* (pp. 106–129). Edward Elgar Publishing.

Chilisa, B. (2020). *Indigenous research methodologies* (2nd ed.). Sage Publications.

Chrzanowska, J. (2002). *Interviewing groups and individuals in qualitative marketing research*. Sage Publications.

Cohen, I. G., Amarasingham, R., Shah, A., Xie, B., & Lo, B. (2014). The legal and ethical concerns that arise from using complex predictive analytics in health care. *Health Affairs, 33*(7), 1139–1147.

Cohen, J., Cohen, P., West, S., & Aiken, L. (2003). *Applied multiple regression/correlation analysis for the behavioral sciences* (3rd ed.). Lawrence Erlbaum Associates.

Cohen, L., Manion, L., & Morrison, K. (2018). *Research methods in education* (8th ed.). Routledge.

Cook, T., & Campbell, D. (1979). *Quasi-experimentation: Design and analysis issues for field settings.* Rand McNally.

Cooksey, R. (1996). *Judgment analysis: Theory, Methods, and Applications.* Academic Press.

Cooksey, R. (2000). Mapping the texture of managerial decision making: A complex dynamic decision perspective. *Emergence: A Journal of Complexity Issues in Organizations and Management, 2*(2), 102–122.

Cooksey, R. (2020). *Illustrating statistical procedures: Finding meaning in quantitative data* (3rd ed.). Springer-Nature.

Cooksey, R., & McDonald, G. (2019). *Surviving and thriving in postgraduate research* (2nd ed.). Springer Nature.

Cooksey, R., & Soutar, G. (2006). Coefficient beta and hierarchical item clustering: An analytical procedure for establishing and displaying the dimensionality and homogeneity of summated scales. *Organizational Research Methods, 9*(1), 78–98.

Cowton, C. (1998). The use of secondary data in business ethics research. *Journal of Business Ethics, 17*, 423–434.

Crossley, N. (2010). The social world of the network: Combining qualitative and quantitative elements in social network analysis. *Sociologica, 4*(1), 1–34.

Czarniawska, B. (2004). *Narratives in social science research.* Sage Publications.

Ditka, G., & Scheer, M. (2021). *Bootstrap methods with applications in R.* Springer Nature.

Duncan, M. (2004). Autoethnography: Critical appreciation of an emerging art. *International Journal of Qualitative Methods, 3*(4), 28–39.

Durlak, J. (1995). Understanding meta-analysis. In L. G. Grimm & P. R. Yarnold (Eds.), *Reading and understanding multivariate statistic* (pp. 319–352). American Psychological Association.

Eden, C., & Ackermann, F. (2002). A mapping framework for strategy making. In A. Huff & M. Jenkins (Eds.), *Mapping strategic knowledge* (pp. 173–195). Sage Publications.

Eichhorn, J. (2021). *Survey research and sampling.* Sage Publications.

Elliott, J. (2005). *Using narratives in social research: Qualitative and quantitative approaches.* Sage Publications.

EMC Education Services (Eds.). (2015). *Data science and big data analytics.* John Wiley & Sons.

Eppler, M. (2006). A comparison between concept maps, mind maps, conceptual diagrams, and visual metaphors as complementary tools for knowledge construction and sharing. *Information Visualization, 5*, 202–210.

Ericsson, K. (2003). Valid and non-reactive verbalization of thoughts during performance of tasks: Towards a solution to the central problems of introspection as a source of scientific data. In A. Jack & A. Roepstorff (eds.), *Trusting the subject: The use of introspective evidence in cognitive science* (Vol. 1, pp. 1–18). Imprint Academic.

Ericsson, K., & Simon, H. (1993). *Protocol analysis: verbal reports as data* (rev. ed.). The MIT Press.

Esin, C., Fathi, M., & Squire, C. (2014). Narrative analysis: The constructionist approach. In U. Flick (Ed.), *The Sage handbook of qualitative data analysis* (pp. 203–216). Sage Publications.

Fayyad, U., Piatetsky-Shapiro, G., & Smyth, P. (1996). From data mining to knowledge discovery in databases. *AI Magazine, 17*(3), 37.

Field, A. (2018). *Discovering statistics using IBM SPSS statistics* (5th ed.). Sage Publications.

Fielding, J., & Fielding, N. (2015). Emergent technologies in multimethod and mixed method research: Incorporating GIS and CAQDAS. In S. Hesse-Biber & B. Johnson (Eds.), *Oxford handbook of multimethod and mixed methods research inquiry* (pp. 561–584). Oxford University Press.

Finlay, S. (2018). Critical arts-based inquiry: Performances of resistance politics. In N.K. Denzin & Y.S. Lincoln (eds.), *The Sage handbook of qualitative research* (5th ed., pp. 561–575). Sage Publications.

Flick, U. (ed.). (2014). *The Sage handbook of qualitative data analysis.* Sage Publications.
Flick, U. (ed.). (2018a). *The Sage qualitative research kit* (2nd Ed.) (10 volume set). Sage Publications.
Flick, U. (2018b). *Doing grounded theory.* Sage Publications.
Flick, U. (2021). *Doing interview research.* Sage Publications.
Flick, U. (2023). *An introduction to qualitative research* (7th ed.). Sage Publications.
Fredericks, B., Adams, K., Finlay, S., Fletcher, G., Andy, S., Briggs, L., & Hall, R. (2011). Engaging the practice of Indigenous yarning in action research. *ALAR: Action Learning and Action Research Journal, 17*(2), 12–24.
Furr, R. (2021). *Psychometrics: An introduction* (4th ed.). Sage Publications.
Gabriel, Y., & Griffiths, D. (2004). Stories in organizational research. In C. Cassell & G. Symon (Eds.), *Essential guide to qualitative methods in organizational research* (pp. 114–126). Sage Publications.
Galletta, A. (2013). *Mastering the semi-structured interview and beyond: From research design to analysis and publications.* New York University Press.
Gandomi, A., & Haider, M. (2015). Beyond the hype: Big data concepts, methods, and analytics. *International Journal of Information Management, 35*, 137–144.
Geia, L., Hayes, B., & Usher, K. (2013). Yarning/Aboriginal storytelling: Towards an understanding of an Indigenous perspective and its implications for research practice. *Contemporary Nurse, 46*(1), 13–17.
Gibbs, G. (2018). *Analyzing qualitative data.* Sage Publications.
Gilbert, N. (2008). *Agent-based models.* Sage Publications.
Gilbert, N., & Troitzsch, K. (2005). *Simulation for the social scientist* (2nd ed.). McGraw-Hill International.
Glass, G., McGaw, B., & Smith, M. (1981). *Meta-analysis in social research.* Sage Publications.
Gobo, G., Fielding, N., La Rocca, G, & van der Vaart, W. (2022). *Merged methods: A rationale for full integration.* Sage Publications.
Gray, D. (2021). *Doing research in the real world* (5th ed.). Sage Publications.
Grbich, C. (2013). *Qualitative data analysis: An introduction* (2nd ed.) Sage Publications.
Gubrium, J., Holstein, J., Marvasti, A., & McKinney, K. (eds). (2012). *The Sage handbook of interview research: The complexity of the craft.* Sage Publications.
Guest, G., MacQueen, K., & Namey, E. (2012). *Applied thematic analysis.* Sage Publications.
Hackett, P., Suvak, J., & Gordley-Smith, A. (2023). *Projective techniques and sort-based research methods.* Routledge.
Hair, J., Hult, G., Ringle, C., & Sarstedt, M. (2022). *A primer on partial least squares structural equation modelling [PLS-SEM]* (3rd ed.). Sage Publications.
Hand, D. (2004). *Measurement theory and practice: The world through quantification.* John Wiley & Sons.
Hayes, A. (2018). *Introduction to mediation, moderation, and conditional process analysis: A regression-based approach.* The Guilford Press.
Henderson, S., & Segal, E. (2013). Visualizing qualitative data in evaluation research. In T. Azzam & S. Evergreen (eds.), *Data visualization, part 1: New Directions for Evaluation*, American Evaluation Association, *139*, 53–71.
Hine, D., Montiel, C., Cooksey, R., & Lewko, J. (2005). Mental models of poverty in developing nations: A Canada-Philippines contrast. *Journal of Cross-Cultural Psychology, 36*(3), 283–303.
Johnson, B. (2001). Towards a new classification of nonexperimental quantitative research. *Educational Researcher, 30*(2), 3–13.
Jones, J., & Goldring, J. (2021). *Exploratory and descriptive statistics.* Sage Publications.
Jones, K. (2005). The art of collaborative storytelling: Arts-based representations of narrative contexts. *International Sociological Association Research Committee on Biography and Society*, (RC38 Newsletter), Summer.
Jorgensen, D. (1989). *Participant observation: A methodology for human studies.* Sage Publications.

Judd, C., McClelland, G., & Ryan, C. (2017). *Data analysis: A model comparison approach to regression, ANOVA, and beyond* (3rd ed.). Routledge.

Kane, M., & Trochim, W.M.K. (2007). *Concept mapping for planning and evaluation.* Sage Publications.

Kaplan, D. (ed.). (2004). *The Sage handbook of quantitative methodology for the social sciences.* Sage Publications.

Kaptchuk, T. (2001). The double-blind, randomized, placebo-controlled trial: Gold standard or golden calf? *Journal of Clinical Epidemiology, 54*(6), 541–549.

Kaye, M. (1996). *Myth-makers and story-tellers: How to unleash the power of myths, stories and metaphors to understand the past, envisage the future, and create lasting and positive cultural change in your organisation.* Business & Professional Publishers.

Kellehear, A. (1993). *The unobtrusive observer: A guide to methods.* Allen & Unwin.

Kelly, G. (1955). *The psychology of personal constructs.* Norton.

Kilduff, M., & Tsai, W. (2003). *Social networks and organizations.* Sage Publications.

King, N. (2004). Using templates in the thematic analysis of data. In C. Cassell & G. Symon (Eds.), *Essential guide to qualitative methods in organizational research* (pp. 256–270). Sage Publications.

Kirk, R. (2013). *Experimental design: Procedures for behavioral sciences* (4th ed.). Sage Publications.

Knoblauch, H., Tuma, R., & Schettler, B. (2014). Video analysis and videography. In U. Flick (Ed.), *The Sage handbook of qualitative data analysis* (pp. 435–449). Sage Publications.

Konstantopoulos, S., & Hedges, L. (2004). Meta-analysis., In D. Kaplan (ed.), *The Sage handbook of quantitative methodology for the social sciences.* Sage Publications, pp. 281–300.

Kovach, M. (2021). *Indigenous methodologies: Characteristics, conversations, and contexts* (2nd ed.). University of Toronto Press.

Kozinets, R., Dolbec, P-Y., & Earley, A. (2018). Using netnographic analysis: Understanding culture through social media data. In U. Flick (ed.), *The Sage handbook of qualitative data analysis* (pp. 262–276). Sage Publications.

Kuckartz, U. (2014). *Qualitative text analysis: A guide to methods, practice, and using software.* Sage Publications.

Kurtz, D. (2013). Indigenous methodologies: Traversing Indigenous and Western worldviews in research. *AlterNative: An International Journal of Indigenous Peoples, 9(*3), 217–229.

Laukkanen, M. (1994). Comparative cause mapping of organizational cognitions. *Organization Science, 5,* 322–341.

Laukkanen, M. (1998). Conducting causal mapping research: Opportunities and challenges. In C. Eden & J. Spender (Eds.), *Managerial and organizational cognition: Theory, methods and research* (pp. 168–191). Sage Publications.

Lee, A. (2005). *Flight simulation: Virtual environments in aviation.* Routledge.

Lee, S., & Skog, A. (2014). Kobayashi Maru in medical simulation: The educational utility of the no win patient scenario (entry no. 20). *Annals of Emergency Medicine, 64*(4), S8–S9.

Lewis, R., & Maas, S. (2007). QDA Miner 2.0: Mixed-model qualitative data analysis software. *Field Methods, 19*(1), 87–108.

Liamputtong, P. (2019). *Qualitative research methods* (5th ed.). Oxford.

Little, T. (ed.). (2013). *The Oxford handbook of quantitative methods in psychology: Vol 2: Statistical analysis.* Oxford University Press.

Liu, B. (2012). *Sentiment analysis: Mining opinions, sentiments, and emotions.* Morgan & Claypool.

Löblich, M., & Pfaff-Rüdiger, S. (2011). Network analysis: A qualitative approach to empirical studies on communication policy. *The International Communication Gazette, 73*(7), 630–647.

Louw, S., Todd, R., & Jimarkon, P. (2011, April). Active listening in qualitative research interviews. In *Proceedings of the International Conference: Research in Applied Linguistics* (pp. 71–8).

Maani, K., & Cavana, R. (2007). *Systems thinking, systems dynamics: Managing change and complexity* (2nd ed.). Pearson Education New Zealand.

Maeder, C. (2014). Analyzing sounds. In U. Flick (Ed.), *The Sage handbook of qualitative data analysis* (pp. 424–434). Sage Publications.
Margolis, E., & Zunjarwad, R. (2018). Visual research. In N. K. Denzin & Y. S. Lincoln (Eds.), *The Sage Handbook of Qualitative Research* (5th ed., pp. 600–626). Sage Publications.
Marotzki, W., Holze, J., & Verständig, D. (2014). Analyzing virtual data. In U. Flick (Ed.), *The Sage handbook of qualitative data analysis* (pp. 450–464). Sage Publications.
Martin, P. (2021a). *Linear regression: An introduction to statistical models.* Sage Publications.
Martin, P. (2021b). *Regression models for categorical and count data.* Sage Publications.
May, T., & Perry, B. (2018). Reflexivity and the practice of qualitative research. In U. Flick (Ed.), *The Sage handbook of qualitative data analysis* (pp. 109–122). Sage Publications.
McBee, M. (2021). *Statistical approaches to causal analysis.* Sage Publications.
McCoach, D., & Cintron, D. (2021). *Introduction to modern modeling methods.* Sage Publications.
McDonald, S., Daniels, K., & Harris, C. (2004). Cognitive mapping in organizational research. In C. Cassell & G. Symon (Eds.), *Essential guide to qualitative methods in organizational research* (pp. 73–85). Sage Publications.
McKeown, B., & Thomas, D. (2013). *Q methodology* (2nd ed., Vol. 66). Sage Publications.
Meadows, D. (2008). *Thinking in systems: A primer.* Chelsea Green Publishing (edited by Diana Wright, Sustainability Institute, upon the death of Donella Meadows).
Mennecke, B., McNeill, D., Ganis, M., Roche, E., Bray, D., Konsynski, B., Townsend, A., & Lester, J. (2008). Second life and other virtual worlds: A roadmap for research (follow-up article to the 2007 International Conference on Information Systems panel). *Communications of the Association for Information Systems, 22*, 371–388.
Mesías, F., & Escribano, M. (2018). Projective techniques. In G. Ares & P. Varela (Eds.), *Methods in consumer research, Volume 1: New approaches to classical methods* (pp. 79–102). Woodhead Publishing.
Mikos, L. (2014). Analyzing film. In U. Flick (Ed.), *The Sage handbook of qualitative data analysis* (pp. 409–423). Sage Publications.
Miles, M., Huberman, A., & Saldaña, J. (2019). *Qualitative data analysis: A methods sourcebook* (4th ed.). Sage Publications.
Miller, J., & Page, S. (2007). *Complex adaptive systems: An introduction to computational models of social life.* Princeton University Press.
Mooney, C. (1997). *Monte Carlo simulation.* Sage Publications.
Mooney, C., & Duval, R. (1993). *Bootstrapping: A nonparametric approach to statistical inference.* Sage Publications.
Musson, G. (2004). Life histories. In C. Cassell & G. Symon (Eds.), *Essential guide to qualitative methods in organizational research* (pp. 34–44). Sage Publications.
Neuman, W. (2020). *Social research methods: Qualitative and quantitative approaches* (8th ed.). Pearson.
Nisbett, R., & Ross, L. (1980). *Human inference: Strategies and shortcomings of social judgment.* Prentice Hall.
Noblit, G., & Hare, R. (1988). *Meta-ethnography: Synthesizing qualitative studies* (Vol. 11). Sage Publications.
Nooraie, R., Sale, J. E., Marin, A., & Ross, L. E. (2020). Social network analysis: An example of fusion between quantitative and qualitative methods. *Journal of Mixed Methods Research, 14*(1), 110–124.
Olaru, D., Purchase, S., & Denize, S. (2009, July). Using docking/replication to verify and validate computational models. In *Proceedings of the 18th World IMACS/MODSIM Congress* (pp. 4432–4438)
Oliveira, M., Bitencourt, C., Teixeira, E., & Santos, A. (2013, July). Thematic content analysis: Is there a difference between the support provided by the MAXQDA® and NVivo® software packages. In *Proceedings of the 12th European Conference on Research Methods for Business and Management Studies* (pp. 304–314)

Onwuegbuzie, A., & Collins, K. (2007). A typology of mixed methods sampling designs in social science research. *The Qualitative Report, 12*(2), 281–316.

Onwuegbuzie, A. & Combs, J. (2010). Emergent data analysis: Analysis techniques in mixed methods research. In A. Tashakkori & Teddlie, C. (Eds.), *Sage handbook of mixed methods in social and behavioral research* (2nd ed., pp. 397–430), Sage Publications.

Paterson, B., Thorne, S., Canam, C., & Jillings, C. (2001). *Meta-study of qualitative health research: A practical guide to meta-analysis and meta-synthesis.* Sage Publications.

Payne, J., Bettman, J., & Johnson, E. (1993). *The adaptive decision maker.* Cambridge University Press.

Peräkylä, A., & Ruusuvuori, J. (2018). Analyzing talk and text. In N.K. Denzin & Y.S. Lincoln (Eds.), *The Sage Handbook of Qualitative Research* (5th ed., pp. 669–691). Sage Publications.

Pidd, M. (2009). *Tools for thinking: Modelling in management science* (3rd ed.). John Wiley & Sons.

Pierce, C., & Aguinis, H. (1997). Using virtual reality technology in organizational behavior research. *Journal of Organizational Behavior, 18*(5), 407–410.

Plowright, D. (2011). *Using mixed methods: Frameworks for an integrated methodology.* Sage Publications.

Price, L. (2017). *Psychometric methods: Theory into practice.* The Guilford Press.

Proctor, T. (2014). *Creative problem solving for managers* (5th ed.). Routledge.

Provost, F., & Fawcett, T. (2013). Data science and its relationship to big data and data-driven decision making. *Big Data, 1*(1), 51–59.

Ragin, C. (1987). *The comparative method: Moving beyond qualitative and quantitative strategies.* University of California Press.

Railsback, S., & Grimm, V. (2019). *Agent-based and individual-based modelling: A practical introduction* (2nd ed.). Princeton University Press.

Rymaszewski, M., Au, W., Wallace, M., Winters, C., Ondrejka, C., & Batstone-Cunningham, B. (2007). *Second life: The official guide.* John Wiley & Sons.

Rapley, T. (2018). *Doing conversation, discourse and document analysis.* Sage Publications.

Rosenthal, R. (1984). *Meta-analytic procedures for social research.* Sage Publications.

Ross, G. (2001). *Visual methodologies: An introduction to the interpretation of visual materials.* Sage Publications.

Saldaña, J. (2018). Ethnodrama and ethnotheatre: Research as performance. In N. K. Denzin & Y. S. Lincoln (Eds.), *The Sage handbook of qualitative research* (5th ed., pp. 377–394). Sage Publications.

Saldaña, J. (2021). *The coding manual for qualitative researchers* (4th ed.). Sage Publications.

Saldaña, J. (2024). Performance shapes for qualitative inquiry. In N. Denzin, Y. Lincoln, M. Giardina & G. Cannella (Eds.), *The Sage handbook of qualitative research* (6th ed., pp. 437–452), Sage Publications.

Schmidt, F., & Hunter, J. E. (2014). *Methods of meta-analysis: Correcting error and bias in research findings.* Sage Publications.

Schreier, M. (2012). *Qualitative content analysis in practice.* Sage Publications.

Schreier, M. (2014). Qualitative content analysis. In U. Flick (Ed.), *The Sage handbook of qualitative data analysis* (pp. 170–183). Sage Publications.

Scott, J. (2017). *Social network analysis: A handbook* (4th ed.). Sage Publications.

Senge, P., Kleiner, A., Roberts, C., Ross, R., & Smith, B. (1994). *The fifth discipline field book.* Nicholas Brealey.

Siegesmund, R. (2024). The arts as research: Nomadic materiality and possible futures. In N. Denzin, Y. Lincoln, M. Giardina & G. Cannella (Eds.), *The Sage handbook of qualitative research* (6th ed., pp. 453–465). Sage Publications.

Silge, J., & Robinson, D. (2017). *Text mining with R: A tidy approach.* O'Reilly & Associates Inc.

Silvester, J. (2004). Attributional coding. In C. Cassell, & G. Symon (Eds.), *Essential guide to qualitative methods in organoizational research* (pp. 228-241). Sage Publications.

Small, S. D., Wuerz, R. C., Simon, R., Shapiro, N., Conn, A., & Setnik, G. (1999). Demonstration of high-fidelity simulation team training for emergency medicine. *Academic Emergency Medicine, 6*(4), 312–323.

Smith, A., & Humphreys, M. (2006). Evaluation of unsupervised semantic mapping of natural language with Leximancer concept mapping. *Behavior Research Methods, 38*, 262–279.

Sprent, P. (2011). *Applied nonparametric statistical methods*. Chapman & Hall.

Sterman, J. (2000). *Business dynamics: Systems thinking and modeling for a complex world*. McGraw-Hill/Irwin.

Symon, G. (2004). Qualitative research diaries. In C. Cassell & G. Symon (Eds.), *Essential guide to qualitative methods in organizational research* (pp. 98–113). Sage Publications.

Tabachnick, B., & Fidell, L. (2018). *Using multivariate statistics* (7th ed.). Pearson.

Taber, C., & Timpone, R. (1996). *Computational modeling*. Sage Publications.

Tachine, A., Bird, E., & Cabrera, N. (2016). Sharing circles: An Indigenous methodological approach for researching with groups of Indigenous peoples. *International Review of Qualitative Research, 9*(3), 277–295.

Tannen, D., Hamilton, H., & Schiffrin, D. (Eds.). (2015). *The handbook of discourse analysis* (2nd ed.). Wiley-Blackwell.

Tansy, O. (2007). Process tracing and elite interviewing: A case for non-probability sampling. *Political Science and Politics, 40*(4), 1–23.

Teddlie, C., & Yu, F. (2007). Mixed methods sampling: A typology with examples. *Journal of Mixed Methods Research, 1*(1), 77–100.

Toerien, M. (2014). Conversations and conversation analysis. In U. Flick (Ed.), *The Sage handbook of qualitative data analysis* (pp. 327–340). Sage Publications.

Trenor, J., Miller, M., & Gipson, K. (2011). Utilization of a think-aloud protocol to cognitively validate a survey instrument identifying social capital resources of engineering undergraduates. In Electronic Proceedings of the *American Society for Engineering Education Annual Conference and Exposition* (pp. 22.1656.1–22.1656.15). Vancouver, CA. (https://peer.asee.org/utilization-of-a-think-aloud-protocol-to-cognitively-validate-a-survey-instrument-identifying-social-capital-resources-of-engineering-undergraduates. Accessed 29 June 2023).

Trochim, W., & McLinden, D. (2017). Introduction to a special issue on concept mapping. *Evaluation and Program Planning, 60*, 166–175.

VanderPlas, J. (2022). *Python data science handbook: Essential tools for working with data* (2nd ed.). O'Reilly Media Inc.

Veal, A. (2005). *Business research methods: A managerial approach* (2nd ed.). Pearson Education.

Waddington, D. (2004). Participant observation. In C. Cassell & G. Symon (Eds.), *Essential guide to qualitative methods in organizational research* (pp. 154–164). Sage Publications.

Walker, S. (2014). Big data: A revolution that will transform how we live, work, and think. *International Journal of Advertising, 33*(1), 181–183.

Webb, E., Campbell, D., Schwartz, R., & Sechrest, L. (2000). *Unobtrusive measures* (rev. ed.). Sage Publications.

Wernet, A. (2014). Hermeneutics and objective hermeneutics. In U. Flick (Ed.), *The Sage handbook of qualitative data analysis* (pp. 234–246). Sage Publications.

Wilensky, U., & Rand, W. (2015). *An introduction to agent-based modelling: Modelling natural, social, and engineered complex systems with Netlogo*. The MIT Press.

Williams, M., Wiggins, R., & McCoach, D. (Eds.). (2021). *The Sage quantitative research kit* (11 volume set). Sage Publications.

Willig, C. (2014). Discourses and discourse analysis. In U. Flick (Ed.), *The Sage handbook of qualitative data analysis* (pp. 341–353). Sage Publications.

Windchief, S., & San Pedro, T. (eds.). (2019). *Applying Indigenous research methods: Storying with peoples and communities*. Routledge.

Wolfram, S. (2002). *A new kind of science*. Wolfram Media, Inc.

Wolfram, S. (2023). *An elementary introduction to the Wolfram language* (3rd ed.). Wolfram Media, Inc.

Woolsey, L. (1986). The critical incident technique: An innovative qualitative method of research. *Canadian Journal of Counselling and Psychotherapy, 20*(4), 242–254.

Chapter 3
Systemic Contextualisation: Situating and Positioning

Abstract This chapter explores the systemic contextualisation of research in order to situate and position the research with respect to relevant and potentially influential considerations. These contextual considerations include: where the researcher is coming from (e.g., their identity, systems they are embedded within) and what they wish to learn (problems, goals, research questions/propositions/ hypotheses); who/what potential data sources might be and why; what has been learned before (published and unpublished); where and under what conditions the researcher might wish to conduct their research; who relevant stakeholders might be and their expectations; who relevant gatekeepers might be that control access to desired data sources; as well as larger contextual influences which may be relevant to the research (e.g., legal, ethical and regulatory, resource-based, environmental, community-based, market-based, institutional/organisational, governmental, sociocultural, technological, political, economic). Contextualisation provides colour, shape, and texture to the methodological landscape, bringing some potential choices into the foreground and allowing other potential choices to recede into the background. Contextualisation is not just a one-off 'upstream' activity for the researcher, it is continually revisited throughout their research journey. Nothing remains static and the researcher must be actively vigilant for contextual changes that may exert pressure to influence how their journey is progressing.

3.1 Contextualising Research

We think, relate, live, and experience within a complex array of interactive and at least partially embedded systems (recall Table 1.1) and research is simply another human activity that unfolds within that complex array. Such systems define contexts that can exert influences on people's thoughts, beliefs, attitudes, motivations and needs, emotions, relationships, cultures and traditions, behaviours, aspirations, achievements, and failures in various ways as we live our lives and these influences do not disappear when we carry out or participate in social and behavioural research. Understanding what influences are coming from where in relevant and important

R. W. Cooksey, *Unity from Diversity*, https://doi.org/10.1007/978-981-97-3462-7_3

contexts can help a researcher to develop their own positioning for their research. Contextualising research means creating the landscape within which the meanings that the researcher intends readers or users to construct from their research become apparent. Contextualisation involves clarifying guiding assumptions, stating research intentions and goals, and drawing boundaries around the research and its context(s). If a researcher appropriately contextualises their research journey, it should be very clear as to (1) where they as well as their data sources, as participants or information providers, are coming from (what can be called 'positioning' arguments); (2) what assumptions and trade-offs they are making in order to make their research 'do-able'; (3) what they will and will not be addressing in their research (e.g., research goals, questions, propositions, and/or hypotheses), including how they are or are not building on the work of others, who their important stakeholders and gatekeepers are, and their choice of research frame(s); (4) what contribution they see their research making to theory, methodology, practice and/or other areas of potential impact, and, importantly, (5) what their research context(s) are and their implications for what they can do and what they may learn. Contextualisation is a dynamic and evolving exercise that changes shape and composition during the research journey and for which identifying and managing key stakeholders and gatekeepers is an important aspect of that exercise. The research journal is an indispensable resource for recording a researcher's thinking about contextualisations.

In the discussions of research contextualisation to follow, we will build up as well as drill down using the four broad domains represented in the Fig. 1.4 Venn diagram as a starting point. The Researcher, Research, Participants and Other Data Sources, and Research Users/Sponsors/ Communities contextual domains are all embedded within a larger set of contextual influences and dynamics and these contextual domains overlap when a specific research project brings them together. The quadruple intersection of the four Venn diagram circles overlaying the area of larger contextual influences is the region from which a specific research project emerges and evolves and is where contextualising and positioning strategies are shaped. The dynamic interconnectedness of contextualisation domains means that any discussions of them are unavoidably nonlinear. In Chaps. 3 and 4, expect to see early foreshadowing of some contextual issues and considerations which will be discussed in more depth later on. As well, those later elaborations will refer to earlier contextualisation considerations.

Typically, social and behavioural research tends to be under-contextualised, especially if we base our judgments on what researchers convey in their research outcomes. Many aspects of contextualisation remain hidden from those who would wish to fully understand the meanings, implications, limitations, and potential uses of the research. However, pluralist systemic thinking demands that researchers dig deep to understand the contextual forces that they either are or anticipate experiencing during their research journey in order to develop strategies for achieving their research goals in the most convincing yet feasible way. Contextual influences may take the form of constraints which can emerge from any or all these contexts at any time during the research journey, which adds to the complexities that the researcher must wrestle with in deciding how to best overcome, circumvent, or adapt to them. Of course, influences may also take the form of new opportunities, which can emerge from any or all these contexts as well, and the researcher must decide whether or not they wish to or are able to pursue them. Constraints operate to limit perhaps even hamstring the research in some critical way whereas opportunities operate to expand/extend the research. A crucial part of the research journey is thus recognising, balancing, and managing the most relevant and impactful of these contextual influences, dealing effectively with constraints and being opportunistic where the need arises—all in pursuit of carrying out a convincing research project. Thus, an important aspect of contextualisation involves anticipating, as far as possible, the critical constraints that exist or which could occur, the possible opportunities that might present themselves, as well as what resources the researcher might need. It is through contextualisation that critical leverage points may be identified which can facilitate (or hinder) adaptation.

Contextualisation may be facilitated by considering three critical questions:

1. What place(s)will/does the research occupy in the larger scheme of things?
2. Why is the research being done, for whose benefit, and perhaps, to whose detriment?
3. Will the research be feasible and realistic to conduct convincingly given the constraints the researcher faces and the opportunities they take advantage of?

Contextualisation is not just a front-end only researcher activity. Continually revisiting and updating one's knowledge about contextual influences is an important way for the researcher to sense when changes are demanded and to flex and adapt accordingly. Furthermore, reflecting on contextualisation can help the researcher to situate, characterise and qualify what they have learned from their research as well as from their own experiences throughout the journey. So, contextualisation as a planning exercise helps in research conceptualisation and positioning and, as a reflective/reflexive exercise, helps in creating meaning for others about what has been learned and where, for whom, and why such learning is relevant.

Since contextual influences may come from various layers of embedded systems within a variety of domains, employing systemic thinking helps to ensure that researchers remain open to perceiving and addressing the relevant aspects of those systems as well as any interactive dynamics that may be playing out between the domains. Pluralist thinking ensures that the most extensive potential toolkit remains available for the researcher to consider as they work to address relevant contextual influences. Relevant contextualisations should be overtly displayed in research outcomes as they provide the background against which the researcher has made their choices and drawn their conclusions and implications. The more explicit the researcher makes and defends the contextualisation of their research and reflects this in consistent decisions throughout their research journey, the more convincing their research outcomes, conclusions, and implications will be.

Through contextualisation, researchers can identify relevant and important stakeholders and gatekeepers to connect with and making such connections can assist them in their contextualisation processes. As well, contextualisation helps the researcher to identify possible target audience(s) for research sharing (i.e., who needs to be convinced), for research impact (i.e., influences the research could facilitate, changes the research could inform, and under what circumstances), and for guiding/undertaking potential/actual innovation development. Appropriately contextualised research will help sponsors/users/readers/potential adopters form appropriate expectations and subsequently draw appropriate inferences and conclusions from any research outcome (a harbinger of research impact).

Effective contextualisation helps the researcher to:

- identify a research problem and begin to form ideas for what research aims/goals they wish to achieve and what research questions/propositions/hypotheses/propositions they wish to address;
- consider potential research contexts within which to pursue their research agenda;
- consider the implications of their research for potential participants and other data sources;
- identify and devise strategies for dealing with relevant and important stakeholders and gatekeepers;
- identify sources of pressure to do things in particular ways to meet specific expectations or achieve outcomes desired by others (e.g., stakeholders, gatekeepers, consumers and the general public, communities and cultural groups, organisations and institutions);
- identify the assumptions they are willing to make in pursuit of their research aims and goals and the quality criteria and meta-criteria that could be applied to make judgments about how well their pursuit has been undertaken;
- where appropriate, justify trade-off choices the researcher has made;
- identify constraints which seek to limit what the researcher can do;
- identify new opportunities that the researcher could pursue; and
- frame their research and knowledge building approach in an appropriate way.

Furthermore, effective contextualisation:

- provides a background against which choices can be made about how to best navigate the 'Data Triangle' in pursuit of research aims and goals; and
- provides a background against which findings, conclusions, and impacts can be appropriately characterised and qualified.

Figure 3.1 provides an extensive cognitive map of contextualisation considerations for and during the research journey, using a scaffold comprising the four main contextual domains from the Venn diagram presented in Fig. 1.4 as well as the larger contextual systems represented by the rounded rectangle within which that Venn diagram is embedded. We will drill down into each of these major contextual domains as well as explore the links and implications that connect them. Considerations associated with stakeholders and gatekeepers emerge from each of these major contextual domains (symbolised by the orange connecting links). Through contextualisation of the research journey, the researcher positions themselves with respect to their intended journey and anticipates relevant research audiences (e.g., users/sponsors/communities). Contextualisation partly depends upon participants and other data sources and facilitates the emergence of specific research-oriented contextual choices. Prior research/knowledge building efforts (commonly, if overly restrictively, labelled the 'literature') contribute to the contextualisation of the research journey. Larger contextual influences often serve to influence, constrain, or open up opportunities for the research.

Darker thicker links signify the major connections between the contextual domains. Dashed links or arrows represent conditional dependencies where influences may or may not be present depending upon circumstances and it is important to understand that these dependencies may evolve and change over the course of the research journey, i.e., contextualisation never stops. This latter point signals a critical limitation to the two-dimensional nature of Fig. 3.1, namely that it neglects the temporal dimension; it must be considered as only a snapshot in time for contextualisation. The researcher must remain alive to the fact that the only way to fully realise the implications of contextualisation is to continually revisit contextualisation considerations with a view to seeing what has changed and what implications those changes might hold for adaptations in the research process during the journey. Each revisit amounts to examining a new, at least partial, snapshot, perhaps stimulated by contextual signals that something has changed in a specific domain or simply by the researcher checking back in to ensure nothing substantive has changed. The complex and interconnected nature of human existence means that, on occasion, contextual circumstances may change, even slightly, in such a way as to create a cascade or ripple effect changing other things downstream. Chaos theory teaches us that even very small contextual changes can result in massive downstream impacts (technically termed 'sensitivity to initial conditions'; see, for example, Cooksey, 2001a), so if contextualisation influences are not monitored and, where needed, adapted to promptly, such changes can threaten not only the viability and convincingness of the research journey itself, but also the potential journeys of other researchers, the lives of participants, and larger environmental, political, health-related, or economic conditions.

3.1.1 Researcher Contextual Considerations

Researcher contextualisation is essential for positioning the research journey and the researcher's role(s) within that journey. This level of contextualisation explicitly focuses on the researcher's identity as it intersects the evolution of their research journey in terms of clarifying what they expect/want to do/achieve, understanding what they are bringing to the table, grappling with any knowledge gaps and skill deficits, reflecting on what pattern(s) of guiding assumptions, knowledge building modes, and research frame(s) they are willing to adopt, identifying/assembling resources and support systems, making choices, decisions, and trade-offs, managing the dynamic interplay with research context considerations (from which the research problem and attendant research questions/propositions/ hypotheses as well as considerations regarding adopting an insider or outsider role in the research context emerge), managing relationships with key stakeholders and potential gatekeepers, as well as identifying potential influences from larger contexts beyond the bounds of their

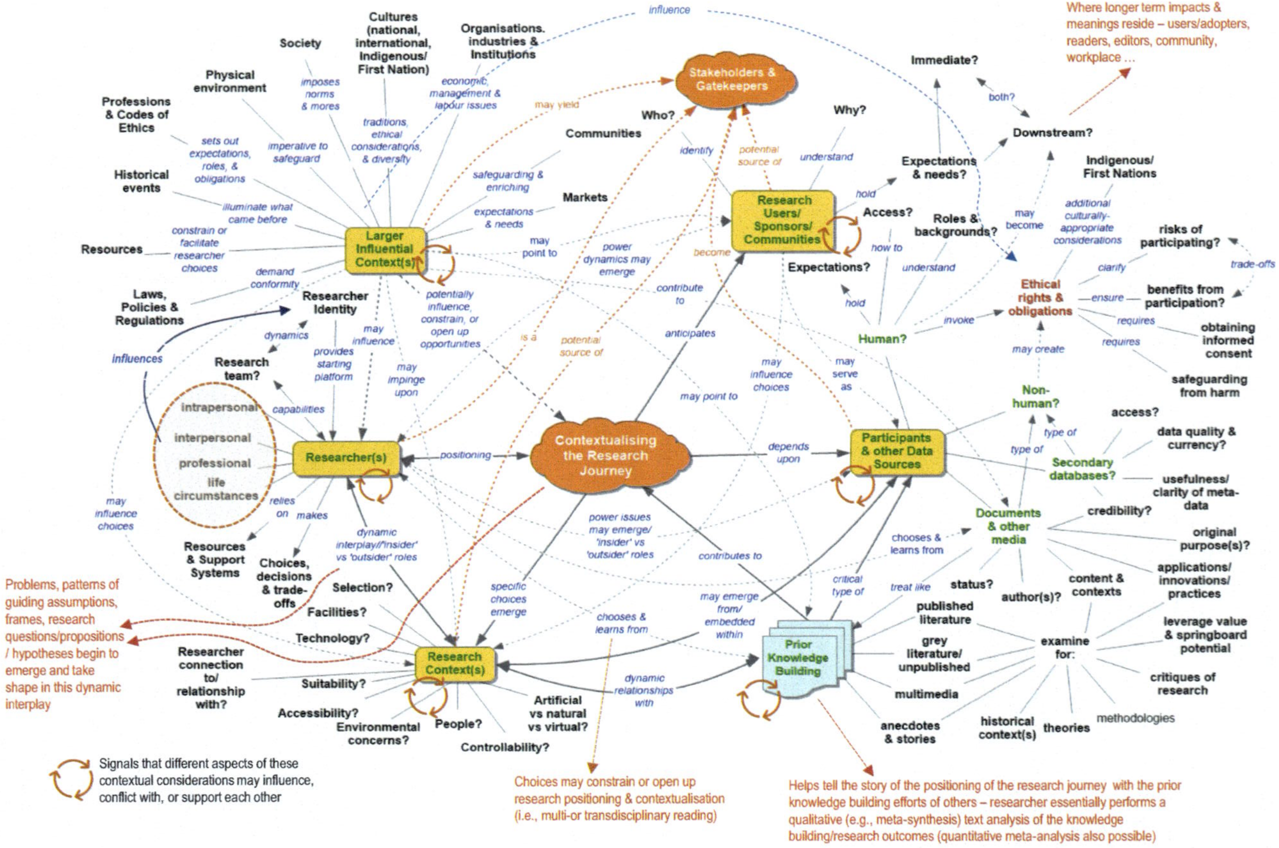

Fig. 3.1 Cognitive mapping of contextual considerations and their potential dynamic interactions for and during the researcher's journey

research journey. Note that if a team of researchers is undertaking the research journey together, then each researcher comes into the journey with their own contextualisations and an additional interpersonal dynamic emerges linked to the nature of the relationships and interactions between the team members.

The systemic thinking perspective embeds an explicit recognition that the researcher is human and that their humanity can influence what they believe in and do. Researcher contextual positioning serves several important purposes, not least of which is ensuring that the researcher understands where they are situated with respect to the research they are conducting. A research problem may surface simply out of researcher curiosity and motivation, perhaps reinforced by other contextual considerations such as prior knowledge building efforts, larger contextual issues, or downstream research users/sponsors/ communities. Relevant aspects of positioning encompass intrapersonal (self), interpersonal (other people the researcher relates to), professional (the researcher's professional identity, networks, and practices) and life circumstances (the researcher's history, state of social and family networks, current events, current activities inside and outside of work, experiences, and roles) domains. All these domains are influential on the nature and development of researcher identity, which in turn forms a core influence on contextualising the research journey. The research journal forms a critical and enduring platform for the researcher to record their deliberations in each of these domains and work through any potential implications for their research.

In the intrapersonal domain, researcher positioning requires critical self-awareness, in terms of knowing their preferences, capabilities and skills, capacities, interests, motivations, needs, goals and expectations. It establishes where the researcher is coming from and what they are trying to achieve. It involves insights into the duality of and tensions between their identities/roles 'as a person' and 'as a researcher'. On the 'as a person' side, considerations emerge that concern:

- basic personality and motivations (e.g., Type A vs Type B personality, introversion/extraversion, open-mindedness, self-awareness, cognitive style, empathy, dogmatism, authoritarianism, curiosity, ability to work with others, preferred types of rewards; motivation to persist; motivation to succeed);
- intellectual (e.g., mathematics, reasoning, critical and systems thinking, logic, language, reading, writing, oral language skills);
- creative abilities (e.g., intuitive thinking, visualisation, artistic capabilities, capacity for lateral thinking);
- beliefs and values (e.g., religious/spiritual beliefs, social/political beliefs, congruence or tensions between work values and values outside of work, ethical stance);
- tolerance for stress/resilience (e.g., ability to rebound when preferred actions are thwarted, constrained, blocked, or inappropriate because of circumstances; tolerance for change);
- family, social, and life history (e.g., who and what their parents/siblings/friends do/did; family support for study, development and professional activities);

- preferences, attitudes, and interests (e.g., likes and dislikes, biases/prejudices, burning questions; comfort zone preferences);
- life goals (e.g., career aspirations, family, reputation, prestige, and status);
- educational history (e.g., schools/university attended; majors and minors in university study; opportunities to pursue interests).

These often provide a backdrop against which some aspects of the 'as a researcher' identity might emerge (e.g., preferences for numbers/mathematics or language, interests in social issues/social justice, long-term career goals, past training in research methods at university). On the 'as a researcher' side, other considerations emerge such as:

- preferred research approaches and patterns of guiding assumptions (which may be shaped by university training, supervisors/mentors, disciplinary norms, and prior research experiences);
- opportunities of interest (e.g., promotion potential and pathways, working with a well-known researcher; bridging the gaps between theory and practice; achieving social justice);
- relevance/sufficiency of prior research training and skill sets/competencies;, and
- access to personal and organisational resources and technology to support their research efforts.

A critical aspect of researcher positioning, on the 'as a researcher' side, is coming to a clear understanding of, and making conscious choices about, the pattern(s) of guiding assumptions that would be most productive/sensible to adopt for their research project. The researcher should be able to clearly unpack their reasoning about adopted guiding assumptions and to convey that reasoning to target audience(s) for intended research outcomes (we will explore these issues more fully in Chap. 4).

In the interpersonal domain, researcher positioning involves considerations of the relationships they have with significant others in their personal and professional lives and how one or more of these might impact on their research journey. Potential relationships to consider might include:

- family (e.g., expectations for work-life balance, time away from home);
- friends (e.g., sounding boards for problems in general; distraction/relief from stress);
- colleagues and peers (e.g., sounding boards for research-related issues; access to professional networks; other members of a research team or collaborative research effort);
- mentors, supervisors and supervisory committees (e.g., for postgraduate students or early career researchers, the people mostly closely related to and influential on their research);
- examiners (e.g., for postgraduate students, professionals who will evaluate their thesis/dissertation/portfolio research outcome, who may or may not be personally known to them);

- reviewers and journal editors (e.g., gatekeepers who pass judgment on the worthiness of a research outcome for publication, who may or may not be personally known to the researcher);
- bosses and/or subordinates (e.g., people/organisation a researcher works for; people who work for you);
- people/gatekeepers who may control valued resources or who might have useful networks to tap into (e.g., people who work with funding organisations; gatekeepers who control access to potential data sources);
- people/groups/organisations who have an interest in and/or who stand to gain or lose by the research the researcher is doing (e.g., key stakeholders, about which more will be said below).

It is in this domain that the researcher begins to unpack just who the most influential people/organisations are on their research journey, how those influences might or do play out and how they might respond or adapt to such influences. Ownership of their research journey and the exercise of power over the shape and conduct of that journey are very important considerations in this domain. It is important to note that positioning with respect to the 'as a researcher' identity may not be fully fleshed out prior to the research journey commencing. In fact, such positioning may continue to evolve over time as research events transpire, constraints are confronted, and opportunities are taken.

In the professional domain, researcher positioning involves understanding what expectations relevant professions, governments and society hold for the researcher. Such positioning encompasses researcher obligations to behave ethically, often formalised in professional standards or codes for ethical research conduct (which may include institutionally- or nationally-mandated policies/guidelines for research conduct and may overlap ethical guidelines for the conduct of research with human participants; see discussion with respect to positioning of data sources in Sect. 3.1.2). Professional codes of conduct typically have sections that detail obligations to deal in good faith with all research participants as well as with other researchers involved in the project and with gatekeepers; deal in good faith with all data gathered, recorded, stored and analysed; obligations to avoid engaging in fraudulent research behaviour (e.g., falsifying data, plagiarising others' work, undisclosed deception of participants, undisclosed use of information about people gathered for purposes other than the research project, or the use of any such information without the consent of the persons concerned) and to avoid subjecting participants to harm, either during the research process itself or downstream as a consequence of their research participation. Note that the distinction between a professional code of conduct and ethical guidelines is often blurred. This is because the difference is really one of perspective: a code of conduct tends to focus on obligations/behaviours as a researcher; ethical guidelines tend to focus on research participants and their rights and protections (obviously, though, it is the researcher who is obliged to ensure that those rights and protections are appropriately safeguarded). Thus, where professional standards/codes of conduct and ethical guidelines overlap lies largely in the relationship between researcher and research participants.

Other aspects of the professional domain in researcher positioning encompass the relationships the researcher has with various professional organisations and networks. The term 'professional' here is intended to encompass all aspects of the researcher's larger professional identity, be it research-based, practitioner-based, service-based, academic, and/or industry-based). Professional networks may be physical, as with peers and colleagues gathered at a conference or virtual with colleagues connected via a web-based service such as *LinkedIn* (https://www.linkedin.com/), *ResearchGate* (http://researchgate.net/) or *Kudos* (https://www.growkudos.com/about). In either case, shared interests and identities are what bring a professional network together and such relationships can facilitate and support a professional's research-oriented activities and may even lead to the emergence of a research team. Professional and academic societies and academies are also sources of professional relationships and/or accreditation for a practitioner/researcher and may help support a professional identity (including, in some cases, offering a code of conduct for members). Many such organisations run annual conferences for their members where professional networks and special interest groups can be created, explored, and extended and where knowledge-sharing is explicitly and regularly encouraged. Finally, the researcher has professional relationships with their employing organisation and co-workers. Any of these relationships can potentially influence the shape of a specific research journey they undertake, perhaps through access to expertise, access to networks that can link them to desired data sources, support for data analysis, collaboration, or access to outlets for research outcomes.

The life circumstances domain of researcher positioning focuses on the intersection between the researcher 'as a person' and the research activities they are undertaking, i.e., the intersection of their life/identities 'as a researcher' and 'as a person'. A researcher's current life circumstances may be closely linked to their motivation for undertaking research. The researcher's cultural background and other aspects of their personal identity (e.g., sexual identity, ethnic identity, religious/spiritual identity) can also play a strong role in motivating research, including whether they are a member of a disadvantaged or marginalised, or oppressed group. They may be motivated to conduct research because they are pursuing a qualification or degree that requires it, because they have a professional or personal interest in or curiosity about a specific issue, problem and/or innovation, because research is a formal expectation of their job role (as it is for university academics seeking publication, market researchers) and/or because they are honouring a commitment or obligation they have made (such as through the receipt of research grant funding, a research consultancy for some organisation or industry or a research team or collaboration they are part of). This is why a researcher is always a stakeholder in their own research. Furthermore, the researcher is the gatekeeper to their own learning and their openness to learning is influenced by both their identity 'as a researcher' and identity 'as a person'. The level of experience they have in conducting research in specific contexts is also relevant. Those experiences may identify additional key stakeholders for their research (e.g., supervisors, bosses, customers, clients) and it may influence the role they undertake in the research context (e.g., educational researcher who is also a teacher; policing researcher who is also a sworn or unsworn police officer; an Indigenous/First Nations or non-Indigenous/First Nations researcher doing research in a specific Indigenous/First Nations context). Finally, the dynamics between the researcher's research identity/role and their other identities/life roles may yield important contextual considerations or sources of pressure. Current and emerging circumstances that arise from sources such as their personal history, family situation, cultural background, work situation, financial constraints, geographical constraints, social network demands inside and outside of work, and any tensions between research identity/role and their other identities/life roles can play a role in contextualising and perhaps constraining one's research journey.

If the researcher is conducting research within their own organisation, community, and/or cultural context, we say that they are an 'insider'. This implies that the researcher possesses more detailed knowledge about the research context and should be able to leverage that insider knowledge in ways that will benefit their research journey as well as identify personal relationships that will require special attention to manage through the research project (e.g., interviewing co-workers or supervisors or subordinates). If the researcher is not conducting research within their own organisation, community, and/or cultural context, we say that they are an 'outsider'. This implies that they will have a greater sense of detachment with respect to the research context but, at the same time, will have to work harder to understand that context in ways that approximate what an 'insider' would already know. There are advantages and disadvantages associated with both the 'insider' and 'outsider' researcher roles as well as potential tensions between them (e.g., see discussions in Breen, 2007; Dwyer & Buckle, 2009; Ergun & Erdemir, 2010; Hayfield & Huxley, 2015; Toy-Cronin, 2018; Smith, 2021, pp. 156–161, discusses insider–outsider tensions from the perspective of Indigenous/First Nations research).

For the 'insider' role, more intimate knowledge of the context may provide insights into who valuable data sources might be, how access to them might be negotiated, and how to contextualise data that are gathered (i.e., contextual sensitivity advantages are conferred). However, the 'insider' role may be accompanied by participant suspicions about what agenda the researcher is really running (e.g., 'is the researcher, who is also an organisational member, simply doing research for management to use against them?'). Furthermore, there are risks that the power relationships the researcher is involved in (e.g., between supervisor and subordinates) could potentially be abused or misused (if they are the supervisor) or could deny them access to certain types of knowledge (if they are a subordinate).

The 'outsider' role comes with the potential benefits of heightened objectivity and perceived neutrality (e.g., perceived to be less likely to be running someone else's agenda). However, drawbacks include lesser capacity to appropriately contextualise what the researcher has learned in the research because of the lack of insider knowledge (i.e., the researcher must work harder to achieve contextual sensitivity and insight) and tougher negotiations to pass gatekeepers for access to desired data sources. It might also mean that the research may need to be reconfigured or augmented to try and gain some relevant 'insider'-level knowledge. If the researcher is conducting their research as part of a consultancy arrangement with the management of an organisation, things get a bit more complicated in that they would still be considered an 'outsider', but not necessarily as someone who would adopt a neutral stance (which could make gaining access to desired sources of data even harder). Even in the context of experimental research in a laboratory, technically, the researcher role would be as an 'outsider' with respect to the contexts that any of the research participants normally inhabit but as an 'insider' with respect to the laboratory, experimental tasks, measurement instruments and conditions the researcher creates and manipulates in order to provide experiences for those participants. Brunswik's (1952, 1956; see also Cooksey, 2001b) concept of 'representative design' outlines a process whereby the researcher can build a meaningful bridge between the 'insider' and 'outsider' perspectives of a laboratory researcher through designing experiences that have meaning for, are relevant to, or are analogues/simulations of experiences that research participants would have outside of the laboratory in their everyday lives. This then enhances the prospects for making convincing generalisations beyond the boundaries of the laboratory experience(s).

Communicating relevant aspects of researcher positioning is often a weak point in many research outcomes, especially published research. Interestingly, different patterns of guiding assumptions take rather different stances with respect to how transparently a researcher exposes their positioning. For some patterns of guiding assumptions (e.g., interpretivist/constructivist, critical, Indigenous/First Nations and systemic othering, to be discussed in more detail in Chap. 4), transparency in researcher positioning as it impacts on the researcher's role in the research context, gaining access to and gathering data from data sources, analysing those data and forming interpretations and conclusions is critical to demonstrate. In contrast, other patterns of guiding assumptions (e.g., positivist and, under certain circumstances, critical realist) downplay the need for the researcher to expose/explore their positioning, instead depending, often uncritically, upon the presumption of objectivity to 'negate' any arguments about influences arising from researcher positioning. However, irrespective of one's adopted pattern of guiding assumptions, it is the case that absence of or thinness in critical details about researcher positioning may have an adverse impact on how convincing the research can be. For example, not making clear that the researcher is an employee of an organisation in which they are doing their research risks raising questions about conflicts between their role as a researcher and their role as an employee. Such conflicts might influence what data sources or information they gain access to, what physical or financial resources they can call upon, and could create problems with data quality arising from power differentials between the researcher and participants (especially if the researcher is in a managerial role). As another example, having a vested interest in a theory the researcher has created and tested or that they strongly believe in can lead to an imbalance in efforts to critically discount or dismiss contrary evidence while simultaneously reinforcing supporting evidence, even to the point of downplaying flaws in their own research (more effort expended) vis-à-vis modifying their theory to accommodate the contrary evidence (less effort expended). Decision researchers refer to this imbalance as 'confirmation bias' (see, for example, Kaptchuk, 2003), where people tend to focus on and seek out evidence that confirms rather than disputes one's perspective; a commonly revealed bias that does not reflect an objective stance. Active avoidance of this problem requires transparent acknowledgement of the risks associated with dogged adherence to one's own perspective. Researchers who adopt the pattern of guiding assumptions for interpretivist/constructivist inquiries refer to this active avoidance process as 'managing or bracketing researcher preconceptions'. Researchers who adopt the pattern of guiding assumptions for positivist inquires use control over experimenter bias (e.g., single- or double-blind studies, see, for example, Kaptchuk, 2001) and other procedures designed to reinforce objectivity to try to combat this problem. Greater awareness of one's own researcher positioning enhances the likelihood the researcher will actively work to avoid such biases.

Another important benefit of awareness of one's own researcher positioning is that key areas requiring further development may be identified. Such insights can add learning value to the research journey which can circumvent potential downstream problems that could emerge. For example, skill deficits in handling certain types of data or data analytic procedures/software support systems may surface and awareness of this can motivate the researcher to seek developmental training or to cultivate a source of expert advice (a human resource). Inexperience with a specific data gathering strategy may motivate the researcher to plan a trialling phase in their research where they obtain practice in implementing the strategy. Awareness of one's lack of access to certain necessary resources, including desired data sources, may motivate the researcher to seek out alternative resources or help them to make a conscious decision to modify their research goals so as to be achievable without the necessity for those resources (one of many trade-off decisions the researcher may have to make during their research journey).

3.1.2 Participants and Other Data Sources Contextual Considerations

Researcher contextualisation and positioning looks at the research journey from the point of view of the researcher, the doer of the research. In contrast, contextual positioning of potential research participants (human) and other data sources (non-human) looks at contextualisation from the viewpoint of the 'objects' of focus in the research. For human data sources, it involves identifying who is desirable to access/connect with and how (including learning which gatekeepers may control such access), understanding what aspects of their roles and backgrounds are important to know about/consider, knowing what ethical protections are required/obligations are imposed (influences coming in from larger contexts beyond the bounds of the research journey), discerning what cultural (e.g., Indigenous/First Nation; minority, multinational) considerations are relevant; understanding where participants are coming from and what they expect, what their communities and cultures expect, what protections and safeguards and quid pro quos they expect if they decide to participate, and what roles they play in relevant (to the research) contexts of their own. For non-human secondary data sources or databases, contextual positioning involves identifying what data sources are relevant and credible and how to access them (including learning which gatekeepers may control such access), learning about data management, quality and maintenance, accessing and understanding any available meta-data, learning about potential gaps and missing data issues. For documents and other media-based data sources, typically produced by humans (e.g., webpages, blogs, YouTube videos, reports, films, newspaper stories, images, social media postings, cartoons, artefacts, music), contextual positioning involves considerations about credibility, status, authorship/ownership, available versions, target audience(s), and the original purposes for which they were produced.

Positioning of Human Participants

Positioning with respect to human participants encompasses considerations associated with the fact that they have their own lives and aspects of their lives may intersect the research journey in a variety of important ways, including how they expect to be/are treated. In short, positioning of participants entails understanding where they come from, what they expect, and what their participation might mean to/for them. For example, the role that the researcher expects them to play in the research may be relatively passive (as in research conducted under the pattern of guiding assumptions for a positivist inquiry, where they are still referred to by the distancing label 'subjects' in some quarters) or active, even participatory or collaborative (as in research conducted under patterns of guiding assumptions for interpretivist/constructivist, critical realist, critical, Indigenous/First Nations or participatory inquiries). This would be something that the researcher should clearly communicate. Irrespective of the pattern of guiding assumptions adopted, understanding the positioning of participants is critical because it may influence the researcher's sampling choices, access to and quality of information gathered, how participants expect to be treated, how experiences might be manipulated, the character of interpretations and conclusions, and the extensional reach of those interpretations and conclusions beyond the research context. Positioning of participants addresses questions such as:

- *Who is participating, from where, and why are they relevant to the research?* This encompasses considerations such as cultural and ethnic background, work, and life experiences relevant to the research or lack thereof, as well as other relevant demographic characteristics. For example, under certain patterns of guiding assumptions (e.g., for positivist or critical realist inquiries), this helps the researcher to define the population of interest for sampling and subsequent generalisation inferences. Under other patterns of guiding assumptions (e.g., for interpretivist/constructivist, systemic othering, critical, Indigenous/First Nations, participatory inquiries), answering this question helps the researcher decide who they wish to connect with for data gathering and what they will or can contribute to the research.
- *What role(s) do participants currently play and what are their prior experiences/roles within the research context?* This may help the researcher to identify majority, minority, marginalised, and/or silenced voices within that context that they may wish to connect with. Different roles in the research context (e.g.,

head teacher, CEO, manager, line worker, new employee, union official, shopper or consumer, community member, and so on) may have implications for what information the researcher gains access to and can potentially influence how well-rounded or representative a story they will be able to tell. Participants who are new to a research context would likely have differing perspectives from those with longer associations and a richer set of experiences with the context, so experience does matter.

- *What are participants' relationships to the researcher?* It is possible that there is no relationship between the researcher, and participants (as is quite often the case in research guided by the pattern of guiding assumptions for a positivist inquiry). However, in naturalistic or field research, the researcher may share a personal, professional, organisational (e.g., manager/subordinate), or collegial relationship with at least some participants. In this case, there are potential power dynamics between the researcher and participants that need to be acknowledged and managed (as when researchers do research in their own organisations). This may also be relevant if the research journey results in an innovation that may benefit participants, thereby connecting the researcher to participants through the vehicle of that innovation.
- *What are the ethical expectations regarding how participants and any information/ knowledge/insights they provide are to be treated/protected/safeguarded?* This highlights the power of ethical constraints over the researcher's research activities by creating obligations and responsibilities that, in dealing with human participants, one cannot ignore. This also acknowledges that participants are themselves important stakeholders in the research (i.e., they have potential to benefit from the research whilst minimising risks and avoiding harm), by virtue of their participation. Equally important, participants are their own gatekeepers. It is important to realise that there is a different set of expectations if participants are from Indigenous/First Nations backgrounds and contexts, given the colonising histories they have lived through and continue to experience. Communities and Elders within those communities must be consulted regarding access to Indigenous/First Nations country and knowledge. The researcher must build trust and relationships with relevant communities as well as with individual community members and Elders before access may be granted and, importantly, the impacts of western modes of thinking associated with colonisation and research practices must be simultaneously recognised and actively countered. As well, the researcher typically undertakes an obligation of reciprocity to share their learning/provide benefit back into the community. Research is done with, not on, Indigenous/ First Nations people. Seeking participants in Indigenous/First Nations research is thus a much more relational process which establishes a partnership between the researcher and the people/community with/within which the research will be carried out (see, for example, discussions in Chilisa, 2020; Kovach, 2021; Smith, 2021).
- *Who controls access to participants and what do they expect in return for granting access?* Here, the role of gatekeepers that the researcher must negotiate with to gain access to participants emerges as an important consideration. It also implies

that those gatekeepers are key stakeholders in the research. In certain contexts (e.g., primary and secondary schools, hospitals and other care facilities, or other public sector organisations, for example), there may be several layers of gatekeepers that the researcher must work their way through before gaining access to desired participants (e.g., state government department, school management, teachers, and parents in the case of research involving students in public schools as research participants). Even then, at the last stage, it is important to emphasise that modern ethical principles in the social and behavioural sciences demand that every participant be their own gatekeeper so that they give their informed consent to participate and understand that they have the right to withdraw their participation at any time. Thus, even if all other gatekeeper hurdles are passed, participants have the final say with respect to whether they will participate and, even if they do agree to participate, retain the right to end their participation at any time. Participation and retention cannot be coerced.

It is important to understand that the term 'participants' as we are using it here refers both to potential participants (those the researcher wishes to approach and connect with) as well as to participants who have already given their permission to participate or are already involved in participating. The issue of ethical expectations with respect to research participants is so important that while we provide a helicopter overview here, it is worth a more detailed separate discussion (see Cooksey & McDonald, 2019, Chap. 15; also check out Bell, Bryman & Harley, 2022a, 2022b, Chap. 6, Cohen, Manion, & Morrison, 2018, Chap. 7, and Mauthner et al., 2012). The Australian Institute of Aboriginal and Torres Strait Islander Studies offers a very detailed discussion of an ethical code of conduct for Indigenous research in Australia, see https://aiatsis.gov.au/research/ethical-research/code-ethics; the Canadian Government offers a similar discussion with respect to Indigenous, Inuit, and Métis peoples, see https://ethics.gc.ca/eng/tcps2-eptc2_2018_chapter9-chapitre9.html. These expectations relate closely to codes of conduct for researchers but focus on the participant's rights. Ethical expectations evolve over time are often codified in ethical guidelines or codes produced by government bodies, professional associations, funding bodies, or special interest groups. The following constitute basic rights for any research participant.

- *Participants have the right to have their informed consent sought and obtained for their research participation.* We flagged this right earlier but unpack it more fully here. Potential participants must be fully informed about what they can expect to happen to them during their participation, what is expected of them, how long their participation could last and how involved it could be, what potential benefits they might realise, what risks they might be exposed to, and what each of their rights are as participants. Participants would generally be asked to sign a statement affirming that they understand this information when they agree to participate. Any promises made by the researcher must be honoured. Research based on observable recordings of individuals obtained for other purposes (such as using camera recordings of shoppers in a store, done to deter shoplifting, to study consumer behaviour in context) or covert observations (as when a researcher goes undercover to study a social group like a gang or a union) is ethically problematic because those shoppers/gang/union members have not given their informed consent to participate in that research. For certain types of research (psychological or sociological research, for example, conducted especially under the pattern of guiding assumptions for a positivist inquiry), the right to informed consent may conflict with the need for the researcher not to signal what they are looking for (because that knowledge risks changing the very behaviours they are studying, something psychologists call demand characteristics). The participant's right to informed consent supersedes the researcher's right to control what participants know about the research. The researcher needs to plan additional steps to offset the possibility that the knowledge conveyed by obtaining informed consent influences the data being gathered (an example of a research journey trade-off being imposed on the researcher by societal concerns).

- *Participants have the right to expect that their privacy and identity will be protected by the researcher and that any information they might provide will be appropriately safeguarded from access by others.* This means that participants have the right to privacy with respect to all aspects of their participation. Their identity must not be linked in any way to any information they might provide (anonymity). Assurances of confidentiality may not always be sufficient, because it begs the question 'confidential to whom?'. The researcher is generally obliged to keep any information provided by participants in a secured location, usually for a specified period of time. If the researcher is conducting longitudinal research over time and needs to track data gathered on separate occasions from each participant over that time period, this needs to be signalled during the informed consent process and each participant's identity and contact details (for tracking purposes) must be kept separate from their actual data records through the use of an anonymising identifier, such as an identification number or pseudonym.
- *Participants have the right to be protected from exposure to unnecessary risks, harm and/or deception, but if such risks are incurred the researcher has the obligation to rectify any adverse consequences.* Participants should expect not to be deceived (i.e., not be lied to) with respect to what the researcher is actually interested in and what will happen to them during their participation. Participants should also expect not to be placed at risk of harm or embarrassment (including not being involved in any circumstances that might cause them to behave uncharacteristically or illegally, i.e., entrapment) nor should they be harmed in any way (physically, socially, or psychologically) during the course of their participation. For certain types of research where deception is absolutely necessary in order to study the phenomena of interest (this includes the researcher saying they are investigating one kind of behaviour when they are really studying some other kind of behaviour in psychological experiments and using placebo (i.e., fake) drugs/treatments in medical drug trials or therapeutic/education/training research), the researcher must take all reasonable steps to ameliorate any damage done/disadvantages created via the deception. Use of deception generally means that true informed consent is difficult to obtain as many participants would not agree to be deceived. Researchers conducting medical drug trials or tests of other

kinds of interventions often get around this constraint by signalling to potential participants that there is a chance they may receive a placebo treatment as part of their participation and that, if they do, they will be given the opportunity to undergo effective treatment after the trial or research is concluded.
- *Participants have the right withdraw their participation in the research at any time without consequence.* Once a participant has given their informed consent to participate, this in no way commits them to go the distance with their participation. They retain to right to withdraw at any time for any reason. What would concern researchers most about this is where their withdrawal is due to something that happens to the participant during the course of the research (recall the previous right regarding exposure to unnecessary risks and harm). The right to withdraw can create an unfortunate but unavoidable feedback effect on the research in that withdrawal of a participant means loss of access to any data they might provide (as well as negating the use of any data they have already provided), which effectively reduces or alters the sample.

Participants' ethical rights may conflict with the researcher's goals and intentions and, if such conflict eventuates, those ethical rights supersede the researcher's right to good data. Ethical dilemmas surface all the time in social and behavioural research and these dilemmas need to be resolved within the localised context of the research. In many cases, ethical guidelines are adopted by research organisations (including universities and funding bodies, such as the Australian Research Council, the National Science Foundation in the U.S., and the Social Sciences & Humanities Research Council in Canada) and compliance is often overseen by some type of ethics committee or governing body. This means that an outside group has the power to influence what the researcher can and cannot do in research involving human participants, often requiring that the proposed research be fully vetted and approved by the overseeing body before the research can commence. Virtually all ethics committees for universities work in this fashion. Such outside research vetting bodies, by virtue of their power, become de facto stakeholders in the research and may actually be held accountable if breaches in research ethics occur within their purview.

Positioning of Indigenous/First Nations Participants

In several countries, including Australia, New Zealand, Canada and the U.S., there are distinct ethical principles and guidelines for research involving Indigenous/First Nation peoples. Such guidelines exist because of their colonised histories, differences in worldviews, beliefs about the nature of knowledge and relationships between the physical, mental, and spiritual worlds, as well as the role of culture and tradition as it bears on learning, knowledge, and experiential truth relative to non-Indigenous cultures (see, for example, the excellent discussions in Chilisa, 2020; Chap. 4; Kovach, 2021; Chap. 4 and pp. 215–218). Critical to these codes of ethics is the fostering of relationships and building of trust. Some codes (e.g., codes adopted by the Australian and Canadian Governments) generally modify and/or augment codes of ethics that apply to non-Indigenous people. For example, the NHMRC document, entitled *Ethical Conduct in Research with Aboriginal and Torres Strait Islander*

Peoples and Communities: Guidelines for Researchers and Stakeholders, built their guidelines around six distinct values: Spirit and Integrity; Cultural Continuity, Equity, Reciprocity; Respect; and Responsibility (NHMRC, 2018, p. 3). These values are continually being reaffirmed through time from the past to the future and Spirit and Integrity forms the integrating central value weaving together the remaining five values. The code shows how each value might be demonstrated in/reflected by research and each is linked to specific ethical requirements for researchers from the National Statement on Ethical Conduct in Human Research 2007 (updated 2018) (NHMRC, ARC & UA, 2018).

In contrast, some codes of ethics for research on Indigenous/First Nations peoples are created by Indigenous/First Nations people themselves (e.g., Aboriginal and Torres Strait Islanders in Australia, Maori in New Zealand, Inuit in Canada) and these codes take a distinctly different approach, seeking instead to construct a specific 'decolonised' ethical framework for research involving Indigenous/First Nations people within that country and establish Indigenous/First Nations people as a collective and self-determining stakeholder in research. For example, the Guidelines for Ethical Research in Australian Indigenous Studies, produced by the Australian Institute of Aboriginal and Torres Strait Islander Studies (AISTSIS), details the following four central principles for ethical research involving Aboriginal and Torres Strait Islander people:

- Principle 1—***Indigenous self-determination***. "The recognition of, and respect for, Aboriginal and Torres Strait Islander peoples' right to self-determination is fundamental to all research conducted in Australia" (AIATSIS, 2020, p. 12). This encompasses the key responsibilities of *recognition and respect* for Indigenous people, their experiences and worldviews, their traditions, and their knowledge and innovations; meaningful *engagement and collaboration* to build trust and ensure the interests and well-being of Indigenous people are protected; ensuring that free and *informed consent* is obtained and maintained at **both** the community and individual level throughout the duration of the research; and ensuring that those who do research with Indigenous people/communities demonstrate sufficient *cultural capability and learning* to ensure that relationships and ethical protections can be respectfully and transparently maintained (AIATSIS, 2020, pp. 13–17).
- Principle 2—***Indigenous leadership***. "To demonstrate merit, Aboriginal and Torres Strait Islander research should be led by Indigenous people" (AIATSIS, 2020, p. 17) This encompasses the key responsibilities of *Indigenous led research* where Indigenous people have collaborative input into decisions about the conceptualisation and implementation of research; incorporating and appropriately acknowledging *Indigenous perspectives and participation* throughout the research process; and appropriate negotiation and control over how *Indigenous knowledge and data* are treated, accessed, and maintained (AIATSIS, 2020, pp. 18–19).
- Principle 3—***Impact and value***. "A shared agreement about the benefit, impact and value of research is of particular significance in Aboriginal and Torres Strait

Islander research" (AIATSIS, 2020, p. 20). This encompasses the key responsibilities of ensuring that appropriate *benefit and reciprocity* as well as avoidance of disadvantage are achieved in research outcomes; and ensuring transparency in understanding any *impact and risk* associated with involvement in the research project (AIATSIS, 2020, pp. 20–21).
- Principle 4—***Sustainability and accountability***. "Respect for Aboriginal and Torres Strait Islander peoples and their worldviews in the conduct of Aboriginal and Torres Strait Islander research requires researchers and institutions to have accountability, over the long term, for the impacts of their actions" (AIATSIS, 2020, pp. 21). This encompasses the key responsibilities of support, respect, and protection for *Indigenous lands and waters* and any knowledge and innovations developed with respect to Country; maintenance of *ongoing Indigenous governance* with respect to storage of, access to, and metadata regarding Indigenous research data; and ensuring culturally appropriate and transparent *reporting and compliance* with respect to research outcomes and the ethical standards that apply to achieving such outcomes, including any changes over the life of the research project (AIATSIS, 2020, p. 22–23).

The ethical dimension of positioning of human participants is probably the most challenging aspect for researchers to grapple with. It raises dilemmas that need to be resolved, creates tensions between what the researcher would ideally like to do and what is ethically appropriate to do and reinforces the need to plan, modify, and extend or augment research activities so that ethical expectations can be met. This means that participants are always the most immediate stakeholders in the research. Doing research involving Indigenous/First Nations people magnifies the complexity and sensitivity of the issues the researcher must deal with; issues created by the historical actions and consequences of colonisation of Indigenous/First Nations peoples by non-Indigenous/non-First Nations people and its on-going ripple effects. The complexities multiply even more if the researcher is a non-Indigenous/non-First Nations person seeking to do research with Indigenous/First Nations people or in an Indigenous/First Nations community because of the extra burdens associated with the building of trust and decolonising of worldviews (e.g., truth-telling about history). It is the ethical dimension that truly brings home the point that social or behavioural researchers do not have power over every aspect of their research journey; some of their power must be deliberately and willingly sacrificed to meet the demands of larger more systemic concerns.

Positioning of Non-human Data Sources

The positioning of non-human data sources looks somewhat different but is no less important to address. If the researcher wishes to access or draw upon a non-human data source for their research, they need to understand:

- the status of the data source (e.g., non-fiction or fiction, fact-based or opinion; public-domain or private);
- whether the data source is formal (e.g., aligned with some institution or group of interest) or informal (e.g., aligned with a personal point of view);

- whether the data source has a regulatory or policy focus (e.g., a secondary database used for the management of a public policy or to meet regulatory requirements);
- whether the data source is sacred and/or of high or symbolic cultural value/ associated with a specific ritual or object (e.g., associated with cultural and community (including Indigenous/First Nations peoples) traditions);
- the credibility of the data source with respect to the currency, quality, and character of its contents;
- the relevance of the data source to the research context;
- who created the data source and why;
- who owns the data source;
- who maintains/error checks the information embedded in the data source and how frequently?
- who controls access to the data source, why and how (points to relevant gate-keepers) and does obtaining such access create ethical obligations for the researcher;
- when and where the data source was created, in what context, and for what purpose(s);
- what the intended audience is/was for the data source;
- what purposes the contents of the data source may be used for (such purposes could create ethical dilemmas for the researcher);
- the availability/accessibility/composition of any metadata associated with the data source (such metadata may hold answers to many of the above questions); and
- the nature of the contents of the data source (e.g., first-hand reflection/recording of events, experiences, images, creative ideas, recordings of performances, thoughts/observations; second-hand reflection/recording of events, experiences, images, creative ideas or thoughts/observations, measurements, financial, demographic, ecological, or economic data associated with other people/organisations/ communities).

Some non-human data sources, such as secondary organisational, institutional, economic or financial databases, policy documents, academic research literature databases, and social media postings, have dynamic content, which means it is important to understand who maintains the data source, how frequently content is updated, whether or not that content has been filtered or vetted in some way and by whom, and how the quality/accuracy/currency/consistency of content is monitored and ensured. Without a clear understanding of the positioning of non-human data sources, coupled with a clear reflection of that understanding in any research outcome, it is much more difficult to be convincing about how relevant or meaningful any content from that data source is for the researcher's intended purposes. As an added complexity, if the researcher wants to access certain institutional or organisational data sources (e.g., internal reports, strategy documents, long-range forecasts) that have been flagged as commercial-in-confidence, this places ethical constraints upon their use of the information provided by or within those data sources. The researcher may, in fact, either be denied access to those documents, even if they would be valuable for their research, or constrained as to how they use such information and whether they can summarise or otherwise represent the information in any research outcome they produce. Gatekeepers may require the researcher to sign a non-disclosure agreement, which will basically mean that they cannot use the information provided by the data source in any way, even as notes in their research journal. In short, such data sources may simply not be feasible to include in the research and the researcher would have to be transparent about any gaps in their knowledge or learning that this constraint might create for them.

3.1.3 Prior Research/Knowledge Building Contextual Considerations

A critically important aspect of research contextualisation concerns the positioning of the research with respect to prior knowledge building research work done by others in other times, places, and contexts (including prior work by the researcher themselves). The context of prior knowledge building, in whatever form the research outcomes take and by whatever medium the outcomes are being shared/communicated, is an essential one for the researcher to continually grapple with throughout their journey. It is the one major context that is especially and often uniquely relevant to a researcher's life. Prior knowledge building contextualisation in the social and behavioural sciences is typically, but not wholly, associated with the systematic reviewing of relevant research literature. If done effectively, a systematic and critical literature review can help the researcher achieve other levels of contextualisation such as formulating research questions/propositions/hypotheses, identifying and offering insights and caveats with respect to relevant/potential research contexts and stakeholders, clarifying the researcher's own contextualisation, and identifying potential outlets for research outcomes. Note that we said 'systematic and critical literature review'; 'systematic' because the review should be structured and oriented toward helping the researcher to build the case/ rationale for their own research journey and 'critical' because the knowledge building work done by others should not just be descriptively summarised, but also critically evaluated, which helps to determine how convincing the work is and how influential that work should be in/on the researcher's current journey. However, and quite importantly, pluralist systemic thinking demands that the researcher, where appropriate, consider other forms of prior knowledge building efforts to tap into besides what appears in written literature as well as knowledge building outcomes that express alternative, minority, or silenced perspectives (e.g., moving beyond discipline boundaries and research guided only by certain patterns of assumptions; considering stories, performances, art, and other creative modes of expression). Pluralist systemic thinking should also motivate the researcher to consider an appropriate balance to achieve between their choices (e.g., published literature, grey literature, trade/professional literature, unpublished work, media-based, anecdotal, stories from cultures with oral traditions of knowledge sharing (such as Indigenous/First Nations cultures), performances, dances, art and other creative works as well as creative forms of texts). In short, focusing solely on written literature may not provide the level of contextual/conceptual depth the researcher needs to convincingly position their current knowledge building effort.

Positioning with Prior Research/Knowledge Building

Positioning one's research with respect to the prior knowledge building efforts and outcomes produced by others in/for other times, places, and contexts, involves trying to establish where the researcher's current research journey fits into the landscape of what has come before. Such positioning might involve, amongst other things, building or testing a theory, creating knowledge to fill a perceived gap, doing something different so as to overcome limitations they see in previous knowledge building work, seeing how previous knowledge applies in a new context, working to provide evidence for practical decision making or change, or trying out a new methodological approach or new forms of knowledge building/representing activities. Positioning with prior knowledge building requires the researcher to logically argue for how their current research fits into, extends, or alters that knowledge building landscape. Positioning arguments with respect to prior knowledge building efforts might adopt:

- a *discovery* stance—seeking to create or learn something new and insightful (e.g., new theory, test new propositions, combining prior knowledge building efforts in new/insightful ways);
- a *cumulative* stance—adding incremental value to or taking the next step in contributing to an existing body of knowledge;
- a *critical* stance—critiquing, correcting, redirecting, re-interpreting, or circumventing previous ambiguities, abuses, or mis-applications of prior knowledge;
- an *application* stance—showing how prior knowledge building efforts may translate into useful actions/applications;
- an *innovation* stance—seeking to translate prior knowledge building efforts into a concrete innovation and/or create new knowledge regarding the value of a pre-existing concrete innovation;
- an *evaluative* stance—seeking to use prior knowledge building efforts to inform a current effort to produce judgments of worth or value regarding a policy, intervention, social program, or other type of development/change process;
- a *greenfield* stance—pushing out into new frontiers of knowledge building that have received minimal or no attention in the past; and/or
- a *replication/extension* stance—testing prior knowledge building efforts in new contexts to see if existing knowledge translates or has meaning for those new contexts.

Positioning with prior knowledge building efforts has a bi-directional dynamic relationship with choices of patterns of guiding assumptions, modes of knowledge building, and research frames such that positioning may influence choice of pattern, mode, and frame and choice of pattern, mode, or frame may influence positioning. Carrying out and presenting a coherent, well-organised and critically executed review of prior knowledge building efforts and outcomes forms the essential foundation for positioning one's current research/knowledge building effort and can serve several useful purposes:

- providing the researcher with contextual insights into what other researchers/knowledge producers have done and why, including some insight into their various contextualisations, positionings, patterns of guiding assumptions, and research frames;
- helping the researcher to achieve new learning insights into essential concepts, theories, and patterns of relationships;
- providing the researcher with methodological insights into how other knowledge builders accomplished their learning goals;
- providing the researcher with practical/professional insights showing how learning might have meaning or be useful beyond specific research contexts; and
- helping the researcher to develop new insights into future directions for knowledge building efforts by suggesting new pathways to follow or pathways to avoid.

Prior knowledge building efforts and outcomes may be encountered in a variety of forms by the researcher and, depending upon the researcher's orientation and needs, those forms will vary in terms of credibility and convincingness for supporting arguments for their current research journey. For example, academic researchers will generally find peer-reviewed published journal articles to be most convincing and useful, followed by research books; media stories and anecdotes, stories, and yarnings will generally be considered least convincing and credible. In contrast, cross-cultural and Indigenous/First Nations researchers may find culturally-based stories and yarnings to be more credible, convincing, and useful than academic peer-reviewed published journal articles. Researchers developing or evaluating an innovation may find trade/professional/practice-oriented publications and particular kinds of grey literature to be as or even more useful and convincing, relative to peer-reviewed published journal articles. In short, there is no single credibility ranking available for forms of prior knowledge building outcomes; what counts as credible and convincing depends upon the researcher's purposes and intended outcomes for their own knowledge building effort. It is fair to say that a major influence on openness to alternative forms of knowledge building outcomes are the pattern(s) of guiding assumptions, mode of knowledge building, and research frame(s) adopted by the researcher (to be explored in more depth in Chap. 4). Those patterns of guiding assumptions that purport to adopt a more 'objectivist' stance with respect to research (e.g., positivist inquiry pattern; possibly the critical realist pattern) will tend to favour peer-reviewed research outcomes as the preferred source of prior knowledge building outcomes; whereas those patterns that purport to adopt a more 'subjectivist' stance will tend to be more open to alternative forms of prior knowledge building outcomes (e.g., for interpretivist/constructivist, critical, Indigenous/First Nations, participatory, postmodern inquiries). As well, research that adopts an exploratory, case study, descriptive, othering/emancipatory, cross-cultural, Indigenous, developmental evaluation, pan-disciplinary, or action research frame will tend to be more open to alternative forms of prior knowledge building outcomes, relative to research that adopts an explanatory (coupled with the pattern of guiding assumptions for a positivist inquiry), evaluation, or survey frame.

- *Peer-reviewed published journal articles.* Peer-reviewed research published in academic journals constitute primary sources of research information that have been formally vetted by expert research peers and judged for their quality and contribution value. However, even within this category, there is a rough hierarchy of credibility, where some journals have a reputation of publishing what is perceived to be very high-quality research (the so-called A* and A journals in many journal ranking systems) whereas others have poorer reputations. Journal ranking systems attempt to rank published research outlets with respect to some combination of citation rates, impact factors and expert perceptions. However, it is important to realise that these ranking systems are not infallible and that different systems may provide different rankings of the same journal. Online and open access journals as well as journals where authors pay a submission and publication fee to have their research published, have been generally perceived as publishing research of lesser quality, although this perception is evolving as publishing systems and the vetting of submitted content become more sophisticated and transparent. An emerging problem in the world of online research publishing is the growth in the number of so-called predatory publishers and journals, where making money without regard to the quality of what is published dominates (see, e.g., https://www.enago.com/academy/avoid-paying-predatory-publishers-journals-for-publication/). Researchers should also be aware that even articles published in peer reviewed journals may not be all that convincing as publication practices are very diverse as well as being highly selective about what is considered publishable in the context of the journal. Researchers need to be aware that many journals prefer to publish, or have a record of publishing, research predominantly produced under specific patterns of guiding assumptions (e.g., specific management, marketing, psychological, sociological, medical/nursing, and educational journals). This trend may reflect editorial policies and preferences as well as larger discipline preferences/expectations. Other journals encourage a more diverse range of submissions, and this trend is slowly spreading.
- *Research books.* A research book may constitute a primary source of information if it focuses on synthesising research done by a specific researcher or team of researchers (see, for example, Beach, 1998; Gigerenzer, Todd and the ABC Research Group, 1999) or a secondary source of information if it brings together research from a range of researchers, perhaps even a range of disciplines (see, for example, Cooksey, 1996; Hammond & Stewart, 2001; Hattie, 2009). Research books will have typically been professionally vetted as a holistic concept in terms of the value they would add to the field. Edited research books are generally produced by well-respected researchers in a particular discipline and often have their individual chapter contributions from different authors specifically invited and peer-reviewed (e.g., Allen, Maguire & McKelvey, 2011; Schuelka & Maxwell, 2016; Sinclair, 2014; Tashakkori & Teddlie, 2010). An edited research book may also be a systematically assembled compendium of previously published

research in an area (e.g., Lichtenstein & Slovic, 2006). The value of research books is that they may help the researcher to organise their thinking about specific research areas and may also provide theoretical, methodological, and practical guidance. Certain publishers (for example, Sage, Oxford University Press, Routledge, Cambridge University Press, John Wiley & Sons, Edward Elgar) routinely engage internationally-known researchers in a specific field to assemble a handbook in a specific research area and the word 'handbook' will generally feature in the book title. One drawback for many research books is that they may not reflect pluralist systemic thinking in their choice of content, thereby providing only a selective view of some field of research endeavour.

- *Grey literature.* "Grey literature can include government reports, committee reports, academic papers, theses, bibliographies, conference papers and abstracts, discussion papers, newsletters, PowerPoint presentations, conference proceedings, program evaluation reports, standards/best practice documents, technical specifications and standards, and working papers" (Alberani et al., 1990, cited in Benzies et al., 2006). However, they are much less likely to have been peer-reviewed for quality and contribution. Some of these, such as working papers, academic conference papers, and Powerpoint presentations, might be interim research outcomes from an on-going research project at a university or research centre. Postgraduate theses, dissertations, or portfolios will have typically been vetted by examiners or a supervising committee, but this vetting is from the point of view of judging the convincingness of the outcome with respect to achieving some credential, such as a PhD or professional doctorate, rather than suitability for publication. Other types of grey literature documents and reports may include: company annual reports; statistical/financial/ economic/educational databases; secondary database information produced by organisations such as the Australian Bureau of Statistics; periodic and final reports to granting/funding bodies; reports submitted to meet the expectations of government regulations (such as equity and access reports for the education sector in Australia); and position and consultation papers (e.g., commissioned by a government body, organisation, or institution). Note that grey literature outcomes may be produced with political, social or other goals (e.g., meeting educational requirements, providing data for other researchers, providing policy advice, enhancing public image, providing information for potential investors) in mind and the researcher must be careful to identify and understand what these other agendas might be and how they might have influenced the content and quality of the outcome as part of their critical evaluation. Grey literature should not be ignored as it can provide contextual as well as theoretical and/or methodological input into research arguments. However, the researcher must take what they say/argue for in context as they will vary greatly in credibility and convincingness.
- *Academic textbooks.* Academic textbooks constitute secondary sources of research information, reporting on and interpreting research from primary sources, primarily for teaching purposes. These secondary sources may look attractive in terms of bringing a lot of research literature together in a straightforward manner.

However, such textbooks organise their understanding of selected research literature in specific ways to enhance their pedagogical goals and those goals may not intersect a research quality or contribution argument or reflect pluralist systemic thinking.

- *Trade/profession/practice-oriented publications and materials.* Trade/profession/practice-oriented publications and materials also constitute secondary sources of research information, often mixing information from primary sources with anecdotal information, and contributions are typically not peer-reviewed. In some cases, the authors attempt to build a useful, potentially marketable, conceptual framework or set of tools/training materials that can be used by managers or other professionals to enhance professional development, produce change, or achieve some other valued result (for example, Hammond, 1996; Senge et al., 1994). In other cases, trade and professional publications may report on the state of play with respect to some issue of concern to the trade or profession, perhaps with the goal of highlighting an important dilemma or policy issue (see, for example, magazines like *Human Capital*, the *Professional Educator*, *Professional Nursing Today*, *Psychology Today*, *The Economist*). These types of publications focus on industry/professional practices, often with the intent of influencing those practices. Such literature can be quite useful ways of learning about communities of practice, thus providing the researcher with practice-focused insights.
- *Media reports and internet materials.* Media reports and internet materials constitute a more informal outlet for research knowledge, with content typically produced by non-researchers. Such outcomes may be research-focused but are not peer-reviewed and may represent third-party perspectives on the research (meaning they are perhaps best viewed as anecdotal stories about research). However, within this category of literature, these sources can themselves vary from more credible, such as Wikipedia and some newspaper/magazine/media stories, academically-oriented multimedia feeds (such as Ted talks: see https://www.ted.com/) and online newsfeeds) to less credible (such as blogs, Facebook/Twitter/X postings, popular magazine articles, YouTube feeds, and manifestos), which are considered to be more in the domain of popular literature or as expressions of public opinion. Such sources of literature may help the researcher to contextualise/position some aspects of their research but are generally far too anecdotal and opinionated to be considered as providing credible research evidence.
- *Anecdotes, stories, and yarns.* Anecdotes, stories, and yarns especially where culturally-based, provide insights into worldviews and perspectives that academic researchers may dismiss or ignore. Indigenous/First Nations researchers would place great importance on such stories or yarns as exercises in truth-telling within their culture, to counter the colonising stories that persist outside their culture. Stories may be written or, in many cultures, oral in form. Anecdotal stories provide rich contextualising information and may reflect culturally-privileged knowledge and learning. Yarns emerge from an Indigenous approach to honest, open, and inclusive conversation called 'yarning' (e.g., see Geia et al., 2013; Kovach, 2021,

pp. 162–164) and provide a way to engage with Indigenous knowledge and traditions. Western researchers have been taught to devalue such sources of information as second-hand or hearsay, but such a perspective cannot easily be decoupled from the cultural practices of an oppressing or colonising group or nation. The value accorded to anecdotes, stories, and yarns would need to be transparently argued for and contextualised by the researcher; pluralist systemic thinking would demand that clear reasoning be applied to judgments of value rather than leaping to immediate rejection of the stories. At the very least, anecdotes, stories, and yarns may provide the creative impetus for an evolving, perhaps greenfield, research project, providing not research data per se, but signs, pointers, and contextualising substance for future directions for research.

When studying and critiquing prior knowledge building outcomes, the researcher could aim to learn from a number of different angles and each angle may help them to establish where their research is situated with respect to what has been done before.

- *Content and context*. The researcher could understand what the prior knowledge builders did and judge whether they did it well (convincingness judgments). What intentions, contexts, and positionings were evidenced in their article? Who were they speaking to in their communications? What were they trying to learn and why?
- *Theories*. The researcher could examine any theoretical positions and propositions that the prior knowledge builders explored and perhaps served as the focus for their knowledge building efforts.
- *Methodologies*. The researcher could learn how the prior knowledge builders obtained the evidence they used to draw their conclusions and set out implications. This should include understanding the research frame and pattern(s) of guiding assumptions that the knowledge producers adopted/reflected. The researcher should try to glean what things seemed to work well and what things did not. The researcher look for practices/strategies they could take on board that might assist in their own research approach.
- *Applications/practices/innovations*. The researcher could learn how prior knowledge builders developed, implemented, or learned anything new and novel, relative to what had been done before. The researcher could look for things such as signalling or evaluating practical applications of what prior researchers learned, showing how what others had previously developed or learned worked in a new context, or signalling potential for any innovation stemming from their work.
- *Leverage value*. The researcher could decide how best to leverage what has been claimed, shown, or learned from the prior knowledge builders to contextualise and position their own research. Here, the researcher could look for things like convergences and/or divergences between what the prior knowledge builders have learned and what others have learned, gaps or unanswered questions that remain after reading what the prior knowledge builders have produced, blind alleys, limitations or obstacles the prior knowledge builders encountered during their journey and any directions for further research signalled by the prior knowledge builders.

- *Culturally-privileged knowledge.* The researcher could decide what can be learned from culturally-privileged knowledge that could be brought to bear on their current research intentions. Here, the researcher looks to broaden their perspective by incorporating/building upon knowledge from non-western sources. Gaining insights into such knowledge will require culturally-appropriate access and be accompanied by culturally-appropriate safeguards and acknowledgment of knowledge sources.
- *Springboard potential.* The researcher can also use what they review as a springboard to discovering other relevant instances of prior knowledge development. Here, the researcher can look at how the prior knowledge builders themselves have positioned their own work with the prior knowledge building of others and search their citations and references for possible new sources of prior knowledge building to investigate. Finding one or more meta-analyses (for quantitative research) or meta-syntheses (for qualitative research) relevant to their research area can also provide springboard potential for a researcher's review of prior knowledge building efforts, particularly those outcomes in written form. One useful benefit of a meta-analysis or meta-synthesis is that its goal is to critically assess patterns of findings or emergent themes/theory in a sample of prior knowledge building outcomes.

Doing a review of prior knowledge building efforts for research purposes can be characterised using the metaphor of a funnel. The search is generally very wide ranging when the researcher starts (at the top of the funnel), where they are just scratching the surface of what is important to examine, and where it is easy to get distracted/diverted by tangential ideas. As the researcher progresses, the search/review becomes much narrower and more focused where they dig deeper into specific aspects of the most relevant prior knowledge building efforts that provide specific foundations for their research (at the bottom of the funnel). The key is for the researcher to appropriately manage the funnelling process in order to add value, logic and coherence to their arguments, always with the goal being to identify and work through prior knowledge building efforts that are most relevant for developing the positioning logic for their own research. The researcher should not prematurely focus their review so tightly that they neglect to consider possible learning and approaches taken to their subject matter/issue from other disciplines and perspectives. Employing pluralist systemic thinking to engage an eclectic multidisciplinary/multi-paradigm approach to research can help to offset the tendency for premature focusing. The researcher should use their research journal to record all summaries, thoughts, critiques, quotes, and ideas that emerge as they examine prior knowledge building efforts and outcomes.

3.1.4 Research Contextual Considerations

When planning and shaping a research journey, the researcher must confront the potential choices they can make regarding the site(s)/context(s) where they will conduct their research. Whether their intended journey takes place in a laboratory setting, a naturalistic field setting/site, or a virtual/simulation setting, every research context choice each has accompanying contextualisation considerations. Research context considerations share dynamic relationships with researcher contextualisation (giving rise to choices regarding patterns of guiding assumptions, research frames, and research questions/ propositions/hypotheses), contextualisation of participants and other data sources (that may emerge from or be embedded within research contexts), and contextualisation associated with prior knowledge building efforts (implications of research context(s) associated with prior knowledge building efforts for the current effort). Thus, the proper positioning of research depends, in part, upon effective contextualisation of the focal research context(s). A range of contextual issues are associated with choices of research context(s), including issues associated with selection, controllability, environment, researcher connection to/relationships with, suitability, artificiality, accessibility, available technology, facilities for data gathering, and emergence of research questions/propositions/hypotheses.

Various contextualisation strategies and positionings may influence the choice of and considerations associated with specific research context(s) and choice of specific research context(s) may shape various positionings in a kind of 'which-comes-first, chicken-or-egg' dynamic. This dynamic continues throughout the duration of the research journey and cannot be fully anticipated and planned for prior to research commencement, because constraints, surprises, and opportunities emerge over time. The best the researcher can hope to do is to try to anticipate and plan for as much as possible, while acknowledging the uncertainties associated with those activities and preparing themselves to be flexible and adaptable, if unforeseen circumstances arise.

- *Research goals, objectives, questions/propositions/hypotheses.* The researcher's goals, objectives, questions/propositions/hypothesis typically emerge from the confluence of contextualisation considerations between researcher positioning and intended or actual research context(s), all of which will be tied to the researcher's general conceptualisation of the research problem. The researcher must form a clear idea of what they want/intend to learn and in what context(s) and, associated with that, what objectives they want to achieve, what questions they want to address, what propositions they want to explore or test, and/or hypotheses they want to test against the background of their adopted pattern(s) of guiding assumptions and choice of research frame. This will then facilitate the researcher's downstream decision making regarding research configuration, data gathering and analysis strategies, and the creation of research outcomes.
- *Positioning with respect to patterns of guiding assumptions and frames.* The researcher's choices with respect to pattern(s) of guiding assumptions and research

frames emerge from the dynamic interplay between the researcher's contextualisation and research context(s) contextualisation. Some patterns and frames may signal the need for a controllable research laboratory or perhaps a virtual/simulated environment, whereas others may signal the need to work out in a field setting, where control is not possible or desirable. Furthermore, this positioning with respect to assumptions and frames will influence the nature, intent, and even the wording of research questions/propositions/hypotheses. Much more will be said about this in Chap. 4.

- *People.* This involves considerations of (1) who and/or what is relevant to connect with in the research context (which can help the researcher to identify not only potential participants, but also potential stakeholders and gatekeepers, as well as non-human data sources, such as contextualising documents); (2) when do people become relevant, including relevant historical details about relevant people in the context; (3) where are relevant people located, which may encompass cultural considerations as well; and (4) why are specific people relevant to/suitable for participation in the research. For example, if the research context is a naturalistic setting, such as a farm, classroom, emergency room, council meeting, or office, then addressing who, when, where and why issues may lead to identification of key people to connect with for gathering data, to negotiate with for gaining access to people, and/or to ensure that any learning that emerges from the research benefits rather than harms others not directly involved in that research. Even if the research context is a laboratory setting, people-oriented considerations still emerge. Here, undergraduate or postgraduate students may constitute the potential participants to enlist, which can mean they have a vested interest in gaining course credit or remuneration in exchange for their participation, may have minimal experience in any tasks the researcher may require them to carry out, and may constrain the extent to which the researcher might wish to extend the learning found in their research beyond the boundaries of the laboratory.
- *Prior relationships.* If there are pre-existing relationships between the researcher and people in the research context (as when the researcher does their research within their employing organisation or in one they worked in/for previously; or a context where the researcher has contacts/friends/prior relationships; or where the researcher has a contractual/paid arrangement with management to conduct research, as an outsider consultant), the potential implications of those relationships for the research must be explored and managed. Prior relationships can also influence participant positioning, especially if people in those relationships are gatekeepers controlling access to other people or to documents or if the researcher wishes to involve people they have a known relationship with as participants (e.g., when a manager/researcher interviews one of their own subordinates).
- *Selection.* If research is to be done in naturalistic settings, it begs the question 'how does the researcher choose which settings/sites to focus on for their research?' Research sites can be sampled in the same way as participants may be sampled. If the researcher chooses a research site with which they have a prior relationship of some sort, this creates a cascading series of considerations about how they should manage their relationships in the research context in such a way as to minimally

impact on data access and quality. If the researcher randomly selects research sites that they have no prior relationship with, then they need to determine how much additional contextual information they need to have about those sites in order to engage in effective research activities.

- *Controllability*. This involves considerations about the controllability/malleability of the specific research context, i.e., the choice between doing the research in a laboratory or in the field under naturalistic circumstances. This consideration depends heavily upon the researcher's positioning with respect to adopted pattern of guiding assumptions. Laboratory-based, highly controlled research is most closely associated with the pattern of guiding assumptions for a positivist inquiry, whereas naturalistic field research is more closely associated with patterns of guiding assumptions for interpretivist/ constructivist or other non-positivist inquiries. However, this is not a one-to-one relationship, as positivist research (even experimentation, via the Manipulated experience-focused data gathering strategy) can be done in the field in naturalistic settings, yielding what is referred to as a quasi-experiment. If research is done in a laboratory under controlled conditions, the design of experimental laboratory tasks to be undertaken by or conditions to be experienced by participant becomes an important consideration, i.e., the representative design of tasks and conditions (Brunswik, 1952, 1956). If this consideration is ignored, the researcher's capacity to generalise beyond the laboratory context is greatly constrained. For example, if participants in a laboratory experiment are asked to do a task they have never done in their lives before or to participate under environmental conditions they have not experienced before, then generalisations based on those tasks and conditions are much harder to make convincing. Brunswik (1956) suggested that this problem could be easily solved by focusing as much effort on sampling the life spaces of potential participants to design tasks and conditions as on sampling of people/participants from some larger population (the latter being a statistical concern in terms of achieving a representative sample).
- *Facilities*. The research context may or may not offer the physical space (e.g., a room for conducting interviews, an office for the researcher to work in), financial or in-kind support (e.g., for research activities such as photocopying, support for travel expenses, or for additional research personnel), technological support (e.g., internet access for the researcher, space on the organisation's website for questionnaire distribution, access to email for soliciting/securing participants; permission to digitally, photographically or video record aspects of the research context), and/or personnel support (e.g., persons to liaise with and organise meetings with gatekeepers, schedule participants for data gathering sessions). Where certain necessary facilities are not available, the researcher will need to adapt their research activities so that the deficit has minimal impact.
- *Environment*. It is becoming more and more important these days for researchers to have to be concerned with the environmental implications of what they do in their research for the context(s) in which the research is conducted. It would not be appropriate for the researcher to conduct research that inadvertently damages the environments of the sites in which they do research (e.g., through creation of

waste, wastage of energy, environmental damage created by travel). Equally, the researcher must be mindful of not creating or conducting their research within an environmental setting that is hostile, in some way, to potential participants (e.g., exposing participants to pollutants or allergens, excessive environmental conditions, such as crowding, heat, humidity and noise, or to risk of physical harm).

- *Accessibility and suitability*. Accessibility to research context(s) can be examined from two perspectives. One perspective considers accessibility from the researcher's point of view raising questions about whether and how they might gain access to the context and who the relevant gatekeepers might be for negotiating such access. A suitable (as in fit for the researcher's purposes) site for the researcher's purposes may not be an accessible site (or vice-versa). In either case, the net effect is that the researcher's choice of research context is constrained. The other perspective is that of the participants raising questions about whether the research site is accessible by participants who might have mobility restrictions or potential psychological/physiological reactivity. The researcher would need to make choices to enhance the prospects that such people could participate in the research if selected. Failure to do so could result in inadvertent sampling constraints or worse, risk exposing the researcher to claims of ethical violation of participant's right to freedom from harm. For participants, accessibility and suitability are connected considerations, one cannot be achieved without the other.
- *Stakeholder/gatekeeper expectations*. Here, considerations revolve around localised emerging constraints within the research context itself and expectations that arise from key stakeholders or gatekeepers in the context. For example, gatekeepers within an organisation may require things from the researcher in return for granting research access, such as regular updates, a report to the organisation, the right to view/vet any report before public release (which can create an ethical dilemma for the researcher if their research findings are unfavourable to the organisation), recommendations for change, insights into who said what during data gathering (creates another ethical dilemma for the researcher, vis-à-vis prioritising protection of privacy/identity of participants over what the gatekeepers in the research context are demanding). An emergent constraint from a research context could be something as drastic as the organisation withdrawing its permission for the researcher's access, once their research has commenced or the organisation going into bankruptcy during their research journey (which means the sampling strategy has to change) or something less drastic (but still requiring some adaptation on the researcher's part), such as when a school states that students in the same classroom should all be treated in the same way, meaning that students who are in the same class cannot be assigned to different experimental conditions, even if that is what the researcher wishes to do.
- *Contextual layers*. Laboratory, virtual/simulated, and naturalistic research contexts are all embedded within layers of contextual settings (e.g., a behavioural laboratory may be owned and resourced by a university; a simulated operating theatre may be jointly owned by a medical school and a large metropolitan hospital); an organisation may be part of a larger public sector government context;

a local council is embedded with the larger community that they serve; a group being studied may be part of a larger department; organisations may have to abide by regulatory/governance requirements; organisations have competitors, markets/ consumers/ clients that they can be influenced by/answer to/need to deal with). Consequently, there may be constraints and opportunities that emerge from these contextual layers in which the research context(s) are embedded. For example, an organisation or institution being researched may be currently experiencing or has very recently experienced a major historical event such as increased pressure from new competitors in the market, imposition of new regulatory or legal requirements, being taken over or absorbed by another company, or suffering a reduction or scoring an increase or in government funding, and such an event may alter conditions of access or expectations of the researcher and may even fundamentally change the context so that it is no longer suitable for the researcher's purposes. Adaptation will almost always be imperative in such circumstances, not least putting in place data gathering strategies that will provide the researcher with insights into these events and their impact, so that they can use that knowledge to qualify any research findings and perhaps redirect or reshape some data gathering activities. A facility in which the researcher wishes to conduct their research, such as a laboratory or meeting room, if it is owned by a larger organisation, may be subject to scheduling (to manage competing pressures from other quarters) and resource-based (e.g., fee/costs for use; limitations on time of use) constraints that the researcher will have to manage. Even if the researcher is conducting research only using data gathered from secondary economic or financial databases (i.e., where no human participants are involved at all), there will be considerations to manage such as knowing who owns the databases and controls access to them so that access can be negotiated or paid for. Those database owners may have their own expectations about what can/should be done with the data they collect, curate, and store and how that source of data should be referenced/cited in any research outcome will need to be managed.

3.1.5 Research Users/Sponsors/Communities Contextual Considerations

Some of the trickier aspects of contextualisation involve considerations associated with potential research users/sponsors/communities, who comprise the audiences for any research outcome. This level of contextualisation requires the researcher to anticipate who might or will be interested in/influenced by/impacted by the outcomes produced during their research journey and then to judge whether those considerations should influence what/how/where they do their research and how those considerations should influence the shaping of any research outcomes. In effect, this involves thinking about the downstream implications of their research then working backwards to see how their research might be shaped so as to mitigate or amplify those implications. Academics who read research publications, practitioners who use research to justify changes in practices, people who buy products or adopt innovations informed by research, communities/workplaces/institutions and organisations who are affected by research that informs policy changes, funders who grant large sums of money to support research endeavours all share an important common interest: forming or holding expectations about what the researcher's research and the outcomes it leads to means for them. Such expectations may not manifest themselves until after a research outcome is shared/communicated, at which time, it will be too late for the researcher to change what they do to better meet those downstream expectations. In some cases, prior knowledge building efforts and consequences that may have flowed from them may expose some user expectations that could be helpful to the researcher planning and undertaking their own journey. In other cases, it is a matter of making a series of educated guesses about what the potential impacts of their research might be and working backward from those anticipations to alter what they do. Anticipating the expectations and needs of research users/sponsors/communities may lead to the identification of a research problem and/or may lead to more nuanced and contextually-sensitive ways of navigating the 'Data Triangle'.

In order to unpack this level of contextualisation effectively, the researcher needs to identify their target audience(s) (i.e., downstream stakeholders) for any research outcomes, anticipate their potential expectations, fears, misgivings, and needs, as relevant, and understand why they are important for the researcher to consider. Researchers need to think about immediate as well as downstream stakeholders, which means differentiating between the needs and expectations of stakeholders who stand to be directly and immediately impacted by the research (i.e., the researcher, participants in the research, other non-participating members from a research site, immediate judges of the quality of research outcomes such as postgraduate supervisors, doctoral examiners/committees, and journal reviewers/editors) from those of more remote and downstream stakeholders (e.g., consumers, members of the general public, innovation adopters, policy decision makers). Immediate stakeholder interests should generally be of higher priority to the researcher and the shaping of their research journey, except and unless there are likely to be strong ethical, societal, cultural, or environmental implications or reactivity to what the researcher has done or has learned. In this latter case, offsetting those anticipated downstream concerns may become imperative for the researcher to deal with if a convincing and societally acceptable research outcome is to be achieved. Enhancing research benefit and minimising potential downstream harm or unintended side-effects have evolved into over-arching general expectations for social and behavioural research.

3.1.6 Larger Contextual Considerations

Larger contextual considerations focus on the larger systems in which the researcher and their research are embedded or connected to, and which may exert influences on the research journey. Such larger contextual concerns may lead to the identification of, or create the imperative for, addressing a research problem, which the researcher then takes on board as the focus for their research journey. Larger contextual forces often serve to constrain what the researcher does and with whom. For example, laws, policies, and regulations must be obeyed by the researcher, the research must be culturally-sensitive, the research must adhere to a professional code of ethics, societal interests in the research must be acknowledged, organisational/institutional/industry, community and market interests should be safeguarded, where relevant, the physical environment must be protected during the conduct of the research, and the research should be sensitive/responsive to relevant historical events and precedents. In some cases, larger contextual considerations may reveal resources that the researcher may be able to tap into to support their research or may reveal resource constraints that will limit what they do. Some larger contextual concerns may exert their influence (even overt pressure) directly or indirectly on the researcher themselves (e.g., university supervisors influence the research of their students; government or social policy constrains university policies which then constrain researchers (e.g., in the realm of ethics, for example); the researcher may belong to a professional organisation which may create expectations for how they do research and even what research they do). Other larger contextual concerns may exert their influence on the systems that researchers must access or navigate in order to get their research done (e.g., cultural expectations such as those held by Indigenous/First Nation communities, Elders, and leaders; funding resources; institutional/organisational policies; legal requirements).

It will not be possible for the researcher to fully anticipate or consider the myriad of potential larger contextual influences and some influences may emerge later in the research journey, which then demands researcher adaptation and flexibility. The best a researcher can hope for is to identify the most crucial larger contextual influences they need to deal with and focus their planning and adaptations on those. In some instances, focusing on dealing with one major source of larger contextual influence may have a beneficial cascading effect in automatically dealing with other related considerations. Equally though, dealing with one major source of larger contextual influence may create the need to address subsequent emergent issues. It is important not to forget that the researcher, participants and other data sources, users/sponsors/communities as well as the research context(s) themselves are all embedded within, connected to, and potentially influenced by these larger contextual systems. Social and behavioural research truly does not occur in a vacuum.

3.2 Managing Stakeholders and Gatekeepers

Every research journey has potential stakeholders who stand to gain or lose depending upon how that journey plays out and what outcomes are generated and for whom. Figure 3.2 presents a cognitive mapping of the issues and considerations associated with managing stakeholders and gatekeepers. A stakeholder may be an individual, a group, a government agency, an institution or organisation, a community, or even some segment of society at large. Key or active stakeholders can be distinguished from ordinary or passive stakeholders by virtue of the direct (stakeholder to researcher) or indirect (stakeholder to researcher via a third party) power they can potentially exert on the researcher to influence the nature and shaping of their research journey. It is the key stakeholders in their research journey that the researcher needs to identify and manage. This does not mean that researchers should ignore the needs/expectations of passive stakeholders, rather it points to the need for the researcher to prioritise their stakeholders as they do not have the resources or capacity to meet the expectations and needs of all. In many cases, immediate stakeholders, who are directly involved with the researcher's journey, are key stakeholders and downstream stakeholders are generally passive (at least until research outcomes have been achieved, at which time, reactivity might be displayed) but may be implicated as a consequence of the researcher's journey.

Gatekeepers are a specific sub-category of stakeholder that have the power to control access to something the researcher will require in order to achieve their research goals, such as contextual knowledge potential research context(s)/sites, facilities, technologies, communities, potential participants, resources and funding, documents and other information, and outlets for research outcomes, and/or, for certain kinds of research, culturally-privileged knowledge. Gatekeepers will always be key stakeholders in a researcher's journey and the researcher may need to devote significant time and effort to passing them (for example, see discussions in Crowhurst & Kennedy-Macfoy, 2013). Gatekeepers may be passed (i.e., desired access granted) using a variety of strategies (e.g., offering payment or other recompense, giving assurances, persuasion, entering into a quid pro quid agreement, offering reciprocity to bring value back to specific constituencies of interest to the gatekeeper), but always upholding key principles (honour ethical obligations and any promises made, respect cultural traditions, maintain research quality/convincingness, and dealing honourably with everyone implicated in the research, i.e., no deception or inappropriate/illegal/unethical actions/behaviours).

When dealing with key stakeholders and gatekeepers, the researcher should build trust and establish a productive relationship for information exchange and enhanced likelihood of gaining access by:

- adopting the attitude that they are genuinely interested in learning from, and, where appropriate, perhaps even collaborating with, the stakeholder/gatekeeper;
- managing their communications in a respectful and transparent manner so as to minimise potential misunderstandings;
- negotiating in good faith so that any agreements reached are honoured with good humour;
- ensuring they are seen to make ethical decisions and undertake actions to safeguard what they learn, who they learn from, and who they learn about;
- ensuring that when/if they respond to demands or expectations from the stakeholder/gatekeeper, they do so in a transparent and respectful way;
- appropriately acknowledging/resolving issues associated with any intellectual property concerns the research may raise;
- inviting stakeholders/gatekeepers to participate actively in and contribute to the research, where appropriate to do so (see, for example, Crowhurst, 2013);

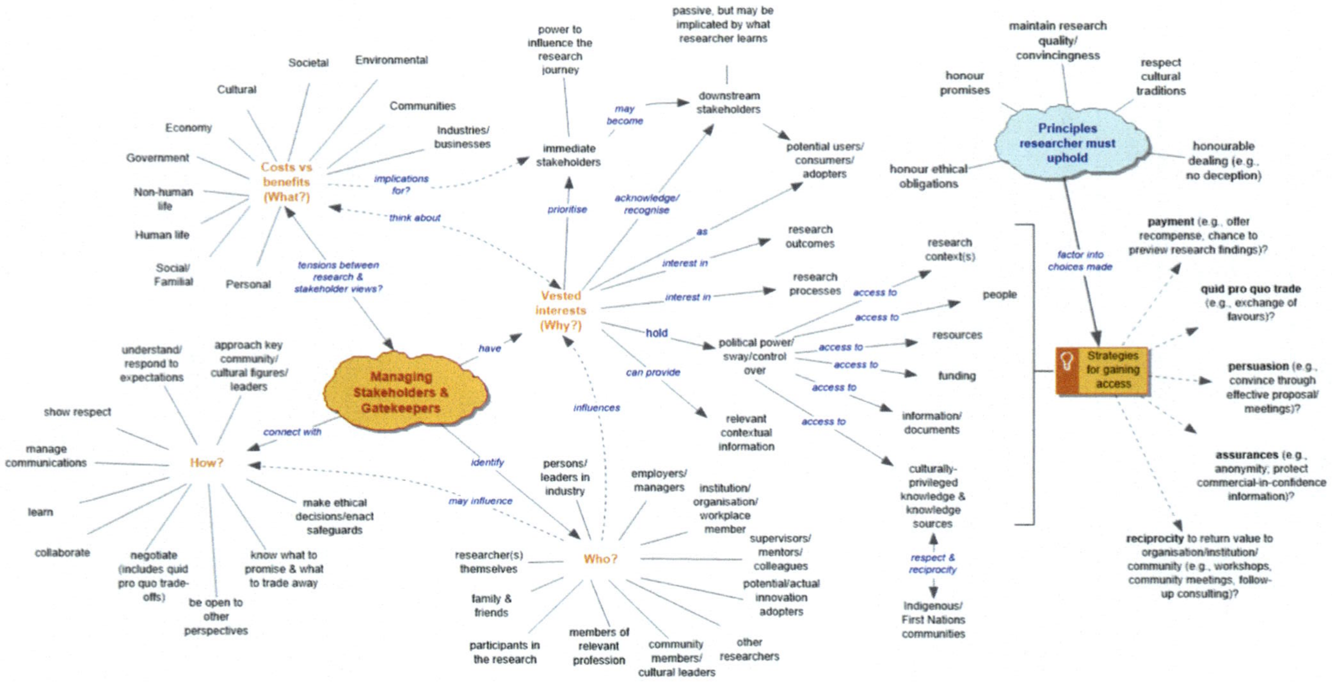

Fig. 3.2 Contextual considerations associated with managing stakeholders and gatekeepers

- ensuring that, where appropriate, they undertake proper cultural protocols and maintain respect for traditions when they are building relationships with cultural (e.g., Indigenous/First Nations Elders/leaders and communities) groups; and
- ensuring that when they are commencing the development of relationships with stakeholders/gatekeepers, they know what they are willing to promise and what they will be unwilling to promise or trade away.

Research where the researcher has an insider role or an outsider role creates different stakeholder/gatekeeper relationships, implications, and power dynamics. A researcher who has an insider role (either currently or formerly) has insider knowledge, may be suspected of trying to further a personal agenda, such as advancement within the company or gaining extra perks, is obliged to obey company policies, is expected to respect the organisation's hierarchy (which may limit access to certain types of information), and may create worries about misuse of power with employees. A researcher who has an outsider role may come with suspicions of an ulterior motive or agenda (e.g., there are there as an outside agent management is using to make changes), can work outside the organisational hierarchy, may be seen as more independent and unbiased, and could be seen as an untainted advocate for marginalised groups within the organisation. In an insider role, the researcher must carefully navigate the tensions between what their research is demanding and what their employer or former employer is expecting. Such navigation can be much more difficult if the researcher is in a supervisor-subordinate relationship with at least some members of the organisation and some of those members are to be participants. The insider researcher must work hard to achieve a balance in their perspectives, minimise conflicts between their roles, and ensure that attributions of personal motives behind their research are dispelled. In an outsider role, the researcher must grapple with learning new jargon, understanding organisational relationships and history, and forming new relationships within the organisation that will be supportive of the research effort. They must devote time and energy to learning as much about the organisation as possible so as to approximate insider knowledge, ensure that any negative perceptions about ulterior motives (about them being in the organisation to do 'research') can be dispelled, and ensure that any findings are appropriately contextualised with respect to the organisation.

3.2.1 Political Nature of the Stakeholder/Gatekeeper Role

Politics comes into play in research in the relationships between the researcher and key stakeholders and gatekeepers with respect to specific research contexts the researcher is interested in. Politics is about the exercise of power to achieve specific goals and is a dynamic that emerges when contextualising and positioning research. From the researcher's perspective, they have certain goals which require that they connect or negotiate with key stakeholders and gatekeepers, so their goals can be achieved. This means the researcher tries to benignly exercise power over key stakeholders and gatekeepers to gain access to specific research contexts, to resources within those contexts, to participants and other data sources, as well as to background information about those research contexts. Such power, from the researcher perspective, needs to be of a persuasive rather than coercive nature, oriented toward achieving their desired ends without creating bad blood, negative impressions, or antipathy in the process. The political process usually involves negotiations or exchanges of some description. For example:

- promising ethical protection for participants in exchange for their agreement to participate and hopefully to stay the course with their participation;
- promising a tangible outcome, such as payment or other incentive (like a chance to win a holiday), to participants in exchange for their agreement to participate and hopefully to stay the course with their participation;
- promising a report on the research results or a set of recommendations for action to key gatekeepers in exchange for access to a research context and perhaps to potential participants and other data sources; this could also be in exchange for allowing the researcher to publish research outcomes;
- highlighting the potential benefits might be realised for potential participants (such as improved health and well-being, better working conditions) in exchange for their agreement to participate and hopefully to stay the course with their participation;
- highlighting the potential generalised benefits that might be realised (such as an economic, public, or social/human resource benefit for a company or community like improved educational, health, legal, or environmental circumstances or outcomes) in exchange for providing information about or allowing access to a research context and perhaps to potential participants and other data sources; or
- providing a well-developed research proposal suggesting positive outcomes and benefits from a specific project to a funding organisation or another relevant organisation or institution in exchange for a research grant, institutional funding, or a consultancy contract.

From the perspectives of stakeholders and gatekeepers, they have organisational/institutional interests that they wish to safeguard as well as their own goals to achieve. However, power leveraged by some stakeholders and gatekeepers may be more coercive than persuasive in nature. For example, gatekeeper power may be leveraged in terms of conditions to be met before the researcher is permitted to pursue their goals, and, in some instances, this may create ethical dilemmas for the researcher (see, for example, McAreavey & Das, 2013). Stakeholder power may be leveraged in terms of concrete expectations of the researcher in the conduct of their research or dissemination of its findings and such expectations may be anchored in the knowledge that the stakeholder will be providing resources (financial, physical, technological) to support the research or is safeguarding larger contextual interests (such as national security, respect for community, or a way of life).

- Gatekeepers who control access to research contexts and/or potential participants or other data sources may demand a report on the research findings or, more seriously, access to who said what to the researcher during their data gathering activities (thereby creating a severe ethical dilemma if the researcher wishes to promise anonymity or confidentiality to potential participants).

- An individual, organisation, or institution may require the researcher to focus their research on a specific innovation or on a specific problem or certain aspects of a problem as a condition of providing access, typically though a consultative arrangement (which may have financial implications as well that could exert additional leverage on the researcher).
- A funding organisation (e.g., Australian Research Council, National Science Foundation) may demand that the researcher provide it with periodic progress reports as well as a final report on their research project as a condition of providing and/or continuing funding and, in addition, may expect proper public acknowledgement of their support in any research outcome.
- An individual, organisation, or institution may expect to dictate or control the dissemination and distribution of research benefits and outcomes, especially where they have provided resources to support the research and, in certain cases, this could lead to difficulties in allocating and managing intellectual property rights.
- An individual, organisation, or institution may expect enhancement of their reputation or public stance, a political/lobbying advantage, or a marketing advantage for an innovation as an outcome of the research, creating pressure on the researcher to deliver such outcomes.
- A supervisor or advisor on a postgraduate research project may demand, as an expectation of their provision of supervisory input, that they be listed as an author (in some cases and, in certain disciplines, as first author) on any research publications that flow from that research project.
- An individual, organisation, or institution may demand that any commercial-in-confidence information uncovered during the research be suppressed in any research outcomes, so that their reputation, market, or political advantage and/or customer relationships are not threatened, and additionally, may demand that their individual or institutional identity and/or participants' identities be hidden or anonymised.
- An individual, organisation, or institution may request or demand changes to the research process or to the content of disseminated research outcomes (so-called 'whitewashing') in exchange for providing access or funding support (e.g., drug trials funded by pharmaceutical companies may be expected to favour a new drug the company wishes to market; cancer research funded by the tobacco industry may be expected to be conducted in such a way as to favour a tobacco-industry perspective; the federal government may put pressure on the researcher to soften or alter their research conclusions/recommendations in contentious areas related to national policy or security (e.g., climate change, trade, terrorism)).
- Indigenous/First Nations Elders may expect reciprocity or return of value to the community from the research as a condition of granting the researcher access to

that community, to community members, and/or to culturally-privileged knowledge, often with the additional expectation that their cultural traditions must be respected.

- A journal editor may ask the researcher to make specific changes to their research manuscript to meet concerns expressed by reviewers as well as the editor in order to increase the chances that the manuscript will be accepted for publication by that journal.
- A university's human ethics committee may demand that the researcher change their plan for sampling participants so that no participant is asked to provide any names of or contact details for other potential participants (i.e., snowball sampling; recall discussion in Chap. 2).

The researcher cannot and should not ignore or neglect dealing with stakeholders and gatekeepers. Instead, they must actively plan for how to deal with those influences that can be anticipated and adapt where their influences emerge downstream. Some demands made by key stakeholders or gatekeepers may place the researcher in an ethical bind that they will need to resolve. It is important to recognise that such dilemmas can create emotional as well as cognitive stress for the researcher as they work through whether they will comply with the demand being made. It is here where ethical as well as personal codes of conduct and beliefs may conflict, and the researcher will be forced to choose a pathway forward. Acceding to an ethically problematic demand may create internal stress and regret for the researcher (as well as public condemnation and potential legal penalties), even if they achieve proper professional recognition for their research outcomes. Not acceding to an ethically problematic demand may mean that the researcher does not get to do the research they wanted to conduct, and this could have downstream effects on the researcher's career or future research prospects. Certainly, not all demands made by stakeholders and gatekeepers create ethical dilemmas or coercive pressure and many requests may actually be made in a collaborative spirit. Many demands can easily be managed through effective negotiation (e.g., considering what the researcher is willing to exchange or promise in order to gain access; considering how to best express research outcomes to meet specific stakeholder needs), but the researcher needs to recognise that this adds to the timeline for the research and it may affect various facets of the research journey, maybe even the nature and convincingness of the research itself.

3.2.2 Key/Immediate Stakeholders

There is a diverse range of key and immediate stakeholders in social and behavioural research, each with their own views as to the advantages/benefits and drawbacks/risks associated with the researcher's journey within in their spheres of interest.

- *Researcher*. The researcher is always an immediate and key stakeholder in their own research journey. They have a keen interest in conducting convincing research that produces valued outcomes that may help them to achieve personal goals (such as achievement, success, prestige, social status, doing something worthwhile in their life) as well as professional goals (such as publication, career advancement, institutional or professional recognition, a degree, access to further research funding, professional connections/enhancement of their professional network, pleasing a mentor or supervisor). The researcher is also the gatekeeper to their own learning, which is why making pluralist systemic thinking part of their researcher identity is so useful. When the research journey is a team effort, each member is their own stakeholder and gatekeeper and there are relational issues between each other to consider as well, such as ensuring they all share the same research goals, understanding what each member brings to the table in terms of personal and profession abilities and expertise, and understanding the relationships each member has with other team members, research context(s), participants, and other stakeholders.

- *Researcher's family & friends.* The researcher's family and friends may become immediate and key stakeholders if the researcher's involvement with their research journey detracts too much from time and attention to them. This can be especially problematic for early career researchers and researchers-in-training (e.g., postgraduate students) because they have less experience managing their research-life balance. Family can be an especially potent source of influence on a research journey in terms of lengths and depth of involvement in the research, scheduling of research activities, travelling to/from research sites, and the like. For the researcher, tensions between family commitments and research commitments can create powerful dilemmas for the researcher when planning, configuring, and implementing their research activities and may stimulate trade-offs that better meet family needs but detract from research convincingness. Friends/social circles, especially if they include work colleagues, can create tensions between research commitments and commitments/expectations for involvement in social activities and networking. If friends are also work colleagues, care must be taken not to allow research-related commitments to affect other work-related interactions and relationships. The gatekeeper role that family and friends may play in a research journey focuses mainly on access to happiness, contentment, and successful relationship, things which certain choices, if made during the research journey, could threaten.
- *University/Research Institution.* Universities and research institutions are stakeholders for the academic and professional researchers they employ and these agencies may express their stakeholder views and political power over the researcher's journey through a variety of gatekeeping means: ethics committee review, resource availability and distributions to support research, control over facilities access and infrastructure (e.g., laboratories, video conferencing and meeting rooms, technology, computing facilities, and software (access and currency)), employment contracts (including tenure), expectations regarding contribution to institutional reputation and to relationships that the organisation has with other institutions and organisations, work allocation policies that affect the balance between teaching/non-research and research time, and/or career progression opportunities. It can be difficult for the researcher to disentangle their personal and career needs from

the necessity to meet institutional/employer policy requirements and work expectations. The employment relationship may create tensions for the researcher that could adversely trade-off research convincingness for research utility or vice-versa (e.g., pressure to publish in A* ranked academic journals, reliance on citation rates for promotion, expectations to bring in research grant money). A lot of the gatekeeper influence from universities and research institutions is indirect, where power and influence are wielded through the organisation's chain of command and accountability for the researcher's research-related activities starts with the researcher but percolates upward and outward within and, potentially, beyond the organisation.

- *Participants.* Participants in the research are always immediate and key stakeholders, not least because of the emphasis and demands that ethical guidelines and codes place on safeguarding their personal, psychological, physical, social, cultural, and workplace well-being. They are also gatekeepers for their own individual participation as well as for their continuation of participation. No matter how many gatekeeper levels there are above the individual, in the end, the participant has the right to control their own participation. That right cannot generally be abrogated or usurped by a higher-level gatekeeper; that is, participation in research cannot be coerced. However, under certain circumstances, participation in research can be incentivised (e.g., by offering recompense, payment, or some other reward, by offering course credit, by promising personal benefit will/could be gained from the research) and, for certain types of institutional-level research (e.g., to ensure the organisation is complying with government policies and regulations, to inform organisational and strategic change initiatives), participation may be expected or mandated. Note that some immediate stakeholders may become research participants if the researcher decides that the contextual information they can provide is central to the progression of their research journey.
- *Environment.* The physical environment may become an immediate and key stakeholder if the research journey requires or utilises resources in ways that are contrary to safeguarding environmental quality and ecological relationships. Research activities that levy heavy energy, fuel use, or physical resource demands, produce excessive resource wastage, and/or add to adverse impacts on climate and ecological integrity (e.g., threats to wildlife, plant life, genetic content) are becoming more difficult to justify. The environment is always an indirect key stakeholder, where power is exerted on its behalf by actors with the authority to direct actions on its behalf.
- *Communities/cultural groups.* For some kinds of research (e.g., Indigenous/First Nations or cross-cultural research, local government research), the community or cultural (where cultural may be taken to encompass religious orientation, ethnic background, minorities, other special interests) group from which participants are or will be drawn, may become an immediate and key stakeholder in the researcher's journey. This is especially the case for Indigenous/First Nations research, where there is an expectation that community/cultural Elders/leaders will be consulted before any research is done in or on their country. In this case, the community/cultural/tribal Elders/leaders serve as gatekeepers for access to

country as well as access to any culturally-privileged knowledge. This does not remove the rights of individual community or cultural group members to control their own participation, it instead adds an extra layer of gatekeeping and an extra layer of accountability to the research process. For research in public areas of various geographic sizes (e.g., village, town, city, county, shire) and demographic compositions as well as under various types of local and governmental control, the governing bodies for those areas may become an immediate and key stakeholder, especially if the research could enhance or threaten public policies, public works, community integrity and vitality, and the like. Finally, the community or cultural group of which the researcher is a member may become an immediate and key stakeholder if the researcher's journey risks creating adverse impacts back into their own community.

- *Supporting organisations/institutions.* The organisations/institutions/industries that commission, fund, pay, or provide in-kind support (including access to their members for research purposes) for the researcher to conduct their research become immediate and kay stakeholders in that research. They will have an obvious vested interest in ensuring that their money/time is well-spent and that the goals they had in mind when they gave their support to the researcher are met. Their vested interest may even extend to expectations as to the research outcomes they would like to see. Such organisations/institutions/ industries can wield a great deal of political power over the research journey by virtue of their gatekeeping role controlling access to supporting funds or other resources. This type of stakeholder can create the most difficult relationship for the researcher to manage, not least because some of the demands they may make in return for their support could create personal or ethical dilemmas for the researcher. As well, the political pressure that such organisations could bring to bear on the researcher may not always be felt up front. For certain kinds of research (e.g., funded by pharmaceutical, chemical, fossil fuel, tobacco and other types of industries), there may be a risk that political pressure on the research journey may not be felt until the time that research outcomes are produced and the content that the researcher wishes to incorporate runs contrary to what the industry expected or hoped to see. If this happens, the gatekeeping role for the supporting organisation transforms from a positive emphasis, namely granting access to research-supporting resources, to a negative emphasis, namely avoiding reputational damage and legal/compensation costs to the industry traded-off against damage to the researcher's reputation and increased risks to future support from the same industry.
- *Mentors/advisors/supervisors.* Mentors/advisors/supervisors represent a deeper stakeholder role than just friends or colleagues because of their potentially significant influence on the development of the researcher's identity and upon at least some of the research journeys that the researcher undertakes. They are more likely to be immediate and key stakeholders for early career and postgraduate researchers-in-training. Mentors/advisors/supervisors may have university/ institution affiliations, perhaps at the same university/institution as the researcher and their influence can encompass contextual, implementational, and outcome production input into a researcher's journey, provision of access to facilities and

data sources and/or resources, as well as drawing upon their greater level of experience in both conducting research and reviewing prior knowledge building efforts. Their relationship with the researcher will generally involve a power differential, especially if the relationship is between advisor/supervisor and postgraduate researcher where the postgraduate researcher is expected to heed and respond to advice and instruction from the advisor/supervisor (similar to the relationship between a professional and a novice or apprentice). Thus, an advisor/supervisor has significant journey shaping and constraining power. A mentoring relationship may start out with power favouring the mentor but may evolve to a point where power is evenly distributed or even inverted in favour of the mentee. Of course, advisors/supervisors may evolve into a mentoring role for the researcher. Generally speaking, mentors/advisors/supervisors become immediate and key stakeholders largely because of the researcher's likely psychological/social need to please or at least not disappoint them.
- *Specific stakeholders.* There may be one or more specific stakeholders who become immediate and key by virtue of the kind/depth of contextual information they can provide to the researcher, especially if there is no other way the researcher can obtain such information. The vested interests of such stakeholders might be utilitarian in nature, expecting a quid pro quo in exchange for the information they provide, or protective in nature, ensuring that the information they provide is not mis-interpreted or mis-used or that the fact that they provided the information does not come back to bite them (which may be a risk if the stakeholder is an insider to an organisation the researcher wants access to and providing the information negatively paints them as an informant). In either case, such stakeholders are gatekeepers of the information, perhaps even documents, they can provide. Where non-human data sources are accessed, the authors/maintainers of the documents become immediate and key stakeholders to consider as well as any people, groups and/or organisations to which such documents belong.

When gauging how best to manage immediate and key stakeholder relationships, the researcher needs to assess the potential benefits and costs that their research journey imposes or will impose upon those stakeholders as well as the potential benefits and costs that the stakeholders might create for the researcher. Here it is clear that there is a bi-directional benefit–cost dynamic: one direction coming from the stakeholder's perspective and the other from the researcher's perspective. This dynamic may create strong opposing forces for the researcher to manage since what stakeholders may want from the research may not accord with what the researcher feels they can or should provide. Relying on key stakeholders/gatekeepers to help contextualise research or provide access is a two-step process: first the researcher must identify and engage with those stakeholders in order to build trust and ensure clear communication, then they can move on to learning things that are important to the contextualisation of their research or working to gain access to something the researcher wants or needs that they control as gatekeeper.

3.2.3 Downstream Stakeholders

While downstream stakeholders are typically less influential in the researcher's journey, the implications of the research journey and its outcomes for such stakeholders should not be ignored. In some cases, where strong reactivity or benefits might be anticipated, the researcher may make certain adaptations to their approach to address/offset/mitigate such reactivity or to crystalise the benefits. In such cases, it is not the stakeholder themselves exerting the power, but the researcher indirectly giving that stakeholder the power to influence what they do by anticipating what their feedback might be then responding to that anticipation. Generally speaking, actual (as opposed to anticipated) feedback loops between downstream stakeholders and the researcher have much longer delays than for immediate and key stakeholders and the feedback may be direct (targeting the researcher) or indirect (targeting what the researcher has learned or produced). Downstream stakeholders who might gain benefit or suffer adverse effects arising from the research could include:

- consumers/users/readers/adopters who need to be convinced about the meaning/value of the research either personally or on behalf on some constituency;
- the general public whose behaviours might be influenced by what the researcher has learned or not learned;
- communities whose practices or facilities might be improved or threatened;
- cultural groups whose practices/traditions might be improved or threatened, whose privileged knowledge might be used beneficially or mis-used, and/or whose beliefs/history/truth may be threatened or given wider public recognition;
- the environment, which may suffer longer-term damage or benefit from the research;
- participants in the research with respect to longer-term after-effects stemming from their participation, or who, having been members of a placebo control group, become benefactors of the researcher's promise to share a beneficial treatment or intervention with them upon conclusion of the research;
- organisations/institutions that support or supported the research, who may have expected certain outcomes to be realised, may have wanted certain things to be said (or avoided); and/or may have gained or lost competitive advantage attributable to the research (consequences that may have intellectual property implications);
- journal editors, reviewers, examiners, advisory committees, and funding organisations who vet outcomes produced from the research journey (such as submitted manuscripts, final reports, theses, dissertations, doctoral portfolios);
- users/adopters of an innovation that was the focus of, and perhaps even developed through, the course of the research, which may have intellectual property implications;
- policy- and decision-makers who might make choices/decisions based upon recommendations arising from the research; and
- more general societal/global interests that might be reinforced or threatened by the research outcomes.

Note that some types of stakeholders may be both immediate as well as downstream stakeholders. For example, participants are immediate and key stakeholders when considering and undertaking their participation and become downstream stakeholders when one considers potential long-term benefits or harm stemming from their participation in that research. Organisations/institutions that support or supported the research are immediate and key stakeholders by virtue of their provision of access and/or resources for the researcher and become downstream stakeholders when research outcomes emerge.

3.3 Overall Research Positioning

Overall research positioning emerges from, as well as being embedded within, contextualisation domains. In holistically positioning their research, the researcher signals the local, general, and wider intents of their journey. To achieve such positioning, the researcher harnesses pluralist systemic thinking in all aspects of contextualisation and extends and integrates that thinking to bring contextualisation considerations, pattern(s) of guiding assumptions (and associated considerations of research quality), and research frame(s) together to form the background for stating the research problem, specifying overall research goals/aims, and evolving the focal research questions/hypotheses/propositions (refer to the emergent and dynamic interplay shown in red font in the lower left quadrant of Fig. 3.1a).

3.3.1 Holistic Positioning Strategies Associated with Contextualisation

There are a large number of potential holistic positioning strategies a researcher could adopt. Establishing a holistic positioning for the research journey helps to provide a preliminary foundation for choices regarding pattern(s) of guiding assumptions and mode of knowledge building. Those choices, when coupled with contextualisations, constitute a research frame, which provides an overall picture of the research journey, within which the researcher's questions/propositions/ hypotheses are embedded. Given this, the following discussions will foreshadow connections to patterns of guiding assumptions and research frames (discussed in Chap. 4), which essentially means that the contextualisation issues explored in this chapter and Chap. 4 are complexly entangled with and dependent on each other, creating a nonlinear relationship between the two chapters.

- *Exploring relationships*: This can mean the researcher will be examining different kinds of social, managerial, or collegial relationships between/among participants in the research. This could also mean examining relationships with the public, with consumers, or with communities, organisations, or institutions. Alternatively, it could mean exploring relationships between concepts, ideas, and/or constructs. Different patterns of guiding assumptions will shape the nature and specificity of the relationships studied. For example, under the pattern of guiding assumptions for a positivist inquiry, the researcher will be interested in testing specific hypotheses about the nature of relationships, through the clear specification of social and relational constructs. Under the pattern of guiding assumptions for an interpretivist/constructivist inquiry, the researcher will be more interested in the meanings inherent in various relationships and in how they have evolved. The patterns of guiding assumptions for a critical inquiry will be more interested in the power dynamics that can be surfaced in various relationships.

- *Extending existing theory*: Here, the research may be providing a new twist or take on a pre-existing theory and/or incorporating new or different constructs or relationship connections. This could also include extending theory through integration of two or more theoretical perspectives. This positioning strategy will primarily be associated with the pattern of guiding assumptions for a positivist inquiry.
- *Developing a new theory*: This positioning strategy involves developing a new theory. However, this type of positioning is intimately linked with the researcher's choice of pattern of guiding assumptions and consequent methodological choices. Under the pattern of guiding assumptions for a positivist inquiry, a new theory is proposed and developed through leveraging what the researcher learns from a coherent review of prior research and knowledge building efforts and is set out for explicit testing before any data are actually gathered. Here, theorising is apriori, driving the research agenda and methodological choices required to appropriately test deductions from that theory. It is also possible for a new theory to be produced at the conclusion of a positivist investigation as a way of inductively pulling exploratory research findings together in a coherent way to provide guidance for future research. Under the pattern of guiding assumptions for an interpretivist/constructivist inquiry, using methodologies such as grounded theory, theory is developed by working iteratively through the data as they are gathered. Thus, theorising is emergent, dynamic, and heavily dependent upon methodological choices for its shape and sufficiency. Furthermore, early emergent forms of a theory are then evaluated and elaborated through more focused data gathering in later stages of the research. Developing and evaluating/testing theories is a key feature of the explanatory research frame.
- *Replicating a test of theory or other research outcome in a new context*: This positioning strategy focuses on showing whether a theory, and its associated constructs and relationships, or some other research outcome, transcend or generalise across specific types of boundaries (e.g., social, cultural, organisational, industrial, political) using a process of replication. A good deal of cross-cultural research implicitly requires such positioning. Replication requires a fairly deep knowledge of the original contextualisation of the research being replicated, to ensure all relevant aspects can be captured. In general, this positioning strategy will likely be associated with the pattern of guiding assumptions for a positivist inquiry.

- *Exploring boundaries*: This positioning strategy brings to the surface an intention to test or examine the generality, transportability, or uniqueness of a phenomenon, perspective, or impact. It seeks to establish the contextual reach of some type of understanding or explanation. For example, under the pattern of guiding assumptions for a positivist inquiry, this goes to the heart of external validity. However, it extends beyond simple considerations of sample size and sample representativeness, to involve considerations of the relationship between the settings, tasks, and circumstances directly observed in the research and those not directly observed but about which something is to be said (an aspect of external validity referred to as representative design by Egon Brunswik (1956; see also Hammond & Stewart, 2001). Under the pattern of guiding assumptions for an interpretivist/constructivist inquiry, the positioning strategy of exploring boundaries may or may not be a viable or desirable strategy to follow. This pattern of guiding assumptions tends to favour the contextual situatedness and potential uniqueness of every individual's or group's perspective rather than the achievement of understandings or theories that can be meaningfully transported across various types of boundaries. However, if the researcher's goal is to explore not only what is unique in particular contexts, but also what emerges in common across them, then it may make sense to invoke this positioning strategy.
- *Describing phenomena in context*: Often, one intention of research is to provide a description of some process, state of affairs, events, perspectives, situations, groups or organisations in the context in which it is being observed. The goal here is to understand what happens and who is involved when with respect to phenomena, behaviour, and events in a specific context, which is the basic purpose of the descriptive research frame. This may be especially relevant for research underpinned by the pattern of guiding assumptions for an interpretivist/constructivist inquiry. In such research, the description will tend to be qualitative, very rich, highly contextualised, and subjective (i.e., anchored in the perspectives of the research participants). Such description is often pursued as a precursor to interpretation and theorising. It is possible that certain types of research guided by the pattern of assumptions for a positivist inquiry can also be positioned as descriptive. In particular, cross-sectional questionnaires often have a descriptive intent. What is less well-understood is that such questionnaires should be regarded as embedded within specific contexts and those contexts will ultimately serve to delimit the boundaries and applicability of the description.
- *Critiquing phenomena in context*: In research guided by patterns of guiding assumptions for a critical, Indigenous/First Nations or systemic othering inquiry, the intent of the study may be to critique a state of affairs, situation, policy, group or organisation with respect to power relationships, constraints and impediments with a view toward overcoming them and thereby achieve improvements or a better relational balance. The goal here is to understand not only what happens and who is involved when with respect to phenomena, behaviour, and events in a specific context, but why and whether such things are good or detrimental, useful, or harmful, oppressing or empowering with respect to one or more groups of interest. Such interest explicitly overlays a power dimension on the research

(i.e., who has power to do what to whom, how do they do it, and does this need changing in order to achieve some societal or cultural goal?). This positioning strategy has very explicit ties to other stakeholders who have a vested interest in the problem being investigated, and those stakeholder interests will need to be addressed as an expected part of the strategy. Such stakeholders will often be those who do not enjoy power, favour, or equity in a current situation (what the purists in critical research might call 'emancipation of the oppressed' classes or people). These dynamics will be well-captured in the othering-emancipatory research frame paired with the pattern of guiding assumptions for a systemic othering inquiry.

- *Evaluating phenomena in context*: In certain types of applied research, the impacts of interventions, programs, or policies may be evaluated in terms of effectiveness, efficiency, acceptance, durability, compliance, and the like. This positioning strategy is explicitly oriented toward gathering evidence to support evaluative judgements, perhaps with a view toward making concrete recommendations about adoption and implementation. Such research may be guided by the patterns of assumptions for a positivist, interpretivist/constructivist, participatory, or critical inquiry, embedded within the evaluation research frame. Thus, arguments for this position must necessarily be derived from and be consistent with the researcher's chosen pattern of guiding assumptions.
- *Understanding perspectives in context*: This positioning strategy is almost always associated with patterns of guiding assumptions for an interpretivist/constructivist, critical, Indigenous/First Nation, or systemic othering inquiry. The broadest view of context may be taken here, including cross-cultural or multicultural contexts. However, the central principle is that the researcher cannot look for commonalities across contexts until they first understand meanings within each context. Thus, the stories sought here are highly situated and this contextualisation will lead relevant audiences to expect to achieve a deep contextual understanding and perhaps even a contextualised theoretical account. This positioning strategy will often be associated with 'describing phenomena in context' positioning strategy.
- *Exploring/filling a perceived gap in knowledge*: Often, when the researcher conducts a comprehensive and critical review of prior research and knowledge building efforts, they will find a gap or an unexplored avenue. This gives rise to a positioning strategy where the researcher explicitly sets out to conduct research to fill the gap or explore that avenue. Gaps or unexplored avenues may be theoretical, methodological, conceptual, contextual, assumption-based, community-based, or application-based.
- *Creating/validating an instrument or measuring process*: This positioning strategy may be closely associated with developing or testing a new theory or with replicating a test of theory in a new context. It is often linked to the pattern of guiding assumptions for a positivist, or in some cases, critical realist inquiry and is typically a preliminary step in a larger study, rather than being an end in itself (which it certainly can be as well). Setting out this positioning strategy will lead relevant audiences to expect a research investigation that has at least two sequential

stages: instrument/measurement development and validation, followed by hypothesis and/or model testing. More rarely but still quite acceptable, the research may be concerned only with instrument/measurement development and validation.

- *Solving a problem and/or informing a decision in a specific context*: This is another explicitly applied research positioning strategy where the research is intended to provide solutions and/or recommendations in a specific context. It may emerge in conjunction with the evaluating phenomena in context positioning strategy. This is a common positioning strategy in action research, for example, where the researcher's intention, in conjunction with a community or organisational group, may be to formulate and perhaps even evaluate solutions and potential decision choices. This positioning strategy has very explicit ties to other stakeholders who have a vested interest in the problem or decision being investigated, and those stakeholder interests will need to be drawn upon and addressed as an expected part of the strategy. Note that this is a hallmark of critical engagement, which is an inherently participatory research process, intended to break down silos between academic research and communities of practice (see, for example, Fear et al., 2006).
- *Improving a process or outcome*: This positioning strategy starts from the premise that processes or outcomes need to be improved and that the research is intended to facilitate the achievement of that improvement as well as providing the necessary evidence to demonstrate that improvement has occurred. The focus could be on improving a change program or process, product or process quality, employee or managerial performance, training quality, and the like. This strategy is often underpinned by the pattern of guiding assumptions for a critical inquiry and is associated with an action research approach, well-captured within the action research frame. In many cases, this positioning strategy may be linked to an explicit expectation of participant involvement in the research process itself. If this is the case, then the pattern of guiding assumptions for a participatory inquiry may be relevant.
- *Achieving cultural insight/understanding*: Here, the researcher seeks to understand and gain insights into other culturally-privileged ways of learning, knowing about, and living in the world, either on its own, or juxtaposed with the ways of other cultures. It is especially relevant for Indigenous/First Nations research but may also be more generally relevant for cross-cultural research (within the cross-cultural research frame). We say culturally-privileged because such knowledge is deeply embedded/held/lived within a specific cultural context and/or specific cultural groups and following traditional western research pathways for accessing such knowledge is not possible or appropriate. In some cases, where the culture concerned has been colonised, subjugated, diminished, marginalised, or dominated by another, e.g., western, culture, uncritically following traditional western research assumptions and practices will offend and negate rather than facilitate access, particularly and especially if the researcher is a member of the colonising culture. Such research calls for the decolonisation of methodology (see, for example, Smith, 2021) and is a hallmark of the Indigenous/First Nations research frame.

- *Developing/evaluating a new social, technological, or process-based innovation*: This positioning strategy focuses on the development, evolution, evaluation, and/or adoption of a specific social, technological, or process-based innovation, which may or may not have been invented/developed by the researcher. This positioning has an explicit impact intent, namely, to produce an innovation and research surrounding that innovation that is useful to or for some purpose and intended users/adopters and is well-captured within the developmental evaluation research frame. The ultimate research goal may be to facilitate the spread of adoption for that innovation.
- *Facilitating learning and reflection*: Sometimes, the researcher may conduct research in order to demonstrate how theory might be translated into action in a group or an organisation, i.e., a strong practice/practitioner focus. This suggests a positioning strategy directly linked to stimulating organisation learning and critical thinking. An expectation of this strategy would be for the research to provide concrete and constructive feedback to a group or organisation in order to facilitate learning and change. Another expectation could be for the research to facilitate the actual learning and reflection process, which would align this strategy very closely with the 'improving a process or outcome' positioning strategy.
- *Creating/combining/synthesising/hybridising positioning strategies*: Various combinations of the above types of positioning are certainly possible. In some cases, positioning strategies will be brought together by virtue of the configuration and conduct of the research. This will be especially if the research employs pluralist logic, the pattern of guiding assumptions for a critical realist inquiry, and/or applied problem-solving research frames such as action research. In other cases, combining or hybridising positioning strategies may occur in order to meet the needs and expectations of various stakeholders in the research, thereby allowing the research to potentially speak to multiple audiences. In greenfield research or in research that confronts complex so-called 'wicked' societal problems, it may be that an entirely new holistic transdisciplinary positioning strategy will need to be created (see, for example, Pohl et al., 2017).

3.3.2 Positioning Strategy Connections

Holistic positioning strategies that arise from contextualisation may be adopted from the outset or may emerge as a research project evolves. Each strategy has potential connections to or with a number of contextual considerations and those connections can often be conceptualised as specific queries that the researcher could address as their research journey unfolds. To clarify, positioning of research can potentially connect, either explicitly or implicitly, to or with:

- *The previous 'literature' in the area*: Where does the research sit in the context of previous knowledge building efforts? What is the research contributing to the 'literature' and/or to what is currently known (or unknown) in the area?

- *Previous events*: Where does the research sit with respect to specific and relevant previous events, including historical events, in the research context?
- *Performance criteria, including regulations and policies*: Where does the research sit in the context of ways of assessing and controlling organisations in the larger environmental context in which the organisation is situated?
- *Market/consumer/client needs*: Where does the research sit in the context of the markets or clientele that organisations or institutions must relate to?
- *Other researchers*: Where does the research sit with respect to other researchers (including mentors and colleagues)?
- *Participants and other data sources*: Where do research participants and other data sources sit with respect to their relationship to the researcher as well as to the research context more generally?
- *Researcher career or educational progression*: Will the research facilitate realising important career or educational goals or desired outcomes for the researcher?
- *Disciplinary, multi- or cross-disciplinary needs and expectations*: Is the researcher committed to one discipline, or are they trying to work beyond the boundaries of a single discipline, using a pluralist or pan-disciplinary logic?
- *Researcher's own connection to the research context*: What is the researcher's particular relationship with the context of the research, and is this relationship changing or evolving?
- *Researcher expectations of self*: How is the researcher situated with respect to possessing the necessary skills, support, life space, and motivation for successfully completing their research journey? If there are any deficits, are there support systems the researcher can draw upon to address?
- *Expectations of significant others* (e.g., people, family, communities, organisations): What might relevant stakeholders expect to see in terms of research processes and outcomes? How might they expect to benefit from the research? How might those expectations translate into obligations the researcher must meet in order to gain access to the data sources and information they need (i.e., where stakeholders become gatekeepers)?
- *Expectations of other, perhaps competing, perspectives* (e.g., disciplinary, methodological, professional, cultural, demographic, social/interpersonal, group, or theoretical perspectives or traditions): Is the research appropriately sensitive to, or does it have meaning for, other relevant perspectives or groups that hold such perspectives?
- *Expectations of decision makers*: How might the research be relevant to people who need to make specific decisions?
- *Possible futures*: How might the research be relevant to organisational investment, choices and strategies for the future? How might the research be relevant to further research in the area?
- *Expectations/needs of a professional organisation*: Does the research appropriately meet the expectations of a relevant professional body, which may encompass, for example, ethical issues, how research outcomes are disseminated/shared, and how research outcomes might be relevant to professional practice?

- *Expectations of a specific research outlet* (e.g., journal, conference, or other vehicle/medium for sharing/communicating research): Does the researcher hold expectations for what constitutes an appropriate outlet for their research? Is the researcher willing and able to meet the expectations of an intended research outlet?

3.4 Integrative Foreshadowing

We have seen that systemic contextualisation facilitates the situating and positioning of a researcher's researcher journey. Contextualisation juxtaposes researcher identity with research contexts, larger contextual concerns, the positioning of participants and other data sources, and anticipation of end users/sponsors/communities and others implicated in or by the research outcomes. Contextualisation builds on prior knowledge building efforts, helps the researcher to identify and engage with key stakeholders and gatekeepers, in order to help them make convincing downstream choices about shaping, implementing, and communicating their research. Contextualisation not only assists the researcher in positioning their research, it also helps them to reflect that positioning to others and to anticipate how those others might respond to such positioning. Effective contextualisation is what gives meaning, colour, and texture to an otherwise sterile landscape of methodological choices for navigating the 'Data Triangle' by narrowing some avenues of potential choice while opening up other potential avenues. It provides the 'lay of the land' for the research journey.

However, we have not yet completed our exploration of the role of contextualisation. What has been discussed in this chapter are higher-level contextual concerns which situate and position the research as a human activity in its time and physical, social, cultural, and other spaces. Such higher-level contextual concerns must, however, feed into and inform thinking that evolves a specific focus and frame for the research. Think of the higher-level contextual activities as operating at the wide end of a funnel, helping the researcher to narrow possibilities and provide meaning and context for choices. This narrowing has its limits, though, in the absence of considerations about what assumptions the researcher is willing to adopt to guide their research journey, what criteria and meta-criteria should be applied to make proper judgments about the quality of their research, how best to frame their research so as to more clearly communicate intentions, and what specific learning/knowledge building will be pursued by the researcher, expressed in the form of research questions, propositions, and/or hypotheses. These latter considerations will assist the researcher in making the final choices needed to complete their research journey successfully and convincingly. In short, they help the researcher to situate and position more precisely the research choices they make in methodological, learning, and impact spaces.

One very important thing to note is that the contextualisation considerations we have explored in this chapter have dynamic interconnections with the considerations about patterns of guiding assumptions, quality criteria/meta-criteria, frames, and research questions, propositions, and hypotheses that provide the focus for the next chapter. The separation of these levels of contextual concerns into two chapters for organisational purposes in this book should in no way be read as these concerns being necessarily sequential in nature and disconnected from each other. On the contrary, some aspects of higher-level contextualisation, say, for example, an interest in Indigenous culture and knowledge, may predispose the researcher to prefer a certain research frame and pattern of guiding assumptions, namely an Indigenous/First Nations frame and a pattern of guiding assumptions for an Indigenous/First Nations inquiry. Choice of a certain research frame and pattern of guiding assumption, say, for example, a case study frame and a systemic othering pattern of guiding assumptions may create points of focus for higher-level contextualisation, namely identification and contextualisation of a particular focal case and use of a critical stance to identify key stakeholders and gatekeepers, relationships and power dynamics within that case context. Dynamics in pluralist systemic contextualisation thinking for how a researcher gets to their research questions, propositions, and/or hypotheses mean that the researcher's final positioning for the research is more likely to be achieved in a nonlinear, perhaps somewhat chaotic, manner than a linear stepwise manner. In metaphorical terms, the chicken may come before the egg, the egg may come before the chicken, the egg and the chicken may emerge simultaneously, or the egg and chicken may be indistinguishable/inseparable.

References

AIATSIS. (2020). *AIATSIS code of ethics for Aboriginal and Torres Strait Islander research*. Australian Institute of Aboriginal and Torres Strait Islander Studies.

Allen, P., Maguire, S., & McKelvey, B. (Eds.). (2011). *The Sage handbook of complexity and management*. Sage Publications.

Beach, L. (Ed.). (1998). *Image theory: Theoretical and empirical foundations*. Lawrence Erlbaum Associates.

Bell, E., Bryman, A., & Harley, W. (2022). *Business research methods* (6th ed.). Oxford University Press.

Benzies, K. M., Premji, S., Hayden, K. A., & Serrett, K. (2006). State-of-the-evidence reviews: Advantages and challenges of including grey literature. *Worldviews on Evidence-Based Nursing, 2*, 55–61.

Breen, L. (2007). The researcher 'in the middle': Negotiating the insider/outsider dichotomy. *The Australian Community Psychologist, 19*(1), 163–174.

Brunswik, E. (1952). *The conceptual framework of psychology*. University of Chicago Press.

Brunswik, E. (1956). *Perception and the representative design of psychological experiments* (2nd ed.). University of California Press.

Bell, E., Bryman, A., & Harley, B. (2022). *Business research methods* (6th ed.). Oxford University Press.

Cohen, L., Manion, L., & Morrison, K. (2018). *Research methods in education* (8th ed.). Routledge.

Cooksey, R. (1996). *Judgment analysis: Theory, methods, and applications*. Academic Press.

Cooksey, R. (2001a). What is complexity science? A contextually-grounded tapestry of systemic dynamism, paradigm diversity, theoretical eclecticism, and organizational learning. *Emergence: A Journal of Complexity Issues in Organizations and Management, 3*, 77–103.

Cooksey, R. (2001b). Brunswik's 'Conceptual Framework for Psychology': Then and now. In K. Hammond & T. Stewart (Eds.), *The essential Brunswik: Beginnings, explications, applications* (pp. 225–237). Oxford University Press.

Cooksey, R., & McDonald, G. (2019). *Surviving and thriving in postgraduate research* (2nd ed.). Springer Nature.

Chilisa, B. (2020). *Indigenous research methodologies* (2nd ed.). Sage Publications.

Cohen, L., Manion, L., & Morrison, K. (2011). *Research methods in education* (7th ed.). Routledge.

Crowhurst, I. (2013). The fallacy of the instrumental gate? Contextualising the process of gaining access through gatekeepers. *International Journal of Social Research Methodology, 16*(6), 463–475.

Crowhurst, I., & Kennedy-Macfoy, M. (2013). Troubling gatekeepers: Methodological considerations for social research. *International Journal of Social Research Methodology, 16*(6), 457–462.

Dwyer, S., & Buckle, J. (2009). The space between: On being an insider-outsider in qualitative research. *International Journal of Qualitative Methods, 8*(1), 54–63.

Ergun, A., & Erdemir, A. (2010). Negotiating insider and outsider identities in the field: "Insider" in a foreign land; "outsider" in one's own land. *Field Methods, 22*(1), 16–38.

Fear, F., Rosaen, C.L. Bawden, R., & Foster-Fuishman, P. (2006). *Coming to critical engagement: An autoethnographic exploration.* University Press of America.

Geia, L. K., Hayes, B., & Usher, K. (2013). Yarning/Aboriginal storytelling: Towards an understanding of an Indigenous perspective and its implications for research practice. *Contemporary Nurse, 46*(1), 13–17.

Gigerenzer, G, Todd, P. & the ABC Research Group. (1999). *Simple heuristics that make us smart.* Oxford University Press.

Hammond, K. (1996). *Human judgment and social policy.* Oxford University Press.

Hammond, K., & Stewart, T. (Eds.). (2001). *The essential Brunswik: Beginnings, explications, applications.* Oxford University Press.

Hattie, J. (2009). *Visible learning: A synthesis of over 800 meta-analyses relating to achievement.* Routledge.

Hayfield, N., & Huxley, C. (2015). Insider and Outsider Perspectives: Reflections on Researcher Identities in Research with Lesbian and Bisexual Women. *Qualitative Research in Psychology, 12*(2), 91–106.

Kaptchuk, T. (2001). The double-blind, randomized, placebo-controlled trial: Gold standard or golden calf? *Journal of Clinical Epidemiology, 54*(6), 541–549.

Kaptchuk, T. (2003). Effect of interpretive bias on research evidence. *BMJ, 326*, 1453–1455. https://doi.org/10.1136/bmj.326.7404.1453

Kovach, M. (2021). *Indigenous methodologies: Characteristics, conversations, and contexts* (2nd ed.). University of Toronto Press.

Lichtenstein, S., & Slovic, P. (Eds.). (2006). *The construction of preference.* Cambridge University Press.

Mauthner, M., Jessop, J., Miller, T., & Birch, M. (2012). *Ethics in qualitative research* (2nd ed.). Sage Publications.

McAreavey, R., & Das, C. (2013). A delicate balancing act: Negotiating with gatekeepers for ethical research when researching minority communities. *International Journal of Qualitative Methods, 12*(1), 113–131.

NHMRC, ARC, & UA. (2018). *National Statement on Ethical Conduct in Human Research 2007 (updated 2018).* The National Health and Medical Research Council, the Australian Research Council and Universities Australia. Canberra: Commonwealth of Australia.

NHMRC. (2018). *Ethical Conduct in research with Aboriginal and Torres Strait Islander Peoples and communities: Guidelines for researchers and stakeholders.* Canberra: The National Health and Medical Research Council.

Pohl, C., Truffer, B., & Hadorn, G. (2017). Addressing wicked problems through transdisciplinary research. In R. Frodeman, J. Klein & R. Pacheco (Eds.), *The Oxford handbook of interdisciplinarity* (2nd ed., pp. 319–331). Oxford University Press.

Schuelka, M., & T.W. Maxwell, T.W. (Eds.). (2106). *Education in Bhutan: Culture, Schooling, and Gross National Happiness*. Springer Science.

Senge, P.M. (1994). *The fifth discipline fieldbook: Strategies and tools for building a learning organization*. Crown Business.

Sinclair, M. (Ed.). (2014). *Handbook of research methods on intuition*. Edward Elgar Publishers.

Smith, L. (2021). *Decolonizing methods: Research and Indigenous peoples* (3rd ed.). Zed Books.

Tashakkori, A., & Teddlie, C. (Eds.). (2010). *Sage handbook of mixed methods in social & behavioral research* (2nd ed.). Sage Publications.

Toy-Cronin, B. (2018). Ethical issues in insider-outsider research. In R. Iphofen & M. Tolich (Eds.), *The Sage handbook of qualitative research ethics* (pp. 455–468). Sage Publications.

Chapter 4
Systemic Contextualisation: Guiding Assumptions and Framing

Abstract This chapter concentrates on the focusing and framing aspects of systemic research contextualisation. To provide focus, the researcher chooses one or more *pattern(s) of guiding assumptions* to underpin their intended research project/journey. To establish one's guiding assumptions, we downplay the more restrictive notion of 'paradigm' (Kuhn, *The structure of scientific revolutions*. The Folio Society, 2012) and rely instead on the patterning of answers to pivotal questions about the specific orientation and intentions of the research. A pattern typically rules some methodological choices in and other choices out. We will see that every pattern of guiding assumptions invokes specific criteria for evaluating research quality. We also discuss meta-criteria for evaluating research quality, centred around the concept of *convincingness*, which transcend specific guiding assumption patterns (yet without ignoring or dispensing with pattern-specific considerations) and facilitate comparisons of research processes and outcomes arising under different patterns. Finally, we consider choices of *research frame* as a holistic and systemic way of capturing the essence of the research journey, encompassing situating, positioning, and guiding assumption contextualisations, choices of mode of knowledge building/production, and learning focus for intended audiences. Within the ambit of a research frame, downstream methodological choices are scoped, configured, and implemented and approaches to communicating and sharing research learning are enacted.

4.1 Pivotal Questions for Establishing Guiding Assumptions

Discussions about guiding assumptions in social and behavioural research are heavily influenced by philosophical discourse (e.g., discussions of 'ontology'—the nature of the world and reality, and 'epistemology'—the nature of knowledge) and have traditionally been associated with Thomas Kuhn's (2012) concept of a research 'paradigm'. A research paradigm is a system of commonly shared (at least amongst a cohort of researchers/peers) assumptions and expectations for helping researchers decide what counts as the type of knowledge one seeks, what counts as something

R. W. Cooksey, *Unity from Diversity*, https://doi.org/10.1007/978-981-97-3462-7_4

worthy of research in pursuit of such knowledge and what constitutes 'good research' (i.e., quality criteria) for generating that knowledge. [For example, in the natural sciences (e.g., geology, chemistry, physics, astronomy, biology, botany, zoology, and so on), the dominant research paradigm is called 'positivist' (sometimes 'post-positivist' or 'normative') and embodies what most people have learned, through their early years of education, to refer to as the 'scientific method' (sometimes called the 'hypothetico-deductive approach').] However, in social and behavioural research, it turns out that there are potentially many different paradigms researchers could adopt in pursuit of relevant knowledge and understanding (see Neuman, 2020, Chap. 4, for a very clear discussion of certain paradigms and their implications; Crotty, 1998, provides a very coherent discussion of the philosophical underpinnings of social research paradigms; Lincoln et al., (2024) provide an excellent critical discussion of paradigm assumptions and controversies).

Paradigms are typically accompanied by a covering label, such as interpretivist, constructivist, positivist, critical, feminist, critical realist/postpositivist, Indigenous, postmodernist, and so on; each label conveying something central about the paradigm. For a very long time, paradigms were viewed as being in competition with and critical of each other, often to the point where one paradigm defined the dominant approach taught and implemented within specific disciplines. For example, in the disciplines of medicine, psychology, marketing, and economics, the positivist paradigm remains the dominant paradigm implemented by many researchers. The interpretivist or constructivist paradigms tend to dominate much educational and sociological research. However, in these as well as other disciplines (e.g., management, nursing and health, criminology), a shifting balance in paradigm preferences can be observed. The thing about a paradigm is that it is a closed system, oriented toward self-maintenance in the face of critique and anomalies, which is frequently achieved by ignoring, denigrating, or otherwise devaluing research done/conclusions drawn within other paradigms (i.e., paradigm blindness). [According to Kuhn (2012), paradigm shifts only occur when enough anomalies have been observed to stimulate a 'revolution' to embrace an alternative paradigm in a cohort of researchers.] Paradigm diversity is often hard-won simply because of the competitive drive to maintain/reify paradigm boundaries. Academic journal editing and research funding practices, as historical gatekeepers of research scholarship, have done their share in the past to maintain such boundaries, by tending to favour research conducted within particular paradigms for decision outcomes. This state of affairs is slowly changing, faster in some disciplines, more slowly in others, as paradigm diversity becomes more important to recognise in social and behavioural research. However, in my view, what continues to slow progress is the power of paradigm labels, which can sometimes become associated with a researcher's identity (e.g., 'X is a dyed-in-the-wool positivist', 'Y only does qualitative constructivist research') and can have the effect of dampening diversity in research outcomes, especially if the researcher internalises the label as part of their identity.

What I want to do in my discussion here is to remove the power of paradigm labels by focusing on the pattern of guiding assumptions implied by the label. Instead of construing an approach to research as 'positivist', we instead refer to a pattern of guiding assumptions for a positivist inquiry. ‘Pattern’ points us to pivotal questions that must be answered about guiding assumptions underpinning the research and ‘inquiry’ attaches the name for a pattern to the research investigation, rather than the researcher. In this way, I advocate a more systemic and deliberately pluralist perspective, one that values the potentials as well as acknowledges the limitations of a range of patterns of guiding assumptions. I emphasise complementarity and synergy rather than competition or rivalry. I will also look behind the labelling of paradigms by focusing on six pivotal questions the researcher needs to address when trying to pin down the guiding assumptions they will adopt. It is the pattern of choices made in their answers to these pivotal questions that creates the association with a specific paradigm label. I will argue that there is no one correct or dominant pattern of guiding assumptions to implement, it is a choice the researcher deliberately makes and, in making that choice, there then follows a cascade of more focused methodological implications and consequences. We will see that achieving clarity in guiding assumptions has clear benefits in terms of narrowing and focusing the range of choices confronting the researcher and offering some criteria relevant for assessing the quality of research guided by that pattern. We will also see that a researcher’s adopted pattern of guiding assumptions may introduce blind spots that they must grapple with. Thus, researchers must take the good with the bad with each pattern of guiding assumptions and, for this reason, research anchored by only a single set of guiding assumptions must be viewed as necessarily partial. Importantly, as a consequence, researchers need not be restricted to a single pattern of guiding assumptions for the entirety of their research journey, multiple patterns could potentially be adopted, either simultaneously or sequentially. This is one important aspect of the concept of ‘pluralism’ that we employ throughout this book. As I argued above, guiding assumptions are those assumptions the researcher explicitly or implicitly adopts when undertaking their research journey. In Table 4.1 (revised and extended from Cooksey & McDonald, 2019, Table 9.1, pp. 249–251), the six pivotal questions, focusing on *Control, Perspective, Locus, Stance, Theory*, and *Learning*, are presented along with a range of choices or positions researchers could adopt in response to each question. The choices shown should not be considered exhaustive of the possibilities as there is always room for growth, evolution, and creative insight with respect to patterns of guiding assumptions.

4.1.1 Control

The first question concerns Control: ‘*to what extent does the researcher need to be able to exercise control over the research context(s)?*’. When we consider the connection between the researcher’s context(s) and research context(s), one key assumption is the extent to which the researcher needs to be able to influence or change what happens in the research context prior to or contiguous with data gathering. For certain types of research, it is essential that the researcher be able to control what happens to data sources in a research context, particularly if they want to gather evidence for

Table 4.1 Six pivotal questions, and associated potential choices of response, for establishing a pattern of guiding assumptions

Pivotal question						
CONTROL: To what extent does the researcher need to be able to exercise control over the research context(s)?						
Possible Response	***High***: The researcher requires high or full (depth and breadth) control over what happens in the research context, as in a laboratory	***Partial***: The researcher requires at least partial control over what happens in the research context as with a field experiment, social/policy intervention, or evaluation of an innovation	***Minimal***: The researcher has minimal or no control—control over what happens in the research context is infeasible, undesirable, or to be actively avoided; contexts and the events to unfold within them are to be taken as given without researcher intervention	***External source***: The researcher accepts, either by choice or circumstance, the loss of control over what happens in the research context to an external agency/force, as when a policy change is enacted, or a natural or societal event/disaster occurs	***Cultural source***: The researcher actively cedes control over the research context to particular cultural interest groups or Indigenous/First Nations communities; the researcher has no power or right to exert contextual control	***Contextually-fluid***: The researcher chooses what level of control they need over context after first assessing the nature and character of the research context against demands of their research questions and, perhaps, the expectations of key stakeholders

(continued)

Table 4.1 (continued)

Pivotal question						
PERSPECTIVE: Whose perspective(s) provides the primary focus for the research process?						
Possible Response	***Researcher-as-Outsider***: The researcher's perspective provides the primary focus for the research process (i.e., The researcher's ideas are privileged and will be centrally tested) meaning that power over conceptualisation and theorising is concentrated in their hands. As an *outsider*, the researcher's life does not intersect the lives of research participants in any meaningful ways except within the context of the research. An outsider researcher possesses little or no knowledge about the research context and has no prior relationships or engagement with any of the people/data sources in it	***Researcher-as-Insider***: The researcher's perspective provides the primary focus for the research process (i.e., the researcher's ideas are privileged and will be centrally tested) meaning that power over conceptualisation and theorising is concentrated in their hands. However, as an *insider*, the researcher's life intersects the lives of research participants in potentially many ways via cultural, social, and/or work groups, including within the research context. An insider researcher possesses lived knowledge about the research context and has at least some engagement with people/data sources who are of interest in the research. An outsider researcher would not be able to easily access or leverage such knowledge	***Participants***: The perspectives of participants and other data sources provide the primary focus for the research process, and therefore the balance of power over conceptualisation and theorising about what is learned through the research process is shifted toward participants (i.e., the researcher's perspective becomes subordinate to the perspectives emerging from or reflected by their data sources); the researcher engages much more deeply with the lives of participants, seeking their worldviews, rather than imposing their own	***Balanced***: The researcher's perspective, either as an insider or outsider, as well as the perspectives of participants and key stakeholders/users all provide some focus for the research process (i.e., power over conceptualisation and theorising about what they learn through the research process is distributed); the researcher-participant engagement process is roughly egalitarian with power shared nearly equally amongst all involved	***Community***: One or more relevant communities (e.g., Indigenous/First Nations or other specific cultural community, queer community, marginalised community) perspectives provide the primary focus for the research process where the researcher engages with participants at a community-level, less as individuals and more as members of that community; any contrary perspectives or alignments (e.g., western, colonising, or other culture, non-marginalised) the researcher shares or may be perceived as being part of, are to be actively withheld (an important aspect of the act of 'decolonising', for example)	***Contextually-situated***: The researcher decides whose perspectives should provide the primary focus for the research process after first assessing the nature and character of the research context against demands of their research questions and, perhaps, the expectations of key stakeholders; this means that researcher's depth of engagement with participants is contextually-determined and emergent

(continued)

Table 4.1 (continued)

Pivotal question						
LOCUS: Where will the locus of 'reality' or 'truth' be considered to reside for purposes of the research?						
Possible Response	***Externalised reality***: The locus of 'social/behavioural reality' resides where data content, patterns/relationships/meanings (including any causal inferences) are exterior to and generalisable from any specific individuals/groups (i.e., a nomothetic or 'average person' interest)	***Internalised realities***: The locus of 'social/behavioural reality' resides where data content/patterns/relationships/meanings (including any causal inferences) are interior and relative to individuals (i.e., an idiographic or individualist interest, sometimes embedded within underlying socio-political structures)	***Community realities***: the locus of 'reality' resides chiefly within the community in which research participants live, invoking a more social/cultural/relational focus	***Hybridised loci***: There are internalised views of realities that can be mapped, at least to some extent, onto phenomena external to the individual or community, which means that some elements of an externalised reality can be at least partially understood and that it is possible for internalised views to be fallible	***Chaotic realities***: The existence/locus of a one or more realities/truths is difficult to pin down because they are inherently chaotic and unstructured; what matters are the idiosyncratic reflections of experiences and relationships by individuals and groups	

(continued)

Table 4.1 (continued)

Pivotal question						
STANCE: What value-orientation, i.e., stance, will the researcher adopt with respect to the research problem, context, and data sources?						
Possible Response	***Objectivist***: The researcher assumes a stance as an 'objective' disinterested scientist who is value-neutral; that is, where the researcher, as far as possible, cannot influence the data collected and where no findings are seen as inherently good or bad, right or wrong. Here, the researcher is seen as having a passive role once data gathering has commenced	***Subjectivist***: The researcher adopts a subjectivist stance, that of an interested and empathic 'passionate participant' who is value-neutral themselves but interested in the value positions of others; in other words, where the researcher, as far as possible, attempts to keep their own preconceptions from influencing their understanding of other's perspectives and where no perspective is seen as better or worse than any other or as being right or wrong. Here, the researcher is seen to be a co-creator of data in conjunction with participants	***Subjectivist—'Decolonised'***: The researcher adopts a subjectivist stance that is 'decolonised', a 'trusted confidant' relative to western or other perceived oppressive/marginalising influences and subjectivities, through privileging Indigenous/First Nations or other worldviews and seeks to understand those views and the contexts in which they evolved. The ethical/value-oriented position here revolves around respect, relational accountability and authenticity and reciprocity. There are important yet different tensions to address if this stance is combined with an insider researcher versus an outsider researcher role	***Critical/Emancipatory/*** **Transformative:** The researcher adopts a Critical/Emancipatory/Transformative (i.e., a change/liberating/action-oriented) stance; they are an active, often participatory, change agent/'transformative intellectual' whose position is value-laden and conscious that some value positions are more moral or correct than others; the researcher deliberately seeks to influence, facilitate, or foment change in power relationships (hence the relevance of emancipation) for specific groups or society at large and/or intends to show that particular viewpoints are privileged, oppressed, or marginalised (reflects an Indigenous/First Nations orientation if melded with the *Subjectivist-Decolonised* stance; systemic othering orientation if other kinds of groups are of interest)	***Interested audience***: The researcher adopts the stance of 'interested observer/audience member' trying to gain insights into the absolutely subjective and relativist experiences of the world expressed by others; the researcher's values and knowledge systems and perspectives are no more or less important than those of others	***Pluralist***: The researcher adopts a position of 'eclectic critical engagement' which acknowledges that there is an important interplay to be understood between the objective manifestation of reality (e.g., cause-effect relationships) and the subjective perceptions of that reality. Thus, both stances need to be explored in order for a more complete understanding to be achieved

(continued)

Table 4.1 (continued)

Pivotal question						
THEORY: What is the role/nature of theory with respect to the researcher's research intentions?						
Possible Response	**A priori**: Theory takes the form of logical and systematic arguments that identify key constructs and the relationships between those constructs of interest to the researcher; the researcher sets out these arguments prior to the conduct of their research, thereby providing the basis for deducing propositions, and/or hypotheses to be tested. Primary logical processes are deduction (from theory to testable hypotheses) followed by induction (from observations to refining the more general theory). Constructs must be operationally defined before the research commences and measurement/quantification processes are the most important avenue for doing this. Constructs must then be validated before any relationships between them can be evaluated. To be considered appropriate, a theory must be translated into a testable model. A testable model should facilitate explanation and prediction. A priori theory can be falsified/shown to be wrong but never proven to be right	***Emergent***: Theory takes the form of an explanatory story, account, or some other description/ representation of how meanings, themes, patterns, and relationships are socially created/constructed, sustained, and communicated in context; theory is intended to emerge as a consequence of the research instead of being formulated a priori. The primary logical process is induction from data to more general theory. Here, theorising occurs a posteriori rather than a priori. Abduction is used to build up accounts for anomalous findings, in order to potentially incorporate them into the emergent theory	***Description***: Theory has minimal formal role in the research; the researcher's goal is to provide characterisations and descriptions of events, persons, groups, organisations, institutions, governments, and/or communities. Description is an a posteriori process whose goal is to observe what (e.g., patterns, events, perspectives) occurs or is encountered, not build an explanation for why	***Critique***: Theory takes the form of a story, performance, or critique that tries to show true conditions and power relationships in a context and provide guidance for people to act to produce change for a better world. The primary logical process is critical argumentation which can emerge in either an a priori or a posteriori manner. Induction and abduction play larger logical thinking roles in building up a critique	***Re-storying***: Theory implies a specific form of emancipatory/ transformative critique that takes the form of stories, characterisations, and/or re-presentations, utilising language and actions to convey culturally-centred meanings, as well as social, time, space, place, spiritual, and cosmic relationships, and that embed/reflect histories/ accounts that contradict, modify, qualify, or negate pre-existing stories that have emerged from other more historically or contextually dominant sources. From a cultural perspective, this view of theorising is a priori in nature, i.e., the culturally-centred stories exist prior to the researcher encountering them	***Performance***: Theory takes the form of a performance or work of artistic expression that can amuse, shock, or stimulate others as a way of conveying particular viewpoints and interpretations of experiences, social structures, and relationships

(continued)

Table 4.1 (continued)

Pivotal question						
LEARNING: What are the researcher's downstream intentions for the learning and knowledge produced or yielded by their research?						
Possible Response	***Explanation and Prediction***: Learning/knowledge from the research will enable explanations of observable (past and present) phenomena and predictions of future phenomena	***Deep understanding***: Learning/knowledge from the research will enable deep and contextually-anchored understanding and sharing of worldviews	***What works/What and How to change***: Learning/knowledge from the research will enable actions, changes, and improvements to be planned, assessed, and evaluated, and/or decisions to be made	***What is/Patterns of association***: Learning/ knowledge from the research will enable insights into what is or has happened to people, groups, organisations, and so on, often in the context of a specific situation, as well as identifying patterns of associations between/ within key entities	***'Decolonisation' of knowledge***: Learning/ knowledge from the research will enable/ teach others to appreciate, evaluate, and perhaps act upon knowledge that comes via other culturally-anchored pathways and histories (e.g., Indigenous/First Nations) and juxtaposes or privileges that knowledge with or over knowledge that has been produced via historically and culturally-dominant (e.g., western) research pathways and histories	***Innovation***: Learning/ knowledge from the research will facilitate and/or reflect the development, evaluation, distribution, and adoption of a technological, social, or cultural innovation intended to change or improve some aspect of people's lives (e.g., individually, socially, culturally, technologically, economically)

externalised cause-effect relationships (i.e., causal relationships that are externally verifiable rather than perceived). The issue of control is of paramount concern in natural sciences research and when we import this natural sciences logic into the social and behavioural sciences, control becomes of paramount concern as well. It is through exercising control that the researcher attempts to ensure that only the causal factors of interest have changed when they observe or measure the effects of interest. Full researcher control is feasibly achievable only in a laboratory in social and behavioural research, yet such control cannot be absolute where humans and their idiosyncrasies are concerned. As soon as one moves out of a laboratory, capacity for the researcher to control the research context declines dramatically, which makes externalised cause-effect relationships that much harder to convincingly demonstrate. The problem becomes more acute in cases where some external agency controls/determines what happens in the research context, rather than the researcher. Table 4.1 more fully details six potential responses to the Control question: *High*, *Partial*, or *Minimal* responses focus on the continuum of potential depth/breadth of control available to the researcher; alternatively, the researcher may recognise that they could or must, by circumstances, consciously cede control to *External* or *Cultural* sources; or the issue of control may be *Contextually-Fluid*, dependent upon known or emergent issues in a research context as they intersect the research questions.

4.1.2 Perspective

The second question concerns Perspective: '*whose perspective(s) provides the primary focus for the research process?*'. Here, we consider the major driving perspective behind the research. The researcher may choose to conceptualise everything about the research and pre-configure all aspects of their research to be consistent with that conceptualisation prior to gathering data. Alternatively, the researcher can choose to remain in the background, to some extent, opting instead to try and learn about the perspectives of the participants (i.e., data sources) in the research. These are diametrically opposing approaches to addressing the perspective question: in one, the researcher's preconceptions count for everything; in the other, the researcher tries to withhold their own perspective so that they can form empathetic insights into/understanding of the perspectives of others (something that the philosopher Max Weber referred to as *Verstehen*; see Crotty, 1998, pp. 67–71). Of course, the researcher can opt to try and maintain a more balanced or egalitarian approach to managing perspectives; a choice that is more challenging to implement, but in the end, may be what is required in order for the research to be convincing (as in, for example, research into the developmental evaluation and adoption of innovations, see Patton, 2011 and Cooksey, 2011, for discussions of this approach). Building upon arguments advanced by Rowan (1981), we can see that this question addresses the connection between researcher and those researched. The more tenuous the connection, the more 'alienated' (in Rowan's words, "treating people as fragments", 1981, p. 93) the researched feel from whatever is being researched about them and the less engagement there is between the researcher and the lives of those researched. Table 4.1 more fully details six potential responses to the Perspective question: *Researcher as outsider* where the researcher perspective dominates but with no connections to/privileged knowledge about the research context or engagement with participants (i.e., greater 'alienation'); *Researcher as insider* where the researcher perspective dominates but with some connections to/privileged knowledge about the research context and at least a shallow perhaps uneven level of engagement with participants; *Participants* where individual participant perspectives dominate through much greater/deeper engagement with their lives by the researcher (i.e., lesser 'alienation'); *Balanced* where researcher and participants perspectives are in held in tension and power in the engagement process is approximately equally shared; *Community* where a community or group perspective dominates and researcher engagement is focused more on individuals as members of that community; or *Contextually-situated*, where perspective dominance and depth of researcher-participant engagement is dependent upon known or emergent issues in a research context as they intersect the research questions.

4.1.3 Locus

The third question concerns Locus: '*where will the locus of 'reality' or 'truth' be considered to reside for purposes of the research?*'. Here, we intersect the philosophical concept of 'ontology' or the nature of the world and reality. However, we need not deeply engage with this concept in order to make use of it. Basically, the researcher can attack the issue of reality by asking where they are assuming it is located, for purposes of their research. If the researcher assumes that the story of reality they are pursuing lies exterior to any particular individual or group, then they are considering reality to be externalised. The implication here is that if the researcher wishes to look for externalised cause-effect relationships, they are interested in a generalisable picture, independent of any particular person. If, on the other hand, the researcher adopts a more relativist positioning on locus by assuming that any story of reality they might gain access to is interior to each individual or group, then they are considering locus to be internalised. The follow-on implication is that any question of cause and effect will always be relative to an individual's or group's perspective, i.e., their construction of cause and effect in their own life space. It is also possible to adopt a hybridised loci perspective whereby both external and internalised worlds and their interplay in context are of interest. For example, such a perspective is a key feature of the patterns of guiding assumptions for a critical realist inquiry (Mingers, 2014; Sayer, 2000). Table 4.1 more fully details five potential responses to the Locus question: *Externalised reality; Internalised realities*; *Community realities*, where locus resides primarily at a community- or group-level; *Hybridised loci*, where internalised and externalised versions of realities are held in tension; or C*haotic realities*, where versions of reality or truth are ephemeral, chaotic, and unstructured.

4.1.4 Stance

The fourth question concerns Stance: '*what value-orientation, i.e., stance, will the researcher adopt with respect to the research problem, context, and data sources?*'. This is probably the most complex and multidimensional of the six pivotal questions, as it invokes considerations of values (i.e., axiology) and observational/psychological distance (from data sources) in pursuit of knowledge creation/development. An objectivist stance means that the researcher tries to maintain distance from, and minimal connectivity with, data sources, so as not to influence or taint the data they may be providing. At the same time, an objectivist stance implies value-neutrality in that the researcher intends to make no judgments about the rightness, wrongness, morality, goodness, or badness of whatever they observe. In contrast, a subjectivist stance means that connectivity between the researcher and data sources is explicitly sought so that understanding of different points of view can be achieved. This connectivity means that any data that result are essentially co-created between the researcher and data sources. Simultaneously, the subjectivist stance also implies value-neutrality in that the researcher intends to make no judgments about the rightness, wrongness, morality, goodness, or badness of whatever they learn, but they are interested in the value positions held or reflected by others. Other potential stances are explicitly value-laden where transformation, change, and improvement are desirable outcomes to pursue, where there are good and bad ways for social systems to work, all centring around how power dynamics are enacted. With a decolonised subjectivist stance, Indigenous/First Nations values are privileged over western values and the power dynamics and positional superiority associated with western colonisation and knowledge are upended or inverted (see, for example, Chilisa, 2020; Kovach, 2021; Smith, 2021, especially Chap. 3). With a critical/emancipatory stance, the researcher's role is to help bring true social conditions and power relationships into the light (revealing illusions) so that people can be liberated from restrictive social conditions (hence, the emancipatory aspect) and ultimately improve their situation. This is a key feature of critical social science (Lincoln et al., 2024). The stance of interested audience is absolutely subjective in that the researcher is an interested watcher of social events and interactions whose value and knowledge systems and perspectives are no more or less important than those of others—a key feature of postmodernist assumptions (for example, see Alvesson & Deetz, 2021, pp. 87–100; Crotty, 1998, Chap. 9). A pluralist stance is one where the researcher accepts that there is value in examining phenomena from an objective stance as well as from subjective stances in order to understand the interplay between them—another key feature of critical realist assumptions (Sayer, 2000). Intrinsic to the pluralist stance is that objective reality is only imperfectly perceivable and partially understandable by individuals or groups and that those subjective understandings and perceptions may be flawed. Table 4.1 more fully details these six potential responses to the Stance question: *Objectivist*; *Subjectivist*; *Subjectivist—Decolonised*; *Critical/Emancipatory*; *Interested Audience*; or *Pluralist*.

4.1.5 Theory

The fifth question concerns Theory: '*what is the role/nature of theory, with respect to the researcher's research intentions?*'. Here, we focus on the conceptualisations the researcher develops and/or uses during the course of their research. Theory has different meanings under different patterns of guiding assumptions. One way of looking at theory is as a network of constructs and relationships between constructs that the researcher assembles prior to the gathering of any data. Theory in this sense is called a priori and provides the means for deducing specific propositions or hypotheses that the researcher wishes to test. A priori theory generally reflects the researcher's preconceptions about what they expect to show in the research. Once data have been gathered and relationships and patterns assessed, the researcher then uses logical induction to infer back to the theory and possibly refine it. Another sense in which theory may be taken is as emergent. Here, the researcher does not start with a theory, rather theory emerges from gathered data as a way of accounting for why the data and attendant meanings and patterns unfolded as they did. This is a process of induction from specific data to a more general theoretical account (in grounded theory, early data-driven theory development can lead to subsequent deduced hypotheses which can be tested by gathering more contextualised data, see discussions in Charmaz, 2014). Thus, the emergent choice for theory means that theory is constructed from the data, not prior to the data. Theory may also emerge as a collaborative phenomenon between researcher(s) and participants, a key aspect of the participatory inquiry paradigm (Heron & Reason, 1997; Lincoln et al., 2024). Abduction, as a logical thinking process, may assist the researcher in incorporating anomalous or surprising findings into their emerging theory. Of course, it is entirely possible that the researcher does not intend for theory to play a role in the research, especially if their intention is straightforward description. Another possible sense in which theory may be taken is as a critique, which is a story or critical argument that illuminates the true nature of social conditions and power relationships with the intention of providing guidance for further action. Here, the researcher may rely on both induction (building up a more general critique from specific observations) and abduction (using anomalous or surprising observations as a basis for critique). Theory may also take on the role of re-storying or cultural re-tellings of events and relationships in a richly layered, interconnected, and spiritual universe, in ways that do not reflect, and may actively reject, western-centred values and knowledge (see, for example, Archibald et al., 2022; McGrath et al., 2023; Ullrich et al., 2022). Finally, theory may take the form of a performance as a vehicle for artistic expressions about meanings, structures, and relationships in a social world (see, for example, Auslander, 2004, in the context of postmodernist research; Saldaña, 2024; Siegesmund, 2024). Table 4.1 more fully details these six potential responses to the Theory question: *A priori*; *Emergent*; *Description*; *Critique*; *Re-storying*; or *Performance.*

4.1.6 Learning

The sixth question, pertinent to identifying the researcher's desired pattern of guiding assumptions, concerns Learning: '*what are the researcher's downstream intentions for the learning and knowledge produced or yielded by their research?*'. Here the focus is on what the researcher intends to do with what they learn and with the knowledge they assemble from their research. One orientation is for using the knowledge the researcher assembles and learning they achieve to enable explanation and prediction of social and behavioural phenomena. This is most often associated with a strong emphasis in showing externalised cause-effect relationships as once a generalised cause-effect relationship is understood, it can be used as a predictive rule as well as serving as an explanation for why things happen as they do (i.e., nomothetic or generalised inferences). A second possible orientation for using the knowledge researcher assembles and learning they achieve is to provide and share a deep understanding of individual worldviews (i.e., idiographic or individual-level inferences). A third possible orientation is to use the knowledge the researcher assembles and learning they achieve to inform themselves and others about further actions and decisions that might/should be taken; thus, learning is to be transformed into action. Another potential orientation is to use the knowledge the researcher assembles and learning they achieve to show others what patterns and relationships have been discovered in what circumstances and contexts. Learning intentions may take on a cultural orientation where western or dominant cultural knowledge is decolonised to reveal other types and forms of knowledge about the world and the place of human beings within it. Finally, learning and knowledge may facilitate and/or reflect the development, evaluation, and/or dissemination of a specific technological, social, or cultural innovation. Table 4.1 more fully details six potential responses to the Learning question: *Explanation and Prediction*; *Deep understanding*; *What works/What and how to change*; *What is/Patterns of association*; *'Decolonisation' of knowledge*; or *Innovation.*

4.2 Choice of Patterns of Guiding Assumptions

In the interest of furthering pluralist systemic thinking, it is important to explore the diversity of possible choices with respect to patterns of guiding assumptions for social and behavioural research. Part of the thinking that contextualises a research investigation must involve deciding upon which pattern or patterns of guiding assumptions the researcher will adopt for purposes of their research. In pluralist systemic thinking, all potential patterns should remain under consideration until the researcher has explicitly considered their relevance for the research journey at hand. The driving logic for choice(s) of pattern should be the contextual fit of one or more patterns to the researcher's needs and intentions; it is a fundamental aspect of a researcher's positioning for the research. Once a pattern of guiding assumptions has been adopted, it brings into focus relevant downstream methodological choices and assists in focusing and shaping one's research questions, propositions, and/or hypotheses. Furthermore, adopting and following a specific pattern of guiding assumptions also implies adopting and utilising criteria for evaluating research quality that are specific to that pattern and knowing these helps to further narrow and focus downstream methodological choices. These criteria are intended to be used to judge the character and quality of research guided by that pattern of assumptions, i.e., what constitutes 'good' research under that pattern. It is important to understand such criteria as the researcher can use them to judge the quality of other's work under a specific pattern of guiding assumptions and others can and will judge the researcher's work in the same way.

A summary table template for identifying a pattern of guiding assumptions is presented in Table 4.2, which condenses the potential responses to pivotal questions to their short yet meaningful labels. In addition to displaying each pattern of guiding assumptions, each pattern will be given a meaningfully descriptive label for the type of inquiry the pattern supports (signalled by ???, which may or may not correspond with the more traditional 'paradigm' name for the pattern). The goal is to associate pattern labels with types of inquiry, not types or identities of researchers. In the summary table template, a specific pattern of guiding assumptions will be revealed

Table 4.2 Summary table template for capturing/displaying patterns of guiding assumptions for different types of inquiries

Pattern of Guiding Assumptions for *???* Inquiry						
Pivotal Question	*Potential Choices for Each Question*					
Control?	High	Partial	Minimal	External source	Cultural source	Contextually-fluid
Perspective?	Researcher-as outsider	Researcher-as insider	Participants	Balanced	Community	Contextually-situated
Locus?	Externalised reality	Internalised 'realities'	Community realities	Hybridised loci	Chaotic realities	
Stance?	Objectivist	Subjectivist	Subjectivist – 'decolonised'	Critical/ Emancipatory/ Transformative	Interested audience	Pluralist
Theory?	A priori	Emergent	Description	Critique	Re-storying	Performance
Learning?	Explanation & Prediction	Deep understanding	What works/ What & how to change	What is/ Patterns of association	'Decolonisation' of knowledge	Innovation

by shading one or more relevant table cells representing typical (but not necessarily exclusive) responses to each pivotal question (row in the table), using darker shading for a single dominant response and lighter shading for potential multiple responses, made either concurrently or sequentially, yet remaining consistent with the overall intent of the pattern. Note that there is nothing that prevents a researcher from making an argument for a non-typical response, if the arguments are consistent with the overall intentions of the pattern. Potential multiple responses also signal the possibility for forming hybrid responses to the pivotal question. Pattern-specific research quality criteria will be discussed where appropriate.

4.2.1 Pattern of Guiding Assumptions for a Positivist Inquiry

Table 4.3 displays the pattern of guiding assumptions for a positivist inquiry (see, for example, discussions in Armstrong & Green, 2022; Crotty, 1998, Chap. 2; Neuman, 2020, Chap. 4). The pattern of guiding assumptions for a positivist inquiry has a long history in the context of the social and behavioural sciences and in the context of the natural sciences before this (in addition to positivism, it is associated with labels such as 'the scientific method', the 'hypothetico-deductive' method, and 'normative science'). In fact, this pattern for social and behavioural research constitutes a straightforward importation from the natural sciences so that, under these guiding assumptions, we study human behaviour and systems in the same way that we study other natural phenomena in the world. This pattern is heavily invested in control processes, the higher the level of control over contexts available to the researcher, the better. However, different degrees of lack of contextual control must also be catered for. In a positivist inquiry, the researcher's perspective dominates proceedings, reflected especially in the a priori nature of theorising about the relationships between constructs/variables (which leads to strong but not absolute preferences for measurement and quantification of data) from which testable propositions or hypotheses can be deduced.

A positivist inquiry focuses on an externalised reality wherein the researcher attempts to arrive at generalisable conclusions (a process of logical induction) about cause-effect relationships through adopting an objective value-free stance. In a positivist inquiry, explanation and prediction are often sought, these being hallmarks of effective generalisation. However, a positivist inquiry may also seek evaluative conclusions about what works or doesn't work in a program or intervention, and what to change to improve effectiveness (such judgments being empirically based upon evidence, not value-based judgments of moral worth) and such goals may extend to the development and evaluation of a specific innovation. Finally, a positivist inquiry may seek to identify/characterise previously unknown relationships, patterns and associations amongst constructs/variables.

Table 4.3 Pattern of guiding assumption responses to pivotal questions for a *positivist* inquiry in social and behavioural research

Pattern of Guiding Assumptions for a *Positivist* Inquiry						
Pivotal Question	*Potential Response Choices for Each Pivotal Question*					
Control?	High	Partial	Minimal	External source	Cultural source	Contextually-fluid
Perspective?	Researcher-as outsider	Researcher-as insider	Participants	Balanced	Community	Contextually-situated
Locus?	Externalised reality	Internalised 'realities'	Community realities	Hybridised loci	Chaotic realities	
Stance?	Objectivist	Subjectivist	Subjectivist – 'decolonised'	Critical/ Emancipatory/ Transformative	Interested audience	Pluralist
Theory?	A priori	Emergent	Description	Critique	Re-storying	Performance
Learning?	Explanation & Prediction	Deep understanding	What works/ What & how to change	What is/ Patterns of association	'Decolonisation' of knowledge	Innovation

Quality Criteria Associated with the Pattern of Guiding Assumptions for a Positivist Inquiry

For research guided by the pattern of guiding assumptions for a positivist inquiry, quantification of observations in the form of measurements is considered to be the most appropriate type of data to gather and analyse (qualitative data can also be used, but typically only after they are transformed into a form amenable to quantitative analysis, i.e., quantitised via the Transformative data-shaping strategy). Since measurement is considered central, it is important to recall the discussion of the Measurement data-shaping strategy in Chap. 2 regarding aspects of measurement quality that are important to evaluate: validity, reliability, and sensitivity (these measurement quality criteria will be further discussed in Chap. 5). However, beyond these specific criteria for measurement quality, it is important to understand the more systemic quality criteria associated with the pattern of guiding assumptions for a positivist inquiry. In social and behavioural research, Campbell and Stanley (1966) and later Cook and Campbell (1979) and Cook et al. (1990) crystallised these standards into a set of four essential types of validity at the research level, namely, construct validity, internal validity, external validity, and statistical conclusion validity. To this list, we will add a fifth criterion, replicability/robustness/reproducibility (see, for example, discussions in Armstrong & Green, 2022; Nosek et al., 2022).

Construct Validity

Construct validity is linked to the quality of the processes the researcher uses to translate theoretical ideas and concepts (called 'constructs') into concrete quantitative measurements (a process called 'operational definition'; also discussed in Chap. 5). Construct validity concerns the appropriate inferences that can be drawn from measurement processes giving rise to the quantitative data used to test propositions and hypotheses. It points to the quality of a researcher's a priori theorising. Construct validity is an important quality criterion under the pattern of guiding assumptions for a positivist inquiry because it focuses on how the researcher attempts to objectively quantify and externally represent some aspect of human behaviour they are theorising about.

Internal Validity

Internal validity refers to the researcher's capacity to appropriately claim that they have found specific and unambiguous relationships between constructs of interest. External causal relationships are especially, but not exclusively, of interest under the pattern of guiding assumptions for a positivist inquiry. In order for an external causal relationship to be claimed to exist between two or more constructs, all three of the following conditions must be met: (1) the causal construct must precede the effect construct in time (*temporal priority of cause*); (2) when the causal construct changes, the effect construct must be observed to change (*covariation*) and (3) alternative plausible cause constructs for the effect construct must be ruled out (*alternative plausible causes are controlled*). A weaker form of relationship, called an association or a correlation, may also be of interest, in which case, only condition 2 (*covariation*) must be met. Internal validity asks whether the configuration and execution of the research warrants the relational conclusions being drawn, whether causal in nature or not. Internal validity is arguably the most important quality criterion under the pattern of guiding assumptions for a positivist inquiry because it focuses on how well the researcher objectively tests for the existence of theorised relationships between constructs by controlling, where possible, for alternative plausible explanations for the relationships observed. Internal validity is generally enhanced through researcher control over events and over the experiences of participants. Such control is a hallmark of the Manipulated experience-focused data gathering strategy and can be achieved through (1) the configuration of research conditions and experiences at the time of data gathering, (2) the use of procedures, prior to data gathering, for reducing or preventing the influence of unwanted causal factors and/or (3) the use of statistical analyses to control for unwanted causal influences during the process of data analysis.

External Validity

External validity refers to the researcher's capacity to appropriately generalise the conclusions about relationships (including any related to external causes and effects) substantiated in their research to other contexts, people, places, and times. Especially important is the capacity for the researcher to generalise what is learned about constructs, patterns, and relationships from the sample of participants employed in their research to the population from which they drew the sample. If any biases occur in the sampling process and/or if issues arise associated with demand characteristics, external validity is reduced. Demand characteristics, in particular, may emerge with human participants and refer to aspects of the research context (e.g., 'Hawthorne effect', where something different happening in the research context changes behaviours), research processes (e.g., 'Guinea pig effect', where something in the way the researcher is gathering the data signals what the researcher is looking/testing for), or the researcher's own behaviours (e.g., 'experimenter bias', where the researcher inadvertently reveals, through how they talk and/or behave, what they expect to find) that change in such a way as to render what the researcher has measured to be idiosyncratic and only locally relevant (see discussion in Rosenthal & Rosnow, 2008, Chap. 7). Sampling biases can be controlled through use of an appropriate (i.e., random) sampling strategy and demand characteristics can be explicitly controlled by the researcher through effective purpose- or treatment-masking procedures. It is important to note, however, that external validity is about more than generalising from a sample to a population. It also concerns how well the research results can be generalised to other situations relevant to the members of the sample. Here we are talking about whether the experiences that members of the sample have while participating in the research bear any relationship or relevance to their lives in outside the research context. If those experiences are relevant only to the research context itself, then it is likely that external validity is lower. This problem is especially relevant for laboratory-based research where participants may experience artificially-controlled environments and tasks that bear little or no relationship to anything they might experience outside of the laboratory. Egon Brunswik (1952, 1956) drew attention to this problem when he developed the concept (and the process) of representative design as a way of trying to ensure that what is learned in highly-controlled laboratory research has some relevance to the larger life space of those who participate in the research as well as to those who do not participate in the research. In short, external validity concerns the meaningfulness of claims about the generality of the relationships found in quantitative data in order to create opportunities for explaining and predicting the phenomena of interest at a more general 'law-like' level.

Statistical Conclusion Validity

Using inappropriate statistical procedures to analyse data or applying statistical procedures to improperly structured, ill-formatted, or erroneous data can lead to misleading and indefensible conclusions. Thus, statistical conclusion validity concerns the appropriateness of the outcomes/interpretations/conclusions resulting from applying statistical procedures to any quantitative data gathered in the research. This is an important quality criterion under the pattern of guiding assumptions for a positivist inquiry because it focuses on how well the researcher employs appropriate statistical procedures to objectively test for the existence of patterns and relationships (e.g., connections between constructs, differences between groups) by statistically controlling, where possible, for alternative plausible explanations for the observations and relationships observed and by providing a logical basis for making statistical inferences from sample to population for purposes of explaining and predicting observations and relationships. 'Appropriateness' also resides in the character of the data themselves with respect to whether they meet the statistical assumption requirements (e.g., normality of data distributions, linearity of relationships, homogeneity of variance, absence of outliers and anomalous data values, independence of errors) associated with analysis procedures they are employing and whether patterns of missing data can interfere with achieving interpretable analysis outcomes.

Relationships Amongst the Validity Criteria

Figure 4.1 (from Cooksey & McDonald, 2019, Fig. 9.3, p. 279) depicts the relationships between the four types of research-level validity for research guided by the pattern of assumptions for a positivist inquiry, conceptually mapped against the timeline for a research project. Construct validity is the first type of validity to focus on as it concerns the nature and quality of the measurements that give rise to the quantitative data being gathered. Construct validity (supported by reliability of measurements) then underpins the achievement of internal validity (arguably the most important type of validity, hence its emphasis in Fig. 4.1) in association with the nature and quality of the research configuration (what is traditionally called 'research design' under the pattern of guiding assumptions for a positivist inquiry). Statistical conclusion validity, through choices of appropriate statistical analysis procedures and their application to appropriate data, facilitates the achievement of both internal validity (as a pathway to establishing covariation, i.e., correlation or association, between constructs) and external validity (facilitating generalisations to relevant populations) and, where appropriate, may assist in supporting claims about construct validity. Equally, construct validity and reliability can help support statistical conclusion validity through reduction of measurement errors. Finally, after statistical conclusions have been drawn, external validity becomes relevant as the researcher makes their claims for the generality of their findings (which themselves may be supported by internal validity). The small opposing red arrows, superimposed on the two-way link between internal and external validity, convey the idea that these two types of validity tend to work in opposition to each other. When the researcher attempts to enhance one type of validity, the tendency will be for the other type to be lower. That is, pursuing high internal validity (typically, through the exercise of more control over context) will generally mean trading off or sacrificing some degree of external validity (usually because samples or participant experiences become less representative of 'real-life' populations and/or contexts). On the other hand, pursuing high external validity (typically, through obtaining larger and more representative samples, using more naturalistic contexts, and representatively designing participant experiences) will generally mean trading off or sacrificing some degree of internal validity (as some control over context is usually lost).

Replicability/Robustness/Reproducibility ('3R Criterion')

Replicability/robustness/reproducibility ('3R criterion', for short) refers to the expectation that whatever the researcher does and learns in under the pattern of guiding assumptions for a positivist inquiry should be able to be replicated, reproduced, or

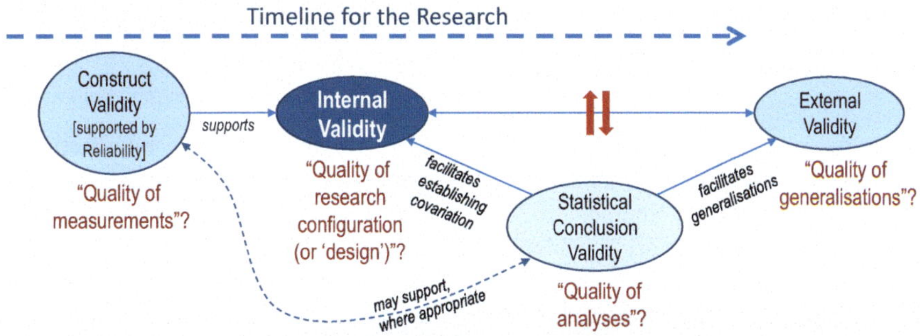

Fig. 4.1 Relationships between validity criteria under the pattern of guiding assumptions for a positivist inquiry

shown to be robust by other researchers. Nosek et al. (2022) characterised these three aspects of research credibility as follows:

> Replication refers to testing the reliability of a prior finding with different data. Robustness refers to testing the reliability of a prior finding using the same data and a different analysis strategy. Reproducibility refers to testing the reliability of a prior finding using the same data and the same analysis strategy (p. 721)

The prospects for success with respect to the '3R criterion' are enhanced through clear and appropriately detailed explication of theory, logical deductions of propositions/hypotheses, descriptions of research contexts (and, where appropriate, research tasks for participants), effective descriptions of measurement processes, access to and sampling of data sources, appropriate details about the data gathering processes, effective utilisation of data analytical procedures as well as clarity/openness about potential conflicts of interest the researcher may have had to deal with when undertaking the research investigation, collectively providing feedforward for future investigations. However, since perfect replication, reproduction, or demonstration of robustness is not possible, empirical success against the '3R criterion' can only be judged retrospectively on the basis of the balance between congruencies and incongruencies of the outcomes of previous research with current research that has a replication, reproducibility, or robustness intent (a feedback-oriented judgment). An interesting tension exists in that getting published in academic journals generally imposes page- or word-limits on submitted manuscripts and a common impact of this constraint is reduction in the level of detail about the researcher's methodological processes (i.e., descriptions/discussions about how the researcher navigated the 'Data Triangle' tend to be one of the first areas to be edited down when trying to meet manuscript submission length requirements). One could argue that academic publishing actually impedes success with respect to the '3R criterion'.

4.2.2 *Pattern of Guiding Assumptions for an Interpretivist/ Constructivist Inquiry*

Table 4.4 displays the pattern of guiding assumptions for an interpretivist/ constructivist inquiry (see, for example, discussions in Crotty, 1998, Chaps. 3, 4, and 5; Lincoln et al., 2024; Neuman, 2020). Under this pattern of guiding assumptions, knowledge and meaning are considered to be socially constructed and vested in individual perspectives and the goal of the researcher is to achieve a deep understanding of those perspectives without superimposing their own viewpoints and potential biases. Under the pattern of guiding assumptions for an interpretivist/constructivist inquiry, researcher control over the research context is not sought or desirable; contextual control may reside in other agencies which then become a focus for perspective seeking from research participants. In saying this, we are explicitly privileging the individual perspectives (i.e., the 'internalised realities') of participants and, where appropriate, communities, over the perspective of the researcher. This pattern of guiding assumptions is thus explicitly subjective in orientation. Theory is emergent in that it is iteratively developed from the data that are gathered (achieved through identifying, capturing, and displaying meanings, patterns and associations embedded within data), rather than being posited a priori by the researcher. Theory may also assume the character of a description (ethnographers, for example, refer to 'thick description' to characterise contextualised theory development from observations over time).

Table 4.4 Pattern of guiding assumption responses to pivotal questions for an interpretivist/ constructivist inquiry in social and behavioural research

Pattern of Guiding Assumptions for an *Interpretivist/Constructivist* Inquiry						
Pivotal Question	*Potential Response Choices for Each Pivotal Question*					
Control?	High	Partial	**Minimal**	**External source**	**Cultural source**	Contextually-fluid
Perspective?	Researcher-as outsider	Researcher-as insider	**Participants**	Balanced	**Community**	Contextually-situated
Locus?	Externalised reality	**Internalised 'realities'**	Community realities	Hybridised loci	Chaotic realities	
Stance?	Objectivist	**Subjectivist**	Subjectivist – 'decolonised'	Critical/ Emancipatory/ Transformative	Interested audience	Pluralist
Theory?	A priori	**Emergent**	**Description**	Critique	Re-storying	Performance
Learning?	Explanation & Prediction	**Deep understanding**	What works/ What & how to change	**What is/ Patterns of association**	'Decolonisation' of knowledge	Innovation

Quality Criteria Associated with the Pattern of Guiding Assumptions for an Interpretivist/Constructivist Inquiry

For a time, researchers working under patterns of guiding assumptions for interpretivist/constructivist and other non-positivist inquires had to adapt positivist validity quality criteria in order to get their research published (see, for example, Golafshani, 2003; Healy & Perry, 2000; LeCompte & Goetz, 1982). This was problematic for two reasons: (1) it privileged 'validity' as the gold standard for all research, meaning that validity criteria always had positivist overtones associated with them, and (2) validity criteria ultimately proved to be ill-suited to the task of assessing the quality of interpretivist/constructivist research. As a consequence, new quality criteria evolved to provide a clearer, better-fitting, and fairer approach to judging the quality of interpretivist/constructivist research as a pathway for "deconstructing truths" (Patton, 2015, Exhibit 9.7). Not all authors agree on these new criteria, so there remains some contention in the field. For most purposes, the following four quality criteria, crystallised by such authors as Fossey et al. (2002), Golden-Biddle and Locke (1993) and Lincoln and Guba (2024), are appropriate for gauging the quality of research guided by the pattern of guiding assumptions for an interpretivist/constructivist inquiry: transparency, authenticity, sufficiency, and transportability.

Transparency

The criterion of transparency focuses on the openness with which the researcher displays all aspects of the research journey they have undertaken. This encompasses how they made choices about who they spoke to or observed or what they read, how the data were gathered, recorded, prepared, and analysed, what they did and thought in the context of their data gathering activities (perhaps including how they felt), and how they might have impacted on the context during the course of their research. One of the key guiding assumptions in the pattern for an interpretivist/constructivist inquiry is subjectivity where the researcher seeks to understand the perspectives/worldviews of the participants in their research rather than imposing their own perspective (as is done under an objectivist guiding assumption). Transparency is an important quality criterion under the pattern of guiding assumptions for an interpretivist/constructivist inquiry because it focuses on the actions the researcher takes to gain access to, record and interpret the meanings and the perspectives of the participants and/or other data sources as well as to manage their own beliefs, preconceptions and emerging ideas to keep them from influencing the perspectives they are gaining access to (via data source choices, contextual choices, and data gathering and analysis processes) and allow deeper understandings and insights to emerge. One of the best tactics the researcher can use to enhance transparency under the pattern of guiding assumptions for an interpretivist/constructivist is to maintain a complete research journal that can also double as an additional data source for their research. Transparency helps others, who might read or use the research, to understand exactly what the researcher did every step of the way and helps to maintain closeness between the data and the interpretations that emerge from those data.

Authenticity (or Trustworthiness)

Authenticity (sometimes referred to as trustworthiness, genuineness, or credibility) is arguably the most important quality criterion under the pattern of guiding assumptions for an interpretivist/constructivist inquiry because it focuses whether the researcher actually gains access to the genuine meanings and perspectives held by their participants and/or reflected in other data sources in the data they have gathered in order to allow deeper understandings and insights to emerge, untainted by their potential biases and preconceptions. Without authenticity, the researcher cannot claim to have understood their participants' points of view or gained insights into their worldviews. Authenticity must be reflected not only in data gathering and recording processes but also throughout data analysis and reporting processes. To enhance authenticity, the researcher must seek, as far as possible, to keep faith with and appropriately and accurately reflect the genuine words, voices, images, and views of their participants and keep their analyses as close to the data as possible. Technological support can usually help with this process by recording verbatim the words and images put forward by participants as well as questions and commentary from the researcher for use in analyses (typically after appropriate and careful transcription, which preserves the verbatim content wherever possible).

Sufficiency (or Completeness)

Sufficiency is an important quality criterion under the pattern of guiding assumptions for an interpretivist/constructivist inquiry because it focuses whether the researcher gains access to genuine voices, meanings, and perspectives from a sufficiently diverse set of relevant data sources in the research context to an appropriate depth in order to develop and convey well-rounded, deeper, and convincing understandings and insights. This may include seeking out non-mainstream, minority, marginalised, disenfranchised, or disempowered perspectives, where necessary, to tell a more complete story. Sufficiency concerns not only who the researcher talked to or observed or what was read or experienced, but also why and about what and to what depth. It focuses on data source choices, of course, but also has wider concerns as well, such as which topics of conversation were followed and the degree or depth of engagement the researcher displayed with each of their data sources.

Transportability

Transportability is an important quality criterion under the pattern of guiding assumptions for an interpretivist/constructivist inquiry only when and where the researcher intends it to be relevant, because it focuses whether they are able to sensibly argue that what they have learned from the voices, meanings, and perspectives of the participants and/or other data sources within the specific research context is relevant for and provides potentially meaningful insights into other contexts for other people or groups at other times and in other circumstances. Thus, transportability is the only quality criterion that may be considered optional. Often, for interpretivist/constructivist inquiries, the researcher may not seek transportability; rather, they seek to tell the story relevant only to the research context at hand and they must keep faith with this intention throughout their processes of assembling and presenting that research story. If transportability is the researcher's intention, this intention should be clearly reflected in their research configuration (using such tactics as multiple case studies or multi-site data gathering). Often, a case for transportability can be made if the researcher makes a deliberate attempt to illuminate convergences and divergences (i.e., compare and contrast) between different contexts being researched (such as comparing the stories and interpretations from two or more different companies or schools).

Relationships Amongst Interpretivist/constructivist Quality Criteria

There are synergies between the various quality criteria for research guided by the pattern of guiding assumptions for an interpretivist/constructivist inquiry, depicted in Fig. 4.2 (from Cooksey & McDonald, 2019, Fig. 9.4, p. 283). Throughout the entire research process, authenticity (arguably the most important quality criterion, hence its emphasis in Fig. 4.2) and sufficiency feed mutually off each other (seeking sufficient diversity of authentic perspectives) and both may be further enhanced by close attention to transparency. Equally, transparency can be enhanced through clarity in conveying what the researcher has done to enhance authenticity and sufficiency. Transportability, if intended, will consist of arguments made toward the end of the research journey, but the researcher's intention also needs to be reflected earlier on in choices of research configuration, data gathering and data source choice strategies, which will in turn make sufficiency a more relevant criterion to meet.

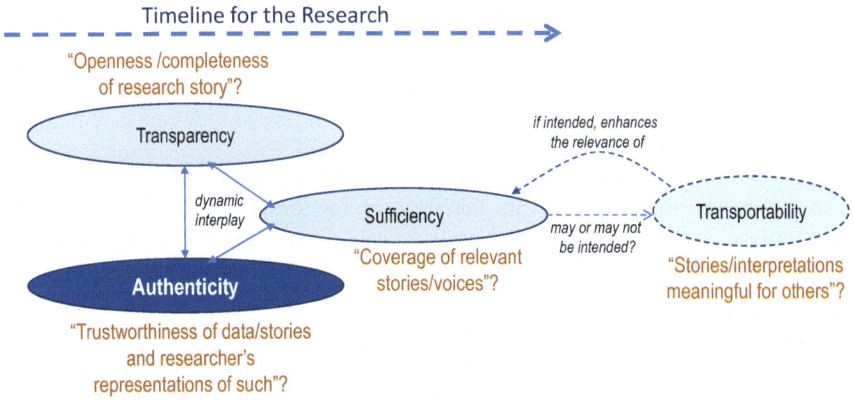

Fig. 4.2 Relationships between quality criteria under the patterns of guiding assumptions for an interpretivist/constructivist inquiry

4.2.3 Pattern of Guiding Assumptions for a Critical Realist Inquiry

Table 4.5 displays the pattern of guiding assumptions for a critical realist (sometimes called postpositivist) inquiry (see, for example, Bhaskar, 2016; Fleetwood, 2014; Lincoln et al., 2024; Maxwell & Mittapalli, 2010; Mingers, 2014; Sayer, 2000; Scott, 2007). This pattern is inherently pluralist in orientation (perhaps with a slight leaning towards a more objectivist stance), stemming from its differentiation between the real world (the world of objects, possessing structures and probabilistic, not deterministic, causal powers, that exists independent of human perception), the actual world (where the structures and causal powers in the real world are activated to produce events, relationships, and consequences) and the empirical world (human experiences of the real and/or actual worlds) (see discussions in Sayer, 2000, Chap. 1). Thus, there is an independent 'externalised' reality, but our understanding of that reality is fallible and probabilistic in nature (embedded within fallible and partial 'internalised realities') and the relationship between the two serves as an empirical focus of interest, entailing a hybridised locus for the research. Researcher control over context may be partial, minimal, or attributed to external sources. The researcher's perspective (whether as insider or outsider) tends to dominate the research agenda, but participants' perspectives are also of interest. Theory is less a priori and more emergent since human experiences are fallible when it comes to patterns in the real or actual worlds. Learning may be focused on probabilistic explanation and prediction, on what works or what needs to change, on patterns, or on innovation. While subjective understanding is pursued, it would not be to the depth that the pattern of guiding assumptions for an interpretivist/constructivist inquiry would demand.

Quality Criteria Associated with the Pattern of Guiding Assumptions for a Critical Realist Inquiry

The predominant quality criteria for the pattern of guiding assumptions for a critical realist or postpositivist inquiry are the same as for the positivist inquiry pattern. Essentially, as Lincoln et al. (2024) have indicated, the researcher is a "'disinterested scientist' as informer of decision makers, policy makers, and change agents" (p. 79). Validity, however, is taken less as an absolute but more as probabilistic given that people's understanding of an 'externalised' reality, including that of the researcher, is not infallible or complete. When the researcher is pursuing an understanding of 'internalised realities' (i.e., social perceptions/understandings of the world including causality) and experiences of participants, then quality criteria for an interpretivist/constructivist inquiry would come into play.

Table 4.5 Pattern of guiding assumption responses to pivotal questions for a *critical realist* inquiry in social and behavioural research

Pattern of Guiding Assumptions for a *Critical Realist* Inquiry						
Pivotal Question	*Potential Response Choices for Each Pivotal Question*					
Control?	High	**Partial**	**Minimal**	**External source**	Cultural source	Contextually-fluid
Perspective?	**Researcher-as outsider**	**Researcher-as insider**	**Participants**	Balanced	Community	Contextually-situated
Locus?	Externalised reality	Internalised 'realities'	Community realities	**Hybridised loci**	Chaotic realities	
Stance?	Objectivist	Subjectivist	Subjectivist – 'decolonised'	Critical/ Emancipatory/ Transformative	Interested audience	**Pluralist**
Theory?	A priori	**Emergent**	Description	Critique	Re-storying	Performance
Learning?	**Explanation & Prediction**	Deep understanding	**What works/ What & how to change**	**What is/ Patterns of association**	'Decolonisation' of knowledge	**Innovation**

4.2.4 Pattern of Guiding Assumptions for a Critical Inquiry

Table 4.6 displays the pattern of guiding assumptions for a critical inquiry (see, for example, Alveeson & Deetz, 2021; Crotty, 1998, Chaps. 6 and 7; Lincoln et al., 2024; Myers & Klein, 2011; Neuman, 2020, Chap. 4; Vaditya, 2018). This is a generalist pattern that commences with the researcher adopting a value-laden position which holds that expressions of and outcomes resulting from power dynamics in the world, which may be widespread, historical, and structural, are largely asymmetrical and while some dynamics are right, just, and fair, other dynamics are wrong, abusive, and oppressive, thereby creating a demand for those who are dominated to be emancipated and for conditions that facilitate/reinforce/maintain the asymmetry to be transformed or changed. Thus, the researcher sets out to produce change acting as both "advocate and activist" (Lincoln et al., 2024, p. 79). Researcher control over context is minimal; most of the control over context can be attributed to external or cultural forces. Perspectives in the research context tend to be balanced between researcher and the researched, but in general, seek to expose social and power dynamics that oppress the researched. An 'externalised reality' is assumed but community-realities (where a broad view of community is intended) may also be relevant. Theory tends to take the form of a critique, a discourse that exposes oppressive power dynamics and proposes rectifying actions to emancipate the oppressed (or help them to emancipate themselves) but may also take the form of performances or re-storying, where marginalised narratives and historical revisions are brought to the surface and shared.

Table 4.6 Pattern of guiding assumption responses to pivotal questions for a critical inquiry in social and behavioural research

Pattern of Guiding Assumptions for a *Critical* Inquiry						
Pivotal Question	*Potential Response Choices for Each Pivotal Question*					
Control?	High	Partial	Minimal	External source	Cultural source	Contextually-fluid
Perspective?	Researcher-as insider	Researcher-as outsider	Participants	Balanced	Community	Contextually-situated
Locus?	Externalised reality	Internalised 'realities'	Community realities	Hybridised loci	Chaotic realities	
Stance?	Objectivist	Subjectivist	Subjectivist – 'decolonised'	Critical/ Emancipatory/ Transformative	Interested audience	Pluralist
Theory?	A priori	Emergent	Description	Critique	Re-storying	Performance
Learning?	Explanation & Prediction	Deep understanding	What works/ What & how to change	What is/ Patterns of association	'Decolonisation' of knowledge	Innovation

The learning pursued by the researcher focuses on understanding what is occurring in terms of patterns of behaviour and on identifying what and how to change things for the better.

Quality Criteria Associated with the Pattern of Guiding Assumptions for a Critical Inquiry

One important quality criterion for the pattern of guiding assumptions for a critical inquiry concerns how effectively the researcher illuminates/convincingly exposes counter-productive/oppressive power dynamics and conveys their historical/structural situatedness (i.e., successful "erosion of ignorance and misapprehensions", Lincoln et al., 2024, p. 79; or "speaking truth to power", Patton, 2015, Exhibit 9.7). This criterion of *illumination* lays the foundation for transformative and emancipatory actions. A second related quality criterion concerns how clearly the researcher signals or undertakes *actions* to effectively rectify problematic power dynamics with a view achieving social justice, equality/equity, and fairness. When/if the researcher interacts with the less powerful participants, who are caught or trapped within problematic power dynamics, in search of their stories, experiences, and perceptions, the *transparency*, *authenticity*, and *sufficiency* quality criteria for an interpretivist/constructivist inquiry pattern also become relevant. Myers and Klein (2011) noted, however, that, unlike the interpretivist/constructivist pattern, critical research focuses on social categorisations that exist a priori and outside the specific context of the research.

4.2.5 *Pattern of Guiding Assumptions for a Systemic Othering Inquiry*

Table 4.7 displays the pattern of guiding assumptions for a systemic othering inquiry. We refer to this pattern as systemic othering because, while it is somewhat closely related to the critical pattern, it is distinguished by a particular focus on the perspectives, cultures, experiences, historical, and present-day circumstances of people/groups/minorities sharing specific identities or characteristics, such as women (feminist orientation, see, e.g., Crotty, 1998, Chap. 8; Hesse-Biber, 2012, 2014; Hesse-Biber & Griffin, 2015; Davies, 2024), people with different racial/ethnic/cultural/national/religious/political backgrounds/affiliations (see, for example, Donnor & Ladson-Billings, 2024), people who differ in terms of one or more specific demographic characteristics, such as age, socioeconomic status, skin colour, sexual, or gender orientation (e.g., LGBTQIA+ identities and orientations; see, for example, Alexander, 2024; Warner, 2004), and people who have been marginalised or oppressed in or by society because of some other targeted characteristic(s) (e.g., disabilities; see, for example, Nusbaum & Lester, 2024). Systemic refers to the idea that othering frequently reflects institutionalised hegemony (e.g., at a community, state, national, international, or global level) creating a 'them and us' set of attitudes, beliefs, motivations, and behaviours that reflect an exclusionary logic to the exercise of power. Intersectionality is also encompassed by the systemic othering pattern, in that marginalising or oppressive treatment may be experienced by an individual or group possessing a more complex combination of different social categorisation identities (e.g., black women, refugees or migrants with a specific religious or cultural background; see, for example, Haynes et al., 2024). The systemic othering pattern is explicitly interested in power dynamics and how such dynamics privilege a dominant or mainstream group at the expense of or to the exclusion of other individuals/groups characterised as different or 'not one of us'.

From a research perspective, researcher control over context is irrelevant, control is vested in specific external or cultural sources. The researcher's perspective, often as an insider, is important, but so are the perspectives of participants (who typically come from specifically identified marginalised groups) leading to an interest in 'internalised realities', perhaps in hybrid combination with the researcher's perspective.

The critical/emancipatory/transformative stance is shared with the critical pattern in that the research is intended to stimulate change, especially from the perspective(s) of the marginalised or oppressed group. However, as there are strong elements of cultural and group-level identity associated with being marginalised or oppressed, the stance may also be characterised as subjectivist-'decolonised' since the perspective(s) of the dominant or oppressing group are explicitly downplayed, suppressed, cast aside, or, in some cases, rewritten (i.e., a 're-storying' view of theory) by perspectives of the marginalised or oppressed group. In effect, the systemic othering pattern seeks to reverse the traditional backgrounding of the excluded and foregrounding of the dominant. This renders the system othering pattern as heavily value-laden, with a view to righting the wrongs perceived by/associated with the treatment of

Table 4.7 Pattern of guiding assumption responses to pivotal questions for a systemic othering inquiry in social and behavioural research

Pattern of Guiding Assumptions for a *Systemic Othering* Inquiry (encompassing feminist, critical race, intersectional, LGBTQIA+, & disability orientations)						
Pivotal Question	*Potential Response Choices for Each Pivotal Question*					
Control?	High	Partial	Minimal	**External source**	**Cultural source**	Contextually-fluid
Perspective?	**Researcher-as outsider**	**Researcher-as insider**	**Participants**	Balanced	Community	Contextually-situated
Locus?	Externalised reality	**Internalised 'realities'**	Community realities	**Hybridised loci**	Chaotic realities	
Stance?	Objectivist	Subjectivist	**Subjectivist – 'decolonised'**	**Critical/ Emancipatory/ Transformative**	Interested audience	Pluralist
Theory?	A priori	Emergent	**Description**	**Critique**	**Re-storying**	Performance
Learning?	Explanation & Prediction	**Deep understanding**	**What works/ What & how to change**	**What is/ Patterns of association**	**'Decolonisation' of knowledge**	Innovation

marginalised or oppressed groups—a critical/emancipatory/transformational imperative. Theory may take the form of a critique that can inform action and change but may also take the form of an in-depth description of events, viewpoints, actions, and consequences, or, as indicated above, a re-storying form. Learning from the research seeks not only deep understanding of marginalised viewpoints and experiences, but also to highlight patterns of systemic othering, provide information to guide change and emancipatory actions, and, in some cases, creation/sharing of knowledge that has been decolonised or freed from the influence of the dominant or oppressing group.

Quality Criteria Associated with the Pattern of Guiding Assumptions for a Systemic Othering Inquiry

The quality criteria specific to the systemic othering pattern of guiding assumptions are the same as for the critical inquiry pattern, *illumination* and *action*, but with a more intense focus on the specific excluded/othered group(s)/subcultures/communities under study. Frequently, the *transparency*, *authenticity*, and *sufficiency* quality criteria from the interpretivist/constructivist inquiry pattern will apply as well, but with value-neutrality being explicitly denied. These criteria apply, in a more focused way, on how the researcher achieves a deep understanding of the 'othered' (e.g., contextualised histories, perspectives, and experiences) as they are impacted by/contrasted with the those of the 'othering' group/culture. The positioning of the researcher with respect to the 'othered' must also be transparently and authentically revealed, in order to clarify insider or outsider status (systemic othering research is much more difficult to convincingly undertake if the researcher is an outsider relative to the 'othered', i.e., being a member or perceived to be a member of the 'othering' group).

4.2.6 Pattern of Guiding Assumptions for an Indigenous/ First Nations Inquiry

Table 4.8 displays the pattern of guiding assumptions for an Indigenous/First Nations inquiry (see, for example, Archibald et al., 2022; Chilisa, 2020; Cram & Mertens, 2015; Getty, 2010; Hart, 2010; Kovach, 2021; Pidgeon, 2019; Smith, 2021; Windchief et al., 2024). We have identified this pattern of guiding assumptions as being distinct from the systemic othering pattern because, in the case of Indigenous/ First Nations peoples, we are talking about the worldviews, lives, cultures, environments/ecologies of the original inhabitants of a land or country being systematically displaced, marginalised, oppressed, re-interpreted, and/or destroyed by an outsider, more powerful, colonising group/culture moving into their country and taking it over (i.e., substantial or complete cultural obliteration, often accompanied by physical/social displacement and/or harm). With the systemic othering pattern, the 'othered' groups typically exist (or may have always existed) within the same context(s) as the 'othering' group/culture (e.g., different racial, political, or religious subgroups, different sexual or gender identity subgroups) or are an outsider group seeking to move into and live within the 'othering' group/culture (e.g., refugee or migrant groups); i.e., reflecting cultural subjugation/revision. A key principle of the pattern of guiding assumptions for an Indigenous/First Nations inquiry is *decolonisation*. Chilisa (2020, p. 11), for example, describes decolonising as a "process of centering the concerns and worldviews of the colonized Other so that they understand themselves through their own assumptions and perspectives" by "conducting research in such a way that the worldviews of those who have suffered a long history of oppression and marginalisation are given space to communicate from their frames of reference". Thus, the researcher's stance is subjectivist with an intent to decolonise Indigenous/First Nation worldviews, often coupled with an intent to provoke action for social transformation. The researcher does not control context, such control is attributed to external or cultural sources. The worldviews sought reside within Indigenous/First Nation communities as it is there where 'reality' is assumed to reside. However, the researcher's worldviews, either as insider or outsider, also must be revealed (an Indigenous/First Nations inquiry is much more difficult to convincingly undertake if the researcher is a member of the 'colonising' culture and is working without any partnering Indigenous/First Nations researchers). Theory can take a myriad of potential forms as a description or characterisation, a critique, a performance, or as re-storying, rewriting the historical stories and worldviews promulgated by colonisers. The learning sought under the guidance of the Indigenous/First Nations pattern of guiding assumptions seeks deep understanding of Indigenous/First Nations

Table 4.8 Pattern of guiding assumption responses to pivotal questions for an Indigenous/First Nations inquiry in social and behavioural research

Pattern of Guiding Assumptions for an *Indigenous/First Nations* Inquiry						
Pivotal Question	*Potential Response Choices for Each Pivotal Question*					
Control?	High	Partial	Minimal	External source	Cultural source	Contextually-fluid
Perspective?	Researcher-as outsider	Researcher-as insider	Participants	Balanced	Community	Contextually-situated
Locus?	Externalised reality	Internalised 'realities'	Community realities	Hybridised loci	Chaotic realities	
Stance?	Objectivist	Subjectivist	Subjectivist – 'decolonised'	Critical/ Emancipatory/ Transformative	Interested audience	Pluralist
Theory?	A priori	Emergent	Description	Critique	Re-storying	Performance
Learning?	Explanation & Prediction	Deep understanding	What works/ What & how to change	What is/ Patterns of association	'Decolonisation' of knowledge	Innovation

worldviews, patterns and relationships associated with those worldviews, actions to create social change and, in general, to decolonise knowledge associated with coloniser worldviews. The values privileged are explicitly those of the Indigenous/First Nations community participating in the research.

Quality Criteria Associated with the Pattern of Guiding Assumptions for an Indigenous/First Nations Inquiry

There are several quality criteria specific to the pattern of guiding assumptions for an Indigenous/First Nations inquiry. All aspects of the research must cultivate and maintain *mutual respect* for all parties involved, as a pathway toward decolonisation. With an Indigenous/First Nations inquiry, there is normally an expectation of genuine *reciprocity* where the researcher returns value from their research back into the Indigenous/First Nations community. The research must provide effective *ethical protection* of Indigenous/First Nations rights which encompasses safeguarding the knowledge that the researcher gains access to or develops. Indigenous/First Nations research would be expected to demonstrate effective *decolonisation* of the methodologies employed, while still maintaining quality in terms of *transparency*, *authenticity*, and *sufficiency* (from the pattern of guiding assumptions for an interpretivist/constructivist inquiry). Finally, the researcher's positioning as insider or outsider must be *transparent* and *authentic*, especially important for an outsider, who may be seen as part of the colonising community.

4.2.7 Pattern of Guiding Assumptions for a Participatory Inquiry

Table 4.9 displays the pattern of guiding assumptions for a participatory inquiry (see, for example, Fear et al., 2006; Heron, 1996a; Heron & Reason, 1997, 2006; Lincoln et al., 2024; Reason & Rowan, 1981; Thiollent, 2011). Participatory inquiry (also referred to as co-operative inquiry, see Heron, 1996a) has, as a central principle, a collaborative relationship between researcher and participants/groups/communities working together to solve problems and create change. The outcome of such collaboration may be change at a local level (as with participatory action research; see, for example, McIntyre, 2008) or the development of an innovation or practice changes to meet certain needs (as with participatory ergonomics; see, for example, Burgess-Limerick, 2018; Noro & Imada, 1991). In a participatory inquiry, researcher and participants are on an approximately equal footing with respect to power and perspective and together, as co-researchers, work toward defining the goals and research questions and propositions. Control over context is contextually-fluid being decided on the basis of researcher and participant needs. There is a balanced interest in perspective learning and sharing, although the researcher's role, whether as insider or outsider, needs to be clearly identified and explored. Reality may be considered to reside at multiple levels (external as well as internal), constituting a hybridised locus. Participatory collaboration is inherently subjectivist in orientation, but also has a transformational intent, to change practices or improve conditions for participants. Theory is typically emergent, forming propositional knowledge that evolves as collaboration continues, but may also be a priori, if the participatory interactions focus on the merits of a pre-existing framework or practice. Theory may also be embodied in performances (i.e., presentational knowledge) at the group-level. Learning intentions focus on the creation/evaluation of change, usually at a local level where participants work/live but may also take the form of an emergent innovation to support change and improvement. Heron and Reason (1997) argued that such learning evolves from the confluence of four types of emergent knowledge: experiential (what is known from direct interactive/social experience), presentational (symbolic knowledge used to convey meaning); propositional (conceptual knowledge), practical (knowledge about how to do things). Heron (1996b, 1996c) argued for the 'primacy of the practical' where the development of practical knowledge is supported by the scaffolding of experiential, propositional, and presentational knowledge.

Quality Criteria Associated with the Pattern of Guiding Assumptions for a Participatory Inquiry

There are three quality criteria specific to the participatory inquiry pattern of guiding assumptions. Heron and Reason (1997) argue that that effective *congruence* between experiential, presentational, propositional, and practical knowledge (where the first three inform the latter) is necessary for realising change and improvement. For participatory inquiry to be successful, all involved must effectively participate, which means that *inclusivity* is an important criterion for quality. Thus, there is no room

Table 4.9 Pattern of guiding assumption responses to pivotal questions for a participatory inquiry in social and behavioural research

Pattern of Guiding Assumptions for a *Participatory* Inquiry						
Pivotal Question	*Potential Response Choices for Each Pivotal Question*					
Control?	High	Partial	Minimal	External source	Cultural source	**Contextually-fluid**
Perspective?	Researcher-as outsider	Researcher-as insider	Participants	**Balanced**	Community	Contextually-situated
Locus?	Externalised reality	Internalised 'realities'	Community realities	**Hybridised loci**	Chaotic realities	
Stance?	Objectivist	**Subjectivist**	Subjectivist – 'decolonised'	**Critical/ Emancipatory/ Transformative**	Interested audience	Pluralist
Theory?	**A priori**	**Emergent**	Description	Critique	Re-storying	**Performance**
Learning?	Explanation & Prediction	Deep understanding	**What works/ What & how to change**	What is/ Patterns of association	'Decolonisation' of knowledge	**Innovation**

or tolerance for voices to be marginalised or sidelined; the social and physical environment for participation must be safe and supportive for all involved. Finally, for participatory inquiry to succeed, *effective actions leading to effective change* must be realised. Heron (1996b) stated that "to say that practice consummates the prior forms of knowing on which it is grounded, is to say that it takes the knowledge quest beyond justification, beyond the concern for validity and truth-values, into the celebration of being-values" affirming "what is intrinsically worthwhile, human flourishing, by manifesting it in action".

4.2.8 Pattern of Guiding Assumptions for a Postmodernist Inquiry

Table 4.10 displays the pattern of guiding assumptions for a postmodernist (or post-structuralist) inquiry (see, for example, Alvesson & Deetz, 2021; Byrne & Callaghan, 2014, Chap. 3; Crotty, 1998, Chap. 9; Fielding, 2009; Kuntz, 2024; Neuman, 2020, Chap. 4). In a postmodernist inquiry, arts and humanities merge with the social sciences and the focus is on the present and potential future (focusing on 'becoming' rather than 'is'), not the past. A postmodernist inquiry embraces absolute relativism and subjectivity in that no better or worse or true or false explanations or accounts can be identified; truth is absolutely relative. Causality is considered to be too complex to be disentangled. All perspectives are valued, but the researcher's no more so than

Table 4.10 Pattern of guiding assumption responses to pivotal questions for a postmodernist inquiry in social and behavioural research

Pattern of Guiding Assumptions for a *Postmodernist* Inquiry						
Pivotal Question	*Potential Response Choices for Each Pivotal Question*					
Control?	High	Partial	**Minimal**	External source	Cultural source	Contextually-fluid
Perspective?	**Researcher-as outsider**	Researcher-as insider	**Participants**	Balanced	Community	Contextually-situated
Locus?	Externalised reality	Internalised 'realities'	Community realities	Hybridised loci	**Chaotic realities**	
Stance?	Objectivist	**Subjectivist**	Subjectivist – 'decolonised'	**Critical/ Emancipatory/ Transformative**	**Interested audience**	Pluralist
Theory?	A priori	Emergent	Description	**Critique**	Re-storying	**Performance**
Learning?	Explanation & Prediction	**Deep understanding**	What works/ What & how to change	What is/ Patterns of association	'Decolonisation' of knowledge	Innovation

anyone else's; the researcher is considered to be an outsider to everyone else's (e.g., participants') truths. Researcher control over context is minimal and they adopt a stance of interested observer or audience, rather than theoriser. Realities are considered to be complex, chaotic, patternless, in a constant state of flux, and must be deconstructed to remove the baggage of the systemic constructions of others (e.g., master narratives). A postmodernist inquiry is inherently interested in breaking with/'short-circuiting' (as Kuntz, 2024, p. 284, put it) the traditional expectations and practices of academic research and research presentation, arguing that understanding emerges through examining multiple texts/discourses reflected or embedded within multiple genres/mediums of communication. A critical/emancipatory/transformative stance is often adopted where the critique is intended to shock and sweep away any notion of generality, logic, or commonness of meaning, and often has a performative/artistic realisation. Learning seeks only deep understanding, not generalisation or structural knowledge.

Quality Criteria Associated with the Pattern of Guiding Assumptions for a Postmodernist Inquiry

There are two specific quality criteria that are relevant to pattern of guiding assumptions for a postmodernist inquiry: *crystallisation* and *aesthetic value*. Ellingson (2009) argued that effective *crystallisation* (in contradistinction to triangulation) emerging from mixed-genre sources/texts is a hallmark of meaningful and impactful postmodernist research. Richardson (2000, cited in Ellingson, 2009, p. 3) stated that "the central image of 'validity' in postmodernist texts" is the "crystal, which combines symmetry and substance with an infinite variety of shapes, substances, transmutations, multidimensionalities, and angles of approach" and provides "deepened, complex, and thoroughly partial understanding". Effective crystallisation reflects authentic and transparent enactment/communication of personal truths, multiple perspectives, accounts, explanations, and meanings. Neuman (2020, Table 4) suggested that good evidence in a postmodernist inquiry has *aesthetic value* that "resonates with people's inner feelings", which creates impact and stimulates understanding.

4.2.9 *Pattern of Guiding Assumptions for a Pragmatist Inquiry*

Table 4.11 displays the pattern of guiding assumptions for a pragmatist inquiry (see, for example, Biesta, 2010; Kaushik & Walsh, 2019; Morgan, 2014). Pragmatism embodies an imperative that asks the researcher to employ approaches and methodologies (even multiple patterns of guiding assumptions or 'paradigms', see discussions in Greene & Hall, 2010) that they feel will work best for addressing a particular research problem and its attendant research questions in context. In a sense, the pattern of guiding assumptions for a pragmatist inquiry can be seen as more of a meta-pattern in that other patterns of guiding assumptions, each yielding quantitative, qualitative, or hybrid/merged data, may be implicated within it, as research needs dictate. The stance adopted, therefore, is inherently pluralist, with respect to achieving an appropriate balance between objectivism and subjectivism. Pragmatism has typically been seen as most closely aligned with mixed methods research, although this should not be taken as an imperative. Researcher control over context is seen as contextually fluid, with control being exercised when and where it makes sense to do so. Whose perspective guides/drives the research is considered to be contextually-situated and may emerge as a consequence of adopting different patterns of guiding assumptions for different aspects of the research. The locus for reality is hybridised, allowing for interest in either/both externalised and internalised realities, as needed. Theorising may be either a priori or emergent, also as a consequence of adopting different patterns of guiding assumptions for different aspects of the research. The learning intent of the pragmatist pattern is diverse, perhaps seek explanation and prediction (when theories are tested), deep understanding (when participant perspectives are sought), assessing what works and what and how to change (when there is an evaluative intent to the research), identifying/utilising patterns and associations (when the research has an exploratory intent), and/or focusing on the development/evaluation of a specific social or technological innovation. In short, the pragmatist inquiry pattern seeks 'actionable knowledge' (Kelly & Cordeiro, 2020). The pragmatist pattern of guiding assumptions is, by definition, an adaptive pattern, which means that the researcher will be continually challenged by the need to ensure consistency and transparency in all aspects of their research (i.e., demonstrating congruence between patterns of

Table 4.11 Pattern of guiding assumption responses to pivotal questions for a pragmatist inquiry in social and behavioural research

Pattern of Guiding Assumptions for a *Pragmatist* Inquiry						
Pivotal Question	*Potential Response Choices for Each Pivotal Question*					
Control?	High	Partial	Minimal	External source	Cultural source	Contextually-fluid
Perspective?	Researcher-as outsider	Researcher-as insider	Participants	Balanced	Community	Contextually-situated
Locus?	Externalised reality	Internalised 'realities'	Community realities	Hybridised loci	Chaotic realities	
Stance?	Objectivist	Subjectivist	Subjectivist – 'decolonised'	Critical/ Emancipatory/ Transformative	Interested audience	Pluralist
Theory?	A priori	Emergent	Description	Critique	Re-storying	Performance
Learning?	Explanation & Prediction	Deep understanding	What works/ What & how to change	What is/ Patterns of association	'Decolonisation' of knowledge	Innovation

guiding assumptions (where several patterns have been adopted), methodological choices, and contextual constraints and opportunities). Of all the patterns discussed thus far, the pattern of guiding assumptions for a pragmatist inquiry resonates most closely with pluralist systemic thinking, but jettisons the a priori necessity or preference for combining quantitative and qualitative data forms in a single investigation typically espoused by mixed methods researchers.

Quality Criteria Associated With the Pattern of Guiding Assumptions for a Pragmatist Inquiry

O'Cathain (2010) presented a framework of quality criteria specifically pertinent to the pragmatist pattern of guiding assumptions. The three criteria we discuss here are integrative distillations of those presented by O'Cathain. The most important criterion is demonstrating *congruence* between the choices made while navigating the 'Data Triangle' and relevant and appropriate pattern-specific quality criteria, depending upon which patterns of guiding assumptions have been pragmatically adopted. There is no index of such congruence, it is demonstrated through clear, logical, and robust argumentation. A second criterion concerns the *effectiveness of planning* to ensure that research questions drive methodological choices, not vice versa. How and why the researcher made the methodological choices they did needs to be clearly communicated; in short, conveying how the researcher created their own actionable research journey. A third criterion concerns the effectiveness of *integration of learning* which focuses on how well the researcher pulls together the learning achieved from each component of their research into coherent and actionable knowledge.

4.2.10 Patterns of Guiding Assumptions and Pluralist Systemic Thinking

Figure 4.3 brings together all nine broad patterns of guiding assumptions in a qualitative version of an icon plot (see Cooksey, 2020, Procedure 5.3), highlighting the visual pattern differences more so than the substance particulars and thereby showcasing the diversity of possibilities for assumptions to guide social and behavioural research inquiries. There are several considerations to keep in mind with respect to patterns of guiding assumptions and their relationships to specific downstream methodological choices. Many researchers believe that there is a relationship between pattern of guiding assumptions and demand for gathering a specific form of data (e.g., quantitative data gathered under pattern of guiding assumptions for a positivist inquiry; qualitative data gathered under the pattern of guiding assumptions for an interpretivist/constructivist, critical, or a systemic othering inquiry). This, however, is a false set of equivalences. It is entirely possible to gather and make effective use of either quantitative or qualitative data, or perhaps hybrid/merged data under any pattern of guiding assumptions. The data will simply be ascribed different meanings and handled differently, drawing upon the Transformative data-shaping data gathering strategy. Thus, while there might be strong preferences for specific data forms under specific patterns of guiding assumptions, this does not mean that non-preferred forms of data are irrelevant or to be avoided in research conducted under that pattern. [Note that many research methods texts make the mistake of referring to approaches to gathering quantitative data as reflecting a 'quantitative' paradigm and approaches to gathering qualitative data as reflecting a 'qualitative' paradigm. Quantitative and qualitative refer simply to forms of data, not to any specific pattern of guiding assumptions that underpin such data.]

Adopting one specific pattern of guiding assumptions does not negate the possibility of adopting another set of guiding assumptions within the same research journey (e.g., see Bowers, 2011). In the past, there has been such a negation tendency (a non-pluralist logic), leading researchers to adopt a preferred pattern of guiding assumptions and to discount the value of other patterns of guiding assumptions. As a consequence, a kind of pattern-based arrogance built up over time, leading to specific patterns becoming entrenched in some disciplines (i.e., in essence becoming the dominant 'paradigm' for the discipline; e.g., positivist pattern in experimental psychology, economics, marketing, and management science; interpretivist/constructivist or systemic othering pattern in humanistic psychology; postmodern pattern for arts-based research, pragmatist pattern for mixed methods research). The problem with pattern entrenchment is that researchers and reviewers tend to make their judgments about research conducted under other patterns of guiding

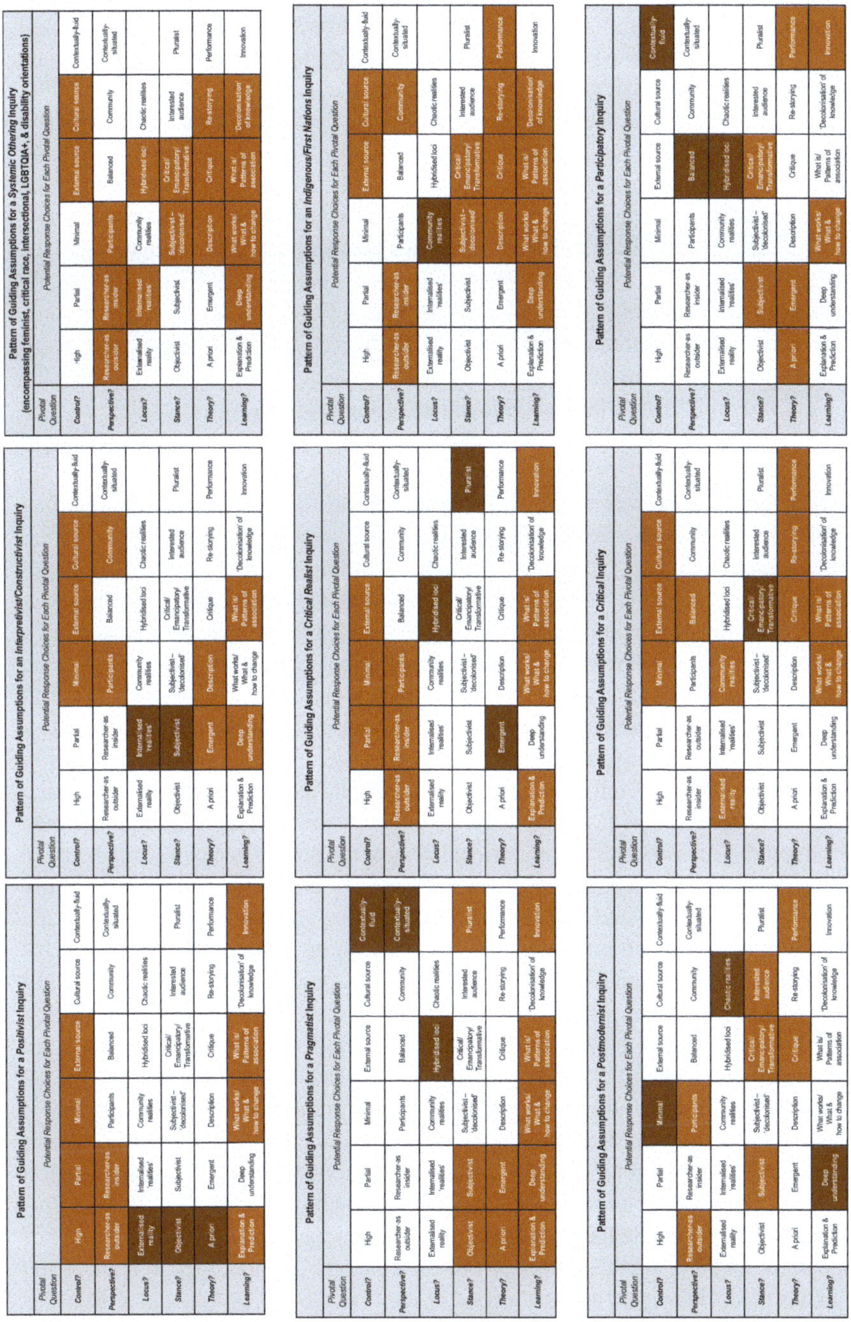

Fig. 4.3 Icon plot showcasing the diversity of patterns of guiding assumptions for social and behavioural research inquiries

assumptions using quality criteria from within their own preferred pattern of guiding assumptions—a confirmational cognitive process that is doomed to making erroneous judgments. Taking a pluralist systemic thinking approach to social and behavioural research helps to overcome such arrogance by asking researchers to keep an open mind about the potential value of working under alternative patterns of guiding assumptions, even within the same research project. The challenge is for researchers to do this in a way where logical inconsistencies don't get introduced into the research processes. Managing to configure one's research to be convincing by harnessing the potential benefits of using different patterns of guiding assumptions and, at the same time, avoiding potential logical inconsistencies is a tough cognitive challenge (see Brocklesby, 1997, for a discussion of some of these challenges). There are a couple of strategies to consider that could help in meeting this challenge: (1) use some type of sequential research configuration (see Chap. 5) where different patterns of guiding assumptions are employed in different stages of the research and/or (2) involve a research team where each member of manages an aspect of the research under a specific pattern of guiding assumptions and the team then works together to integrate what they find.

More and more, researchers are becoming aware that any pattern of guiding assumptions offers both advantages as well as disadvantages (and even blind spots) when adopted. No one pattern of guiding assumptions offers only advantages meaning that no one pattern of guiding assumptions can be promoted as being better in some essential way compared to all others—each simply reflects a researcher's particular pattern of answers to the pivotal questions. Importantly, by harnessing pluralist systemic thinking when configuring research, it may be possible to offset the disadvantages of one pattern of guiding assumptions employed in one part of the research with the advantages of a different pattern of guiding assumptions in another part of the research. It is critical in any high-quality research investigation that the researcher provides clear arguments about the guiding assumptions they have adopted. This is especially important because it displays to anyone interested in learning from the research (the 'audience') that the researcher has thought carefully about what they are assuming, and that the decisions and choices they have made throughout their research process are consistent with those assumptions. This helps the audience to understand where the researcher is coming from and helps them to decide which criteria for judging the quality of the research they should be relying upon. If the researcher leaves their assumptions hidden or undiscussed, they are depriving the audience of critical information they need to correctly understand the choices they have made. Ironically, in practice, researchers will find that a sizeable portion of the published literature they read is associated with authors who do not clearly unpack their guiding assumptions for the audience (perhaps because of length restrictions imposed by journals), and the audience is left to ferret this out for themselves using their knowledge of the discipline, contextual knowledge, the journal of publication, and any signals embedded in the strategic methodological choices the author has made. However, this is not good research practice. The better practice is for researchers to be explicit and transparent about their adopted pattern(s) of guiding assumptions and how they are connected to their research questions/propositions/

hypotheses in order to remove any ambiguity about where they are coming from (Allard-Poesi & Maréchal, 2001: 46–47).

It is also good practice to let the problem and the research context(s) inform the choice of the most suitable and feasible pattern of guiding assumptions and associated practices and strategic methodological choices, and not researcher, stakeholder, or discipline preferences. Too often, researchers will let their paradigm and methodological preferences, capabilities, and comfortable ways of thinking determine which research problems are investigated and how; stakeholders can be guilty of this as well. This amounts to forcing the research problem to fit a perspective, a particular way of thinking, and specific method(s), thereby ruling out other potentially beneficial ways of thinking, other perspectives and/or other methods. For example, there can be a tendency for researchers who do not like quantitative methods/data to automatically and uncritically gravitate toward a pattern of guiding assumptions associated with qualitative data gathering, even when such approaches do not really suit the research question they want to address. Conversely, researchers with a strong affinity for the assumptions associated with the scientific method (i.e., the pattern of guiding assumptions for a positivist inquiry) and its preference for quantitative data and statistical analyses, may express a view that non-positivist patterns of guiding assumptions and their association with qualitative methods/data are 'soft', subjective and imprecise and have no place in their research endeavours even when their research questions may be best addressed by them. The better strategy, in all cases, is to choose a pattern of guiding assumptions to fit the research problem and contexts (note that this is a logic borrowed directly from the pragmatist pattern). If this demands that the researcher develop new skills and capabilities or learn a new way of thinking, then it is better to do so than to force the research problem to fit their current modes of thinking, skills and capabilities. In short, the researcher should learn to question their personal status quo with respect to patterns of guiding assumptions, which is a specific type of reflexivity.

There are several different and logically defensible ways that a researcher could argue for their choice of pattern(s) of guiding assumptions:

- *contrast argument* where the researcher explicitly and defensibly contrasts the pattern of guiding assumptions they will be adopting against other potential patterns of guiding assumptions, highlighting that the benefits of the chosen pattern of assumptions would not be achievable under other patterns of guiding assumptions;
- *assertion of congruence* where the researcher makes a straightforward assertion that the pattern of guiding assumptions they will be adopting is most congruent with their research needs;
- *previous research deficiency* where the researcher argues that previous research guided by a particular pattern of assumptions was flawed or deficient in some way, and, as a consequence, their research is being positioned to overcome or circumvent those flaws or deficiencies by adopting an alternative pattern;
- *need for a pluralist approach* where the researcher argues for using multiple patterns of guiding assumptions, not from a basis of deficit in previous approaches,

but from an assertion that more stands to be gained if a pluralist approach is adopted. By a pluralist approach, we mean that, for distinct aspects or phases of the research, different patterns of guiding assumptions are adopted with the ultimate goal of integrating the partial stories gleaned under each pattern. By doing this, some of the weaknesses associated with one pattern of guiding assumptions may be compensated for by the strengths of another pattern. The pluralist argument is the most complex as well as most demanding one to make in terms of the systemic thinking required for getting it right.

4.3 Pattern-Independent Meta-criteria for Judging Research Quality

Thus far, we have only considered research quality criteria specific to certain patterns of guiding assumptions. The operative word here is 'specific'. For example, quality criteria under the pattern of guiding assumptions for positivist inquiries (i.e., validity criteria) have little meaning under patterns of guiding assumptions for interpretivist/ constructivist inquiries (despite historical attempts to force them to) and vice versa. This makes it difficult, if not impossible, to compare the quality of research conducted under different patterns of guiding assumptions, because no quality criterion is shared with all patterns. Kuhn (2012) identified this as the problem of 'incommensurability' meaning that pattern (or 'paradigm', in his words)-specific quality criteria have difficulty achieving clear meaning and utility outside of their own boundaries. This problem is further exacerbated if one includes practice-based research in the mix, which require a focus on the practical utility of research outcomes (e.g., the critical realist, participatory, or pragmatist inquiry patterns) (see, e.g., Bleijenbergh et al., 2011). A number of researchers have attempted to identify higher-order more inclusive research quality criteria (e.g., Bryman et al., 2008; Caracelli & Riggin, 1994; Dellinger & Leech, 2007; Fàbregues & Molina-Azorin, 2017; Healy & Perry, 2000; Krathwohl, 1985; Marshall, 1990; Murthy, 2000; O'Cathain, 2010; Sale & Brazil, 2004), but none has really taken research quality criteria to the next level as pluralist systemic thinking requires. Additionally, the frameworks that have emerged have tended to be very cumbersome where the same two tactics appear to be in play: either additional conceptual torque is placed on positivist or 'traditional scientific' validity criteria (see, for example, Collins, 2015; Le Compte & Goetz, 1982, as well as discussions in Lincoln et al., 2024) or unique quality criteria are devised for research conducted under different guiding assumptions or for different data forms (also see discussions in Lincoln et al., 2024). For example, O'Cathain (2010) proposed a comprehensive framework of quality criteria, spanning qualitative, quantitative, and mixed methods research, which encompassed 8 different domains and 44 different criteria. Patton (2015, Exhibit 9.7) identified no fewer than seven alternative sets of quality criteria that could be relevant to qualitative inquiry: traditional scientific research criteria, social construction and constructivist criteria; artistic and evocative criteria; participatory and collaborative criteria, critical change criteria, systems thinking and complexity criteria, and pragmatic, utilisation-focused criteria.

The increasing diversity in potential quality criteria in social and behavioural research has value in that it is forcing researchers to think more deeply about what is most important to worry about when conducting their research. However, such diversity is a two-edged sword; it leaves researchers with nowhere to turn for a coherent way to provide holistic, systemic, and comparative judgments about the quality of their own research vis-à-vis the research of others, independently of the pattern of guiding assumptions they adopt and the research choices they make (a point indirectly made by Easterby-Smith et al., 2008, in their discussion of pluralism in qualitative research). This deficit creates difficulties for researchers wishing to work across patterns of guiding assumptions as well as for consumers or reviewers of research who may be tempted to apply criteria from within their own preferred pattern to make potentially ill-founded judgments about the quality of research arising from within other patterns (e.g., Armstrong, 1997; Hojat et al., 2003; Pratt, 2008; Savall et al., 2008). The moral of the story thus far is that no coherent set of meta-criteria for judging research quality has been available (Metcalf, 2006, purported to offer a set of 'guarantors' for knowledge claims, based on Churchman's systems theorising, but even he fell short of offering coherent and paradigm-independent criteria). The survey of social policy researchers in the UK by Bryman et al. (2008) certainly indicated there was a need for criteria that had more universal appeal for use in evaluating mixed methods research. Hammersley (2009) argued that both tactics, in essence, were problematic and that what was needed was to recognise that

> the differences in view cannot be mapped onto the distinction between quantitative and qualitative approaches, even though the argument is often formulated in those terms. It is essential to engage with the complexities of this issue if any progress is to be made in resolving the disputes. (p. 51)

If we think pluralistically, complexly, and systemically about social and behavioural research, then criteria for holistically evaluating research quality must be able to operate effectively in that arena. This cannot happen if the only approaches taken are either linear extensional (from one paradigm to another—a 'paternalistic' approach) or fragmentary (a reductionist contextually-specific approach). Accordingly, researchers still need to deal with the following two questions: (1) how can researchers evaluate the quality of a social or behavioural research outcome independently of the pattern of guiding assumptions that underpinned that research; and (2) how can researchers compare the quality of research investigations that are guided by very different patterns of guiding assumptions? To address both questions, Cooksey (2001, 2006, 2008; see also Cooksey & McDonald, 2019, Chap. 9) developed a simpler logical system of higher-order criteria, centred on the concept of convincingness, for judging the quality, coherence, and value of any type of social and behavioural research, irrespective of assumptions and philosophical alignments; i.e., quality criteria that could be applied across all patterns of guiding assumptions. These higher-order criteria are referred to as meta-criteria in order to clearly signal that they are intended to operate above and beyond where localised pattern-specific quality criteria exert their influence. The criteria for judging research quality held under specific patterns of guiding assumptions are not ignored, rather they are subsumed as a part of what the meta-criteria focus on, especially with respect to how consistently the researcher meets the quality expectations inherent in their adopted choice of pattern(s) of guiding assumptions and produces a research outcome that is congruent with those expectations. The meta-criteria, by their very nature, are explicitly pluralist in intent and invoke systemic thinking about research. They interconnect and operate synergistically (building upon complexity theory and pluralist reasoning) and they can be used to assist in planning and evaluating one's own research (anticipating judgments of convincingness, hopefully building its potential achievement into the research via the choices the researcher makes) as well as to assist in evaluating the planned or actual research proposals or outcomes of others (evaluations of convincingness).

The focal point of the meta-criteria is the assumption that, in the end, research is fundamentally a 'storytelling' activity (Daft, 1983) and that research quality is reflected in the impacts of that story on the consuming audience, interested stakeholder, sponsor, community, culture, or readership. The research act itself focuses on conceiving, constructing, and displaying all the necessary components of that story. This logic is similar to Lincoln's (1995) arguments for the 'inquiry community' as the 'arbiter of research quality' in qualitative and interpretive research but extends those arguments to research conducted under any particular pattern of guiding assumptions. Quality is inherent in the configuring and execution of the research itself in that the researcher can do things as part of their research to improve quality. This is where pattern-specific quality criteria operate. However, and more importantly, quality is vested externally in the value of social and behavioural research as seen by relevant others. Thus, inferences about the quality of research may be drawn by those who review or examine it, read it, fund it, build upon it, change practices based on it, supervise it, use it, apply it, provide input to it, critique it, learn from it, take ownership of it, and share knowledge arising from it with others. In the end, it does not matter if the researcher thinks their research is of high quality, it is the judgments of quality made by others that matter. Thus, the researcher needs to be oriented toward convincing others of the quality and value of their research.

Figure 4.4 displays a mind map of the system of meta-criteria, adapted and expanded from Cooksey and McDonald (2019, p. 286). There are 12 meta-criteria in the system. The 12 meta-criteria are grouped into three higher-order domains, *contextualisation*, *realisation*, and *explication*, which then inform judgments about overall research quality, captured in the concept of *convincingness*. As well, there are interconnections and dynamic interplay between specific meta-criteria as well as between the three domains of meta-criteria, which means that focusing on quality in one area may enhance or have implications for quality in another area.

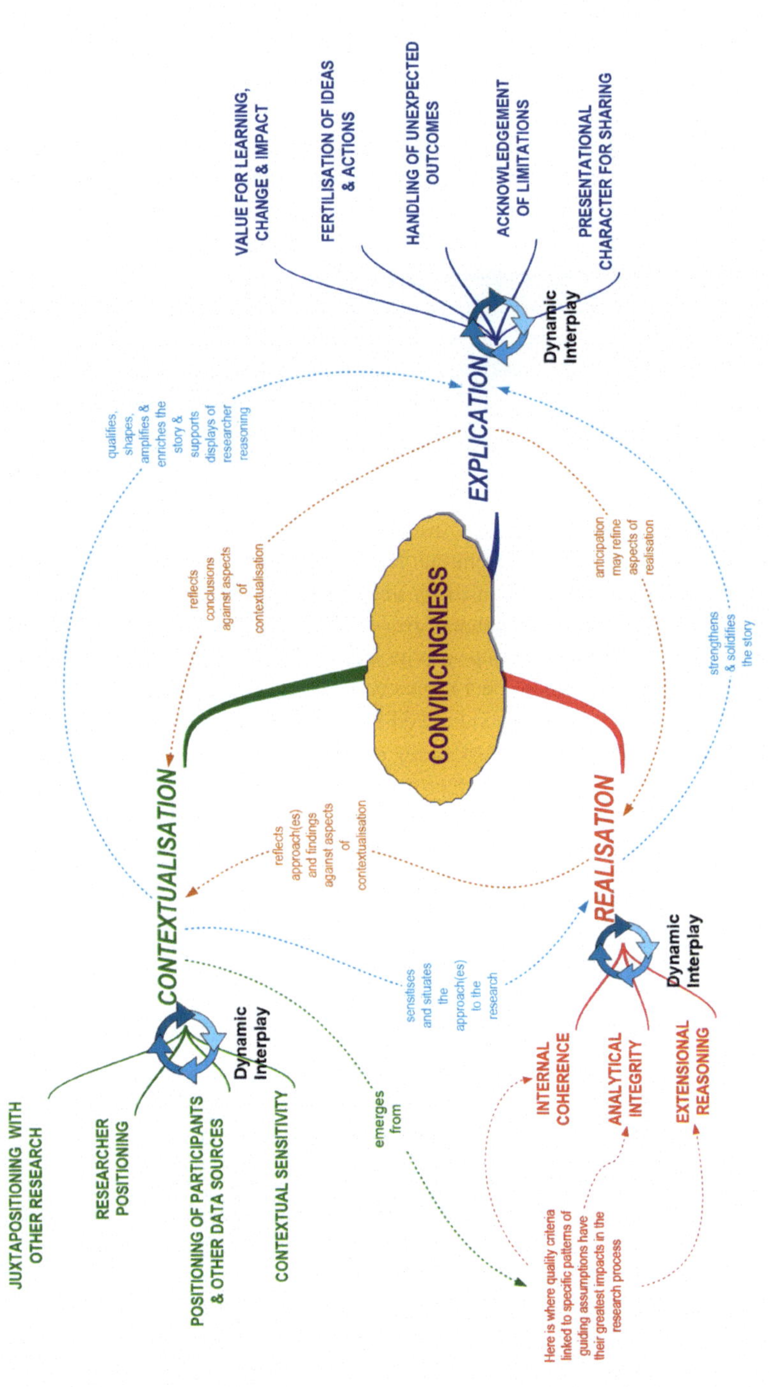

Fig. 4.4 Mind map of 12 meta-criteria and their interrelationships in pursuit of *convincingness* (adapted from Cooksey & McDonald, 2019, Fig. 9.5, p. 286)

4.3.1 Convincingness

Convincingness constitutes the central quality judgment and addresses the most important question regarding research quality: is the research story, as an integrated whole, convincing with respect to the arguments being made and the learning be advocated? That research should be convincing to a relevant audience is not a new idea since it begs the question: convincing to whom? Golden-Biddle and Locke (1993) talked about research as being convincing and that being convincing has various dimensions. However, their focus was solely on research conducted under the pattern of guiding assumptions for an interpretivist/constructivist inquiry. Aguinis et al. (2010) and Duncan and Harrop (2006) both presented 'consumer-centric' perspectives on research quality but offered no meta-criteria for judging quality that would work independently of pattern of guiding assumptions. Consumers may be seen to encompass a diverse range of stakeholders (e.g., communities, cultures, societies, organisations, and other institutions) having an interest in, potentially benefitting from, or potentially adversely affected by the research, potential users of the outcomes from the research (e.g., academics, practitioners, potential innovation adopters), and funding bodies and other types of contractors who provide resources to support the research. Convincingness is influenced by and, in most cases, is explicitly linked to, a specific presentation/performance of the research story (e.g., an outcome such as a proposal, report, thesis, dissertation or portfolio, conference paper or presentation, published article) or artistic performance. Telling a convincing research story comprises attending to those aspects that add power, weight, and value to that story in ways that effectively reach one or more specific audiences while at the same time managing or accounting for any emergent issues or concerns having the potential to confuse or devalue that story. It also requires that the researcher knows who potentially needs to be convinced by their research, i.e., their intended audience(s).

4.3.2 Three Domains of Meta-criteria

An overall judgment about convincingness represents a cognitive synthesis of judgments about the meta-criteria, focusing on answers to three principal questions about research quality, where each question implicates a specific domain of relevant meta-criteria (refer to Fig. 4.4):

- *How well has the research been contextualised?* This question defines the *contextualisation* domain of meta-criteria, which establishes the boundary conditions, research goals, questions, propositions, and/or hypotheses, opportunities, constraints, and assumptions surrounding the research, the research context(s), the researcher, and all relevant data sources. Contextualisation meta-criteria help to situate the research within relevant larger systemic contexts, including the extant literature, prior research and knowledge building efforts, organisations, institutions, and society as well as the researcher's own context (including choices about patterns of guiding assumptions), the context(s) relevant to research participants and other data sources and the context(s) relevant to potential research consumers/audiences (anticipating possible research outcomes). Thus, this domain of meta-criteria captures the dynamics and specifics associated with the contextualised emergent positioning of a specific research journey.

- *How appropriately have the research processes and learning been realised?* This question defines the *realisation* domain of meta-criteria, which focuses on the methodological steps taken, configurations of, and implementation procedures associated with, the data gathering strategies adopted, approaches to identifying, selecting, and safeguarding data sources, and procedures and technologies employed to access, assemble, analyse and establish/construct patterns, relationships, meanings, interpretations and extensions using the information/data collected during the research journey. In short, the realisation domain concerns how well the researcher has navigated the'Data Triangle'. It is in the context of the realisation meta-criteria that we also explicitly look for congruence between researcher decisions and the expectations and 'localised' quality criteria established by adopted pattern(s) of guiding assumptions. Thus, these meta-criteria capture the dynamics and specifics associated with the contextualised emergent execution of the research journey.
- *How appropriately and carefully has the research been explicated?* This question defines the *explication* domain of meta-criteria, which focuses on how and how well the story of the research journey, in the context of a specifically-produced research outcome, is configured and presented, how well the researcher has or will handle unexpected findings, what limitations and caveats apply to the research and what that story might mean for other researchers, for theory, for methodology, for actions, for practitioners and/or for other relevant stakeholder groups. The research outcome may be anticipatory in intent (as with a research proposal) or retrospective (as in a thesis or journal article). The focus here is on meeting the needs and expectations of anticipated research consumers and audiences through the sharing of specific research outcomes. Thus, explication meta-criteria capture the dynamics and specifics associated with the emergent contextualised stories about a specific research journey and their associated implications and, where relevant, applications. This meta-criteria domain is most closely related to and predictive of research impact.

As depicted in Fig. 4.4, the higher-order contextualisation, realisation and explication domains of meta-criteria dynamically interconnect, cross-influence and, often, trade-off with each other to build a convincing case for the research and its findings, conclusions, and implications. The 12 meta-criteria to be discussed in more detail below share important dependencies with each other as they work together to inform judgments about the convincingness of research for given purposes. These meta-criteria should not be expected to be of equal importance/relevance in all research investigations. This means that an important aspect of judging convincingness is reflecting on and applying an appropriate balance in emphasis among the meta-criteria, given the contexts, intentions and emergent textures of the research.

Contextualisation Meta-criteria

Juxtapositioning with Other Research

This meta-criterion focuses on the role(s) that the extant literature and previous research experiences and knowledge building efforts play in informing the research journey. A convincing story will coherently and critically argue for how and where the current research journey is situated relative to previous relevant research with respect to the research topic, perspectives, frames and frameworks, data sources, context(s), theories, practices (including tensions/dynamics between theory and praxis), and methodological choices. This meta-criterion forms an important aspect of research contextualisation because it is often the starting point for the researcher to build the case for their research. Juxtapositioning requires the researcher to engage in critical thinking and evaluative reflection on past research as it intersects their own emerging research agenda. and the system of meta-criteria can actually help in that critical evaluative reflective process. Often, what contributes to convincingness is the emergence, through juxtapositioning with other research, of a niche for the research, an innovative turn or twist, theoretical development or extension, new contextual application, change imperative, or a logical or methodological gap where the researcher can see that research is needed. In this sense, juxtapositioning with other research may reveal or highlight critical leverage points for the current research.

Researcher Positioning

This meta-criterion reflects the importance of understanding where/how/why the researcher is situated in the research context. A research story that displays the researcher's context, background, motivations, intentions, assumptions and values, cultural status, as well as constraints encountered and opportunities pursued, reinforces a more open and convincing account. This meta-criterion demands that the researcher critically reflect on their own role(s) in the research context as well as on their relationships with key/influences of stakeholders. Some roles may need to be negotiated with gatekeepers who control access to research contexts or resources and, where relevant, with others involved in the research, such as postgraduate supervisors, research team members, research participants, institutional or organisational partners, and/or contracting or funding organisations. Researcher positioning also concerns how clearly and logically the researcher argues for their adopted pattern(s) of guiding assumptions, choice of research context(s), as well as the goals of their research and develops the research questions/propositions/hypotheses to be investigated. In situations where the researcher is also associated with the research context in some way (i.e., researcher-as-insider), their positioning needs to include discussion of their researcher role vis-à-vis any organisational/institutional/managerial/cultural roles they hold in the research context (e.g., the researcher would need to clarify any tensions between demands as a researcher and demands/responsibilities as a research context member and how this tension may influence their researcher role). If the researcher's role is as an outsider, the nature and implications of that role should be explored. Researcher positioning shares an important dynamic with the juxtapositioning with other research meta-criterion when the extant literature and previous research experiences form an important basis for arguments about patterns of guiding assumptions, research framing, emerging research goals, and research questions/propositions/hypotheses.

Positioning of Participants and Other Data Sources

This meta-criterion reflects the importance of knowing where/how/why the participants and any other data sources (e.g., documents, recordings, images, videos, performances) are situated and relevant with respect to the research context and goals. A research story that explores participants' contexts, expectations, concerns, and vested interests in the research being conducted and/or the contexts, credibility, and positioning of relevant documents and other handiworks of human activity reinforces a more open and convincing account. This meta-criterion requires the researcher to critically reflect on any power dynamics/relationships that may exist between themselves and participants in the research context and the implications that managing these might have for research quality and value. This is especially important if the researcher is conducting research in an organisation they work for or culture they belong to (researcher-as-insider). Positioning of participants and other data sources will often have a dynamic relationship with choice of pattern of guiding assumptions (e.g., the positivist, interpretivist/constructivist, participatory, systemic othering, or Indigenous/First Nations inquiry patterns) and this dynamic would need to be fully explored (for Indigenous/First Nations research, this would extend to considerations about the ownership of knowledge). Working with this meta-criterion explicitly acknowledges that participants also have differing roles, backgrounds, and expectations about the research, thus recognising that they have voices to be heard. This creates an important dynamic or dependency with the contextual sensitivity meta-criterion (discussed below) concerning the choices of organisations, participants, and other data sources for inclusion in the research along with the ethical and social responsibility considerations that such choices may evoke. Downstream, this dynamic may be reflected in how the researcher makes sense of/interprets their findings vis-à-vis participants.

Contextual Sensitivity

This meta-criterion captures the extent to which the methodological choices, implementation, findings, and conclusions of research are appropriately sensitive to locally-relevant contextual choices and considerations. Openly reflecting such sensitivity will enrich both the contextualisation and the realisation of the research, may facilitate relationships with key potential gatekeepers, and may enhance the prospects for meeting extensional reasoning, value for learning, change, and impact, handling of unexpected outcomes, and acknowledgement of limitations meta-criteria expectations (discussed below). Contextual sensitivity should be taken to encompass any personal, ethical, political, cultural, and organisational considerations associated with configuring any data gathering protocols, tasks, interventions or treatments and identifying/choosing data sources in specific contexts. Contextual sensitivity means being responsive to contextual nuances for data gathering, such as physical and social environment, time and place (including, where relevant, historical or cultural background), and identifying/accessing data sources. It also plays a major role in helping the researcher to appropriately situate, characterise, and qualify anything they learn from the data they have gathered through critically reflecting on the contexts in which the data were obtained. Even where contextual constraints are beyond researcher control, through external imposition (such as where a school principal rules that the researcher may not treat students differently within a classroom cohort), a convincing research story should be open about the perceived potential impacts of such constraints.

Realisation Meta-criteria

Internal Coherence

This meta-criterion focuses closely on how well the research works as an integrated set of activities designed to address the researcher's specific research goals, questions, propositions, hunches or suspicions, anticipations, or hypotheses/predictions. One important consideration is how well the entire research process works as a logical sequence of choices and actions to both inform and support researcher inferences, interpretations, conclusions, and implications. Congruence between pattern of guiding assumptions and the data gathering configurations and strategies the researcher has implemented is another important consideration under this meta-criterion, making this a meta-criterion where pattern-specific quality criteria are recognised and have an explicit role to play. For example, under the pattern of guiding assumptions for a positivist inquiry, internal coherence would depend, in part, upon meeting the expectations of measurement quality, construct validity, and internal validity. Under the pattern of guiding assumptions for an interpretivist/constructivist inquiry, internal coherence would depend, in part, on transparency, authenticity and, in certain cases, sufficiency. Under the pattern of guiding assumptions for an Indigenous/First Nations inquiry, effective 'decolonisation' of all data gathering strategies would be essential to demonstrate. Against this meta-criterion, a convincing study should provide, from a range of possible alternative accounts, the strongest warrant for the conclusions drawn, given the quality of the positioning, configuration, and execution of the research. Deductive reasoning plays an important role in the realisation of internal coherence as pluralist systemic thinking, contextualisations, guiding assumptions, and research framing synergistically combine to inform a general research approach within which the researcher 'deduces' and specifies the questions to be addressed. The shape and precision of such deductions is strongly influenced by the pattern of guiding assumptions one adopts. There are some critical inter-dependencies between internal coherence and analytical integrity (discussed below) through showing congruence between the researcher's analytical choices and the positioning, configuration, and execution features of their research, presentational character (discussed below) through informing the story the researcher tells about the positioning, configuration, and execution of their study and contextual sensitivity, where influences in or emerging from the research context may add to or detract from the intended positioning, configuration, execution, interpretations, and implications of the research.

Analytical Integrity

Here, we focus on the quality and implementation of the researcher's choices of analytical technologies, pathways, and approaches for making sense of/building meaning from the data that they gather, consistent with their choice of pattern of guiding assumptions and with due attention to the research questions/propositions/hypotheses being addressed. For example, under the pattern of guiding assumptions for a positivist inquiry where quantitative data are gathered, statistical conclusion validity becomes relevant. Under the pattern of guiding assumptions for an interpretivist/constructivist inquiry, keeping interpretations closely connected to the data and to the voices of participants, while managing the researcher's own preconceptions becomes relevant (reflections of authenticity). Convincing research will show how and why the researcher made specific analytical choices, argue for their appropriateness/suitability to the task at hand, and sensitise what they have learned against the quality and character of the available data and the context(s) in which they were gathered. Analytical integrity will dynamically interact with presentational character (discussed below) in that analytical choices, particularly those related to displays of data themes, patterns and relationships, should be sensitive to audience expectations and competencies with attention to how the researcher should communicate findings, meaning and interpretations most clearly and extensional reasoning (discussed below) in that the researcher's analytical choices may inhibit or enhance the reach of the stories emerging from their quantitative, qualitative, or hybrid/merged data analyses.

Extensional Reasoning

The final realisation meta-criterion focuses on the researcher's intentions and reasoning for extending the findings from localised research context(s) to other contexts of potential interest, including other participants, groups, organisations, cultures, tasks, times, situations, and places. Here, the researcher establishes the defensible boundaries and extent of legitimate reach for meanings and implications of their research findings, against the background of the contexts, constraints, configuration, and execution of their research. This is another meta-criterion where pattern-specific quality criteria have an explicitly recognised role to play. For example, under the pattern of guiding assumptions for a positivist inquiry, external validity and representative design of tasks and conditions becomes relevant. Under the pattern of guiding assumptions for an interpretivist/constructivist inquiry, sufficiency and, where appropriate, transportability, become relevant. Extensional reasoning shares an important inter-dependence with the value for learning, change, and impact meta-criterion (discussed below) in that the boundaries and reach of conclusions help to shape where value for learning from the research might be realised and where change and impact might occur. Extensional reasoning also establishes the nature and role of any forecasts, explanations, and speculations that the researcher may produce to help make sense of their data patterns, explain why things happened the way they did, and argue for use and applications by others in the future. Both inductive and abductive reasoning play important roles in the meeting the extensional reasoning meta-criterion: inductive reasoning to pull together the separate threads of stories that emerge from the data into a more coherent higher level set of conclusions aided by abductive reasoning to help build accounts and explanations for emergent and possibly anomalous data patterns. Explaining why things happened the way they did creates important interdependencies with the meta-criteria for handling of unexpected outcomes (discussed below) in terms of how the researcher handled data anomalies during their analysis processes and acknowledgement of limitations (discussed below) in terms of what limits and caveats apply to the extensions and implications of their conclusions.

Explication Meta-criteria

Value for Learning, Change and Impact

The value for learning, change, and impact meta-criterion captures the extent to which the researcher makes the meanings and implications of their findings/conclusions logical, transparent, and closely connected to their goals and questions/propositions/hypotheses. Convincing research will clearly indicate where its important contributions are intended to be and what knowledge it is contributing to. Value for learning, change and impact needs to focus on relevant context(s) for relevant research audiences, highlighting the nature of the intended learning, changes and/or impacts from the research and where and how the research may be relevant to academics, other researchers, professionals, managers, organisations, communities, cultures, markets, potential/actual innovation adopters and/or practitioners. In short, value for learning, change and impact is all about the emergent 'take-home' messages from the research story and it is responsibility of the researcher to make the connections needed for conveying those messages to their intended audience(s) or, where appropriate, to facilitate co-constructing those messages with their intended audience(s). The goal is to answer the 'so what' question for the audience. The processes undertaken by the researcher to meet the expectations of the internal coherence, analytical integrity, and extensional reasoning meta-criteria provide the central platform for value for learning, change, and impact. Value for learning, change and impact may be displayed for theory, methodology, disciplinary advancement, practice, change, action, emancipation, policy, education and development, management and many other points of focus and may convey important concrete lessons for development, application, innovation, improvement, and change in social and organisational contexts. In action learning research, for example, value for learning, change and impact may stimulate shifts in the positioning of both the researcher and research participants (creating dynamics with researcher positioning and positioning of participants & other data sources) as well as reflecting contextual sensitivity to relevant groups, communities, cultures, institutions or organisations with respect to how change unfolds and how and where impacts may be realised.

Fertilisation of Ideas and Actions

The fertilisation of ideas and actions meta-criterion is explicitly forward-looking in intent. It involves an element of futures-oriented systems thinking, extrapolating from the processes undertaken to support extensional reasoning. Fertilisation of ideas and actions explicitly captures the notion that convincing research should stimulate interest in and possible guidance for future research and signal, where relevant, any potential innovations, evaluations, decisions, and/or follow-up actions, applications or interventions by practitioners that might flow from the research (in this way, it shares a dynamic with value for learning, change and impact). Convincing research should also leverage possible or actual blind alleys and dead ends the researcher encountered in ways that may assist future researchers to avoid or circumvent the traps that the current research may have fallen into. Thus, we can see that there is a dynamic interplay with the acknowledgement of limitations meta-criterion (discussed below). Furthermore, convincing research poses unanswered questions arising from the research which can point to directions for future research or future actions to take, a key role played by abductive reasoning, often building upon the handling of unexpected outcomes. This meta-criterion has a built-in delayed feedback loop linked to whether the research story is experienced and cited/used/applied/adopted by future relevant others and whether it has an influence on some future relevant context(s). Meeting expectations against this meta-criterion can therefore provide the researcher with opportunities to learn about the 'impact' of their research (another dynamic link to value for learning, change and impact).

Handling of Unexpected Outcomes

For research to be convincing, the researcher must openly discuss and actively deal with any unintended or unanticipated, i.e., surprising, outcomes arising from their research. Discussion should logically set out how the researcher handled such outcomes without being dismissive, defensive, or apologetic. These outcomes may range from anomalies in the data (e.g., quantitative data that did not meet appropriate statistical assumptions; qualitative stories that contradicted previous accounts or revealed a surprising or very unique perspective) to unanticipated constraints imposed on the research that affected data quality or quantity (e.g., an organisation withdrew its participation from the research or influenced data gathering activities in ways that were not anticipated; poor questionnaire response rates). How the researcher deals with such outcomes will very much depend upon their adopted pattern(s) of guiding assumptions and the methodologies and analytical strategies they employed (creating dynamics with all three realisation meta-criteria). Handling of unexpected outcomes can have a particularly powerful positive or negative influence on convincingness depending on whether the researcher attempts to logically and openly account for such outcomes or instead attempts to explain them away, ignore them or gloss over them. Here is where abductive reasoning can be utilised and displayed by the researcher to develop accounts and explanations for unexpected outcomes and patterns that may lead to future testable deductions and theoretical directions to pursue. Handling of unexpected outcomes shares important dynamics with the acknowledgement of limitations meta-criterion discussed below (e.g., recognising where limitations created unexpected outcomes in during the research journey or where contextual insensitivity may have produced unintended side-effects) and the fertilisation of ideas and actions meta-criterion (e.g., using the unexpected outcomes as springboards for identifying directions, new theorising, and methodological improvements for future research).

Acknowledgement of Limitations

The acknowledgement of limitations meta-criterion reflects the expectation that convincing research should incorporate the researcher's critical reflections on the limitations, constraints, and difficulties they encountered through their research processes. Many limitations tend to be dynamically created when the researcher makes trade-off decisions, often on the fly, to move from what might be ideally desirable to implement in the conduct of their research to what is feasible to implement, given the constraints they encounter. Acknowledgement of limitations shares an important dynamic with handling of unexpected outcomes in that limitations that the researcher could not effectively manage or circumvent may have produced some of the unexpected outcomes that they need to address. Like handling of unexpected outcomes, acknowledgement of limitations can have a very powerful positive or negative influence on convincingness depending on whether the researcher actively engages in critical thinking about and reflection on their own research by openly discussing its limitations or instead avoids critical thinking by deflecting, glossing over, ignoring, or discounting them. Effective acknowledgement of limitations can enhance convincingness by providing a backdrop for clearer and more appropriately bounded arguments for value for learning, change and impact and extensional reasoning as well as helping to make a stronger case for fertilisation of ideas and actions. Acknowledgement of limitations also shares an important dynamic with the contextual sensitivity meta-criterion in that limitations may arise because of contextual constraints or unique circumstances that were not feasible or possible for the researcher to overcome.

Presentational Character for Sharing

The final explication meta-criterion is presentational character for sharing research learning, knowledge, and applications (shortened to presentational character for discussion purposes). This meta-criterion focuses on who needs to be convinced by the research, how the researcher might best convince them, and what to take away from the research, all of which contribute to research impact. Here, we explicitly recognise the central role played by audience(s) or intended users in helping the researcher to produce/realise a quality research outcome. A research outcome exhibits presentational character when it ensures that audiences understand the researcher's journey from where they started to where they ended up, within the constraints/guidelines imposed by the medium within which the research outcome is expressed and the setting in which it is experienced. Conveying a story that targets the wrong audience, is pitched at the wrong level, is culturally insensitive, and/or is poorly crafted (i.e., lacking logical coherence and clarity in writing), are some of the quickest ways to damage the convincingness of the research investigation (as well as, quite possibly, the researcher's own credibility). This means that convincingness can be strongly influenced by the strategic choices the researcher makes when writing, delivering, displaying, or performing their story, above and beyond any inherent strengths their research processes might possess and how compelling their investigation was to pursue. Deficiencies in presentational character can have a powerful perceptual as well as cognitive impact on audiences, which may override strong positions on contextualisation and realisation meta-criteria. Even minor conceptual or mechanical difficulties (e.g., using the wrong report format or writing/formatting style, inappropriate modes of expression, poor vocabulary choices, inappropriate representation for a theoretical or conceptual framework, errors in statistical reporting, tables or figures; grammatical errors; lack of proof-reading evidenced by many typographical errors) may detract from convincingness, even where the researcher has correctly targeted relevant audiences (e.g., examiners for a postgraduate thesis, academic readership/ reviewers/editors for a specific journal, research colleagues, managers, employers, practitioners, even laypeople, such as members of a community or workforce). On the other hand, a research process that is weaker against one or more contextualisation or realisation meta-criteria may recover some impact potential and standing with audiences through strong presentational character. There is, however, an important dynamic to recognise here: this 'quasi-recovery' is likely only to happen if the researcher's story is also strong with respect to acknowledgement of limitations, value for learning, change and impact, and handling of unexpected outcomes. Presentational character is enhanced where the researcher pays attention to the appropriate balance in emphasis/weighting on various meta-criteria, depending upon intended audience(s) and the purposes and anticipated uses of their research.

4.3.3 Working with the Meta-criteria

The meta-criteria can be used to enhance the meaningfulness, quality, and cumulative nature of research. Firstly, the meta-criteria reinforce the idea that research processes need to be seen in a more realistic as well as holistic light as a contextualised, constrained, and very human activity, subject to numerous influences. They also promote systemic thinking to help the researcher combat the tendency to reify specific research 'paradigms' and approaches (i.e., negating 'paradigm' arrogance). Secondly, the meta-criteria can facilitate the researcher's identification or recognition of (1) critical leverage points (i.e., places that emerge during their research journey where a choice they made altered the pathway for their research in a productive or positively adaptive way or counter-productive negative way) and (2) constraints within their research context and the wider contexts in which it is situated, that had the potential to impact on research quality in various ways. By anticipating and making decisions on how to best handle emergent constraints and take action at critical leverage points, research quality can be enhanced. Thus, the meta-criteria offer a more systemic and flexible landscape within which to energise the researcher's creative harnessing of a plurality of patterns of guiding assumptions, research frames, processes, and configurations, research entry and exit points, data gathering strategies and method mixtures, and positionings with respect to theory, methods and practice, where the researcher's goal is to enhance the convincingness of their research. The meta-criteria can assist in (1) directly informing appropriate evaluations of research outcomes and products, thereby circumventing problems that can emerge when pattern-inappropriate criteria are applied to judge the quality of specific research endeavours and (2) reinforcing the importance of not only considering pattern-specific research quality criteria but also looking beyond these localised assumption-specific criteria using the meta-criteria when planning and/or evaluating research processes and outcomes.

When focusing on convincingness in research planning, the contextualisation and realisation meta-criteria domains take centre stage in anticipation of how explication might best unfold. How best to meet explication meta-criteria entails the researcher anticipating how relevant audiences might evaluate the study they are planning and adapting their research configuration in accordance with what they can foresee being raised as potential issues by audiences (realising, of course, that they will never be able to anticipate everything). Part of researcher preparation for meeting explication meta-criteria expectations involves (1) being very clear about research configuration and procedural choices and trade-offs they have made to enhance the chances that their research will be feasible and realistic as well as convincing, and (2) knowing the strengths and limitations of their planned research so that they can critically reflect on this knowledge in any research outcome they produce.

Researchers in the context of conducting their review of prior research/knowledge building efforts can employ the meta-criteria as guides for judging the quality of a specific research outcome (journal article, book, consultancy report, conference paper, manuscript for a journal, media report, social media report, thesis, dissertation or portfolio). If the researcher employs meta-criteria for judging research quality, they can avoid (or at least recognise) 'paradigm' confusion, blindness, bias, or arrogance in their evaluations while ensuring that the criteria they draw upon to make their judgments remain transparent, accountable, and comparable across different types of social and behavioural research. When judging convincingness in evaluating others' research, the contextualisation and realisation meta-criteria domains form the reflective backdrop against which the explication domain is contextualised and judged. Furthermore, the meta-criteria can facilitate higher-order 'learning' by helping the researcher to maintain an awareness of the importance of diversity in patterns of guiding assumptions, choices, and approaches to social and behavioural research and of the need to employ criteria sensitive to that diversity, reflecting what Flood and Romm (1996) called 'triple loop learning'. When reading a researcher's research explication with a critical eye, the reader looks to be convinced by attending to (1) how well that researcher has contextualised and realised their research processes to get to the conclusions they have drawn; (2) evidence of the researcher's awareness of and responsiveness to emergent issues and constraints during their research;

and (3) evidence of the researcher 's critical insights into what they have done and learned and how to move forward. Research proposals can also be evaluated using the meta-criteria but from a prospective ('what the researcher plans to do') rather than a retrospective ('what the researcher actually did') mindset.

4.4 Choice of Mode of Knowledge Production/Building

There are several modes of human knowledge production or knowledge building, and it is important for researchers to be aware of the mode they are engaging with their specific research endeavour. While there is a relationship between mode of knowledge production and patterns of guiding assumptions, the relationship is not necessarily one to one. So, in a manner similar to that used to identify and adopt patterns of guiding assumptions, researchers should reflect on their chosen mode of knowledge building. Such reflection supports pluralist systemic thinking and helps the researcher to place their research within a larger context. The choices for mode of knowledge building reflect more integrative considerations of the researcher's intentions coupled with the extent to which their research needs to connect to a wider range of stakeholder and audience interests with respect to the implications of the research outcome(s) they envisage.

- *Mode 1 knowledge building* refers to traditional academic and scientific knowledge, built up through disciplinary lines and usually with the intention of dissemination to and use by other researchers (Gibbons et al., 1994). Building up such knowledge is largely a cognitive and empirical exercise on the part of the researcher and immediate application of the knowledge may not be of specific interest. Mode 1 knowledge building tends to occur relatively independently of interactions and engagement with or inputs from key stakeholders and end users (prototypical classical discipline-centred academic/university basic research). Mode 1 research typically serves to achieve discoveries and the researcher intends to disseminate what they have learned via peer-reviewed outlets such as academic journals and other academically-respected forms of research outcome (e.g., a thesis/dissertation/portfolio, conference paper). While Mode 1 knowledge building is most strongly associated with the pattern of guiding assumptions for a positivist inquiry, it may also be associated with the pattern of guiding assumptions for a critical realist or interpretivist/constructivist inquiry.
- *Mode 2 knowledge building*, on the other hand, refers to a more engaged form of participatory knowledge building that is transdisciplinary in focus, embedded within social and economic systems, and involves a wide range of researchers and stakeholders (see, for example, Gibbons, 2001; Gibbons et al., 1994; Nowotny et al., 2001, 2003). There is an explicit focus on application for Mode 2 knowledge, often coupled with a focus on participation and action. Such knowledge building depends heavily, if not completely, upon high-level interactions with and inputs from key stakeholders and end-users. Thus, knowledge is co-created, multi/

transdisciplinary and participatory (inputs may be physical, emotional, cultural, intellectual, resource-based, power/control over access to contexts, data sources, and so on). Research has a practical, problem-solving, and user-focused outcome or intention (i.e., applied research), such as facilitating the development and adoption of innovations or changes in practices. This mode of knowledge building may rely upon critical engagement between the researcher and communities for whom the research is intended to have meaning/impact (see, for example, Fear et al., 2006). Mode 2 knowledge building is commonly associated with the patterns of guiding assumptions for a participatory inquiry, a critical realist inquiry, critical inquiry, a pragmatist inquiry, and, occasionally, a positivist or an interpretivist/ constructivist inquiry.

- *Mode 3 knowledge building* is founded in systems theory (creating a knowledge system) and occurs at multiple levels within multiple systems/networks (Carayannis et al., 2016). It is especially relevant to innovation research. Mode 3 knowledge building may also reflect a higher-order hybridisation or synthesis of Mode 1 and Mode 2 knowledge building. Mode 3 knowledge building is associated with the concepts of the Triple Helix (capitalising on relationships between university-industry-government), the Quadruple Helix (Triple Helix relationships embedded within a media-based, arts-based, and culture-based public/ civil society), facilitating a democratisation of knowledge (Carayannis et al., 2016, p. 19) and the Quintuple Helix (embedding the Quadruple helix within a fifth helix, "natural environments of society" (Carayannis & Campbell, 2010, p. 62). Mode 3 knowledge production is more commonly associated with the patterns of guiding assumptions for a participatory, critical realist, or pragmatist inquiry.
- *Culturally-privileged knowledge building* captures an emerging awareness of how knowledge is developed, protected, regarded, and passed on within Indigenous/ First Nations research contexts (emerging from work especially in North American, African, and Australian/New Zealand Indigenous/First Nation cultures; see, for example, Chilisa, 2020; Kovach, 2021; Wilson, 2001; Steinhauer, 2002) as well as in systemic othering research (emerging from work in feminist, intersectional, and queer research; see, for example, Alexander, 2024; Davies, 2024; Haynes et al., 2024; Hesse-Biber, 2012). Knowledge building in this view is culturally privileged, meaning that it is developed, resides in, and is shared within specific groups, and historically-anchored yet freed from colonising or othering influences, oppression, or rewriting, meaning that the knowledge is to be understood on its own terms rather than being dismissed, filtered or edited by the worldviews of other, more powerful, outsider influences. For the pattern of guiding assumptions for an Indigenous/First Nations inquiry, knowledge exists and can be shared (under the right cultural circumstances); it is not owned by an individual and it is not handed down from a colonising group. From an Indigenous/First Nations perspective, knowledge is more about relationships (and being accountable for those relationships) than about objects and things, including relationships between researchers and Indigenous/First Nations people (Wilson, 2001). Specifically, Wilson (2001) said,

> Our systems of knowledge are built on the relationships that we have, not just with people or objects, but the relationships we have with the cosmos, with ideas, concepts, and everything around us. For research it is important to think about our relationship with the ideas and concepts that we are explaining. Because this relationship is shared and mutual, ideas or knowledge cannot be owned or discovered. (p. 177)

Thus, knowledge is seen as culturally and historically anchored, relational, and privileged and permeates all aspect of life, the cosmos, and relationships, thus directly embedding systemic thinking. Researchers work to gain access to this privileged knowledge for potential sharing with the wider community or for creating additional benefits to Indigenous/First Nations communities. The pattern of guiding assumptions for a systemic othering inquiry uses a similar type of logic (but perhaps taking less of a systemic worldview), except that knowledge is typically considered to be other-privileged (where 'othering' category reflects the 'cultural' basis for the privilege, e.g., gender, race or ethnic background, sexual identity, disability). Note that in an Indigenous/First Nations or a systemic othering inquiry, the purpose of the research may be to juxtapose culturally-privileged knowledge held by the colonised or marginalised group(s) with the knowledge held by non-Indigenous/non-First Nations (i.e., colonising) or 'non-othered'/dominant/majority group(s), a focus that is very consistent with an explicitly critical intent.

4.5 Choice of Research Frame

The typical methodological discussions offered in many research methods texts are confusing because the term 'method' is often used not only to refer to specific ways of gathering data (e.g., interview, questionnaire) but also to more general research approaches like action research, transdisciplinary research, survey research, and case study research. The problem is that different levels of conceptual abstraction are at play here, which many authors simply ignore. A data gathering strategy is simply that—a set of actions undertaken in research to collect a specific form of data from a specific type of data source under a specific pattern of guiding assumptions. Action research is not a data gathering strategy, 'method' or a 'methodology', but a way of holistically conceptualising the research. The same is true for 'methodologies' like survey research and case study research. To resolve this confusion, we need a higher-order concept or *frame* for systematically organising research thinking (first introduced in Cooksey & McDonald, 2019, Chap. 11). A research frame captures the dynamic and synergistic intersection of researcher positioning, research contexts, participants' positioning and contexts, mode of knowledge building, and audience contexts, all embedded within the larger social, political, and physical worlds. It helps to contextualise social and behavioural research more fully, providing a more holistic picture of how research purposes can/will be translated into research strategies and tactics that will generate or apply learning and knowledge which will speak to and influence specific audiences. A research frame embeds ways of creating, applying, and disseminating knowledge about the social world that might be relevant to specific stakeholders in that research. Figure 4.5 (significantly updated from Cooksey & McDonald, 2019, Fig. 11.1, p. 350) displays a mind map of different frames that can be identified for social and behavioural research. Each research frame can stand on its own to conceptualise and contextualise research. There are 12 potential frames grouped into four higher-order categories; three frames are categorised as *transformational* (seeking to stimulate or enact change); four frames are categorised as *focused* (pursuing a specific type of research process and story); three frames are categorised as *generalist* (pursuing research processes and stories that have broader intent), and two frames are categorised as *complexity/innovation* (pursuing more complex or innovation-related agendas). Note that is possible to synergistically combine different frames (from either the same category or from different categories) to form hybrids (signalled by the red dashed links in Fig. 4.5 connecting the different frame categories) in order to achieve specific research aims. For example, a researcher could envisage an othering/emancipatory action research frame; a cross-cultural Indigenous/First Nations research frame; a cross-cultural survey research frame; a pan-disciplinary developmental evaluation frame; an evaluative survey frame; an action research developmental evaluation (e.g., participatory action research for social innovation) frame; a descriptive case study frame; an explanatory (theory-testing/experimental) survey frame; or a descriptive/exploratory/explanatory frame (phenomenography, grounded theory, or ethnography).

While a research frame may or may not implicate specific patterns of guiding assumptions or the use of specific data gathering strategies or data sources, they may help the researcher to narrow their choices as well as to identify key areas of prior research and knowledge building efforts they should explore. A research frame may also utilise specific concepts and specialised language which can facilitate communication with certain types of audiences (but could also possibly inhibit communication with other types). As well, each research frame comes with its own in-built constraints, expectations, and assumptions which the researcher needs to be aware of and work with. Quantitative research, qualitative research, and mixed methods research are not considered to constitute research frames because they are far too general to be helpful in focusing and situating research. In addition, they only focus on the form of data gathered, which is a very small part of the much bigger set of considerations associated with the choice and implementation of a research frame. Any of the frames shown in Fig. 4.5 may employ strategies for gathering and interpreting quantitative and/or qualitative (or, in some cases, hybrid/merged) data.

Political dynamics are implicit in certain research frames (e.g., Indigenous/First Nations research, othering/emancipatory research, action research, evaluation research) where patterns of guiding assumptions for a critical, systemic othering, or Indigenous/First Nations inquiry that have an espoused value position have been adopted. In these types of frames, the researcher may, from the outset, be oriented toward revealing disadvantage or correcting social perceptions or injustices. Their political agenda would be one of change and, in some cases, the researcher expects to be an agent of that change, through their expert power as a researcher and, in other cases, those being researched (who may feel politically disadvantaged or silenced,

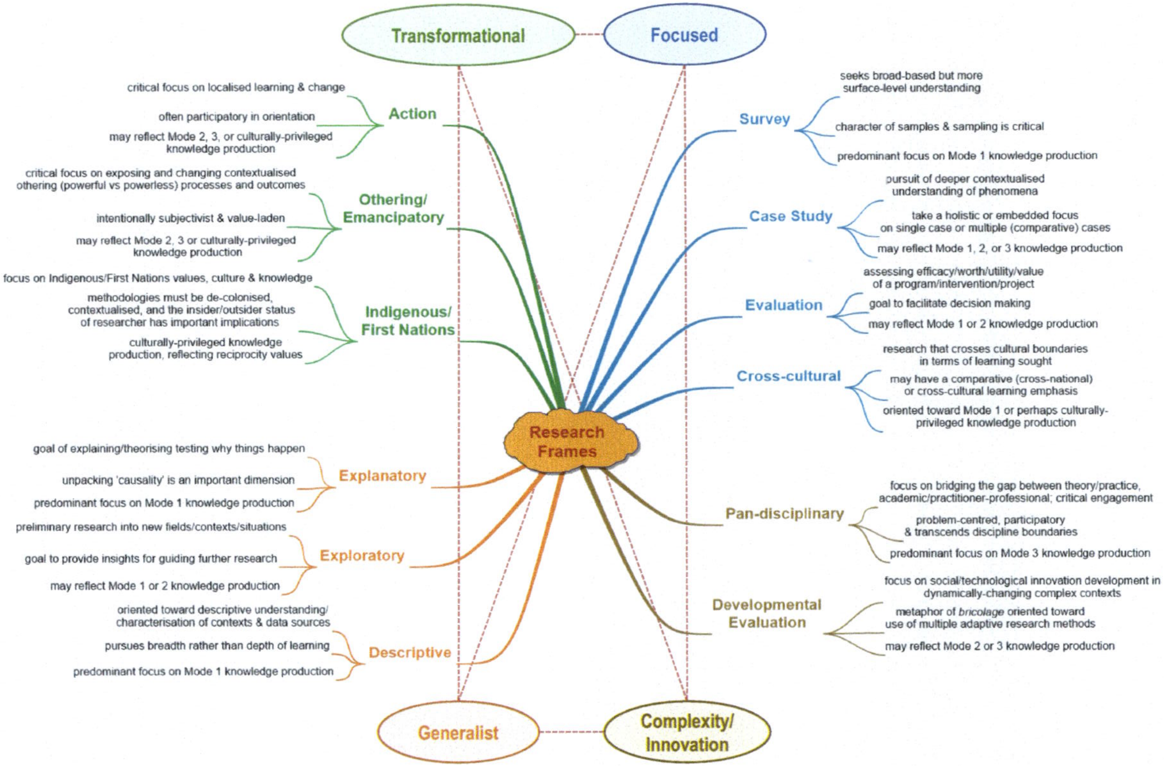

Fig. 4.5 Mind map of important frames for social and behavioural research highlighting some key features of each frame (dashed links signal that there are potential synergistic combinations between frames in different frame sets)

unjustly treated or otherwise held in a position of lesser status in society) may expect the research to aid in carrying their change agenda forward to wider society. Specific research frames may also bring political forces between competing stakeholders to the fore. For example, action research may try to facilitate organisational change in a climate that has been traditionally reluctant to embrace such change. Indigenous/First Nations or othering/emancipatory research may try to expose colonising, oppressive, or discriminating practices attributable to what they view as the dominant cultural milieu. Developmental evaluation may try to bring an innovation into a specific context where adoption of innovations has typically been very slow and/or strongly resisted. Whatever the specifics of the political dynamic associated with a research frame, the bottom line is the researcher cannot ignore it. They must engage and manage it in such a way as to facilitate achieving a convincing research outcome while, at the same time, avoiding the adverse consequences of political alienation (e.g., reputational damage, resistance to future funding requests, whether by the researcher or their colleagues, resistance to requests for access made by other researchers).

4.5.1 Focused Frames

Survey Research Frame

Survey research (see, for example, Eichhorn, 2021; Fowler, 2014; Ruel et al., 2015; Krosnick, 1999; Rossi et al., 2013; Toepoel, 2015) is commonly, simplistically, and mistakenly thought of as research conducted using questionnaires (hence, the reason many people inappropriately equate the term 'survey' with 'questionnaire'). However, survey research is more properly and holistically conceived as a research frame which reflects the pursuit of learning across many issues and spanning a wide range of people. Survey research typically favours breadth of learning and knowledge (i.e., a nomothetic emphasis) at a more surface level, over depth of learning (i.e., an idiographic emphasis) and quantitative data over qualitative data (consistent with the pattern of guiding assumptions for a positivist inquiry). Breadth of learning is generally achieved through accessing large representative samples of participants, the larger and more representative the sample the better. Some depth of learning can be achieved through asking many questions about a specific topic area (a strategy that may be suitable under the pattern of guiding assumptions for an interpretivist/constructivist or critical realist inquiry), but there is often a practical limit to what participants are willing or able to handle.

Research within the survey research frame generally favours Measurement-data shaping strategies (e.g., self-report questionnaires, objective tests, and assessments) and/or structured interviews, organised to yield quantitative or quantifiable data). For survey research conducted under the pattern of guiding assumptions for an interpretivist/constructivist inquiry (not common, but certainly possible), the Transformative data-shaping strategy and the Textual artefact-based strategy are implemented using open-ended (i.e., free-response) questions. Note that it is possible to include both highly structured questions and open-ended questions—a hybrid strategy often employed in mixed methods research, perhaps guided by the pattern of assumptions for a pragmatist inquiry. Self-report questionnaires can be administered face-to-face (individually or in a group) or online in a web-based format. These types of questionnaires require a great deal of front-end design work, validation, and pilot testing to ensure they function as intended. In these cases, the researcher is completely 'hands-off' when it comes time to gather the data, which means the questionnaire must function independently. Structured interviews, conducted via a variety of modes (e.g., face-to-face, phone, or internet), may also be used in the survey research frame where the interview serves as a vehicle for the oral delivery of a questionnaire. This requires interviewers to be trained to a high standard of consistency to ensure that errors are not introduced into the data by virtue of the interview situation itself. For specific purposes, it may be possible to embed an experimental manipulation within the questionnaire itself, which effectively pairs the Manipulation experience-focused strategy with the Measurement data-shaping strategy (e.g., in certain types of marketing research on consumer decision making, such as discrete choice or conjoint measurement experiments).

Typically, the researcher predetermines (i.e., controls) exactly what they wish to learn from participants and will most often also delimit exactly how participants will convey information to them. When data gathering commences, the researcher generally adopts a more hands-off 'objective' approach to data gathering, even where a structured interactive mode of data gathering is employed. For some modes of questionnaire administration (e.g., mail, online), the measurement instrument or questionnaire takes on the de facto role of 'researcher' (albeit an inanimate one), thus making the design of the instrument absolutely critical for the researcher to get right. Generally, survey research is done in the field, away from the context where the researcher works. However, what constitutes the field is open and may range from simply the general public (relatively context-free; may even be at home with an online marketing or local government questionnaire) to shoppers intercepted in a shopping centre to employees in an organisation.

The kinds of research questions that can be addressed in the survey research frame may focus on what participants do, what they think, feel, expect and/or prefer, what their essential characteristics are, and what they know about. Often, research questions take the form of propositions that posit certain patterns of relationships between various constructs (e.g., gender will relate to income level; job satisfaction will relate to salary level or commitment to the organisation) but stop short of stating strong causal inference hypotheses. This is because, unless the researcher is very careful and has sufficient control over context or their measurement processes, survey research will not be able to convincingly demonstrate cause-effect relationships. Survey research typically employs statistical procedures designed to facilitate generalisation from sample to population (accompanied by estimated error limits, which get smaller with larger and larger sample sizes; larger samples yield greater precision). In order for statistical generalisation to work effectively, formal probability sampling procedures need to be used, such as random or stratified random sampling. The goal is to obtain a sample that is representative of the population in all essential characteristics. In many cases, it may not be feasible or even possible to implement a formal probability sampling scheme, in which case, a non-probability sampling scheme must be used. The price paid for this trade-off is less convincing capacity to generalise. Survey research is often oriented toward building Mode 1 knowledge where the researcher, from within their disciplinary background, drives the research agenda, often with the goal of disseminating research outcomes to other researchers. However, it is also common for survey research to have a more applied focus, where the results are intended to provide information for decision and policy-making (e.g., organisational survey research to examine specific issues; community survey research to tap into attitudes toward/preferences for infrastructure priorities). However, this does not constitute genuine Mode 2 knowledge building because the role of participants is generally passive rather than active in the knowledge building process.

In survey research, participants are often referred to as 'respondents' to signal their passive role as responders to questions put to them by the researcher in a prescribed manner (although this labelling trend is slowly changing in favour of 'participant' as 'respondent' implicitly devalues/dehumanises both the participants themselves as well as the information they are providing to the researcher). However, including some open-ended questions to invite participants to express their own views on an issue can be a useful strategy for maintaining their interest in participating. The survey population/sampling frame is often identified according to expectations that members have knowledge, preferences, or perceptions about some issue, context, or process that the researcher wishes to access. In other cases, participants may be targeted simply to learn who they are and what demographic characteristics they possess (as with a population census for a country). Irrespective of the reasons for inclusion in a sample, the researcher will generally be uninterested in what or how participants think outside the bounds of the data gathering strategy being employed. Since the researcher determines the nature and content of any questions asked, the only real power a participant has is over whether they will choose to participate and how they will respond. Modern ethical principles dictate that, aside from survey research conducted for mandated organisational, institutional, or governmental purposes, participants have the right to make all decisions regarding their participation and their responses. This effectively gives participants the power to influence the nature and composition of the sample, which can then affect the researcher's capacity to generalise from sample to population. Furthermore, participants will determine for themselves whether to answer specific questions and the more sensitive the information requested, the more likely it is that at least some participants will decline to respond, potentially creating missing data problems for analysis processes.

Since the survey research frame emphasises breadth (and generalisability) of learning over depth of learning, it gains its power for learning from its focus on large representative samples. Executives, managers, educators, government officials, and other kinds of decision makers may use the results from survey research to inform their decisions about a specific issue (e.g., public policy, community relationships with local government, policing, abortion, climate change), event (e.g., planned infrastructure development, impending election, recreational activity), process (e.g., financial, hospitality, sales, or medical services) or object (e.g., a product, the participant or another person/people, such as one's supervisor or one's work/social network). The most productive learning from survey research concerns distributions of characteristics and patterns of relationships. Survey research, unless very carefully conceived and highly controlled, is a poor research frame for testing causal hypotheses (the explanatory frame, discussed below, is a much stronger frame to work within if strong cause-effect inferences are desired). Population-based survey research such as a census may yield learning that can influence public policy at a national, or possibly international, level. Primary learning outcomes from survey research emerge from patterns in data that are inferred to be present in some population of interest (however widely or narrowly defined).

Case Study Research Frame

Case study research (see, for example, Byrne & Ragin, 2009; Cohen et al., 2018, Chap. 19; Dul & Hak, 2008; Gray, 2021, Chap. 11; Kazdin, 2011; Tight, 2017; Yin, 2017) is a research frame that employs a range of data gathering strategies to achieve the goal of a deep and/or extensive contextual (potentially including historical) understanding of one or more specific phenomena in the context of a 'case' (which may be construed as an event, a program, a project, an individual, a relationship, a decision, a group, a school, an organisation, a community). Case study research overall, or specific data gathering strategies used within it, may follow virtually any of the patterns of guiding assumptions, depending upon the researcher's learning needs and intended impacts. The research goal in a case study typically centres on understanding how and why things happen the way they do in a specific context. A case study may take an inductive theory development trajectory, a deductive theory testing trajectory, or a descriptive trajectory (which means that the case study frame is often combined with the explanatory, exploratory, or descriptive research frames, see discussions of these frames below). The focus in a case study may be on a single level of analysis (holistic or whole-of-case) or multiple levels of analysis (embedded case; e.g., clients, administrators, doctors, nurses, wards, departments in an organisation, groups, programs, components of programs) and single (which could be a single individual) or multiple cases may be of interest. Multiple case studies will usually be required if the researcher intends to generalise or transport their learning across cases or to make explicit cross-case comparisons (identifying commonalities between cases and uniquenesses within each case). In the latter, the researcher will generally have two levels of analyses to conduct and stories to relate: within-case analyses and stories and between-case analyses and stories. Generally speaking, the greater the depth of learning pursued by the researcher in a case study, the more resource-intensive (in terms of time, money and data gathering energy) their research will be.

Case study research is inherently pluralist employing a range of data gathering strategies and the Textual artefact-based strategy (e.g., document analysis) usually forms one important data gathering strategy. A case study guided predominantly by the pattern of assumptions for a positivist inquiry would typically employ quantitative data gathering strategies such as measurement instruments/questionnaires or structured interviews in the context of sampling within a single focal organisation or institution (yielding a so-called 'single case study'). Even though the primary data the researcher collects may be quantitative, there would still be an expectation that they would weave a contextual story around the data from each case and this would normally require access to documents and other data sources and perhaps preliminary interviews to get the picture right. Generalisability in a positivist case study is necessarily highly restricted to the specific case study organisation, but depending upon the sampling scheme, the researcher may be able to generalise within that organisation (possible if the case study configuration is embedded). To overcome this limitation, the researcher could conduct multiple case studies to enhance contextual cross-comparison and generalisability (a strategy called the 'comparative case method'). Such stories would largely be qualitative in emphasis. Alternatively, the researcher could conduct an interpretivist/constructivist case study, where they might employ a range of qualitative data gathering strategies to flesh out a detailed and rich contextual story of the case group, community or organisation (quantitative data might also be used to highlight within case demographic, social and/or behavioural patterns). In single qualitative case studies, the goal would be to achieve a deep contextual understanding, usually assuming that what is learned may not be transportable to other cases but may be transportable within the case study group or organisation. However, the researcher could conduct two or more interpretivist/constructivist comparative case studies, to enhance the chances of demonstrating the transportability of what is learned. The choice of cases needs to be very strategic in order to maximise value for learning, change and impact relative to resources expended. If the researcher adopts a critical, Indigenous/First Nations, systemic othering, or participatory inquiry pattern of guiding assumption, qualitative data collection will typically be emphasised, usually relying on several types of data sources.

It is the researcher's role to explicitly define the case context(s) of interest, be it a person, group, program, organisation, or community, and delineate the boundaries of case focus (what, who, when and where). The researcher is typically, but not always, as an outsider to the case study context, which is why access to multiple types of data sources is crucial for contextualisation. Participants are typically those involved in or with the case context in some way, but some participants may play more of a stakeholder role, helping the researcher to historically situate, contextualise, and understand the case. Some data sources will typically be non-human in nature, comprising case-related documents and other artefacts, which can help to flesh out the researcher's contextual knowledge. Research questions in the case study frame typically focus on learning how and why things happened in the case context. Learning from case study research may have several different intentions for different audiences. The researcher may intend to build theory for other researchers (inductive exploratory case study), test pre-existing theory in a specific case context (deductive explanatory case study), evaluate change or evolution over time, perhaps as a function of an intervention or change event (longitudinal case study, where the outcomes may be of interest to key managers/executives in the case context) or simply provide a descriptive account (descriptive case study). In most cases, the researcher's goal will be Mode 1 knowledge building and learning outcomes primarily emerge from patterns in data that help to inform and convey the story or stories about the case study context. However, if the researcher is pursuing Mode 2 knowledge building, they will need to adopt a pattern of guiding assumptions consistent with that intent, such as for a systemic othering, pragmatist, or participatory inquiry, and their approach will need to be much more inclusive of participant input.

Evaluation Research Frame

The evaluation research frame (see, for example, Alkin & Christie, 2023; Bamberger & Mabry, 2019; Donaldson et al., 2009; Gray, 2021, Chap. 12; Fink, 2014; Patton, 2015; Patton & Campbell-Patton, 2021; Shaw et al., 2006) focuses on making assessments and judgments about the efficacy, worth, utility, and value of a program, project, or intervention in specific contexts with the intention of facilitating decision making about that program or intervention. The research context(s) for an evaluation are defined as any context(s) where, and circumstances in which, a 'program' (for example, a curriculum, medical or psychological treatment, social policy, or community intervention) occurs (e.g., in a school, a classroom, a hospital or ward, an organisation or workplace, an executive team). Key research questions/propositions in an evaluation may include: Did the program realise its goals/objectives? Did the program have an impact and, if so, for/on whom and in what time frame? Did the program work better than other programs or no program? Who benefitted from the program and who did not? Were there any unintended effects of the program? Was the program cost-effective? Did users/participants see the program as effective? Should the program be continued/invested in/modified/extended into other contexts? Evaluation research may have a *formative* emphasis, which focuses on how a program evolves over time on the basis of feedback from participants and other stakeholders (could be tied to key milestones within the program) and/or a *summative* focus, which focuses on the outcomes achieved by the end of the program, perhaps in comparison to where participants were prior to its commencement or to participants in other types of programs. Utilisation-focused evaluation (see, e.g., Patton & Campbell, 2021) tends to take a more holistic view of evaluation, encompassing not only participants and institutional/community stakeholders in the program being evaluated, but also intended users as well as the researchers/evaluators themselves, invoking Mode 2 or perhaps even Mode 3 knowledge building. Utilisation-focused evaluation is thus a more systems-oriented evaluation process.

Evaluation research may be guided by any pattern of assumptions depending upon what the researcher wishes to learn and how deeply the researcher wishes their learning to extend. Summative evaluations will tend to rely more on patterns of guiding assumptions for positivist or critical realist inquiries. Formative evaluations will tend to rely more on patterns of guiding assumptions for interpretivist/constructivist, participatory, or critical inquiries. If evaluations have both a formative and summative orientation, choosing a pattern of guiding assumptions for a critical realist, participatory, or pragmatic inquiry may prove beneficial. A wide variety of data gathering strategies may be used for evaluation research, but experience-focused strategies will likely prove useful since a program, treatment, or intervention provides a concrete focal experience for participants and the Measurement data-shaping strategy is often useful for summative evaluations. Interaction-based, participant observation, artifact-based, and participant-centred data gathering strategies often prove useful in formative evaluations.

The researcher's role assumes an additional dimension when conducting evaluations in that they conduct their research in order to inform and draw conclusions and provide recommendations for decision makers. Thus, the researcher is explicitly expected to make value judgments based on their research and those judgments may carry powerful financial, political, and social implications. The researcher's competencies, positioning, and credibility have key impacts on the convincingness of their evaluation and the likelihood that decision makers will heed any recommendations they put forward. Participants in the evaluation research frame are those who take part in the program, however defined, and may often be self-selected or required (e.g., by their doctor, organisation, community, or institution) to undergo/experience that program. Participants thus constitute a very important set of primary stakeholders in evaluation research. Other participants may include staff/consultants who designed and/or conducted the program as well as the organisation or community that adopted/implemented the program. Another set of primary stakeholders for learning in the evaluation research frame are those who will make decisions (e.g., adoption, investment, resource allocation, removal) concerning the program, based on the researcher's findings and recommendations. The designers and distributors of a program will also want to learn whether and how it works and whether modifications will be needed. Utilisation-focused evaluation explicitly focuses on the longer-term utility/viability of a program. Downstream potential clients may also be considered as stakeholders who could be impacted, depending upon what decision(s) are made. Overall, the primary learning outcomes emerge from patterns in data that signal effectiveness and/or impact (whether beneficial or adverse) of the program being evaluated.

Cross-Cultural (Cross-National or Comparative) Research Frame

The cross-cultural research frame (see, for example, Cleary, 2013; Ember & Ember, 2009; Liamputtong, 2010, 2022; de Vaus, 2008; Harkness et al., 2010; Peterson, 2015; Van de Vijver & Leung, 2021) encompasses any research that crosses cultural boundaries. In the comparative approach to cross-cultural research, the researcher is looking for patterns, similarities, differences, and uniquenesses across the cultural groups they are interested in (e.g., comparing Southeast Asian and Middle Eastern cultures with respect to their attitudes toward risk taking in business). If the comparative inferences are intended to be relevant to or generalised to the national level (e.g., to Vietnam and Saudi Arabia), this gives rise to cross-national research. In a cross-cultural learning approach, the researcher is looking to understand behavioural and social patterns within another cultural group, of which they are not a member (e.g., an Australian researcher studying participation in higher education in the Middle East). Cross-cultural learning research will likely require much longer timelines than research conducted within other frames because some degree of immersion in the cultural context being studied is necessary in order to maximise the depth and value of the learning achieved. It is possible to combine the cross-cultural learning approach with the comparative approach in instances where the researcher wants to compare cultural groups and they are a member of one of the cultural groups involved in the comparisons. In this case, the complexity of the research takes a quantum leap because the researcher must work to offset the potential problems associated with both cross-cultural comparison and cross-cultural learning. Any research context may potentially be involved in cross-cultural research depending upon the nature and sampling of the participants. For some types of cross-cultural research, the researcher may be a visitor in the home nation of participants (as, for example, when an Australian researcher does research in China with Chinese students as participants). For other types (especially studies with comparative intentions), a cross-cultural research team may be involved, where each team member gathers data from a sample of their own national group, either in a laboratory or a field setting. Mode 1 knowledge building is common in cross-cultural research, but, depending upon circumstances, culturally-privileged knowledge building may move to centre stage, which may make convincing comparative research much more difficult to achieve. Mode 2 knowledge building in cross-cultural research would be more difficult to accomplish with a researcher who is an outsider to the culture in which participants may be involved as co-researchers.

The researcher takes on an outsider role if they are gathering data from participants in a sample whose ethnic/national/cultural background differs from theirs. Here they must take steps to ensure they have sufficient understanding of the cultural group and how to behave within it to facilitate effective and meaningful data gathering and this may require one or more preliminary stages in the research. Outsider status will necessitate constant attention to potential power and knowledge sharing issues, especially if the researcher is gaining access to more privileged types of knowledge within the cultural group being studied. Additionally, the researcher must be extra cautious not to fall into the trap of 'going native', which means gradually losing their perspective as a researcher and moving more closely toward adopting and reflecting the perspective(s) of those in the cultural group they are researching (evolving from 'outsider' to 'insider'). If 'going native' occurs, the research may suffer irreparable damage to convincingness. Thus, the researcher must take steps to minimise the risk of this happening, which adds to the overall complexity of the research project. The researcher takes on an insider role if they are gathering data from members of their own cultural group. Much more cultural and local knowledge can be taken for granted in this situation, but care must be taken when they then make comparisons with other cultural groups. The issue here is ensuring, as far as possible, conceptual and contextual equivalence where comparisons are being made. Understanding the positioning of any participants in cross-cultural research demands that the researcher acquires and enact a deeper understanding of the cultural context and backgrounds of those participants. An insider researcher might have an easier time here. In cross-cultural research, there is always the risk that unintentional, possibly negative, power dynamics may influence the conduct of the research and the quality of the data obtained, especially where there has been a past history of such dynamics between the culture the researcher comes from (e.g., the US) and the cultures that participants come from (e.g., participants from Iraq or Cuba). This risk is at its most potent in Indigenous/First Nations research, which is one reason why Indigenous/First Nations research is considered to be a distinct research frame (see discussion below). The risk is also potent if the cross-cultural research also crosses religious boundaries (as, for example, when a Christian researcher attempts to do research in a predominantly Islamic nation). It is important to note that cross-cultural research may be essentially undertaken within the boundaries of a single nation, if that nation is itself multicultural in composition and that composition is reflected in the sample or if the nation has or has had strong historical and/or political divides between geographic areas (such as between northern and southern states in the US) or demographic groups (e.g., a dominant middle- or upper- class group and poor minority groups).

Cross-cultural research may be undertaken under the guidance of a wide variety of patterns of assumptions. Where the pattern of guiding assumptions for a positivist, or perhaps critical realist, inquiry is adopted for comparative research, close attention must be given to the definition and measurement of constructs so to ensure that they have the same meaning (connotations and denotations) across all the cultures being compared. This will normally require additional preliminary stages in the research. Data gathering strategies, in particular, those that pursue connecting with people and those that provide structured experiences for people, have to be tailored to meet the needs, expectations, and conventions of all cultural groups to be compared. Where the pattern of guiding assumptions for an interpretivist/constructivist inquiry is adopted, the researcher will often need to spend more time in context in order to achieve meaningful interpretations of what they hear and see. If the pattern of guiding assumptions for a critical, systemic othering, or Indigenous/First Nations inquiry is adopted, the researcher will need to spend time trying to understand the political, perhaps oppressive contexts, the participants have or may be experiencing before attempting any critique (essentially blending the cross-cultural frame with one of the transformational frames). Unless decolonisation, critique and/or change are to be explicitly pursued, the researcher must take steps to remain value-neutral, non-judgmental, culturally sensitive, and open-minded about what they observe within a cultural group; this is especially important in cross-cultural learning approaches. The most dangerous thing the researcher can do in cross-cultural research is to unilaterally impose their own views, values, ideology on what they are looking for, seeing, and hearing.

Primary learning outcomes from cross-cultural learning research where researcher and participants come from different cultural backgrounds emerge from patterns in data that help the researcher gain/achieve insights into/understanding of life/work/study/relationships and so on in the cultural group they are researching. Primary learning outcomes from cross-cultural research that has an explicit comparative intent (e.g., cross-national research) emerge from patterns in data that effectively and convincingly reveal commonalities between and uniquenesses within different cultural groups. In particular, highlighting commonalities in cross-cultural patterns feeds learning that reinforces what may be generalisable across cultural boundaries (i.e., reinforcing and valuing connectedness). Highlighting contrasting cross-cultural patterns feeds learning that signals diversity and perhaps the perils of assuming commonality (i.e., reinforcing and valuing distinctiveness). Both directions for learning may provide concrete guidance for future actions and applications of knowledge.

4.5.2 Transformational Frames

Action Research Frame

Action research (see, for example, Chevalier & Buckles, 2019; Fine & Torre, 2021; Gray, 2021, Chap. 13; Coghlan, 2019; Greenwood & Levin, 2007; McIntyre, 2008; McNiff, 2013) has the explicit goal of pursuing change to improve social/human conditions, often in localised contexts. While there are many 'flavours' of action research, they are all hallmarks of the same basic frame: produce localised change for the better. Thus, action research is often associated with the pattern of guiding assumptions for a critical inquiry, but where the aim is achievement of productive change/improvement in a specific context, rather than emancipation of an oppressed or disadvantaged group. Within this frame, the researcher can employ a range of data gathering strategies to achieve the goal of targeted improvement and change within a group, institutional, or organisational context. One widely-used variant of action research, particularly in the education, management, and innovation fields, is participatory action research (Chevalier & Buckles, 2019; Fine & Torre, 2021; McIntyre, 2008), where the research is participant-driven and where one of the participants might be the researcher themselves, researching within their own or another group, school, community or organisation. In this particular version of the action research frame, guidance would be offered by participatory inquiry pattern of assumptions, participants in the research context would generally control the research agenda, and all data gathering would be geared toward obtaining evidence that can be used for learning what things need to be improved or changed, for planning how to achieve improvements or change, and for showing the effectiveness of implemented change and improvement processes or interventions. Participatory action research is one pathway toward achieving critical engagement between academic researchers and practitioners/community members and institutions (Fear et al., 2006).

Action research tends to unfold in a cyclical manner moving through a series of stages (diagnose, plan change, implement change/intervene, evaluate change, learn, and modify where necessary), the outcome of which may lead to successive research cycles. The research focus is on a localised context within which changes and improvements are to occur (e.g., a work group, a classroom, hospital ward, department, community group). Research unfolds in an action research cycle as the researcher and participants work hand-in-hand in the localised context to map, produce, evaluate, and further promote change. In this way, the action research frame can be seen to favour Mode 2 knowledge production, knowledge that is co-produced and shared between researchers and participants in localised context(s) where change can be most meaningful and most impactful. Qualitative interaction-based (e.g., focused and group interviews) and observation-based (e.g., participant observation) data gathering strategies tend to be preferred but data-shaping and experience-focused strategies that gather quantitative measurement data can also be used.

The role of the researcher is to act as both a participant (a researcher-as-participant role) and an 'expert-in'-residence' (vis-à-vis research practices) therefore becoming a quasi-insider who is an active/involved resource in the research context rather than maintaining an objective and more distant hands-off stance. The researcher becomes an active part but not the primary driver of localised change and improvement. Participants (e.g., workers, teachers, nurses, managers, community members) act as quasi- or co-researchers (participants-as-researchers role), actively engaging with the researcher in seeking data to solve problems, plan changes, and evaluate change and impact. They stand to be most immediately affected by successful or failed changes; that is, they are the immediate stakeholders/research users. Since action research emphasises Mode 2 knowledge production, research learning and impact initially tends to be localised to the action research context itself. However, changes in practices, new interventions, and/or new innovations may then spread/be distributed to other stakeholders and downstream users and one research goal may concern how best to facilitate such distribution. Thus, similar groups in similar contexts may benefit from what is learned. Knowledge is disseminated primarily through implementation of successful changes, perhaps facilitating the building of communities of practice. An important by-product of action research, in any form, can be the development of new social relationships and technical skills in pursuit of targeted problem solving.

Othering/Emancipatory Research Frame

The othering/emancipatory research frame (see, for example, Browne & Nash, 2010; Davies, 2024; Haynes et al., 2024; Hesse-Biber, 2012, 2014; Letherby, 2011; Nusbaum & Lester, 2024; Warner, 2004) focuses on the othering experiences of marginalised groups, defined in whatever way, in and by society, communities, organisations, schools, and families and the power, political, social, ethnic, educational, religious, and behavioural implications of those experiences. Othering research adopts a critical orientation where there is a presumption of unequal power relationships between marginalised group(s) (the 'othered' group) and the dominant hegemony/majority (the 'othering' group) who, either by direct or indirect action, confers upon them the status of 'other', to be treated differently. These asymmetrical power dynamics need to be understood, exposed, and undone/rectified/transformed so that the marginalised group(s) may be emancipated from the worldviews and actions of their 'otherers'. The othering/emancipatory frame is thus aligned with the pattern of guiding assumptions for a systemic othering inquiry often melded with an interpretivist/constructivist orientation. The pattern of guiding assumptions for a positivist inquiry and associated quantitative measurement are generally held to be incongruent with an othering perspective, unless that perspective is radically modified toward inclusiveness of the experiences and constructs associated with being othered.

Othering research tends to unfold within the natural experiential contexts of the research participants who are associated with a particular marginalised group (e.g., female, ethnic, religious, lesbian, gay, queer, bisexual, intersex, transgender, disabled, poverty-stricken, homeless, and so on). Such research may focus exclusively on the 'othered' or may pursue the perspectives of 'othered' and 'otherers' for purposes of contrast. Inherent subjectivity and biases in knowledge building must be acknowledged and understood which means that othering research must therefore be considered as value-laden rather than value-free. Interaction-based, participant-centred, and artefact-based data gathering strategies that yield qualitative data are considered to be more appropriate for accessing the authentic subjective worldviews, experiences, and stories of people who have been othered in society. Authenticity in the voices and stories being sought and heard is extremely important in all variants of the othering/emancipatory research frame. This imposes a great burden of responsibility upon the researcher to ensure that any data gathered and voices heard are demonstrably authentic and trustworthy. The researcher is not trying to tell their own story but those of the participants in the research (acknowledging, of course, that their own story may interact or resonate with participants' stories). Researchers working within an othering/emancipatory frame may either be insiders (i.e., a member of the 'othered' group being researched) or outsiders. A researcher-as-insider may find it easier to obtain access to authentic perspectives of the 'othered'. However, they need to be constantly aware of the fact that their research may be perceived, especially by the dominant 'othering' group, as simply running a political agenda (and therefore to be disregarded), so the researcher must constantly work to counter this impression through scrupulous choice of, engagement with, and protection of data sources as well as implementation of data gathering strategies and analysis strategies. While a researcher-as-outsider may not run the same risk of running afoul of external political perceptions, they will have to work much harder to withhold their preconceptions from influencing the research process and will likely struggle to gain access to the authentic perspectives of the 'othered' group because the participants themselves may suspect a political agenda. In many cases, research that includes perspectives from both marginalised ('othered') and non-marginalised or dominant ('othering') groups of participants, however defined, can add balance to the research (which helps counter perceptions of political agendas), value to knowledge that is produced, and impetus for social action. In every case, the researcher will need to fully explore what constitutes being 'othered' with the participants in their research.

Primary learning outcomes for 'othering' audiences emerge from patterns in data that open up new perspectives and interpretations of experiences of the 'othered', perhaps leading to transformative change through the unfreezing of entrenched patterns of experience. This may require retelling of stories in ways that will enhance the chances of convincing those who may have not previously shared such awareness by creating/stimulating awareness for why action and change might be needed. Primary learning outcomes for 'othered' audiences emerge from patterns in data that generate new pathways and leverage for achieving transformational and/or political change, build leverage for unfreezing entrenched patterns of experience and relationships, create political leverage for rebalancing/rectifying/emancipating relationships (e.g., enriching the reasoning and logic of the convinced or moving those prepared to be convinced toward being convinced, perhaps sufficiently enough to participate in action and change). Learning for both types of audiences may also emerge from patterns in data that help to share, compare, and contrast perspectives and interpretations of experiences, create leverage to help the 'converted' and 'unconverted' move closer together in perspectives to achieve a common vision for action and change. Speaking to both types of audiences simultaneously poses significant challenges because of the great care needed to craft a story that speaks to both sides of any issue.

Indigenous/First Nations Research Frame

Indigenous/First Nations research is a more recent evolution of cross-cultural research (see, for example, Botha, 2011; Chilisa, 2020; Kovach's, 2021; Smith, 2021; Walter & Andersen, 2013; Windchief et al., 2024). Indigenous/First Nations cultures encompass a myriad of Australian Aboriginal and Torres Strait Islander cultures, New Zealand Maori cultures, Inuit cultures in Canada, Native American cultures in the North, Central and South Americas, African cultures, Scandinavian cultures, Pacific Islander cultures, Southeast Asian cultures, and so on. Indigenous/First Nations research focuses on the development of knowledge of and about as well as understanding knowledge development, implications, and practices within a specific Indigenous/First Nations culture in its native context(s). Indigenous/First Nations research also focuses on understanding cultural knowledge, relationships, history, practices, and potentials from the Indigenous/First Nations perspective, as distinct from western and colonising perspectives. Such research is strongly aligned with the pattern of guiding assumptions for an Indigenous/First Nations inquiry (which effectively melds the pattern of guiding assumptions for an interpretivist/constructivist with the pattern for a participatory inquiry and a strong othering/critical/emancipatory/empowerment orientation; some authors, such as Smith, 2021, argue for researchers who are Indigenous/First Nations people to enact an activist role). Reid (2020) illustrated a different argument that it may be possible to productively integrate western and Indigenous/First Nations approaches through effective collaboration. Indigenous/First Nations research generally favours the use of qualitative data gathering strategies, implemented with a decolonising mindset, with a heavy emphasis on interaction-based (especially Focused, Depth, or Group) strategies, the Participant Observation strategy, and the participant-centred Creative Arts and the artefact-based Creative Works strategies, although there have been some recent developments with respect to the use of the Measurement data-shaping strategy for gathering/interpreting quantitative data from an Indigenous/First Nations perspective (see Walter & Anderson, 2013).

Because of the unique nature of traditions associated with how Indigenous/First Nations knowledge is acquired, accumulated, shared, and 'owned' by the Indigenous/First Nations communities, Indigenous/First Nations knowledge building is not consistent with any of the western concepts of Mode 1, Mode 2 or Mode 3 knowledge building. Research questions cannot be posed in ways that presuppose, reinforce, or reflect western or colonising anchoring of concepts and theories. Instead, Indigenous/First Nations research tends to be undertaken from a critical stance that is anchored in the recognition that the colonisation histories of Indigenous/First Nations peoples and cultures (including the non-Indigenous research practices associated with the colonising culture) has led to Indigenous/First Nations knowledge being stolen, co-opted, usurped, suppressed/hidden, distorted, or completely subsumed or destroyed by the colonising culture. Part of the role of Indigenous/First Nations research is to recover, privilege, and preserve the content, power, strength, and ownership of Indigenous/First Nations knowledge where it resides. Another important purpose for Indigenous/First Nations research is the co-creation and sharing of knowledge about Indigenous/First Nations lifespaces, practices, social mores, cultures, experiences, spiritual and worldviews with wider and likely non-Indigenous/non-First Nations audiences to foster learning about, respect for, and understanding of Indigenous/First Nations ways of knowing, ways of being, and ways of doing (Martin & Mirraboopa, 2003). Given the uniqueness of the human systems and spaces within which Indigenous/First Nations research operates, distinct sets of ethical principles for guiding such research have been devised in several countries. Two examples: in Australia, there are the 2020 Guidelines for Ethical Research in Australian Indigenous Studies established by the Australian Institute of Aboriginal and Torres Strait Islander Studies (AIATSIS; see http://aiatsis.gov.au/research/ethical-research/guidelines-ethical-research-australian-indigenous-studies); in Canada, there is Chap. 9 of the 2022 Tri-Council Policy Statement (Canadian Government's Panel on Research Ethics), entitled Research Involving the First Nations, Inuit and Métis Peoples of Canada (https://ethics.gc.ca/eng/policy-politique_tcps2-eptc2_2022.html).

We can identify two different pathways for conducting Indigenous/First Nations research. The first approach involves an Indigenous/First Nations person studying their own culture or another Indigenous/First Nations culture (this appears to be more consistent with Martin & Mirraboopa's, 2003 Indigenist research concept). If this describes the research situation, this approach would be characterised as 'insider' research in that the researcher shares at least some commonalities with the Indigenous/First Nations cultures they are doing research with. The second approach involves a non-Indigenous/non-First Nations person studying an Indigenous/First Nations culture, thereby constituting 'outsider' research. If this describes the research situation, this approach is much riskier and likely to be much less revealing and trusting if the researcher is working on their own, because of its negative association with historically oppressive colonisation effects, especially if the researcher is a member of the colonising culture (e.g., a white Australian university researcher studying aspects of the Anaiwan Indigenous cultural context in the New England region of New South Wales). The 'outsider' approach has a better chance of success if the researcher works collaboratively with a team that involves Indigenous/First Nations people from the cultural group that is being studied, but it will mean that the researcher will need to set their own predilections and preferences aside in favour of Indigenous/First Nations traditions. Nicholls (2009) argued that this is possible through a multi-layered process of self, interpersonal, and collective reflexivity embodied in a participatory methodology that works to build and maintain relationships and attendant social obligations. Furthermore, such 'outsider' research will have a much longer timeline, simply because it will take that much longer to build up trusting relationships within the focal cultural group.

What is learned in the Indigenous/First Nations research frame needs to be carefully negotiated and knowledge sharing and dissemination carefully and sensitively managed, with continual attention to what that learning and sharing brings back to the Indigenous/First Nations community (i.e., following the principle of reciprocity). It must be understood that social and behavioural research, from the Indigenous/First Nations perspective, is research done *with* not on Indigenous/First Nations people. The research experience will be inherently participatory in nature and affirming of Indigenous/First Nations values and experiences and will demand holistic/systemic thinking about the relationships with/connections between people, culture, physical environment, the cosmos, and the spiritual world (this is what differentiates the Indigenous/First Nations frame from the othering/emancipation frame).

The researcher's positioning in such research may emerge from one of three specific places: (1) insider researcher whose background and experiences are congruent with/same as the Indigenous/First Nations background of participants (e.g., research in or from the same regional group or language background); (2) insider researcher whose background is Indigenous/First Nations but whose experiences are incongruent/not same as the Indigenous/First Nations background of participants (e.g., from a different regional group or language background); or (3) outsider researcher who comes from a non-Indigenous/non-First Nations background. Potential contexts for Indigenous/First Nations research include (a) natural Indigenous/First Nations contexts inhabited by Indigenous/First Nations participants (Indigenous and Torres Strait Islander people in Australia would consider this to be research on Country; see https://aiatsis.gov.au/explore/welcome-country#toc-what-is-country) or (b) non-Indigenous/non-First Nations natural contexts with Indigenous/First Nations inhabitants or where Indigenous/First Nations and non-Indigenous/non-First Nations people are co-mingled (i.e., would also be considered research on Country). The (a) contexts may provide insights into the 'within-culture' aspects of Indigenous/First Nations life as well as into tensions and adaptations that emerge as those contexts interact with/respond to outside pressures whereas the (b) contexts may provide insights into 'between-culture' as well as 'within-culture' tensions and adaptations that occur when Indigenous/First Nations and non-Indigenous/non-First Nations people live together.

Participants in Indigenous/First Nations research may come from shared or different Indigenous/First nations backgrounds. They may occupy traditional knowledge keeping/spiritual/Elder/gender/social or other roles within their Indigenous/First Nations group or society. They may also occupy roles within non-Indigenous society, and this may, in fact, be a point of focus for the research (e.g., Indigenous/First Nations employees of a university; elected Indigenous/First Nations officials (e.g., members of parliament, representatives/senators in Congress, ministers in or councillors for state or local government). Generally speaking, participants in Indigenous/

First Nations research will need to explicitly include a broader range of stakeholders and reach right back into the community with which the research is being conducted.

The primary learning outcomes for non-Indigenous/non-First Nations audiences emerge from patterns in data that provide a strong cultural understanding or highlight social justice/equity/identity/social policy issues. Primary learning outcomes for Indigenous/First Nations audiences emerge from patterns in data oriented toward cultural development/preservation, conflict management, political action, or consensus-building. Primary learning outcomes for a mixed audience of Indigenous/First Nations and non-Indigenous/non-First Nations audiences emerge from patterns in data oriented toward displaying a strong cultural understanding (especially with respect to relations between Indigenous/First Nations and non-Indigenous/non-First Nations people), policy development, reconciliation, conflict management, or social equity. What the researcher learns needs to return value to the Indigenous/First Nations community with which they worked, and that learning must be shared with full respect for Indigenous/First Nations cultural practices and expectations.

4.5.3 Generalist Frames

Descriptive Research Frame

The descriptive research frame (see, for example, Elliott & Timulak, 2021; Gray, 2021; Tripodi & Bender, 2001; Mittal, 2011; Lambert & Lambert, 2012; Siedlecki, 2020; Thomlison, 2001) encompasses any research that purports to describe, summarise, count, or otherwise identify, record, analyse and reflect on important characteristics of and relationships between relevant data sources and groups of data sources, however defined. Most descriptive research is oriented toward Mode 1 knowledge production, producing descriptive stories, and/or databases for other researchers to utilise and build upon. Descriptive research may be population-based (i.e., providing census data) or may focus on characterising specific nations, groups, cultures, organisations/institutions, events, or samples. Typical research questions tend to be of the 'what', 'when', 'where', 'who' and/or 'how' types (e.g., What is observed? What is happening and where? What is the situation? Who lives/works/studies here? How is this outcome being achieved? When does/did event X occur?) rather than seeking deeper meaning and/or patterns of causal relationships. Some types of descriptive research may provide data for secondary databases or an institutional database with the data gathering exercise often being mandated by government or industry regulations or initiatives. Such descriptive databases then can be accessed by researchers for other purposes and types of research. Descriptive secondary databases may also be used to inform the creation/modification of social, institutional, and legislative policies. Thus, in many cases, the production of a descriptive secondary database is done to meet the needs of stakeholders other than the researchers who design, gather, and record the data in the database.

Descriptive research generally pursues breadth rather than depth of learning and may conducted under a wide range of patterns of guiding assumptions, including for a postmodern inquiry, yielding qualitative, quantitative, and/or hybrid/merged data. The general intention is to create and display a 'picture' or story about some context (and participants within that context). Descriptive research, on its own, cannot explain what is observed and/or recorded, it only notes what things are observed and/or recorded. In some cases, descriptive research may only be a part of the researcher's goal, especially if the research is being guided by pattern of assumptions for an interpretivist/constructivist inquiry. Here, description may be an initial goal of the research, namely, to establish observational and interpretational contexts, before proceeding to or as a part of exploratory or explanatory research (another instance where research frames can be blended or hybridised). For example, a descriptive ethnography has a goal to produce what is called 'thick description', which is description coupled with emerging interpretive work being associated with what is being observed. The research context can be widely varying for this research frame, but generally tends toward naturalistic contexts, away from the researcher's own context. For descriptive research feeding into a secondary database, the research context tends to be more broadly defined, say at a community, industry, institutional, or national level of analysis and the data gathered tend to be aggregated to that level.

The researcher's role in the descriptive research frame often tends to be more observational than interactional in nature, especially if their research is guided by the pattern of assumptions for a positivist inquiry. The researcher is there to record what is observed rather than to interpret what is observed. In some cases, the researcher may be contextually embedded with the goal of providing a descriptive story of some research context (as with a descriptive ethnography or case study guided by the pattern of assumptions for an interpretivist/constructivist inquiry). For descriptive research guided by the pattern of assumptions for a positivist inquiry, participants tend to be respondents to pre-designed descriptive questions. Under other patterns of guiding assumptions, participants may be more centrally involved with the researcher, where the research seeks to produce a descriptive and contextualised story. For descriptive research feeding into a secondary database, data sources will tend not to be individual participants, but a particular individual who produces and/or reports data aggregated across many individuals and/or groups within a context of interest.

Descriptive research is often conducted to provide information for decision makers and policy makers. Learning for such stakeholders will hopefully translate into meaningful decisions and policies. As well, descriptive research, and especially research that produces secondary databases, will provide data for other researchers to work with. Learning here will emerge from the research conducted using the secondary database (e.g., a human resources, financial, economic, or medical database) and will generally concern research questions that were not the focus of the originator(s) of the database. Primary learning outcomes emerge from patterns in data that help build up a coherent picture or narrative about some context or population of interest.

Exploratory Research Frame

The exploratory research frame (see, for example, Babbie, 2011 pp. 95–96; Davies, 2006; Flynn & Uttley, 2021; Stebbins, 2001; Swedberg, 2020) encompasses preliminary research into new fields of investigation, situations, and/or contexts (so-called 'greenfield' research). Of the three generalist research frames (descriptive, exploratory, and explanatory), exploratory research can be considered to be the frame most open to building knowledge in new areas and new ways. Depending upon the researcher's goals, they might pursue either Mode 1 or Mode 2 knowledge building; the latter, of course, engaging much more participatory input as well as demanding much more of the researcher's engagement with the context in order to build up perspectives on the issues being explored and to determine the needs to be met by further research. Any context may be useful in exploratory research, but often those contexts closer to where potential participants live, study, and/or and work will be more productive. In many cases, the research context will not have been the previous target of research, which creates the need for exploration.

Exploratory research addresses many of the same types of research questions as descriptive research, but with a more explicit goal of building knowledge to stimulate, guide, and provide input to further research. Theories, propositions, and hypotheses are typically not tested in exploratory research. However, some ways of doing exploratory research may lead to preliminary theoretical ideas, propositions, and hypotheses, reflecting a theory-building emphasis (e.g., providing initial inputs into grounded theory development) consistent with Mode 1 knowledge building. In such cases, research questions that ask 'why' things are happening become important to address. Exploratory research may be conducted under a wide range of patterns of guiding assumptions, including for a postmodern inquiry, but in some disciplines, such as marketing research, may be guided by the pattern of assumptions for an interpretivist/constructivist inquiry in early phases of the research. This signals that many researchers use exploratory research early in their research investigations to gain a deeper understanding of contexts and issues prior to embarking on their main research phase. Exploratory research tends to reflect more depth of focus than breadth of focus, since the learning achieved tends to open up more questions than it resolves and generalisable or transportable findings are not being pursued. In fact, one of the distinguishing features of the exploratory research frame is that it is oriented toward opening up lines of investigation, rather than trying to come to a definitive answer about any context or issue.

The researcher's role is generally to be a learner, to engage at some level with a new or under-explored research context of interest with a view to learning about that context and what might be useful to take forward as input into further research. The researcher generally has a higher level of uncertainty with respect to what might be learned within this research frame in that surprises may lie around every corner. Participants often take on a more active role of helping the researcher to learn about the research context, what the issues are, what is important and so on. Here, the pattern of guiding assumptions for an interpretivist/constructivist or participatory inquiry may be more useful to adopt. In exploratory research, the researcher-participant relationship is somewhat closer than for descriptive research and, in some cases, even explanatory research. Under the pattern of guiding assumptions for a positivist inquiry, the participant takes a more passive role, but one where the researcher cannot/does not anticipate all that might be learned. Exploratory research tends to work best with interaction-based, observation-based, or artefact-based data gathering strategies. In sequential research configurations (to be discussed in Chap. 5), exploratory research often occupies the first phase of the research, providing input into subsequent phases.

Exploratory research can be an end in itself by providing other researchers with signals for new and further directions for research, preliminary ideas for theorising as well as input into how to more effectively contextualise and configure further research. The researcher may be learning in order to satisfy their curiosity or to determine what might be feasible to do in further research within a specific context including learning how best to shape the next phase of their journey. Primary learning outcomes for exploratory research emerge from patterns in data that provide signals and directions for future research endeavours as well as help to build up a coherent picture or narrative about some context of interest.

Explanatory Research Frame

The explanatory research frame (see, for example, Babbie (2011 pp. 97–101); Blatter & Haverland, 2012; Bryant & Charmaz, 2007; Cook et al., 1990; Flynn & Uttley, 2021) encompasses any research that has the goal of explaining and/or theorising why things happen. Explanatory research chiefly pursues Mode 1 knowledge building, seeking, and testing theoretical accounts for events and observations. It has a strong association with the pattern of guiding assumptions for a positivist inquiry where the researcher's goal is to test causal and theoretical propositions, hypotheses, or predictive relationships, deduced from theory, then make inductive inferences back to that originating theory. Explanatory research also has strong ties to the Measurement data shaping strategy and experience-focused data gathering strategies, because quantitative data are often pursued within this frame. Under the pattern of guiding assumptions for a positivist inquiry, explanation equates to understanding the theorised causal linkages between constructs and perhaps making predictions of future behaviours based on those theorised linkages (making predictions is one manifestation of generalisation). To do this convincingly, (1) causal factors must be observed before effects are observed, (2) effects must change if their putative cause(s) change and, importantly, (3) alternative plausible causes for observed effects must be ruled out (via control procedures). The stronger the causal inferences intended, the more tightly controlled the research context(s) must be and the more the research will be focused on breadth of learning (i.e., the pursuit of generalisation) rather than depth of learning. Research questions in the explanatory frame under the pattern of guiding assumptions for a positivist inquiry will tend to use causal language that varies in strength (from anticipating strong causal implications to anticipating simple associations) depending upon the contextual circumstances of the research. Explanatory research may also be conducted under critical realist assumptions, where the goal may be to achieve an understanding not only of external causal patterns but also of people's perceptions and understandings of those patterns.

There are other perspectives available for the explanatory research frame. For example, explanatory research also encompasses inductive theory-building approaches such as grounded theory, which tend to be guided by the pattern of assumptions for an interpretivist/constructivist, participatory, or, in some cases, critical or systemic othering inquiry. The explanatory frame would not resonate well with the pattern of guiding assumptions for a postmodern inquiry. In this case, theoretical explanations and accounts emerge from the data instead of being proposed by the researcher prior to their research being conducted. Furthermore, emerging theoretical propositions are tested by gathering more contextualised data. Here, the focus is more on depth of learning (i.e., immersion in context) rather than breadth of learning and research questions are worded in more open language that suggests what the researcher is interested in learning about without anticipating what they might find. Interaction-based and observation-based data gathering strategies, and strategies for exploring people's handiworks gathering strategies will tend to be preferred. It is fair to say that, of all the possible ways of framing social and behavioural research, the explanatory research frame is the most sensitive to and most strongly shaped by the pattern of guiding assumptions the researcher adopts. This is largely attributable to the differing roles of and perspectives on causality and theory within the different patterns.

The research context for the explanatory research frame may range from a highly controllable experimental laboratory to a naturalistic field setting. Contextual choice is pivotally connected to the pattern of guiding assumptions adopted. Under the pattern of guiding assumptions for a positivist inquiry, the researcher's capacity to influence or manipulate what happens in the context will determine choices. Under a pattern of guiding assumptions for a non-positivist inquiry, the researcher will tend to prefer naturalistic settings, where they have minimal capacity to influence what happens. The researcher's role in the explanatory research frame will also be heavily dependent upon the pattern of guiding assumptions they adopt. Under the pattern of guiding assumptions for a positivist inquiry, the researcher's role will involve prior theorising about what they hope to find coupled with maintaining a more 'objective' distance from what they are observing so they do not contaminate the data being gathered, thereby testing and evaluating theoretical propositions. Under any pattern of guiding assumptions for a non-positivist inquiry, the researcher's role will become embedded in the research context whilst maintaining a 'researcher attitude', in as far as it is possible and feasible to do so. Here, the researcher pursues subjective perspectives that are not their own in order to build up a theoretical account for those perspectives. Under the pattern of guiding assumptions for a critical realist inquiry, the researcher may adopt a more balanced stance, to critically examine both internal (within participants) and external (in the world inhabited by the participants but distinct from any particular participant's viewpoint) perspectives. Where the adopted pattern of guiding assumptions is not value-neutral, theorising may be a priori or conceptually-driven, but will not typically be data-driven in the same way as a grounded theory approach would be.

The positioning of participants, in terms of their role from the researcher's perspective, is also critically dependent upon the pattern of guiding assumptions the researcher adopts. Under the pattern of guiding assumptions for a positivist inquiry, the larger context(s) and life histories of participants are largely irrelevant; participants are simply inter-substitutable data sources whose idiosyncratic differences can hopefully be assumed to contribute nothing more to the research than random errors. Under the pattern of guiding assumptions for an interpretivist/constructivist inquiry or for other non-positivist inquires, who participants are, where they have come from and the role(s) they may play in the context(s) of the research all become relevant and important for the researcher to learn about. Participants thus become a primary source of information for emergent theorising. Adopting the pattern of guiding assumptions for a critical realist inquiry allows the researcher to seek a balance between these two attitudes.

Most explanatory research tends to engage in Mode 1 knowledge building, trying to add to the disciplinary core of knowledge about social and behavioural phenomena, often through publication in peer-reviewed academic journals. Thus, other researchers are the primary learners from such research and that learning may influence what they investigate and how in the future. Explanatory research tends not to ascribe high value to application of what is learned to concrete problems and practices, except in so far as such applications may be envisaged as emerging downstream or where the explicit intent is to construct and test predictive forecasting or decision-making models (in the economic, management, or financial disciplines, for example). Primary learning outcomes emerge from patterns in data that signal defensible inferences about causes and effects or that provide authentic, coherent, and consistent accounts for what has been observed.

4.5.4 Complexity/Innovation Frames

Pan-Disciplinary Research Frame

The pan-disciplinary research frame (see, for example, Aagaard-Hansen, 2007; Bergmann et al., 2012; Choi & Pak, 2006; Fear et al., 2006; Duke, 2011; Fedorov, 2020; Hadorn et al., 2008; Lang et al., 2012; Leavy, 2011; Lury et al., 2018; Padmanabhan, 2019; Repko & Szostak, 2020; Szostak, 2015; Toomey et al., 2015; Vienni-Baptista, Fletcher & Lyall, 2023) encompasses research that is intended to work across or break down the boundaries/barriers between disciplines in addressing complex real-world problems. There is a continuum of pan-disciplinary modes ranging from multi-disciplinary to cross-/inter-disciplinary to trans-disciplinary (see Fig. 4.6), where modes are distinguished by the degree of integration across disciplines and the engagement of knowledge building (as well as directionality of influence) outside the academy (Aagaard-Hansen, 2007; Toomey et al., 2015). The pattern of guiding assumptions for a critical realist, participatory, postmodern, or pragmatic inquiry would all resonate well with different modes in this research frame. However,

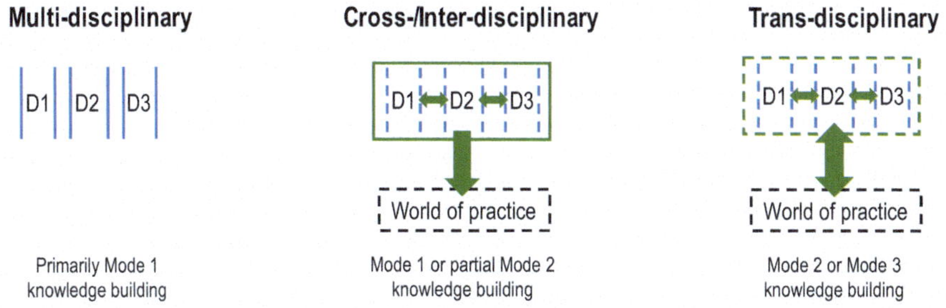

Fig. 4.6 Conceptual illustration of modes of pan-disciplinary research

it is possible to envisage that the critical or systemic othering patterns could be useful, as long as their value-laden orientation is explicitly explored across the boundaries of more than one discipline.

Multi-disciplinary research (e.g., see discussion in Aagaard-Hansen, 2007) employs knowledge building practices from within several distinct disciplines within the academy in situations where a monodisciplinary approach would not be or has not been productive. Such research may be done by a single researcher, using a variety of discipline perspectives, or, more likely, by a team of researchers, comprising different disciplines. Typically, Mode 1 knowledge building is pursued and while the perspectives of multiple disciplines are used, the integrity/boundaries of those disciplines remain intact. It is, in effect, using disciplines as a lens for triangulation on the research problem. Connectiveness with the world of practice tends to be of little interest in multi-disciplinary research, which means that the pattern of guiding assumptions for a participatory inquiry is generally not preferred. In many cases, the patterns of guiding assumptions adopted may coincide with the disciplines within which the research problem is being addressed.

Cross-/inter-disciplinary research (e.g., see discussions in Toomey et al., 2015; Repko & Szostak, 2015) employs knowledge building practices from within several distinct disciplines within the academy but in situations where a monodisciplinary approach would not be or has not been productive and where connection to the world of practice outside the academy is desired. In such research, Mode 2 knowledge building is more likely to be pursued (although Mode 1 is also possible depending upon how tenuous the connection is to the world of practice). Inter-disciplinary research moves beyond multi-disciplinary research by seeking to break down the boundaries and barriers as well as harnessing synergies between disciplines, facilitating learning, innovation, change, and evolution. However, inter-disciplinary research still privileges the academy over the world of practice such that the knowledge built tends to be applied to the world of practice without full participation from that world. Virtually any patterns of guiding assumptions may be adopted, often coinciding with particular disciplines, but the pattern for a participatory inquiry would be unlikely to be adopted.

Trans-disciplinary research (e.g., Cooksey, 2011; Leary, 2011; Padmanabhan, 2019; Szostak, 2015) has an explicit Mode 2 knowledge building focus, but depending upon circumstances and research/innovation needs, may adopt a Mode 3 knowledge building focus. It attempts to bridge the divide between theory and practice and between academic and practitioner/professional. Trans-disciplinary research is typically highly participatory and collaborative in emphasis, drawing upon different academic disciplines and non-academic practitioner and community inputs into the research. Trans-disciplinary research typically reflects the pattern of guiding assumptions for a critical realist or pragmatic inquiry, focusing on problem-solving, but melded with the pattern of guiding assumptions for a participatory inquiry. Such research is ultimately intended to foster innovation and influence decision making in specific contexts. Trans-disciplinary research seeks to break down the boundaries between theory and practice by diversifying the stakeholder influences and inputs to and ownership of the research. Stakeholders thus become active collaborators in the research process. In this way, research moves beyond academic constraints toward stimulating change and innovation where it matters most, be it in a community, a business, the environment, government, non-government organisation, or another focal context. One goal, therefore, is to produce research that is useful to the public/community. The practice of critical engagement (see, e.g., Fear et al., 2006) is one way of approaching the trans-disciplinary mode of the pan-disciplinary research frame.

One creative approach to pan-disciplinary research is arts-based, where non-traditional ways of building, displaying, and applying knowledge are employed; an approach that resonates strongly with the pattern of guiding assumptions for a postmodern inquiry. The goal is to engage thinking and perception (across all sensory channels) through the use of various types of performance (theatre, dance, music, drama), imagery (photography, painting, and sculpture), multiple medias (blogs, film and online) and language genres (narratives/stories, literature, poetry, scripts and screenplays). An arts-based approach to pan-disciplinary research may be especially useful if the research is intended to elicit meaning from, have meaning for, and/or be useful to systemically othered, cross-cultural and/or Indigenous/First Nations audiences.

While it is certainly possible for a single researcher to conduct pan-disciplinary research, it would be an onerous undertaking. In the multi-disciplinary mode, a single researcher approach would likely be more feasible if the disciplinary attacks on the research problem were undertaken sequentially rather than simultaneously. The inter-disciplinary and trans-disciplinary modes will be much more feasible and effective to undertake with a collaborative research team that is willing and able to seek out, learn from, and build upon feedback from various sources, including each other. Appropriate attention must be given to the leadership of this team. If the approach is community-based, multiple stakeholders need to be involved in the research team. The goal is to look for synergies between disciplines and be willing to look beyond the boundaries of one's own discipline (including one's own preferred patterns of guiding assumptions) as well as being willing to be adaptive and flexible in approach. There is little room in inter-disciplinary and no room in trans-disciplinary research for academic 'snobbery' and detachment. However, a Mode 1 knowledge building mindset will actively limit/constrain the potential and effectiveness of Mode 2 knowledge building as well as undermine the building of trust.

Since pan-disciplinary research, particularly inter-disciplinary and trans-disciplinary research, focuses on addressing complex real-world problems, such research is best done in the field where the problem resides, in professional or community contexts where practices can be influenced, and/or where innovation development can occur. In trans-disciplinary research, the research context usually involves the cultivation of a knowledge co-development relationship between researchers and participants which means that a range of interaction-based and participant-centred data gathering strategies will be quite useful. Research participants are those whose are involved with, impacted by or are stakeholders in the research to address the problem. In the trans-disciplinary research, stakeholders may become members of the research team which may mean they take on dual roles: researcher and participant. In a very real way, effective trans-disciplinary research reaches out to the public in general, even if every member of the public cannot be explicitly included in the research itself (they could be considered 'downstream' stakeholders, for example). The research, therefore, should have meaning for/be useful to all relevant audiences in the research context.

Primary learning outcomes from pan-disciplinary research for public audiences emerge from patterns in data that meaningfully connect what has been learned to what must be done. They should provide impetus and invitation for social action, involvement, advocacy, and innovation. Primary learning outcomes from pan-disciplinary research for scientific audiences emerge from patterns in data that signal as well as evaluate useful directions, shapes, and forms for social action as well as for innovations and setting the stage for their effective adoption (Lang et al., 2012). The scientific community would also be looking for potential directions for new research and innovation development. Primary learning outcomes from pan-disciplinary research for public policy audiences emerge from patterns in data that signal needed changes in societal practices/policies relevant to addressing specific problems that are sensitive to the diversity within different public groups/spaces. This then may influence decision making regarding how resources are deployed within communities. Learning outcomes for researchers, especially in the inter-disciplinary and trans-disciplinary modes, will likely include changes to their researcher identities vis-à-vis their discipline, leading them to be more likely to work as 'boundary-spanners' in future research endeavours and enhance their prospects for developing and engaging in pluralist systemic thinking.

Developmental Evaluation Frame

The Developmental Evaluation (DE) frame (see, for example, Leonard et al., 2016; Patton, 2011, 2016; Patton, McKegg & Wehipeihana, 2015; Rey et al., 2014) is most closely associated with Michael Quinn Patton (2011) and focuses on innovation (social, technological, policy) development as an adaptive response to dynamically changing contextual circumstances. DE emerged as a research/innovation process that draws upon multiple disciplines, seeks co-production of knowledge, and explicitly works through and within the complexity and unpredictability of social circumstances, thereby aligning with, but extending beyond, the trans-disciplinary mode of the pan-disciplinary research frame. Patton positioned DE as a type of utilisation-focused evaluation but one where changing situational contexts and needs work to stimulate innovative responses to meet those needs as well as anticipate productive responses to future emergent needs. This means that DE involves a cyclical approach to research coupled with a deep and reflective learning interconnectivity between the researcher/evaluator/consultant, stakeholders, users, and innovators. DE research tends to align with the pattern of guiding assumptions for an interpretivist/constructivist, participatory, critical realist, or pragmatic inquiry but need not be tied to those patterns of guiding assumptions. DE is explicitly focused on Mode 2 participatory knowledge co-production, but, depending upon circumstances, may venture into Mode 3 knowledge building. DE may employ a wide range of data gathering strategies in pursuit of its purposes. In fact, Patton (2011) employs the metaphor of 'bricolage' to describe the process of DE (see also Kincheloe, 2005; Kincheloe & Berry, 2004). Bricolage describes a process of using whatever tools are necessary, even creating new tools, to achieve the deep learning demanded by DE, a logic that resonates with the pattern of guiding assumptions for a pragmatist inquiry.

The researcher, in the DE frame, undertakes the role of a bricoleur or a jack-of all-trades in their approach, accessing the broadest possible toolkit while being open to creating new tools and approaches where necessary. Thus, to meet emerging, perhaps unanticipated needs, methodological innovation is highly likely in DE. Reflective learning capability is demanded by all involved in DE, since the focus is on understanding, monitoring, and modifying innovative adaptations to dynamically changing/evolving situations and circumstances. The dynamically changing situational circumstances for a specific social problem defines the research context for DE. Participants are stakeholders, users, potential adopters, and innovators involved in the research context and stand to benefit from as well as contribute to adaptive innovation to address emergent and complex contextual problems. For DE, a deep level of contextual engagement is sought as innovation-focused knowledge is co-produced by all involved. Developmental evaluation for innovation tends to be inherently participative in nature, so some hybridisation of other patterns of guiding assumptions with the pragmatic and participatory inquiry patterns may be preferable.

The deep engagement demanded of the researcher and the participants, stakeholders, users, potential adopters, and innovators should result in productive and reflective as well as anticipatory learning. This type of learning balances a focus on the past ('what we did?') with a focus on the present ('what is happening now?") as well as a focus on the future ('where do we go from here?"). Contextualised adaptation through innovation is the prime goal for reflective learning in DE. Overall, the primary learning outcomes emerge from patterns in data that provide feedback and input into innovation development, modification, and adoption/use.

4.6 Harnessing Synergies

Patterns of guiding assumptions, modes of knowledge building, and research frames form three higher-level planes for focusing pluralist systemic thinking and contextualisation in social and behavioural research. These considerations emerge from the intersection of more general contextualisation considerations and the researcher's thinking about how best to conceptualise and realise their journey on a specific research project. Contextualisation and thinking each influence and shape each other in dynamic ways over time. The choices of patterns of guiding assumptions, modes of knowledge building, and frames are built upon the researcher's contextualisation activities and learning and those choices can be singular or plural, as needed (Bowers, 2011; Hesse-Biber, 2012; Layder, 1993; Lewis & Keleman, 2002; Mingers & Brocklesby, 2003). Broader research contextualisations get refined, reshaped and more sharply focused once those choices have been made. This is not a linear, deterministic, or one-time process, but an evolutionary dynamic that unfolds in time and space and is subject to the shocks and jolts that unanticipated/emergent constraints and opportunities create throughout the research journey. Planning is the researcher's attempt to anticipate as many aspects of this evolving dynamic as possible, keeping in mind the ultimate goal of producing convincing research outcomes. Flexibility and adaptability are the researcher responses demanded when planning and anticipation fail. What we have seen in this chapter is that patterns of guiding assumptions serve a necessary function in helping the researcher to focus their downstream research choices, modes of knowledge building speak to the relevant audiences for learning from, participating in, and benefitting from research, and frames provide a way of holistically situating the research journey, within which those patterns of guiding assumptions and modes of knowledge building operate. Together, patterns of guiding assumptions, modes of knowledge building, and frames provide necessary inputs for making choices about how best to navigate the 'Data Triangle', further infusing shape, colour and life into the sterile landscape of methodological possibilities.

4.7 The Emergence of Research Questions, Propositions, and Hypotheses

The form of and language used in stating research questions, propositions, or hypotheses are sensitive to the underlying pattern of guiding assumptions, mode of knowledge building, and research frame adopted by the researcher. Higher-level contextual concerns may also contribute to such statements. Figure 4.7 (significantly updated and expanded from Cooksey & McDonald, 2019, Fig. 11.8, p. 399) takes the researcher's emergent research contextualisations and positionings pathway from Fig. 1.4 (and as subsequently elaborated in Figs. 3.1a, 3.2, and 4.3) as the starting point and diagrammatically expands the considerations associated with the identification of the research problem and evolution of the research questions, propositions, and hypotheses. Higher-level contextualisations help to surface the research problem and identify prior knowledge building efforts associated with it. However, a research problem is typically amorphous, perhaps even ambiguous, in nature, lacking specifics as to what is of interest, who or what might be implicated, and who the intended audience(s) might be for research purporting to address the problem. Thus, researcher cannot simply focus on a research problem, they must tear the problem apart, identify the essential aspects of interest, and transform the problem into more focused statements of interest and intention regarding what the researcher wishes to learn and the knowledge they wish to build. Considerations about patterns of guiding assumptions, modes of knowledge building, and frames as well as learning from prior knowledge building efforts assist in this transformational process. Research quality criteria and meta-criteria help the researcher to ensure that anything they do to investigate their research questions, propositions, and/or hypotheses by navigating the 'Data Triangle' is done in a way that enhances rather than detracts from convincingness. Research questions, propositions, and hypotheses provide boundaries for the research, making clear what is of interest and implying that anything not captured in them may be outside the researcher's focus. However, under some patterns of guiding assumptions, research questions and propositions may morph and change over the course of the researcher's journey, sometimes in response to what has been learned in prior knowledge building efforts or phases of the research. This means relaxing the implication that what is not stated in research questions, propositions, and hypotheses resides outside the researcher's interest, leaving room for new and emergent statements of research focus. Research hypotheses generally do not emerge/evolve in the same way; once stated, they must be evaluated as stated and if disconfirming evidence emerges, they must be discarded. If theory changes and new hypotheses are deductively formulated, they are generally tested with new research.

Research questions, propositions, and hypotheses must be stated in a manner consistent with the expectations of the researcher's chosen pattern of guiding assumptions and they must be consistent with the intended positioning and framing of the research. There may be disciplinary, prior research/knowledge building outcomes, and other contextual influences (e.g., communities of practice, researcher's prior experiences and beliefs) on the substance and expression of research questions, propositions, and hypotheses, which the researcher may need to take on board or at least critically examine (see, for example, discussions in Plano et al., 2011). Research questions, propositions, or hypotheses may signal methodological approaches or may be worded in a manner consistent with a particular research approach. Research propositions and hypotheses are generally stated assertively, yielding statements that predict or at least anticipate patterns and relationships between constructs or concepts. Research questions tend to be open-ended and equivocal, avoiding statements about anticipated relationships or patterns and, in many cases, not invoking specific constructs or concepts. However, research questions, propositions, and hypotheses are all more specific than a general statement of the research problem statement and the choice of language for them will be critical decisions the researcher will have to make. Depending upon the research problem, the researcher may have just

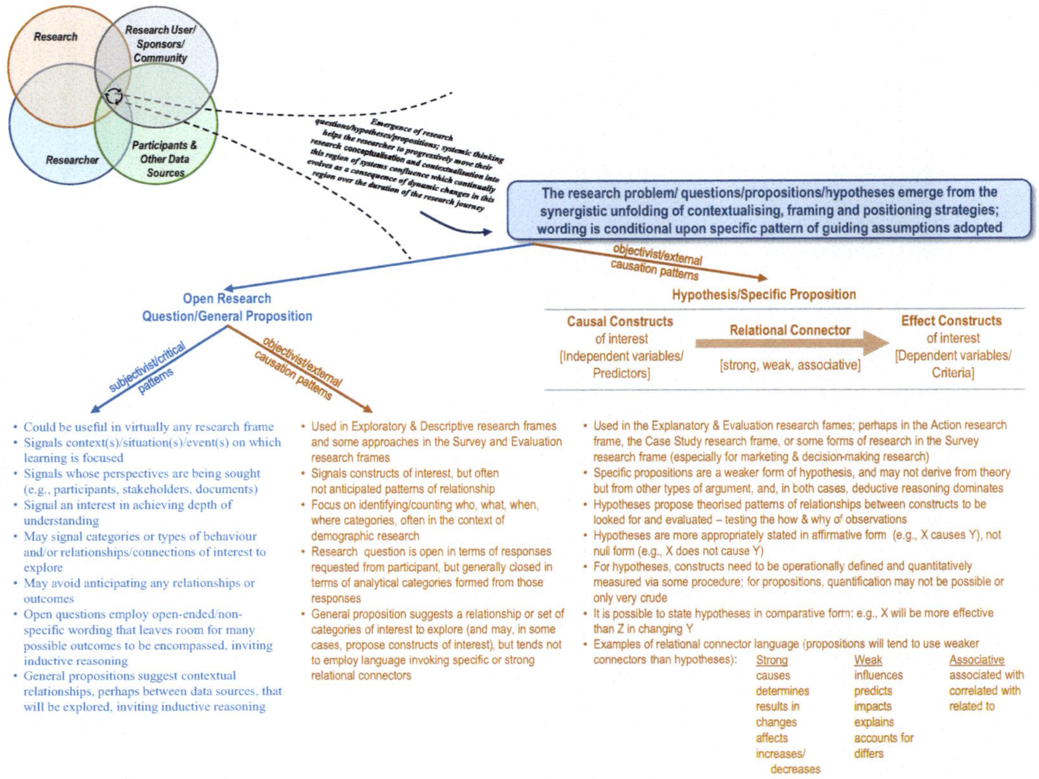

Fig. 4.7 Depiction of the distinctions between types of research questions, propositions, and hypotheses

a single emergent research question, proposition, or hypothesis, or they may have a series of connected research questions, propositions, or hypotheses. In general, it is better practice to have a few well-stated research questions, propositions, or hypotheses than to have many exceedingly precise questions or hypotheses.

4.7.1 Research Questions

A research question (sometimes referred to as an 'open research question') is open ended and does not anticipate any relationship between constructs, concepts, or ideas (some authors use the term 'research question' to encompass propositions and hypotheses, but that is not the use being promulgated here). Research questions invite the use of inductive reasoning to evaluate evidence and draw conclusions. Open research questions (follow the left/blue branch in Fig. 4.7 to the left/blue forked pathway) tend to be associated with subjectivist patterns of guiding assumptions where they do not anticipate what might be found (critically-oriented patterns of assumptions may be adopted, in which case, research questions may signal a specific value position or 'othering' expectation). Typically, an open research question points more toward general focal interests and perhaps to types of behaviour/phenomena of interest and often, the researcher may have just a single general research question. Open research questions use looser language more focused on who's perspectives are to be understood in what circumstances (a subjectivist intent, such as "what do A, B and C think about event or situation P?", "How do A, B and C accomplish tasks 1, 2 and 3?", "What can we learn from A, B and C?", "Why and how has situation P come about in context G?", and tend to shy away from naming specific constructs or relationships of interest. Open research questions set out the contexts(s) and phenomena that are the researcher's focus for learning, without second-guessing what patterns might emerge. Open research questions may also focus on specific participants or types of participants, where relevant. Furthermore, open research questions may evolve or be further refined or shaped during the course of the research journey as a consequence of what the researcher has learned earlier on (something that should not occur under the patterns of guiding assumptions that have objectivist external causation interests, unless the researcher uses a sequential research configuration—see Chap. 5). Thus, there is often an expectation of flexibility and evolution attached to open research questions in that, as analyses progress, further questions may evolve to pursue greater depth and focus. It is possible to employ open research questions in research guided by objectivist/external causation patterns of assumptions (follow the left/blue branch in Fig. 4.7 to the right/orange forked pathway), especially where an Exploratory or Descriptive research frame has been adopted or where census-type questionnaires are being used in the Survey research frame. Constructs are still of interest in such questions, but relational patterns between them are typically not proposed. Such research questions tend to focus on identifying/counting categories of who, what, when and where with respect to some population of interest and they are often employed in demographically-oriented research.

4.7.2 Hypotheses

A research hypothesis is deductively derived from theory that emerges from prior, normally published, knowledge building efforts. It is the most precise form of research statement and is worded in in such a way that it can be empirically disproven (shown to be false). Hypotheses are never confirmed; they can only be disconfirmed. Hypotheses are associated with the objectivist/external causation patterns of guiding assumptions and assert theorised connections between specific constructs that are to be tested. A review of prior research and knowledge building outcomes should clearly lead their audiences to expect the constructs and relational patterns that are implicated in the research hypotheses. As shown in right/orange branch of Fig. 4.7, hypotheses have two key components: (1) identification of key constructs on either side of a (2) relational connector. Constructs generally require operational definition, measurement, and quantification and must be shown or argued to possess construct validity. In hypotheses, constructs play one of five different roles when they are translated into measured variables: (1) *independent* (i.e., reflecting putative causes), (2) *dependent* (i.e., reflecting putative effects), *mediating* (i.e., reflecting putative intermediate causes), (4) *moderators* (i.e., reflecting putative conditional causes) or (5) *extraneous* (i.e., reflecting anticipated alternative plausible causes to be explicitly controlled for). The researcher must keep these roles straight in all their research hypotheses. Hypotheses are typically evaluated using some form of inferential statistical analysis (see, for example, discussions in Cooksey, 2020, Chap. 7), however, they are more productively stated in affirmative language (i.e., stating what the researcher is hoping to show) than null hypothesis wording (which is the base form of hypothesis that inferential statistical testing seeks to reject/disconfirm). Nothing is more boring to read than a list of hypotheses worded as null hypotheses, when it is obvious that the researcher is expecting some kind of relationship or group difference.

Hypotheses may vary in terms of the strength of causality they theorise to exist: strongly causal, weakly causal, or simply associative (refer to Fig. 4.7). Strongly causal hypotheses state the precise causal patterns that the researcher is looking for, clearly pointing to the constructs thought to be causes and those thought to be effects. Such hypotheses are most more likely to be used in the Explanatory and Evaluation research frames. The language used to express strong causal hypotheses signals that strength: e.g., "X causes Y", "X determines Y", "X results in Y"). Such hypotheses require strong theoretical arguments to justify them, with the logic well-anchored in prior research/knowledge building efforts. Choosing wording like "X influences Y" or "X predicts Y" provides weaker causal statements than "X causes Y" and allows the researcher to soften their causal claim while still signalling a directional intent for the theoretical inference. Weak causal hypotheses may be more appropriate if the theoretical justification for strong causal hypotheses is ambivalent or contradictory. Associative hypotheses back away from making causal assertions because the research area may be new (as with the Exploratory research frame), the research focus is case-based (as with the Case Study research frame), the research intent may be descriptive (as with the Descriptive research frame) and/or where theoretical justification is lacking (as with certain ways of implementing the Survey research frame). Such hypotheses instead make claims of association, using language such as "X is associated with Y", "X is correlated with Y", "X is related to Y". Technically, once the researcher sets out a hypothesis to be tested, it should not be changed downstream within the same research project. To do so might expose the researcher to claims of running a fishing expedition.

In cases where the researcher is testing precise theoretical predictions, they would be expected to state hypotheses to reflect the anticipated relationships between two or more constructs or anticipated differences between different groups (e.g., see Tharenou et al., 2007; White, 2009). However, the researcher must be careful in stating anticipated directions of difference or relationship. If the researcher has no theoretical basis on which to say that 'X will increase Y', then they can back away from predicting the direction by simply stating that 'X will be associated with changes in Y'. Similarly, if the researcher has no compelling reason to expect that "Group A will show significantly improved performance relative to group B", then they can soften their expectation by stating "Group A performance will significantly differ from Group B". This is important because stating a direction will often requires the researcher to use a slightly modified form of hypothesis testing called a 'one-tailed test' and, if they hypothesise the wrong direction, then they will be in a real mess with respect to the convincingness of their conclusions. Hence, the researcher should have a strong theoretical basis to make such precise directional predictions.

4.7.3 Research Propositions

A research proposition may be formed deductively or inductively or may emerge abductively as a consequence of some anomalous or unexpected outcome. A research proposition is a more focused statement relative to open research questions but is much less precise or theoretically grounded than a hypothesis. We can differentiate between two types of research proposition: general and specific. A general research proposition is a less open form of research question, whereas a specific proposition is a weaker form of hypothesis. Where research is guided by subjectivist/critically-oriented patterns of guiding assumptions (follow the left/blue branch in Fig. 4.7 to the left/blue forked pathway), general research propositions are often used. For example, for research associated with an othering/emancipatory or Indigenous/First Nations frame, the perspective of a specific group or community may be privileged by the wording of the question (e.g., by flagging the group as colonised, discriminated against, oppressed, marginalised, minority, silenced and the proposition focuses on identifying the conditions or power relationships/dynamics that contributed to that state of affairs). In such cases, patterns of relationship are signalled, and the researcher's intention is to learn why this might be so and what maintains the pattern.

Where research is guided by objectivist/external causation patterns of assumptions but is relatively less tightly conceptualised (follow the left/blue branch in Fig. 4.7 to the right/orange forked pathway), a general proposition suggests a relationship or set of categories/conditions, perhaps even proposes constructs, to explore but avoids language that signals strong relational connectors. For example, for a researcher investigating a new innovation, a general research proposition might be that an innovation improves the experiences of X in the context of Y. Where research is guided by objectivist/external causation patterns of assumptions and is relatively more tightly conceptualised (follow the right/orange branch in Fig. 4.7), specific propositions emerge where stronger theoretically deduced hypotheses cannot be justified, such as when a researcher suspects there is a connection between constructs, but this suspicion emerged from anecdotal or personal experiences (e.g., an intuition or hunch) rather than from theory established by prior knowledge building efforts. Thus, non-theoretical types of deductive argument underpin the statement of specific propositions. The constructs invoked in a specific proposition may be only crudely quantifiable if at all.

4.8 Integrative Foreshadowing

As we have said before, social and behavioural research is a human activity, contextualised in human spaces, and historical timelines. Unlike the physical and natural sciences, the social and behavioural sciences deal with phenomena associated with thinking feeling beings, one of whom is the researcher. While it is certainly possible to learn about and build knowledge regarding human behaviour in some context as if it were a physical or natural phenomenon, this is not the only way that such learning and knowledge building can be achieved. Thus, we must consider the assumptions the researcher is willing to adopt to guide their research activities in order to provide further contextual focus for the research. Those assumptions arise from patterns of answers to pivotal questions about the focus and intentions of the research/researcher. Each pattern of guiding assumptions invokes its own criteria for making judgments about research quality, but, with pluralist systemic thinking, one needs a way to make judgments that are not bound by particular patterns of guiding assumptions, giving rise to meta-criteria for research quality, focusing on research convincingness. Researchers can also benefit from framing their research where a frame holistically encompasses researcher, participant, mode of knowledge building, and intended audience(s) considerations, including adopted pattern(s) of guiding assumptions, as a way of signalling the overall intentions of the research. Adopting a frame makes it easier for a researcher to communicate their research intentions to others. The final step in research contextualisation involves statements about where the researcher wishes to concentrate their learning and knowledge building. Depending upon higher-level contextualisations, choices of pattern of guiding assumptions and frames, these statements will take the form of research questions, propositions, and/or hypotheses. One visualisation tool that could be useful for recording contextualisation thinking is the *contextualised journey map*, first introduced in Fig. 1.13, where the blue boxes overlaying the grey amorphous area prompt deeper and broader contextual thinking. Those contextualising considerations inform would then prompt (refer to the gold box with the journey icon) more focused thinking about relevant choices to be made for the researcher's journey. It could be useful to revisit this map from time to time during the journey to see if and where any changes have occurred that might require an adaptive response or trade-off. The visualisation tool could be embedded in the researcher's journal along with notes/memos that clarify/elaborate on the researcher's thinking.

It is important to note that we should distinguish between 'contextualisation within' and 'contextualisation about'. Any pattern of guiding assumptions that adopts a subjectivist, subjectivist-decolonised, or critical/emancipatory stance and/or any research journey which pursues Mode 2, Mode 3, or culturally-privileged knowledge building routinely uses depth of contextualisation to help understand the perspectives they uncover/discover. This is 'contextualisation within' which means that contextualisation activities follow *as a consequence of making a specific choice* of pattern of guiding assumption or mode of knowledge building (choice of research frame, depending upon circumstances, may also invoke 'contextualisation within'). 'Contextualisation about' concerns the making of choices about patterns of guiding assumptions, modes of knowledge building, and research frames themselves. It is a higher, more systemic, level of contextualisation intended to guide the researcher's choices by clarifying and critically examining potential options, in terms of strengths and weaknesses, losses and gains, what is feasible, and what will be convincing. Thus, 'contextualisation about' thinking precedes and informs other contextualisation choices; 'contextualisation within' follows from the making of specific choices. Pluralist systemic thinking confers its greatest advantages when applied to 'contextualisation about'. Unfortunately, 'contextualisation about' is where a lot of social and behavioural research remains vague. Questions such as 'how and why was a particular pattern of guiding assumptions, mode of knowledge building, or research frame chosen?', 'how well do those choices fit the research context(s)?', and 'what are the immediate and downstream implications of such choices for creating/producing the anticipated research outcomes?' typically remain unanswered. External constraints such as page length restrictions for journal articles, sponsor expectations, and audience expectations certainly contribute to an environment where there is little space or capacity for such questions to be answered. Our message here is that wherever such questions are addressed, the convincingness of research outcomes will likely be enhanced.

Contextualisation at multiple levels helps the researcher to focus their choice-making during their research journey. But, in order to complete their journey, the choices have to be made, research activities have to devised and configured, then implemented. Implementation requires procedural considerations, namely, detailing precisely how will the researcher carry out each of their research activities, for what purpose, in what order, with whom or what as a data source, using what data gathering strategies, and applying what analysis processes. In short, the researcher must make their choices for navigating the 'Data Triangle', then actually implement those choices, in a specified order, to navigate the 'Data Triangle' in real time and space, and in a manner consistent with the expectations of their adopted pattern(s) of guiding assumptions and frame. The research journey, at this point, transforms from an anticipatory perspective (what is planned) to an implementation perspective (what is done) and being convincing to one or more intended audience(s) becomes the overriding imperative for any research outcome.

References

Aagaard-Hansen, J. (2007). The challenges of cross-disciplinary research. *Social Epistemology, 21*(4), 425–438.

Aguinis, H., Werner, S., Abbott, J., Angert, C., Park, J., & Kohlhausen, D. (2010). Customer-centric science: Reporting significant research results with rigor, relevance, and practical impact in mind. *Organizational Research Methods, 13*(3), 515–539.

Alexander, B. (2024). Queer/quare theory: Worldmaking and methodologies [revisited]. In N. Denzin, Y. Lincoln, M. Giardina, & G. Cannella (Eds.), *The Sage handbook of qualitative research* (6th ed.). Sage Publications, pp. 167–208.

Alkin, M., & Christie, C. (Eds.). (2023). *Evaluation roots: Theory influencing practice* (3rd ed.). The Guilford Press.

Allard-Poesi, F., & Maréchal, C. (2001). Constructing the research problem. In R. Thietart (Ed.), *Doing management research: A comprehensive guide* (pp. 41–50). Sage Publications.

Alvesson, M., & Deetz, S. (2021). *Doing critical research.* Sage Publications.

Archibald, J., Xiiem, Q., Lee-Morgan, J., & Santolo, J. (Eds.). (2022). *Decolonizing research: Indigenous storywork as methodology.* Bloomsbury

Armstrong, J., & Green, K. (2022). *The scientific method: A guide to finding useful knowledge.* Cambridge University Press.

Armstrong, J. (1997). Peer review for journals: Evidence on quality control, fairness, and innovation. *Science and Engineering Ethics, 3*, 63–84.

Auslander, P. (2004). Postmodernism and performance. In S. Connor (Ed.), *The Cambridge companion to postmodernism* (pp. 97–115). Cambridge University Press.

Babbie, E. (2011). *The basics of social research* (5th ed.). Wadsworth Cengage Learning.

Bamberger, M., & Mabry, L. (2019). *RealWorld evaluation: Working under budget, time, data, and political constraints* (3rd ed.). Sage Publications.

Bergmann, M., Jahn, T., Knobloch, T., Krohn, W., Pohl, C., & Schramm, E. (2012). *Methods for transdisciplinary research: A primer for practice.* Campus Verlag, [translated from the original German edition by R. Faust].

Bhaskar, R. (2016). *Enlightened common sense: The philosophy of critical realism.* Routledge.

Biesta, G. (2010). Pragmatism and the philosophical foundations of mixed methods research. In A. Tashakkori & C. Teddlie (Eds.), *Sage handbook of mixed methods in social & behavioral research* (2nd ed.). Sage Publications, pp. 95–117.

Blatter, J., & Haverland, M. (2012). *Designing case studies: Explanatory approaches in small-n research.* Palgrave McMillan.

Bleijenbergh, I., Korzilius, H., & Verschuren, P. (2011). Methodological criteria for the internal validity and utility of practice-oriented research. *Quality & Quantity, 45*(1), 145–156.

Botha, L. (2011). Mixing methods as a process towards Indigenous methodologies. *International Journal of Social Research Methodology, 14*(4), 313–325.

Bowers, T. (2011). Towards a framework for multiparadigm multimethodologies. *Systems Research and Behavioral Science, 28*(5), 537–552.

Brocklesby, J. (1997). Becoming multimethodology literate: An assessment of the cognitive difficulties of working across paradigms. In J. Mingers & A. Gill (Eds.), *Multimethodology: The theory and practice of combining management science methodologies* (pp. 189–216). John Wiley & Sons.

Browne, K., & Nash, C. (Eds.). (2010). *Queer methods and methodologies: Intersecting queer theories and social science research.* Ashgate.
Brunswik, E. (1952). *The conceptual framework of psychology.* University of Chicago Press.
Brunswik, E. (1956). *Perception and the representative design of psychological experiments* (2nd ed.). University of California Press.
Bryant, A., & Charmaz, K. (Eds.). (2007). *The Sage handbook of grounded theory.* Sage Publications.
Bryman, A., Becker, S., & Sempik, J. (2008). Quality criteria for quantitative, qualitative and mixed methods research: A view from social policy. *International Journal of Social Research Methodology, 11*(4), 261–276.
Burgess-Limerick, R. (2018). Participatory ergonomics: Evidence and implementation lessons. *Applied Ergonomics, 68*, 289–293.
Byrne, D., & Callaghan, G. (2014). *Complexity theory and the social sciences: The state of the art.* Routledge.
Byrne, D., & Ragin, R. (2009). *The Sage handbook of case-based methods.* Sage Publications.
Campbell, D., & Stanley, J. (1966). *Experimental and quasi-experimental designs for research.* Rand-McNally.
Caracelli, V., & Riggins, L. (1994). Mixed-method evaluation: Developing quality criteria through concept mapping. *Evaluation Practice, 15*(2), 139–152.
Carayannis, E., & Campbell, D. (2010). Triple helix, quadruple helix and quintuple helix and how do knowledge, innovation and the environment relate to each other?: A proposed framework for a trans-disciplinary analysis of sustainable development and social ecology. *International Journal of Social Ecology and Sustainable Development (IJSESD), 1*(1), 41–69.
Carayannis, E., Campbell, D., & Rehman, S. (2016). Mode 3 knowledge production: Systems and systems theory, clusters and networks. *Journal of Innovation and Entrepreneurship, 5*(1), 1–24.
Charmaz, K. (2014). *Constructing grounded theory* (2nd ed.). Sage Publications.
Chevalier, J., & Buckles, D. (2019). *Participatory action research: Theory and methods for engaged inquiry.* Routledge.
Chilisa, B. (2020). *Indigenous research methodologies* (2nd ed.). Sage Publications.
Choi, B., & Pak, A. (2006). Multidisciplinarity, interdisciplinarity and transdisciplinarity in health research, services, education and policy: 1. Definitions, objectives, and evidence of effectiveness. *Clinical & Investigative Medicine, 29*(6), 351–364.
Cleary, L. (Ed.). (2013). *Doing cross-cultural research with integrity: Collected wisdom from researchers in social settings.* Springer Nature.
Coghlan, D. (2019). *Doing action research in your own organization* (5th ed.). Sage Publications.
Cohen, L., Manion, L., & Morrison, K. (2018). *Research methods in education* (8th ed.). Routledge.
Collins, L. (2015). Validity criteria in multimethod and mixed research. In S. Hesse-Biber & R. Johnson (Eds.), *The Oxford handbook of multimethod and mixed methods research inquiry* (pp. 240–256). Oxford University Press.
Cook, T., & Campbell, D. (1979). *Quasi-experimentation: Design and analysis issues for field settings.* Houghton Mifflin.
Cook, T., Campbell, D., & Peracchio, L. (1990). Quasi-experimentation. In M. Dunnette & L. Hough (Eds), *Handbook of industrial and organizational psychology*, (Vol. 4). Consulting Psychologists Press, Inc., pp. 491–576.
Cooksey, R. (2001). What is complexity science? A contextually-grounded tapestry of systemic dynamism, paradigm diversity, theoretical eclecticism, and organizational learning. *Emergence: A Journal of Complexity Issues in Organizations and Management, 3*(1), 77–103.
Cooksey, R. (2006). *Evaluating research quality: Meta-criteria for management and organisational research.* In CD-ROM Proceedings of the 2006 Annual December Conference of the Australian and New Zealand Academy of Management, Yeppoon, Qld.
Cooksey, R. (2008). *Paradigm-independent meta-criteria for social & behavioural research.* CD-ROM Proceedings of the UNE Postgraduate Research Conference, Armidale, NSW.

Cooksey, R. (2011). Yours, mine or ours: What counts as innovation? *Journal of Agricultural Education and Extension, 17*(3), 283–295.
Cooksey, R. (2020). *Illustrating statistical procedures: Finding meaning in quantitative data* (3rd ed.). Springer Nature.
Cooksey, R., & McDonald, G. (2019). *Surviving and thriving in postgraduate research* (2nd ed). Springer Nature.
Cram, R., & Mertens, D. (2015). Transformative and Indigenous frameworks for multimethod and mixed methods research. In S. Hesse-Biber & R. Johnson (Eds.), *The Oxford handbook of multimethod and mixed methods research inquiry* (pp. 91–109). Oxford University Press.
Crotty, M. (1998). *The foundations of social research: Meaning and perspective in the research process.* Allen & Unwin.
Daft, R. (1983). Learning the craft of organizational research. *Academy of Management Review, 8*(4), 539–546.
Davies, B. (2024). Feminist inquiry. In In N. Denzin, Y. Lincoln, M. Giardina, & G. Cannella (Eds.), *The Sage handbook of qualitative research* (6th ed., pp. 123–136). Sage Publications.
Davies, P. (2006). Exploratory research. In V. Jupp (Ed.), *The Sage dictionary of social research methods* (pp. 110–111). Sage Publications.
de Vaus, D. (2008). Comparative and cross-national designs. In P. Alasuutari, L. Bickman, & J. Brannen (Eds.), *The Sage handbook of social research methods* (pp. 249–264). Sage Publications.
Dellinger, A., & Leech, N. (2007). Toward a unified validity framework in mixed methods research. *Journal of Mixed Methods Research, 1*(4), 309–332.
Donaldson, S., Christie, C., & Mark, M. (2009). *What counts as credible evidence in applied research and evaluation practice?* Sage Publications.
Donnor, J., & Ladson-Billings, G. (2024). Critical race theory and the postracial imaginary. In N. Denzin, Y. Lincoln, M. Giardina, & G. Cannella (Eds.), *The Sage handbook of qualitative research* (6th ed., pp. 137–152). Sage Publications.
Duke, P. (2011, July 29). The emergence of pandisciplinarity. *Pandisciplinarity.* http://pandisciplinarity.blogspot.com/2011/07/emergence-of-pandisciplinarity.html.
Dul, J., & Hak, T. (2008). *Case study methodology in business research.* Butterworth-Heineman.
Duncan, S., & Harrop, A. (2006). A user perspective on research quality. *International Journal of Social Research Methodology, 9*(2), 159–174.
Easterby-Smith, M., Golden-Biddle, K., & Locke, K. (2008). Working with pluralism: Determining quality in qualitative research. *Organizational Research Methods, 11*(3), 419–429.
Eichhorn, J. (2021). *Survey research and sampling.* Sage Publications.
Ellingson, L. (2009). *Effective crystallization in qualitative research: An introduction.* Sage Publications.
Elliott, R., & Timulak, L. (2021). *Essentials of descriptive-interpretive qualitative research: A generic approach.* American Psychological Association.
Ember, C., & Ember, M. (2009). *Cross-cultural research methods* (2nd ed.). AltaMira Press.
Fàbregues, S., & Molina-Azorín, J. (2017). Addressing quality in mixed methods research: A review and recommendations for a future agenda. *Quality & Quantity, 51*, 2847–2863.
Fear, F., Rosaen, C.L. Bawden, R., & Foster-Fuishman, P. (2006). *Coming to critical engagement: An autoethnographic exploration.* University Press of America.
Fedorov, S. (Ed.). (2020). *Research trends in multidisciplinary research and development* (Vol. 6). Weser Books.
Fielding, N. (2009). Going out on a limb: Postmodernism and multiple method research. *Current Sociology, 57*(3), 427–447.
Fine, M., & Torre, M. (2021). *Essentials of participatory action research.* American Psychological Association.
Fink, A. (2014). *Evaluation fundamentals: Insights into program effectiveness, quality, and value* (3rd ed.). Sage Publications.

Fleetwood, S. (2014). Bhaskar and critical realism. In P. Adler, P. Du Gay, G. Morgan, & M. Reed (Eds.), *Oxford handbook of sociology, social theory, and organization studies: Contemporary currents* (pp. 182–219). Oxford University Press.
Flood, R., & Romm, N. (1996). *Diversity management: Triple loop learning*. John Wiley & Sons.
Flynn, S., & Uttley, C. (2021). Mixed methods research: Explanatory and exploratory designs. In S. Flynn (Ed.), *Research design for the behavioral sciences: An applied approach*. Springer Publishing, pp. 457–479.
Fossey, E., Harvey, C., McDermott, F., & Davidson, L. (2002). Understanding and evaluating qualitative research. *Australian and New Zealand Journal of Psychiatry, 36*(6), 717–732.
Fowler, F. (2014). *Survey research methods* (5th ed.). Sage Publications.
Getty, G. (2010). The journey between western and Indigenous research paradigms. *Journal of Transcultural Nursing, 21*(1), 5–14.
Gibbons, M. (2001). Mode 2 society and the emergence of context-sensitive science. *Science and Social Policy, 27*(3), 159–163.
Gibbons, M., Limoges, C., Nowotny, H., Schwartzman, S., Scott, P., & Trow, M. (1994). *The new production of knowledge: The dynamics of science and research in contemporary societies*. Sage Publications.
Golafshani, N. (2003). Understanding reliability and validity in qualitative research. *The Qualitative Report, 8*(4), 597–607.
Golden-Biddle, K., & Locke, K. (1993). Appealing work: An investigation of how ethnographic texts convince. *Organization Science, 4*(4), 595–616.
Gray, D. (2021). *Doing research in the real world* (5th ed.). Sage Publications.
Greene, J., & Hall, J. (2010). Dialectics and pragmatism. In A. Tashakkori & C. Teddlie (Eds.), *Sage handbook of mixed methods in social & behavioral research* (2nd ed., pp. 119–143). Sage Publications.
Greenwood, D., & Levin, M. (2007). *Introduction to action research: Social research for social change* (2nd ed.). Sage Publications.
Hadorn, G., Biber-Klemm, S., Grossenbacher-Mansuy, W., Hoffmann-Riem, H., Joye, C., Wiesmann, U., & Zemp, E. (2008). *The Emergence of transdisciplinarity as a form of research*. In G. Hadorn, H. Hoffman-Riem, S. Biber-Klemm, W. Grossenbacher-Mansuy, C. Joye, C. Pohl, C. Wiesmann, & E. Zemp (Eds.), *Handbook of transdisciplinary research* (pp. 19–39). Springer.
Hammersley, M. (2009). Assessing validity in social research. In P. Alasuutari, L. Bickman, & J. Brannen (Eds.), *The Sage handbook of social research methods (Paperback edition)* (pp. 42–53). Sage Publications.
Harkness, J., Braun, M., Edwards, B., Johnson, T., Lyberg, L., Mohler, P., Pennell, B., & Smith, T. (2010). *Survey methods in multinational, multiregional, and multicultural contexts*. Wiley.
Hart, M. (2010). Indigenous worldviews, knowledge, and research: The development of an Indigenous research paradigm. *Journal of Indigenous Social Development, 1*(1), 1–16.
Haynes, C., Stewart, S., & Patton, L. (2024). Intersectionality methodology: A qualitative research imperative for black women's lives. In N. Denzin, Y. Lincoln, M. Giardina, & G. Cannella (Eds.), *The Sage handbook of qualitative research* (6th ed., pp. 153–166). Sage Publications.
Healy, M., & Perry, C. (2000). Comprehensive criteria to judge validity and reliability of qualitative research within the realism paradigm. *Qualitative Marketing Research: An International Journal, 3*(3), 118–126.
Heron, J. (1996a). *Co-operative inquiry and the primacy of the practical*. Sage Publications.
Heron, J. (1996b). *Co-operative inquiry and the primacy of the practical*. https://johnheron-archive.co.uk/primprac.
Heron, J. (1996c). Quality as primacy of the practical. *Qualitative Inquiry, 2*(1), 41–56.
Heron, J., & Reason, P. (1997). A participatory inquiry paradigm. *Qualitative Inquiry, 3*(3), 274–294.
Heron, J., & Reason, P. (2006). The practice of co-operative inquiry: Research 'with' rather than 'on' people. In P. Reason & H. Bradbury (Eds.), *Handbook of action research: Concise paperback edition* (pp. 144–154).

Hesse-Biber, S., & Griffin, A. (2015). Feminist approaches to multimethod and mixed methods research: Theory and praxis. In S. Hesse-Biber & R. Johnson (Eds.), *The Oxford handbook of multimethod and mixed methods research inquiry* (pp. 72–90). Oxford University Press.

Hesse-Biber, S. (ed.). (2012a). *Handbook of feminist research: Theory and praxis*. Sage Publications.

Hesse-Biber, S. (2012). Weaving a multimethodology and mixed methods praxis into randomized control trials to enhance credibility. *Qualitative Inquiry, 18*(10), 876–889.

Hesse-Biber, S. (2014). *Feminist research practice: A primer* (2nd ed.). Sage Publications.

Hojat, M., Gonnella, J., & Caelleigh, A. (2003). Impartial judgment by the "gatekeepers' of science: Fallibility and accountability in the peer review process. *Advances in Health Sciences Education, 8*, 75–96.

Kaushik, V., & Walsh, C. (2019). Pragmatism as a research paradigm and its implications for social work research. *Social Sciences, 8*(9), 255.

Kazdin, A. (2011). *Single-case research designs: Methods for clinical and applied settings* (2nd ed.). Oxford University Press.

Kelly, L., & Cordeiro, M. (2020). Three principles of pragmatism for research on organizational processes. *Methodological Innovations, 13*(2), 1–10.

Kincheloe, J., & Berry, K. (2004). *Rigour & complexity in educational research: Conceptualizing the bricolage*. Open University Press.

Kincheloe, J. (2005). On to the next level: Continuing the conceptualization of the bricolage. *Qualitative Inquiry, 11*(3), 323–350.

Kovach, M. (2021). *Indigenous methodologies: Characteristics, conversations, and contexts* (2nd ed.). University of Toronto Press.

Krathwohl, D. (1985). *Social and behavioral science research: A new framework for conceptualizing, implementing, and evaluating research studies*. Jossey-Bass.

Krosnick, J. (1999). Survey research. *Annual Review of Psychology, 50*, 537–567.

Kuhn, T. (2012). *The structure of scientific revolutions* (rev. ed.). The Folio Society.

Lambert, V., & Lambert, C. (2012). Qualitative descriptive research: An acceptable design. *Pacific Rim International Journal of Nursing Research, 16*(4), 255–256.

Lang, D., Wiek, A., Bergmann, M., Stauffacher, M., Martens, P., Moll, P., Swilling, M., & Thomas, C. (2012). Transdisciplinary research in sustainability science: Practice, principles, and challenges. *Sustainability Science, 7*(1), 25–43.

Layder, D. (1993). *New strategies in social research*. Polity Press.

Leavy, P. (2011). *Essentials of transdisciplinary research: Using problem-centred methodologies*. Left Coast Press.

LeCompte, M., & Goetz, J. (1982). Problems of reliability and validity in ethnographic research. *Review of Educational Research, 52*(1), 31–60.

Leonard, S., Fitzgerald, R., & Riordan, G. (2016). Using developmental evaluation as a design thinking tool for curriculum innovation in professional higher education. *Higher Education Research & Development, 35*(2), 309–321.

Letherby, G. (2011). Feminist methodology. In M. Williams & W. Vogt (Eds.), *The Sage handbook of innovation in social research methods* (pp. 62–79). Sage Publications.

Lewis, M., & Keleman, M. (2002). Multiparadigm inquiry: Exploring organizational pluralism and paradox. *Human Relations, 55*(2), 251–275.

Liamputtong, P. (2010). *Performing qualitative cross-cultural research*. Cambridge University Press.

Liamputtong, P. (Ed.) (2022). *Handbook of qualitative cross-cultural research methods: A social science perspective*. Edward Elgar Publishing.

Lincoln, Y. (1995). Emerging criteria for quality in qualitative and interpretive research. *Qualitative Inquiry, 1*(3), 275–289.

Lincoln, Y., Lynham, S., & Guba, E. (2024). Paradigmatic controversies, contradictions, and emerging confluences, revisited. In N. Denzin, Y. Lincoln, M. Giardina, & G. Cannella (Eds.), *The Sage handbook of qualitative research* (6th ed., pp. 75–112). Sage Publications.

Lury, C., Fensham, R., Heller-Nicholas, A., Lammes, S., Last, A., Michael, M., & Uprichard, E. (Eds.). (2018). *Routledge handbook of interdisciplinary research methods*. Routledge.

Marshall, C. (1990). Goodness criteria: Are they objective or judgment calls? In E. Guba (Ed.), *The paradigm dialog* (pp. 188–197). Sage Publications.

Martin, K., & Mirraboopa, M. (2003). Ways of knowing, being and doing: A theoretical framework and methods for Indigenous and indigenist re-search. *Journal of Australian Studies, 27*(76), 203–214.

Maxwell, J., & Mittapalli, K. (2010). Realism as a stance for mixed methods research. In A. Tashakkori & C. Teddlie (Eds.), *Sage handbook of mixed methods in social & behavioral research* (2nd ed., pp. 145–167). Sage Publications.

McGrath, A., Troy, J., & Rademaker, L. (Eds.). (2023). *Everywhen: Australia and the language of deep history*. University of Nebraska Press.

McIntyre, A. (2008). *Participatory action research*. Sage Publications.

McNiff, J. (2013). *Action research: Principles and practice*. Routledge.

Metcalf, M. (2006). Seeking Churchman's system for guaranteeing knowledge claims. *Systems Research and Behavioral Science, 23*, 51–59.

Mingers, J., & Brocklesby, J. (2003). Multimethodology: Towards a framework for mixing methodologies. In G. Midgley (Ed.), *Systems thinking: Volume IV: Critical systems thinking and systemic perspectives on ethics, power and pluralism* (pp. 227–253). Sage Publications.

Mingers, J. (2014). *Systems thinking, critical realism and philosophy: A confluence of ideas*. Routledge.

Mittal, V. (2011). Descriptive research. In J. Sheth & N. Malohatra (Eds.), *Wiley international encyclopedia of marketing:* Vol. 2. Wiley. http://onlinelibrary.wiley.com/https://doi.org/10.1002/9781444316568.wiem02002/full.

Morgan, D. (2014). Pragmatism as a paradigm for social research. *Qualitative Inquiry, 20*(8), 1045–1053.

Murthy, P. (2000). Complex societal problem solving: A possible set of methodological criteria. *Systems Research and Behavioral Science, 17*, 73–101.

Myers, M., & Klein, H. (2011). A set of principles for conducting critical research in information systems. *MIS Quarterly, 35*(1), 17–36.

Neuman, W. (2020). *Social research methods: Qualitative and quantitative approaches* (8th ed.). Pearson Education.

Nicholls, R. (2009). Research and Indigenous participation: Critical reflexive methods. *International Journal of Social Research Methodology, 12*(2), 117–126.

Noro, K., & Imada, A. (Eds.). (1991). *Participatory ergonomics*. CRC Press.

Kuntz, A. (2024). Poststructural engagements. In N. Denzin, Y. Lincoln, M. Giardina, & G. Cannella (Eds.), *The Sage handbook of qualitative research* (6th ed., pp. 273–291). Sage Publications.

Nosek, B., Hardwicke, T., Moshontz, H., Allard, A., Corker, K., Dreber, A., et al. (2022). Replicability, robustness, and reproducibility in psychological science. *Annual Review of Psychology, 73*, 719–748.

Nowotny, H., Scott, P., & Gibbons, M. (2001). *Re-thinking science: Knowledge and the public in an age of uncertainty*. Polity Press.

Nowotny, H., Scott, P., & Gibbons, M. (2003). Introduction: Mode 2 revisited: The new production of knowledge. *Minerva, 41*(3), 179–194.

Nusbaum, E., & Lester, J. (2024). Critical disability studies and diverse bodyminds in qualitative inquiry. In N. Denzin, Y. Lincoln, M. Giardina, & G. Cannella (Eds.), *The Sage handbook of qualitative research* (6th ed., pp. 209–222). Sage Publications.

O'Cathain, A. (2010). Assessing the quality of mixed methods research. In A. Tashakkori & C. Teddlie (Eds.), *Sage handbook of mixed methods in social & behavioral research* (2nd ed., pp. 531–555). Sage Publications.

Padmanabhan, M. (Ed.). (2019). *Transdisciplinary research and sustainability: Collaboration*. Routledge.

Patton, M., & Campbell-Patton, C. (2021). *Utilization-focused evaluation* (5th ed.). Sage Publications.

Patton, M. (2011). *Developmental evaluation: Applying complexity concepts to enhance innovation and use*. The Guilford Press.

Patton, M. (2015). *Qualitative research & evaluation methods* (4th ed.). Sage Publications.

Patton, M. (2016). What is essential in developmental evaluation? On integrity, fidelity, adultery, abstinence, impotence, long-term commitment, integrity, and sensitivity in implementing evaluation models. *American Journal of Evaluation, 37*(2), 250–265.

Patton, M., McKegg, K., & Wehipeihana, N. (Eds.). (2015). *Developmental evaluation exemplars: Principles in Practice*. The Guilford Press.

Peterson, M. (ed.). (2015). *Cross-cultural research*. Sage Publications.

Pidgeon, M. (2019). Moving between theory and practice within an Indigenous research paradigm. *Qualitative Research, 19*(4), 418–436.

Plano Clark, V., & Badiee, M. (2011). Research questions in mixed methods research. In A. Tashakkori & C. Teddlie (Eds.), *Sage handbook of mixed methods in social & behavioural research* (2nd ed., pp. 275–304). Sage Publications.

Pratt, M. (2008). Fitting oval pegs into round holes: Tensions in evaluating and publishing qualitative research in top-tier North American journals. *Organizational Research Methods, 11*(3), 481–509.

Reason, P., & Rowan, J. (Eds.). (1981). *Human inquiry: A sourcebook of new paradigm research*. Wiley.

Reid, B. (2020). Positionality and research: "two-eyed seeing" with a rural Ktaqmkuk Mi'kmaw community. *International Journal of Qualitative Methods, 19*, 1–12.

Repko, A., & Szostak, R. (2020). *Interdisciplinary research: Process and theory* (4th ed.). Sage Publications.

Rey, L., Tremblay, M. C., & Brousselle, A. (2014). Managing tensions between evaluation and research: Illustrative cases of developmental evaluation in the context of research. *American Journal of Evaluation, 35*(1), 45–60.

Richardson, K. (2010). *Thinking about complexity: Grasping the continuum through criticism and pluralism*. Emergent Publications.

Rosenthal, R., & Rosnow, R. (2008). *Essentials of behavioral research: Methods and data analysis* (3rd ed.). McGraw-Hill.

Rossi, P., Wright, J., & Anderson, A. (Eds.). (2013). *Handbook of survey research*. Academic Press.

Rowan, J. (1981). A dialectical paradigm for research. In P. Reason & J. Rowan (Eds.), *Human inquiry: A sourcebook of new paradigm research* (pp. 93–112). Wiley.

Ruel, E., Wagner, W., & Gillespie, B. (2015). *The practice of survey research: Theory and applications*. Sage Publications.

Saldaña, J. (2024). Performance shapes in qualitative inquiry. In N. Denzin, Y. Lincoln, M. Giardina, & G. Cannella (Eds.), *The Sage handbook of qualitative research* (6th ed., pp. 437–452). Sage Publications.

Sale, J., & Brazil, K. (2004). A strategy to identify critical appraisal criteria for primary mixed-method studies. *Quality & Quantity, 38*, 351–365.

Savall, H., Zardet, V., Bonnet, M., & Péron, M. (2008). The emergence of implicit criteria actually used by reviewers of qualitative research articles. *Organizational Research Methods, 11*(3), 510–540.

Sayer, A. (2000). *Realism and social science*. Sage Publications.

Scott, D. (2007). Resolving the quantitative–qualitative dilemma: A critical realist approach. *International Journal of Research & Method in Education, 30*(1), 3–17.

Shaw, I., Greene, J., & Mark, M. (2006). *The Sage handbook of evaluation research*, Sage Publications.

Siedlecki, S. (2020). Understanding descriptive research designs and methods. *Clinical Nurse Specialist, 34*(1), 8–12.

Siegesmund, R. (2024). The arts as research: Nomadic materiality and possible futures. In N. Denzin, Y. Lincoln, M. Giardina, & G. Cannella (Eds.), *The Sage handbook of qualitative research* (6th ed., pp. 453–466). Sage Publications.

Smith, L. (2021). *Decolonizing methods: Research and Indigenous peoples* (3rd ed.). Zed Books.

Stebbins, R. (2001). *Exploratory research in the social sciences.* Sage Publications.

Steinhauer, E. (2002). Thoughts on an Indigenous research methodology. *Canadian Journal of Native Education, 26*(2), 69–81.

Swedberg, R. (2020). Exploratory research. In C. Elman, J. Gerring, & J. Mahoney (Eds.), *The production of knowledge: Enhancing progress in social science* (pp. 17–41). Cambridge University Press.

Szostak, R. (2015). Interdisciplinary and transdisciplinary multimethod and mixed methods research. In S. Hesse-Biber & R. Johnson (Eds.), *The Oxford handbook of multimethod and mixed methods research inquiry* (pp. 128–143). Oxford University Press.

Tharenou, P., Donohue, R., & Cooper, B. (2007). *Management research methods.* Cambridge University Press.

Thiollent, M. (2011). Action research and participatory research: An overview. *International Journal of Action Research, 7*(2), 160–174.

Thomlison, B. (2001). Descriptive studies. In B. Thyer (Ed.), *Handbook of social work research methods* (pp. 131–141). Sage Publications.

Tight, M. (2017). *Understanding case study research: Small-scale research with meaning.* Sage Publications.

Toepoel, V. (2015). *Doing surveys online.* Sage Publications.

Toomey, A., Markusson, N., Adams, E., & Brockett, B. (2015). Inter-and trans-disciplinary research: A critical perspective. *GSDR Brief*, 1–3.

Tripodi, S., & Bender, K. (2001). Descriptive studies. In B. Thyer (Ed.), *The handbook of social work research methods* (2nd ed., pp. 120–130). Sage Publications.

Ullrich, J., Demientieff, L., & Elliott, E. (2022). Storying and re-storying: Co-creating Indigenous well-being through relational knowledge exchange. *American Review of Canadian Studies, 52*(3), 247–259.

Vaditya, V. (2018). Social domination and epistemic marginalisation: Towards methodology of the oppressed. *Social Epistemology, 32*(4), 272–285.

Van de Vijver, F., & Leung, K. (2021). *Methods and data analysis for cross-cultural research* (2nd ed.). Cambridge University Press.

Vienni-Baptista, B., Fletcher, I., & Lyall, C. (Eds.). (2023). *Foundations of interdisciplinary and transdisciplinary research: A reader.* Bristol University Press.

Walter, M., & Anderson, C. (2013). *Indigenous statistics: A quantitative research methodology.* Left Coast Press.

Warner, D. (2004). Towards a queer methodology. *Qualitative Research in Psychology, 1*(4), 321–337.

White, P. (2009). *Developing research questions: A guide for social scientists.* Palgrave Macmillan.

Wilson, S. (2001). What is an Indigenous research methodology? *Canadian Journal of Native Education, 25*(2), 175–179.

Windchief, S., San Pedro, T., & Kovach, M. (2024). Why we do Indigenous methodologies: Contemplations on Indigenous protocol, theory, and method. In N. Denzin, Y. Lincoln, M. Giardina, & G. Cannella (Eds.), *The Sage handbook of qualitative research* (6th ed., pp. 241–258). Sage Publications.

Yin, R. (2017). *Case study research: Design and methods* (6th ed.). Sage Publications.

Chapter 5
Making Convincing Choices: Scoping, Configuring and Implementing

Abstract This chapter explores the choices a researcher makes to convincingly scope, configure, and implement their activities around the 'Data Triangle'. These enacting choices follow from and are more focused than choices related to contextualisation, patterns of guiding assumptions, and framing, and provide concrete pathways for the researcher to traverse in order to gather and analyse the data they need, from the sources they intend to or need to connect with. Here is where the landscape of methodological choices gets shaped, configured, pruned, and detailed (in short, 'brought to life') to produce the 'Data Triangle' the researcher navigates. We examine the issue of trade-offs, namely the adaptive decisions a researcher must sometimes make during their journey in anticipation of future situations or events (feedforward) or in response to emergent, perhaps unexpected, situations or events (feedback). With respect to configuring research activities, we utilise the concept of a 'Method Unit' (MU, which synergistically combines one data gathering strategy, one form of data, and one type of data source) as the fundamental building block. MUs provide the most flexible way of implementing pluralist systemic thinking. Following on from research configuration, we discuss implementation considerations underpinning each component of each MU that must be addressed.

In this chapter, the 'rubber meets the road'. All the contextualisation work that the researcher does helps them to make preliminary choices to focus their research. This work then provides the background against which the research will be scoped, configured, and implemented. Figure 5.1 provides a concept flow map for scoping, configuring, and implementing research activities. These research activities are centred on successfully and convincingly navigating the 'Data Triangle' to achieve the researcher's intended goals. 'Scoping' provides the boundaries and specific choices, content, and intentions for research activities, 'configuring' arranges those chosen activities in the most meaningful and convincing way, and 'implementing' puts those choices, in their intended configuration and with all necessary supporting procedures, into action. Of course, as events in the research journey unfold over time, choices may yield feedback that causes the researcher to revisit and possible revise their choices. Thus, scoping, configuring, and implementing are not necessarily linearly sequenced

R. W. Cooksey, *Unity from Diversity*, https://doi.org/10.1007/978-981-97-3462-7_5

processes; they will build and feed off each other as the research journey progresses, stimulating the need to make trade-offs. Some trade-offs may be made early on during the research journey as it becomes clear that some choices will or will not be viable for the researcher. Other trade-offs may be needed as a consequence of dynamically changing circumstances that emerge as the research progresses. Each trade-off needs to be evaluated against the often-conflicting demands to maintain convincingness and to make the research journey feasible for the researcher to achieve, within the constraints they confront. We will unpack the implications of Fig. 5.1 through further discussion and elaborations in the chapter.

5.1 Scoping of the Research Journey

Scoping of research is a process of setting requirements and boundaries for research activities, narrowing, and specifying the choices for navigating the 'Data Triangle'. Scoping is a natural consequence of employing pluralist systemic thinking in research contextualisation as well as in making choices of pattern(s) of guiding assumptions, mode of knowledge building, and research frame(s). *Scoping moves the researcher's journey into a space where it is feasible and realistic for the researcher to successfully undertake in pursuit of one or more convincing research outcomes.* In order to make research feasible and realistically achievable, the researcher makes choices relevant to the nature of the research activities that they will use to gather the evidence they need as well as choices relevant to configuring the patterns, connections, and synergies between those research activities. In short, scoping focuses on how the researcher intends to navigate the 'Data Triangle', in real time and space, to address their research questions/propositions/hypotheses within the constraints they confront and given the opportunities they encounter. Appropriately scoping and configuring research is often a very iterative thinking forward-working backward type of process where the 'final' (in quotes to signal that nothing is ever really final in research—adaptations and trade-offs are almost always necessary) research configuration does not emerge until the researcher has explored the implications of earlier choices for their later choices and has revisited some of those early choices in light of impediments or constraints they experience or can foresee down the track. The iterative scoping and configuring process can often be aided by conversations with other people, including peers, colleagues, funding or commissioning organisations, relevant stakeholders/gatekeepers, and communities of practice, as they may bring to light issues that the researcher has not yet considered or has insufficiently considered. Scoping must be done in light of joint contextual and research-focused considerations and in full knowledge of the target audience(s) for whom convincingness will be a matter for their judgment.

Scoping involves synergistic considerations with respect to contextualisation (including stakeholder and gatekeeper positionings) which backgrounds the research, the emerging research problem/goals/questions/propositions/hypotheses (which form the focus/intent of the research), patterns of guiding assumptions (which help

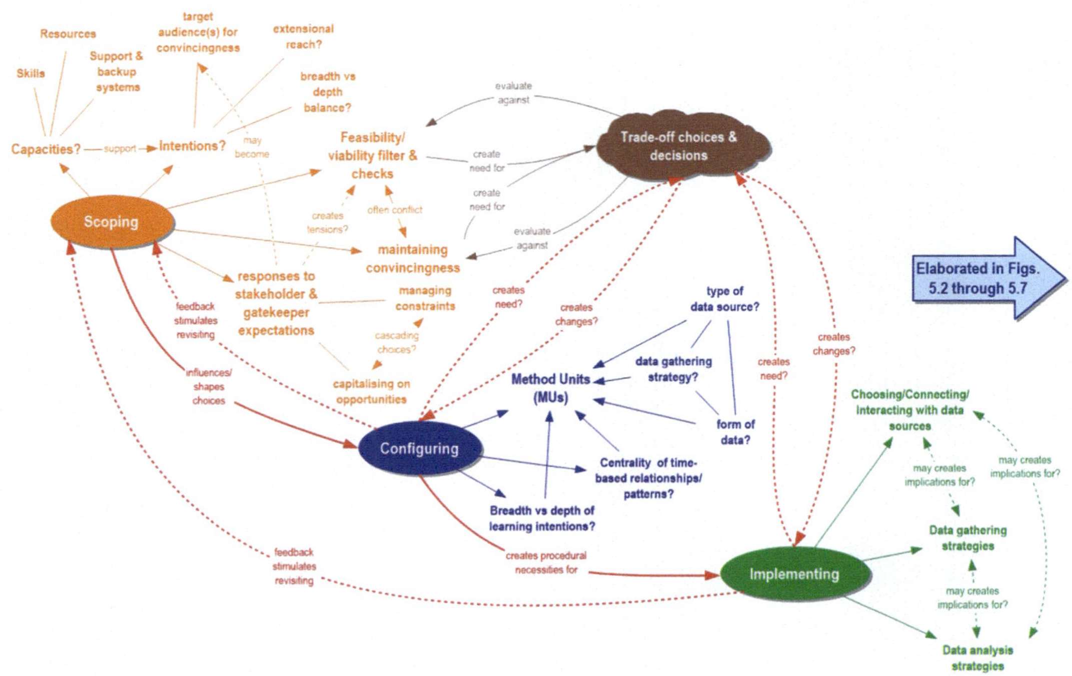

Fig. 5.1 Relationships between scoping, configuring, and implementing activities for navigating the 'Data Triangle'

focus researcher choices), mode of knowledge building (which links the researcher to intended audiences and participants), research frames to capture the intent and positioning of the research, and quality criteria/meta-criteria (which link research, configuration, and implementation choices to convincingness) together with considerations of available and required resources and/or access to those resources to inform choices around the 'Data Triangle' methodological landscape. Here pluralist systemic thinking has one of its greatest impacts. Here also, is where stakeholder and gatekeeper impacts are greatest because what they expect, require, or hope for exerts pressures on the researcher to scope their research accordingly. For example, for Indigenous/First Nations research, negotiation and discussion with relevant communities and their Elders needs to occur prior to any scoping process and such discussions may seek to set limits, re-prioritise, or re-interpret research-related issues for the researcher. Experimental laboratory research may require the researcher to negotiate access to laboratory resources, constrained by university or research organisation policies as well as by considerations regarding sharing access and use with other researchers. Thus, issues about resourcing and access often hinge upon successful negotiation and discussion with relevant stakeholders and gatekeepers. Some of these considerations may then exert downstream influences on the researcher's choices around the 'Data Triangle' methodological landscape. Contextualisation will influence choices of patterns of guiding assumptions (which will then raise considerations about pattern-specific quality criteria). Equally, choice of patterns of guiding assumptions will influence contextualisation and may create demands for certain other choices to be made (e.g., a postgraduate student who wishes to adopt the pattern of guiding assumptions for a systemic othering inquiry within the othering/emancipatory frame may need to seek out a supervisor with specific experience in such research; a researcher who wishes to adopt the pattern of guiding assumptions for a participatory inquiry and Mode 3 knowledge building within a developmental evaluation frame will need to negotiate with their targeted community or relevant professional/industry/government groups with respect to their willingness to support/facilitate such research and intellectual property issues).

The researcher scopes their research when their choices are 'filtered' by feasibility and viability constraints resulting in some choices being ruled out (which delimits research approaches) and other choices being ruled in (which augments research approaches). Pluralist systemic thinking is more focused here with the researcher trying to make choices that will be workable while still meeting research goals, contextual demands, and quality expectations. This is the nexus between what the researcher considers to be ideally desirable and what they consider to be feasible, viable, and realistic to achieve. Choices that survive feasibility and viability 'filtering' should be examined for their downstream implications and subsequent considerations associated with each choice. This process may set up a decision cascade, where making one choice automatically demands other choices. For example, choosing to do semi-structured interviews brings up considerations about where to conduct and how to record the interviews, what the topical landscape looks like, how transcripts will be shaped, and so on. Choosing to conduct an online questionnaire within a survey research frame brings up considerations about the design and presentation of the web pages that present the questions to participants and record their responses, what software would be best to use, how multiple responses will be handled, how errors will be trapped and prevented, and so on.

Self-questioning is one cognitive tactic the researcher can use to work through the scoping process, keeping in mind the resources they have access to and the constraints (e.g., human resources, cultural expectations, ethical requirements, time, and access) they must work within:

- 'if I implement A, what else do I now have to consider?'
- 'if I want to implement A, do I need other human resources (e.g., research team, research assistants, informants) to implement it?'
- 'if I want to implement A, do I need to acquire/practice/build up my skills for implementing it?'
- 'if I implement A, will I achieve the balance between breadth and depth that I desire?'
- 'if tracking the passage of time (e.g., as with evaluating pretests and posttests, following trends across days, weeks, months, years) is a central element in the learning I want to achieve and I implement A, how effectively will that play out?'
- 'if I am pursuing external causal inferences and I implement A, how effectively will that play out?'
- 'if I intend what I learn to have extensional reach/implications (e.g., generalisation; transportability) for other people, groups, communities, organisations/institutions, contexts, cultures, tasks, times/occasions, can I achieve this if I implement A?'
- 'if I implement A, are there likely to be any unintended immediate or downstream consequences?'
- 'what do I need to put in place to ensure that A works effectively?'
- 'does implementing A require technological (e.g., hardware, software, online) support and, if so, what kind and can I access it?'
- 'what source(s) of data (primary, secondary, generated/created by the researcher) do I need when implementing A?'
- 'what form(s) of data (quantitative, qualitative, hybrid/merged) would be best to obtain when doing implementing A?'
- 'if I want to implement A but am denied access to or lack resources to implement it, what are my viable options, given my research intentions, and my desire to maintain convincingness and quality?'
- 'can I compensate for problems I anticipate in implementing A through augmenting my research by doing B or C (an explicitly pluralist systemic thinking question)?'
- 'are there any institutional, cultural, ethical or legal requirements/expectations associated with doing A that I should know and do something about?

Through the scoping process, the nature and structure of the Method Units (MUs) that the researcher wants to employ become clear. Those MUs then need to be configured in a way best suited for achieving the researcher's goals within their adopted pattern(s) of guiding assumptions and frame(s) while maintaining convincingness/

quality, as well as meeting any contextual demands/expectations. The researcher also needs to employ pluralist systemic thinking to identify/plan for the potential synergies, connections and cross-influences, and dynamics that may be relevant for their specific configuration of MUs. The researcher then needs to create/configure all the procedural elements/documents/instrumentation/tools/software/ethical requirements necessary for carrying out each MU. All these elements must be finalised before research implementation can commence. As the research journey unfolds in real time and space, unexpected obstacles, constraints, or opportunities may emerge which may create the demand for trade-offs to be made in the researcher's approach. Such demands will move the researcher back into the wider set of considerations in order to gauge feasibility, viability, and any potential threats to research quality associated with alternative choices under consideration. Trade-offs may impose losses (where emergent constraints are operating) or provide gains (where opportunities have emerged). In either case, maintaining or enhancing research convincingness must always be a focus for choice making. This will often create tensions or oppositional forces to practicality and feasibility that the researcher will have to actively work through.

One useful strategy for research scoping is to work through some thought 'experiment' scenarios, guided by self-questioning at all stages (supported through the use of relevant frameworks/tools, where appropriate; revisit Sect. 1.6.2 to review). This involves trying to anticipate what could happen if certain choices are made. Another aspect of this exercise would involve thinking about what the researcher might do if certain of their desired choices are thwarted by things that emerge which are out of their control (e.g., organisations decline to participate; the researcher doesn't get ethical permission for their sampling plan, questionnaire items create adverse effects for participants from a specific cultural background). As the researcher works through these scenarios, they should record their thinking in their research journal. This will provide the researcher with additional material for defending their choices when it comes time to creating/producing and sharing their research outcomes. Scenario thinking can also help the researcher to figure out what a Plan B might look like for their research—a plan they could use if critical or insurmountable obstacles are encountered to interfere with their initial or ideal choices.

Figure 5.2 provides a more elaborated way to view the relationships between scoping, configuring, and implementing activities. Given the research contextualisations (remembering that these will be continually revisited, updated, and revised as the research journey progresses), scoping focuses choice making on what is necessary to navigate the 'Data Triangle' successfully and convincingly. Such choice making is evaluated using a conceptual feasibility/viability 'filter' that includes choices that would augment and enhance convincingness of the research, excludes choices that would limit the research and lessen convincingness, and adaptive choices that represent trade-offs between what is desired and what is feasible in the hopes of maintaining, enhancing, or at least minimising adverse impacts on convincingness. Choices that survive the feasibility/viability 'filter' are then configured in a specific spatial and temporal arrangement so as to provide the best opportunity to address

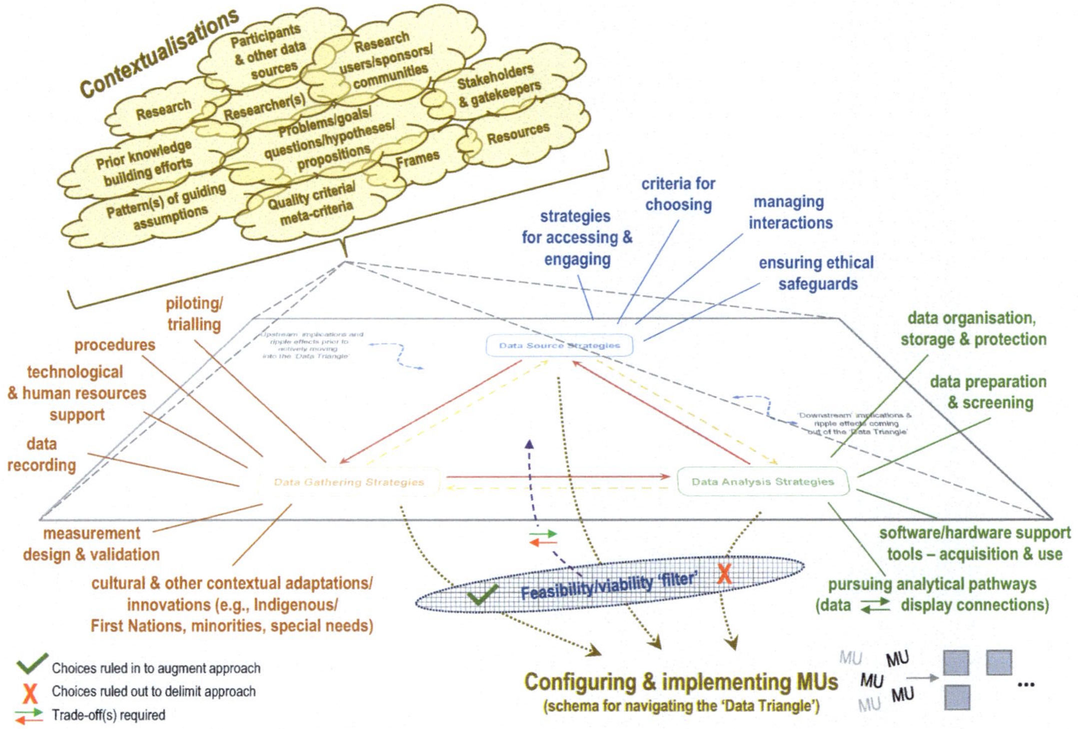

Fig. 5.2 Processes for navigating the 'Data Triangle' in light of contextualisations and emergent feasibility/viability considerations

the researcher's research questions/propositions/hypotheses. Implementation necessities emerge at each vertex of the 'Data Triangle' as a consequence of scoping and configuring processes.

5.2 Making Trade-Off Decisions

As we have previously indicated, contextualising, framing, scoping, and configuring social and behavioural research frequently requires making trade-off decisions, where the researcher gives away, modifies, or augments some valued angle, aspect, or component of their research in order to incorporate another angle, aspect, or component that they value in a different way. All trade-off decisions move aspects of the research journey away from the researcher's ideal or originally most desirable or intended form toward a form that may be more feasible, realistic, acceptable, or useful. All social or behavioural research has flaws, some more serious or damaging than others, created by certain trade-offs the researcher had to make in order to get the job done. This means that *convincingness* (whether maintaining, regaining, or minimising loss of) is a central concern in determining if a trade-off decision constitutes a Loss or a Gain. Trade-off decisions are frequently motivated by actions, feedback, or influence from one or more stakeholders/gatekeepers in a particular context. Making trade-off decisions is the hallmark of the *flexible and adaptive researcher* where the goal is to manage the trade-offs in such a way that they can still lead to a convincing piece of work.

Many trade-off decisions are conscious choices that the researcher makes to give away or modify aspects of their research that they would ideally or initially like to implement but, because of constraints and emerging obstacles, become unrealistic or infeasible to accomplish. These can be characterised as *Loss* trade-offs, and they often, but not always, sacrifice some convincingness in the process. The conditions that demand Loss trade-off decisions generally arise because, in social and behavioural research, humans (in a 'researcher' role) study humans (in a 'participant' or contextual role) and/or the systems and artefacts that humans inhabit and produce. What happens to and with those humans is dependent upon larger circumstances, many of which cannot be predicted or anticipated. Researchers must live, work, prosper, survive, have careers, have families and friends, and so on, as part of their everyday lives (the 'researcher' role/identity being just one of many roles/identities they enact in their lives) and research activities can be influenced by those other aspects of life as well as by aspects of the lives of others. Thus, research cannot be fully compartmentalised from other aspects of one's life and this often creates the need to make Loss trade-off decisions. Such decisions tend to create or impose limitations on the initial research contextualisation and/or configuration that need to be acknowledged and, where possible, actively managed.

Other trade-offs involve conscious choices that the researcher makes to evolve or augment their initial research contextualisation and/or configuration so that additional or different and perhaps more convincing set of valued outcomes, likely for a different audience, might be achieved. These can be characterised as *Gain* trade-offs, and they often, but not always, involve downplaying or reshaping one research goal in order to pursue a new, more useful, or opportunistically emergent goal. Gain trade-offs are typically made when the researcher realises that their initial research contextualisation and/or configuration is missing some important aspect or ignores an opportunity that would make it more convincing and, accordingly, they augment their research in such a way as to facilitate achievement of those new valued outcomes. Gain trade-offs most often lead to a more complex research contextualisation and/or configuration with attendant increases in time, effort, and other resources that such increased complexity demands and they may create new sets of stakeholders/gatekeepers to consider. However, they may also provide ways to overcome or mitigate limitations imposed by the initial research contextualisation and/or configuration or created by previous Loss trade-off decisions (which means that some trade-offs may operate as Loss → Gain chains of connected decisions). In general, Gain trade-off decisions give away convincingness in one aspect of the research in order to achieve an even greater or more focused degree of convincingness in another aspect.

Contextualised Loss/Gain trade-offs refer to trade-offs that can either constitute a loss or a gain depending upon specific contextual circumstances in the research journey, including what the impact on convincingness is likely to be. If convincingness ultimately takes a hit, the trade-off would be considered a Loss. However, if convincingness is ultimately improved (or at least maintained in the face of prior Loss trade-offs), the trade-off would be considered a Gain. Many contextualised loss/gain trade-offs come about because of the researcher's personal and professional investment in their research journey and how that investment dynamically interacts with the investments of others.

Loss or Gain trade-offs (whether contextualised or not) may be called for at any juncture (thereby creating critical leverage points) during the researcher's journey, where the researcher make changes along the way in order that their goals and valued outcomes can be achieved (even though those goals themselves may be modified as a consequence of the trade-offs made). No matter what type of trade-off researchers make, what drives their decision is their need to achieve valued outcomes that their current research contextualisation and/or configuration cannot or is likely not to provide. For Loss trade-offs, the major value of interest is typically enhanced feasibility while minimising damage to convincingness. For Gain trade-offs, the major value of interest is typically enhanced convincingness or usefulness, but likely for a different audience than originally envisaged. Whenever a Loss or Gain trade-off is made, there will almost always be implications for research quality criteria and the meta-criteria. With any trade-off, researchers need to understand what they stand to gain (the 'pros') as well as what they stand to lose (the 'cons') by making the decision. Researchers also need to consider whether potential downstream issues will be created by a trade-off decision and, if so, take steps to address those emergent issues, where possible.

There are several types of Loss, Gain, and contextualised Loss/Gain trade-offs researchers might make, summarised in the mind map in Fig. 5.3. Each type of trade-off accomplishes a different purpose, produces a different valued outcome,

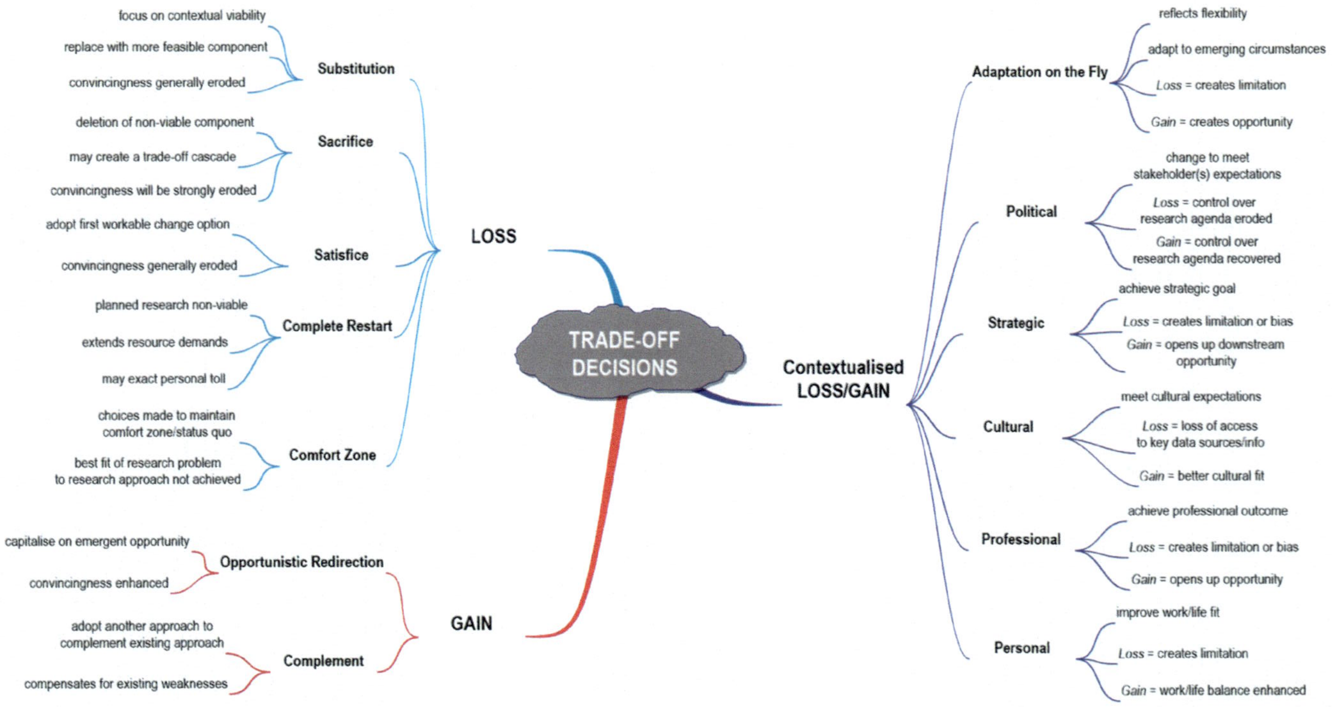

Fig. 5.3 Summary mind map of loss and gain trade-off decisions in research

and the need for it may arise from different sets of circumstances. While some trade-off decisions may yield positive outcomes, others may create ethical dilemmas and other downstream problems for the researcher, which may require them to make further trade-offs, thereby precipitating a decision cascade. Since any trade-off decision affects the very nature of the research that researchers are conducting, it is important for them to record all their thinking about each trade-off in their research journal, including its implications for research quality criteria and meta-criteria, so that its logic, necessity, and nature can be reconstructed, when needed, but especially when communicating the research story or responding to reviews and criticisms. In this regard, Loss trade-off decisions (as well as contextualised Loss/Gain trade-offs that result in a Loss) create the need to acknowledge/discuss the resulting limitations and their implications for any conclusions in research outcomes. Gain trade-offs (as well as contextualised Loss/Gain trade-offs that result in a Gain) create the opportunity to showcase flexibility and adaptability in the face of complex and changing circumstances in pursuit of a more convincing research outcome. However, such decisions also create the need for the researcher to construct a narrative explaining the reasoning behind the trade-off.

5.2.1 Loss Trade-Off Decisions

Substitution

Substitution is a Loss trade-off that involves substituting a research angle, aspect, or component that is more likely to work in the research context for another component or approach that is less likely to work or that has, in fact, not worked. Two examples where a substitution trade-off is made: (1) a researcher replaces one data gathering strategy that they have found to be contextually unviable (in a trialling phase, for example) with another data gathering strategy that is likely to be more contextually viable; and (2) a researcher makes the decision to use a less powerful statistical analysis technique that is more appropriate for handling emergent anomalies in their quantitative data. The potential downside of this trade-off decision is that the substitution may require the researcher to pull back from aspirations to generalise or infer external causality or may be accompanied by changes in their adopted pattern of guiding assumptions, which must then be accommodated. The substitution trade-off generally shifts the balance in emphasis between internal and external validity (and, possibly, statistical conclusion validity) in research guided by the pattern of guiding assumptions for a positivist inquiry or between authenticity and sufficiency for research guided by the pattern of guiding assumptions for an interpretivist/constructivist inquiry. In terms of the meta-criteria, a substitution trade-off may shift the balance in emphasis and character between internal coherence, analytical integrity, and extensional reasoning and may create additional limitations that will need to be acknowledged in the research story.

Sacrifice

Sacrifice is a Loss trade-off that involves the deletion of a research angle, aspect, or component, thus, removing a part of the research because it is no longer viable, or its intent has become irrelevant. This decision can involve removal of a research question from the researcher's goals. The sacrifice trade-off is a more drastic decision and, as a result, can create real consequences for meeting the expectations of specific research quality criteria and meta-criteria. In many cases, this trade-off may be forced on the researcher by changing circumstances in their research context (such as a school pulling the plug on its participation in the research halfway through the data gathering process or a gatekeeper denying access to certain desirable data sources). In other cases, the researcher may come to realise that a particular research goal is not achievable within the resource constraints they confront. For example, a researcher may intend to conduct multiple case studies in the context of their investigation but then discover that they have insufficient time and resources to gather and analyse the type and quantity of qualitative data beyond a single case study (which means sacrificing capacity to draw strong conclusions about transportability of what is learned). In any case, the sacrifice trade-off will create limitations that the researcher will need to address in their research story. Downstream, a sacrifice decision (e.g., losing access to a desired school) may lead to a substitution decision (e.g., selecting a lesser desirable but accessible school) to try and recover from some of the losses incurred by the sacrifice, illustrating a trade-off cascade.

Satisfice

The satisfice decision is a Loss trade-off where a research contextualisation or configuration choice is made immediately if it looks like it will work satisfactorily. The term 'satisfice' was coined by Nobel laureate Herbert Simon (1957, cited in Anderson et al., 1981, pp. 214–215) referring to making a decision by adopting the first solution to a problem that a person thinks will work. It is a hallmark of 'bounded rationality' (see Anderson et al., 1981, p. 30) associated with human decision making in complex and uncertain circumstances. It means one makes a decision without giving rational consideration to all possible options; instead, one makes the first workable (i.e., satisfactory) choice, where satisfactory means the choice will meet their minimum standards. Circumstances for satisficing tend to occur in times of high stress and where insufficient time and resources are available to make a fully rational choice (see, for example, discussions in Janis & Mann, 1977). In terms of a research trade-off, this means deliberately giving away looking for optimal research angle, aspect, or component in favour of the first concrete research angle, aspect, or component that the researcher thinks will get the job done to a minimum set of specifications. This usually means that the researcher does not consider the full range of options for configuring their research in order to choose the best option. An example of the satisfice trade-off in action would be where a researcher adopts the Survey research frame, intending to interview people about racism in their community under tight resource and time constraints, but discovers that building enough trust to gather authentic data from participants takes far too long, is too costly, and is not always successful (perhaps because of the sensitive nature of the topic). The researcher immediately jettisons their interview strategy and decides, instead, to administer a simple anonymous quantitative questionnaire to a convenience sample of undergraduate students because it is the cheapest and quickest way to get workable data from a large sample on the issue. A satisfice trade-off almost invariability means a reduction in the quality of the research and its capacity to convince, especially with respect to its internal coherence and extensional reasoning. It can sometimes be a hallmark of a novice researcher; it can be highly practical (especially where resources are very tight) yet represents the antithesis of thinking systemically.

Complete Restart

This is the ultimate and most drastic Loss trade-off one can make. Quite simply, it means that, in light of emerging circumstances, constraints, and obstacles, the researcher completely abandons their current research problem and decides instead to pursue a completely new problem in a new context. This trade-off decision involves a complete rethink of the research questions and the associated research contextualisations, positionings, and configuration. As an example, suppose a researcher conducts a pilot study to fine-tune a quantitative questionnaire on clinical decision making using a small sample of medical practitioners. As a consequence of their pilot testing experience, the researcher learns that medical practitioners were extremely reluctant to be queried about their clinical decision-making practices and outcomes, giving rise to an extremely low response rate of 2%. As a result, the researcher abandons their intention to study clinical decision making and completely restarted their research in another area of interest (this example was inspired by Ross, 1999, PhD research experience). When a complete restart trade-off is made, the entire research journey commences again, almost from scratch, a consequence that has obvious implications for timelines and resources, as well as, at least in the short term, psychological impacts on the researcher.

Comfort Zone

A comfort zone trade-off is Loss in nature, resulting in less convincing research. As the name suggests, this trade-off seeks to keep the researcher safely operating within their comfort zone, with respect to their research preferences, skills and abilities, and prior experiences. In short, it preserves the researcher's current identity as a researcher. It involves (a) incorporating a specific angle, aspect, or component into the research plan that they prefer, think is most expedient, or feel most comfortable/competent in carrying out or (b) ruling out an approach that they feel is least preferable, expedient, or they feel least comfortable/competent in carrying out. The outcome associated with (a) may reflect a philosophical bias such as adopting only the pattern of guiding assumptions for a positivist inquiry in their research (which itself may reflect disciplinary biases or selectiveness in prior research training). The outcome associated with (b) may reflect a skills deficit or a general bias against quantitative (e.g., the researcher hates or can't work with numbers/statistics) or qualitative (e.g., the researcher thinks qualitative research is not credible) research approaches. This kind of trade-off is thus made to meet personal rather than professional needs, incurring sacrifices in research quality and convincingness, and has the effect of stifling thinking openly, pluralistically, and systemically about research. It substitutes expedience and personal comfort for quality and suitability. The way to negate the pressure to consider this trade-off is to take that pressure as a signal that one needs to develop new skills and new thinking patterns before pursuing the research project, perhaps seeking a mentor who could work with them, or involve more than one researcher (i.e., a team) in the journey.

5.2.2 Gain Trade-Off Decisions

Opportunistic Redirection

Opportunistic redirection is a Gain trade-off that can create unanticipated positive outcomes for the researcher because it is not stimulated by unviability of an angle, aspect, or component of the research but by the potential viability of pursuing a new angle. Here, the researcher decides to pursue a new research pathway within the same broad research context, which can involve a partial or a complete rethink of the research configuration (or even further back, a rethink of the research questions), because a new situation has presented itself or new resources have become available. Thus, an unanticipated emergent opportunity in the research context (e.g., a shift in market forces, due to the emergence of new competitors, may amplify the possibilities for exploring strategic decision making in an organisation, when the original intention may have been to study decision making in performance reviews) provides the impetus to shift the research focus. Part of what may feed this decision is the researcher's interest and curiosity, stimulated by the new opportunity. However, the downside to this decision might be that the timeline and resource demands for the research blows out as the researcher needs to plan and reconfigure their research activities to follow the new direction. There may also be some consequences for research quality criteria and meta-criteria (such as the need for re-contextualising the research problem, need to become familiar with new areas of prior research, etc.) that the researcher would need to address. The distinction between this trade-off and the complete restart trade-off, discussed earlier, is that the opportunistic redirection trade-off remains broadly focused on the research context and problem area the researcher commenced working within, whereas with the complete restart trade-off decision, even the research context and problem changes.

Complement

Complement is a Gain trade-off that involves adding an angle, aspect, or component to the research contextualisation and/or configuration whose strengths can at least partially compensate for the weaknesses of another angle, aspect, or component. Thus, the researcher extends/augments their research through adding a complementary facet. This, of course, adds to the complexity and resource-intensiveness of the research, but with the potential to yield a more convincing story. In terms of research configuration, to be discussed in more detail below, this trade-off decision would be signalled in a shift from a *simultaneous MU* configuration to a *sequential MU* configuration. The trade-off achieved by the complement decision trades away resource efficiency in exchange for enhanced research effectiveness. This particular trade-off decision may actually make the research stronger and more convincing (against certain research quality criteria and meta-criteria) than its original version (where resource efficiency is valued). The complement trade-off frequently motivates the conduct of mixed methods research through, for example, adding a quantitative component to a predominantly qualitative investigation or vice-versa.

5.2.3 Contextualised Loss/Gain Trade-Off Decisions

Adaptation on the Fly

Adaptation on the fly may constitute a Loss trade-off or a Gain trade-off depending upon contextual circumstances. This trade-off involves a mid-stream correction, redirection, or reconfiguration of the research and it reflects flexibility and nimbleness in action. In certain situations, circumstances may emerge during the data gathering process, for example, where the researcher must do something different in order to proceed. For example, a researcher may be using a cause mapping data gathering strategy in an interpretivist/constructivist investigation, asking participants to produce a visualisation (i.e., a cause map) of their understanding of the causes of particular events they have experienced. However, part way through data gathering, the researcher encounters participants who are uncomfortable with or incapable of producing such a visualisation. An adaptation-on-the-fly trade-off might be to ask the participant to think aloud (i.e., verbally explaining their thinking) while they drew the visualisation the best they could (this example was inspired by Sandall's, 2006, PhD research experience) and the researcher then using the think-aloud transcript to help flesh out the visualisation. The implications of this adaptive decision are, for this participant and any others encountered who were in the same situation, their data do not take the same form as the data produced by other participants, meaning that internal coherence and analytical integrity could be affected, especially since the researcher would have played a role in generating some visualisations but not others. In this situation, the trade-off would represent a Loss for the research. Another example would be where a researcher conducts econometric research using a secondary national database of institutional economic, financial, and labour statistics, but upon commencing their sampling of the database, they realise there are gaps in the information that is stored, creating a missing data problem. This generates the need to adapt their research approach to either estimate the missing data values by using more complex statistical processes (creating implications for analytical integrity because missing value estimation is a contentious problem) or to remove any entity from the sample that has missing data and choosing a replacement entity (creating implications for extensional reasoning because the sampling plan cannot be fully realised). Here again, the trade-off would represent a Loss to the research.

The adaptation-on-the-fly trade-off decision is also a hallmark of the *evolutionary* MU configuration, to be discussed below, where subsequent MUs are not known or planned for until a critical choice point has been reached in the research journey. The trade-off being made here is giving away certainty about the research configuration and living with the uncertainties of what is to come, i.e., short-term Loss. In the longer term, though, the trade-off is likely to represent a Gain for the research as the choice of what to do next will have been fully informed by what the researcher had previously learned. Flexibility is maximised and convincingness will likely be enhanced, but at the cost of not knowing what downstream research activities will be and whether they will be viable until it comes time to make the choice (i.e., after the first MU has been completed). An example might be where the researcher conducts qualitative semi-structured interviews about remuneration practices with a sample of employees of a large company and what is signalled to the researcher in these interviews is a need to interview union officials inside and outside the organisation (who had previously argued and won a legal case against the organisation for paying entry-level employees below current minimum wage standards), as well as senior managers within the organisation. The short-term Loss would be the researcher not knowing up front that they would need to conduct additional unanticipated interviews with the concomitant resource demands that would entail. The longer-term Gain would be the researcher achieving a much more well-rounded and likely more convincing story, with more diverse information/perspectives to build interpretations upon.

Political

A political trade-off involves the researcher adapting their planned research by adding a specific research angle, aspect, or component in order to meet a specific stakeholder's (or group of stakeholders) expectations. One might make this trade-off because it will help the stakeholder achieve a valued goal in their own context (such as gaining information that will help them to manage a change process within a community or evaluating an innovation for use in a particular industry). Generally, a researcher might consider making this trade-off if the stakeholder offered to provide resources or access (if they are also a gatekeeper) in exchange for adapting the research to meet their needs. The quid pro quo nature of politically motivated decisions, in many cases, results in a Loss for the researcher, specifically a loss of some control over their research agenda. In making this kind of Loss trade-off, the researcher must be very careful not to 'sell their soul' as it can potentially bring certain ethical as well as intellectual property concerns into the foreground. For example, the CEO of an organisation may be willing to provide the researcher with access to their organisation for data gathering purposes as well as provide resources to support those data gathering activities as long as they are willing to share what they learn from every research participant. This would create an ethical dilemma for the researcher in the case where they need to promise full anonymity to every participant. The ethical decision would be to withdraw from pursuing the research in that organisation, i.e., to make a sacrifice trade-off instead. The practical decision would be to agree in general to the CEO's request but negotiate strongly for doing it in such a way as to protect individual participant identities, i.e., a compromise political trade-off.

However, not every political trade-off should be seen as constituting a Loss. In situations where the research focuses on the development and evaluation of an innovation, a political trade-off decision could be made to move away from an academic researcher-dominated innovation process (i.e., Mode 1 knowledge building) toward a more resource-intensive yet participatory research process (i.e., Mode 2 or Mode 3 knowledge building) involving potential adopters of the innovation participating in the research and development for that innovation (along the lines suggested by Cooksey, 2011). In the longer term, this political trade-off may constitute a Gain for the research and would likely enhance the convincingness of the knowledge produced as well as the chances of successful and perhaps more widespread, innovation adoption.

Strategic

A strategic trade-off occurs when the researcher alters or adds a research angle, aspect, or component in order to create opportunities to achieve longer-term strategic research goals (e.g., enhancing grant or publication opportunities, creating good will for future research in a specific context, adding value to the researcher's institutional profile, developing/extending a network of research colleagues through forming or joining a research team). The strategic trade-off can constitute either a Loss or Gain depending upon circumstances. The trade-off would constitute a Loss if the researcher's current research contextualisation and/or configuration is modified simply to achieve the strategic goal; it would constitute a Gain if the researcher's research contextualisation and/or configuration is augmented to achieve the strategic goal. This trade-off strategy can create new ethical concerns for the researcher, if their valued goal becomes more personal than professional and/or can introduce biases into decisions they make while navigating the 'Data Triangle', if those decisions run contrary to certain stakeholder expectations or sound research practices (in which case, it may backfire, and the researcher may lose everything). For example, suppose a researcher wishes to conduct a longitudinal year-long study to implement and evaluate an innovative classroom teaching technology at a particular high school but is denied access by the principal. After some negotiation with the principal, the researcher agrees to first conduct a small-scale short-term study within the school and workshop the results with staff. If successful, the principal will agree to grant the researcher permission to conduct their longer term, more involved, study in the school (note that strategic and political trade-offs may coincide, as in this example). As another example, a researcher may modify their research in ways they perceive as consistent with the goal of ultimately achieving a research grant from a specific funding institution because they see this as a way of helping their institution lift its public research profile. Whether this strategic trade-off constitutes a Loss or a Gain can only be judged downstream: the ultimately desired research grant is obtained (Gain) or not (Loss).

Cultural

The cultural trade-off may be either Loss or Gain depending upon research-specific circumstances. Some of these circumstances may exist from the commencement of the research (e.g., within an Indigenous/First Nations, cross-cultural or othering/ emancipatory research frame), requiring trade-off decisions at the very start of the research. However, other circumstances may emerge during the course of the research (e.g., during a piloting or trialling phase, or arising from a specific stage in a sequential or evolutionary MU configuration). A cultural trade-off decision might entail changing data gathering strategies or approaches to desired data sources in order to better meet cultural expectations or address culture-specific issues. A cultural trade-off decision may represent a Loss if it creates incompatibilities between the researcher's positioning/capabilities and the cultural fit of the research. For example, in cross-cultural research, a researcher may plan to adopt the pattern of guiding assumptions for a positivist inquiry to underpin the conduct of a cross-cultural survey. However, language translation problems associated with specific concepts to be addressed via questionnaire items may demand that the researcher explore these issues under the pattern of guiding assumptions for an interpretivist/constructivist inquiry using an interactive data gathering strategy, but the researcher may lack the appropriate skills/experiences/resources to do so convincingly. A cultural trade-off decision may represent a Gain if the changes made improve the cultural fit/ suitability of the research through enhanced access to data sources and/or desired information. For example, a researcher conducting classroom-based research may discover that some classes have a significant number of Indigenous/First Nations children and decide that, instead of treating classes as holistic units, value might be gained from focusing more closely on the perspectives of those children and their parents/community as relates to practices in those particular classrooms. The Gain here would be an enriched, productively nuanced, and better-informed story about classrooms practices than might otherwise have been achieved.

Professional

The professional trade-off may be either Loss or Gain, depending upon the researcher's circumstances and intentions. A Loss professional trade-off is strictly utilitarian in nature and involves adding or modifying an angle, aspect, or component of the research in order to enhance the researcher's chances for achieving a valued career-oriented goal, such as a postgraduate degree, a promotion, a publication, tenure, a research grant, working with a valued colleague or mentor, more consulting work, more professional prestige, opening up a new job opportunity. In a sense, professional trade-off decisions can be seen as a sub-variant of strategic trade-offs. The Loss version of the professional trade-off will generally tend to detract from convincingness as its tighter focus on achieving a professional gain may inhibit pluralist systemic thinking about what is best for the research project itself. An example of a Loss professional trade-off would be where the researcher aspires to get their research published in a well-respected marketing journal but learns that their chances of acceptance would be greatly enhanced if structural equation modelling is used as the data analytic theory-testing approach. The researcher may not have initially planned to use this procedure for their analysis but decides to use the approach in hopes of gaining an edge for the publication, even though their data may be of insufficient quality to warrant the use of the procedure. The Gain version of the professional trade-off has a learning application focus and involves the researcher adding or modifying an angle, aspect, or component of their research so that they can create an opportunity to learn new knowledge or acquire/apply new skills (such as tapping into new and unexpected data sources, using a new data gathering or data analytic technique, enhancing the extensional reach of their research), in ways not envisaged in their initial research plan. This version of the trade-off thus seeks to add value and convincingness to the research. An example of the Gain-oriented version of this trade-off would be where the researcher learns about a new analytical procedure and sees value for enhancing analytical integrity as part of convincingness by using that new procedure in their current study, when they had initially planned to use other more customary approaches.

Personal

The personal trade-off is not necessarily Loss or Gain in nature. Rather, it is an adaptive response to the researcher's circumstances and intentions. This trade-off explicitly connects with the researcher's larger life space and may involve adding or modifying an angle, aspect, or component of their research in order to accommodate family and lifestyle needs (e.g., minimise time away from family, minimising travel, avoiding conflicting work/life commitments) or to accommodate limitations imposed by physical impairments such as blindness, deafness, or mobility restrictions. While the personal trade-off may create difficulties in terms of the convincingness of the research that need to be dealt with, its overall priority is to improve the degree to which the research project can fit within the researcher's larger life and circumstances—the epitome of trying to achieve a realistic and feasible project. For example, if the researcher has a very young family and a working spouse, they may not wish to conduct their research away from their hometown. This might lead them to make a personal trade-off to conduct their research using a local convenience sample of participants, rather than a geographically dispersed (but more representative) sample, accepting the limitations on generalisation that this choice would impose. As another example, suppose the researcher had a severe vision impairment such that they are legally blind. Such personal circumstances might lead the researcher to adopt the pattern of guiding assumptions for an interpretivist/constructivist inquiry and a semi-structured qualitative interview data gathering strategy rather than adopting the pattern of guiding assumptions for a positivist inquiry and using a more broad-reaching quantitative internet questionnaire (which a vision-impaired person would find very challenging to design and construct). In the end, this could actually enhance the convincingness of the research because of their heightened listening skills, but would mean a shift in the research focus, from a breadth to a depth emphasis (see Smith-Ruig & Sheridan, 2009, for a more detailed case study exploring the research adaptation and accommodation experiences of a vision-impaired researcher, including an emergent tendency to favour electronic literature over paper-based printed literature because of how adaptive technologies for the blind functioned).

5.3 Configuring Research Activities Using Method Units (MU)

Effectively and convincingly navigating the 'Data Triangle' requires a pluralist systemic framework for configuring data gathering activities; a framework that unifies the diversity of research methods/approaches under a single umbrella. This unifying configuration framework was first introduced in Cooksey and McDonald (2019, Chap. 12). It embodies methodological diversity, being robust enough to accommodate a wide variety of research approaches, including quantitative, qualitative, mixed, and hybrid/merged methods. It is also sufficiently open-ended to permit the creative expansion and evolution of new data gathering activities and approaches. The framework is also suitable for accommodating a wide variety of patterns of guiding assumptions and is adaptable to contextual influences, where required. The unifying configuration framework builds upon, expands, and extends the design frameworks argued for by a number of mixed methods researchers (e.g., Cameron, 2023; Creswell & Plano Clark, 2018, Chap. 3; Plano Clark & Ivankova, 2016, Chap. 5, Hall & Howard, 2008; Leech & Onwuegbuzie, 2009; Loomis & Maxwell, 2003; Nastasi, Hitchcock & Brown, 2010; Poth, 2023). While those design frameworks provided a useful platform to build upon and appear in the unifying configuration framework in some form, the systemic restriction imposed on them by the 'mixed methods' logic of requiring both quantitative and qualitative data in some mixture, as traditionally depicted in the literature (e.g., Azorín & Cameron, 2010; Creswell & Plano Clark, 2018; Tashakkori & Teddlie, 2010) has been rejected. Therefore, the unifying configuration framework is not artificially constrained to just mixing quantitative and qualitative data forms in a research investigation. Instead, researchers can use the unifying framework to combine particular choices of *any* data gathering strategies, data forms, and types of data sources into specific patterns to create a complete research plan. Note that the unifying configuration framework explored here, when considered in conjunction with research contextualisation and scoping, reflects many of the complexities and possibilities for innovating in 'mixed methods' research flagged by Poth (2018) as well as Poth (2023).

In order to implement the unifying framework effectively and ensure its pluralist systemic intent, we need to discard some pretty hefty historical research methods baggage, namely the term 'research design'. The term 'research design', in most discussions about research methods, has been the most traditional way to refer to patterns of connected data gathering activities. However, 'research design' carries strong positivist connotations that require some degree of researcher control over context; the researcher is, therefore, to be understood as the 'designer', evoking connotations of contextual manipulation and a power relationship that favours the researcher These are connotations consistent with experimental and quasi-experimental research, so we will use the term 'design' only with reference to those particular approaches. Since our goal here is to invite pluralist systemic thinking with respect to different patterns of guiding assumptions and some patterns explicitly do not seek researcher control over context, in some cases, actively giving participants power in the research process, the connotations of 'design' are far too limiting and invite biased thinking. Instead, we use the more neutral term 'research configuration' to refer to the patterns and associated connections between activities to be undertaken when navigating the 'Data Triangle'. Thus, configuring research activities means working out the pattern whereby data gathering activities will be implemented, the sources of data from whom information/data will be sought, and the form(s) such information/data will take. Depending upon the research frame and associated pattern of guiding assumptions, a configuration, when implemented, will likely implicate all three facets of the 'Data Triangle': data source strategies, data gathering strategies, and data analysis strategies.

5.3.1 The Method Unit (MU)

The **Method Unit** (**MU**) is the basic conceptual building block for research configurations in the unifying framework. A MU is defined as a 'triple composite' comprising choices of:

- **one data gathering strategy** [(e.g., self-report questionnaire, semi-structured interviews, cause mapping, meta-analysis). However, certain strategies may be naturally combined and work as a single MU choice, under specific patterns of guiding assumptions (e.g., researcher-manipulated experiments/quasi-experiment, systematic observation, or meta-analysis combined with measurements under the pattern of guiding assumptions for a positivist inquiry); various merged methods (e.g., Delphi method, Q methodology, text mining, intervey) are also possible which yield merged data (see, e.g., Gobo et al. 2022)], combined with
- **one data form** [which may, in some cases, be determined by the choice of data gathering strategy in light of a specific pattern of guiding assumptions (e.g., measurement strategy yielding quantitative data; unstructured interviews yielding qualitative data). A hybrid/merged data form is also possible yielding data where quantitative and qualitative distinctions are blurred or so tightly integrated that they cannot be effectively distinguished)], combined with
- **one type of data source** [(human (e.g., individuals or groups) or non-human (e.g., document, multimedia recording, artistic performance) accessed, selected, or sampled in some way].

Every research configuration constitutes an amalgam or pattern for combining/connecting MUs. Hybrid combinations of MU configurations are also possible (e.g., simultaneous with sequential; longitudinal with case-based). Research frames, various contextualisations, positionings, and patterns of guiding assumptions may have dynamic influences on the researcher's choice and patterning of MUs, which may reduce or even determine the available/desirable choices at any one point in a configuration. Configurations unfold in both in time and space. For example, with the Manipulative data gathering strategy, traditional experimental and quasi-experimental designs (defined as researcher-assembled/manipulated combinations of hypothesised causal factors or conditions, consistent with positivist inquiry guiding assumptions) constitute a single data gathering strategy, where the design, no matter its complexity, is explicitly embedded within the MU. The only exception is where the passage of time is a dimension/independent variable of interest. In such cases, the original experimental or quasi-experimental MU is repeated on one (as with pretest–posttest experiments) or more occasions (giving rise to longitudinal MU configurations). Repeated measures experiments where participants are measured under a number of experimental conditions, but where the passage of time is **not** considered an independent variable of interest (at most it is a nuisance or extraneous variable), do not give rise to a longitudinal configuration; instead, these are sequential MU configurations. Sequential MU configurations may unfold in different spaces (e.g., contexts, locations, organisations) as well as different times for each MU in the sequence. Casy study configurations drill down deeply into a case context, often in both space (e.g., environment, geography, demography, culture, current events) and time (e.g., historical timelines and changes). Sequential and longitudinal MU configurations may require data analysis activities between each stage or time period of the configuration, especially where qualitative or hybrid/merged data forms are being gathered. Certain longitudinal and sequential MU configurations, involving quantitative measurements, may require data analysis to be carried out once the longitudinal or sequential series has been completed.

Figure 5.4 presents a summary mind map showing eight broad classes of MU configurations. The definition of an MU is embedded in the diagram at the top right-hand corner. Configurations sharing a central common feature (grouped by dashed line borders) differ with respect to the researcher's intentions they are intended to address, signalled by the label attached to each configuration. This mind map is by no means exhaustive of the possibilities for configurations. There is ample room for researcher creativity to be manifested in new hybrids and evolutionary patterns. What are not adequately captured in Fig. 5.4 are the nature, synergies, interactions, and dynamics that the researcher envisages between MUs. This means that, for any configuration that involves more than one MU, the researcher must specify the relationships and emphases that exist between the various MUs and reflect those considerations in how they implement 'Data Triangle' activities as well as how they present research outcomes to intended audience(s). With respect to sequential configurations, the four configurations presented are differentiated by intent and emphasis between the MUs in the sequence. For example, the explanatory sequential configuration gives primary emphasis to data gathering with the first MU and secondary emphasis to follow-up data gathered in the second MU as a way of explaining observed patterns revealed in the first MU. The exploratory sequential configuration gives primary emphasis to the second MU in the sequence, whose structure and content has been influenced

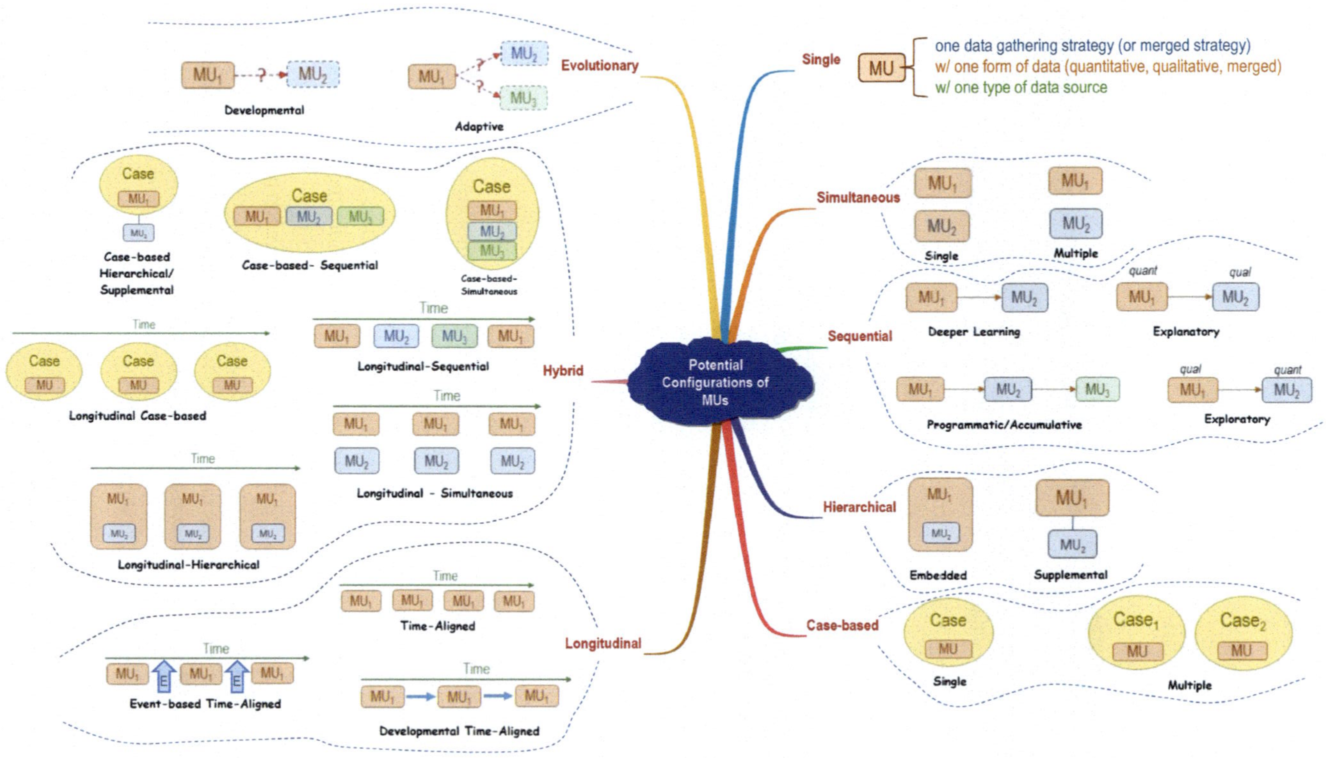

Fig. 5.4 Summary mind map of potential MU configuarations and sub-varieties (adapts and integrates elements from Figs. 12.1 through 12.23 in Cooksey & McDonald, 2019, pp. 439–467)

by what was learned in the initial exploratory MU in the sequence). In the discussions of potential MU configurations below, refer to Fig. 5.4, for a diagrammatic representation of the configuration.

5.3.2 Single Configuration

The Single configuration category is characterised by a single MU, comprised however the researcher chooses. This is the simplest and most basic research investigation that can be conceived; what some researchers have referred to as 'monomethod' research (see, for example, Johnson & Onwuegbuzie, 2004; Spector, 2006). It is thus a non-pluralist configuration. Examples could include in-depth semi-structured qualitative interviews with a sample of teachers from metropolitan high schools, quantitative questionnaires administered to a sample of shoppers in a shopping mall, or structured quantitative observations of children on a playground at a preschool.

The biggest advantages of a single MU investigation are its simplicity and, in most cases, its resource (time, effort and money) efficiency. The biggest disadvantage is that convincingness, in the form of a well-rounded story, is difficult to achieve. Too many questions are often left unresolved in single MU research. Consequently, it is becoming harder to defend the use of the single MU configuration in most research frames. However, it should be noted that the single MU configuration is more likely to be seen in the Survey research frame, especially if the data gathering strategy involves using a questionnaire to simply gather one form of data from a specific constituency, often for very applied purposes such as decision- or policy-making. For example, a local city council may use a short questionnaire to gather quantitative ratings of satisfaction with a range of council-provided services from members of the community. This information may then be used to help inform certain resource allocation decisions. Some types of laboratory experiment investigations, conducted within the Explanatory research frame, may use a single MU configuration to gather quantitative measurements from a sample of undergraduate student participants under specific manipulated experimental conditions. However, in more recent years, quantitative monomethod research guided by the pattern of guiding assumptions for a positivist inquiry (which historically was the prototypical pattern in many social and behavioural disciplines) has come in for criticism, due to what has been termed 'common method variance', a bias that potentially contaminates research outcomes because the same strategy has been used to gather all quantitative measurements (see Spector, 2006).

5.3.3 *Simultaneous Configurations*

The basic Simultaneous MU configuration involves implementing different MUs at the same time or nearly the same time. This configuration is sometimes classified as a 'mixed methods' approach, if the researcher explicitly includes multiple forms of data or as a 'triangulation' (or 'convergent') approach if the researcher uses multiple data gathering strategies coupled with multiple forms of data and/or types of data sources (see discussion in Creswell & Plano-Clark, 2018, Chap. 3). Generally, the researcher carries out the 'Data Triangle' activities within each MU independently of the other MUs (i.e., cross-influences between simultaneous MUs are minimised), so the focused stories arising from each MU have a chance to emerge.

Some advantages of the Simultaneous configuration include:

- resource (particularly time) efficiency in sampling, data gathering, and analysis activities;
- the opportunity to access a deeper or a broader level of learning in a short space of time;
- the opportunity to pursue convergence in the stories arising from the different MUs; and/or
- the opportunity to achieve cross-sectional (a 'slice of life' emphasis) learning, where appropriate for the researcher's goals (especially where the researcher has adopted a survey or exploratory research frame).

Some disadvantages of the Simultaneous configuration include:

- the configuration is much more challenging for a single researcher to implement if different patterns of guiding assumptions are used for each MU (i.e., a much broader pluralist skill set is required of a single researcher);
- convergence in the stories that emerge from each MU may not occur, leaving the researcher with a more challenging integration to achieve;
- the learning focus is episodic and cross-sectional in nature (focusing on a single time period, i.e., a 'slice of life' emphasis, rather than longitudinal), so this is not the most convincing configuration to use if change/growth over time is important to understand/demonstrate; and
- reduced usefulness if finding evidence for external causality is a goal of the research since temporal priority of cause before effect is much harder to arrange/ensure/argue for in the Simultaneous MU configuration.

A Simultaneous configuration can prove useful in:

- the survey, descriptive or exploratory research frames as a pathway to assembling a time-efficient cross-sectional story, through employing more diverse data gathering strategies and/or forms of data and/or types of data sources.

- the case-study research frame as a pathway to enhancing learning in the context of a single visit to the case context through employing more diverse data gathering strategies and/or forms of data and/or types data sources; or
- the cross-cultural (comparative emphasis) research frame, where it can be useful in facilitating an efficient comparison of samples from different cultural contexts using the same data gathering strategy (say, a questionnaire constructed to have culturally equivalent forms).

There are two variants of Simultaneous configuration possible: Single and Multiple. The choice of variant to use is generally informed by the researcher's resourcing/feasibility constraints and research goals as well as by their scoping choices with respect to breadth-depth balance for learning and extensional reach. The Single Simultaneous configuration is more efficient in terms of resource demands, but favours breadth over depth of learning and offers limited extensional reach for generalisations. The Multiple Simultaneous configuration is more demanding in terms of the researcher's time, effort, and financial resources, but favours depth over breadth of learning and offers the researcher the opportunity to achieve a more convergent and convincing story that has broader contextual/extensional reach in conclusions. For either variant, implementing more than two MUs at the same time would be very onerous, bordering on being infeasible for one researcher. A research team, however, could do this.

Single Simultaneous Configuration

The Single Simultaneous configuration involves using a single data gathering strategy in conjunction with multiple types of data sources and/or data forms. Thus, it is basically a monomethod approach gathering quantitative and/or qualitative data from one or more types of data source. Three examples would be (a) where the researcher administers a quantitative questionnaire to a sample of managers in a large organisation (MU_1) and, at the same time, administers a similar quantitative questionnaire to a sample of non-management employees from the same organisation (MU_2); (b) where the researcher conducts semi-structured qualitative interviews with teachers sampled from high schools in rural/regional Australia (MU_1) and, at the same time, conducts semi-structured qualitative interviews with the parents of students sampled from high schools in rural/regional Australia (MU_2) or (c) where the researcher uses a document-based archival sample of government reports to gather quantitative financial data (MU_1) and, as well, uses the same document-based archival sample of government reports to gather qualitative data, focusing on the language used in the reports (MU_2).

Multiple Simultaneous Configuration

The Multiple Simultaneous configuration involves using multiple data gathering strategies in conjunction with multiple data sources and/or data forms. Triangulation of different data gathering strategies (generally considered to be a stronger as well as more popular form of triangulation, see Jick's, 1979, discussion of 'method triangulation'; see also Flick, 2018) is one distinct advantage of this particular configuration.

Two examples would be: (a) where the researcher administers a quantitative questionnaire to a sample of employees in the manufacturing industry (MU_1) and, in the same time period, conducts qualitative semi-structured interviews with a sample of senior managers within that same industry (MU_2) or (b) where the researcher conducts a study of emergency room procedures in a training simulation context using a sample of nurse practitioners to gather quantified behavioural observations (MU_1) and, at the same, conducts qualitative focused interviews with a sample of emergency room doctors reflecting on specific issues associated with trauma management procedures (MU_2).

5.3.4 Sequential Configurations

The basic Sequential MU configuration involves sequentially implementing different MUs but where mapping data patterns/meanings against the passage of time is not of explicit interest in the research. This configuration is also sometimes classified as a 'mixed methods' approach, if the researcher explicitly incorporates both quantitative and qualitative data (Creswell & Plano Clark, 2018; Teddlie & Tashakkori, 2011). In this configuration, each MU stage involves a complete progression through the 'Data Triangle' from sampling to data gathering to analyses. The outcomes from the analyses in the previous MU(s) then provide input into the next MU stage.

Some advantages of this configuration include:

- learning from earlier research stages can provide feedforward to later stages to help shape and refine what is done and perhaps how;
- different stages may be guided by different patterns of guiding assumptions, allowing the weaknesses of one pattern to be offset or mitigated by the strengths of another pattern; and
- a single researcher can feasibly cope with a sequential MU configuration.

Some disadvantages of this configuration include:

- it is more time and resource-intensive;
- generally, analysis of data gathered in the first MU stage must be completed before the second MU stage commences (if learning from the first MU is to help in shaping what is done in the second MU) adding to the timeline for the research; and
- integrating stories between stages can be more challenging, especially where patterns of guiding assumptions differ for each stage.

A Sequential configuration may be useful in a range of research frames:

- the Explanatory research frame where the researcher is interested in constructing grounded theory under the pattern of guiding assumptions for an interpretivist/

constructivist inquiry, first using a set of interviews of key managers in an organisational context context and then participant observations of group meetings in that context;

- the Exploratory research frame where the initial exploratory MU phase uses a wide lens to explore a new contexts, which then sets the stage for a subsequent research phase that employs a narrow lens for further exploring issues of interest that arose in the first phase;
- the Survey research frame where a quantitative questionnaire MU stage is followed by semi-structured qualitative interviews;
- the Action research frame where each MU phase embodies an aspect of the cycle of action learning intended to change or transform practices/behaviours in a specific context;
- the Indigenous/First Nations research frame where early MU stages can provide the researcher with the foundations for building trust to pursue deeper learning in later MU stages; or
- the Pan-disciplinary research frame or Developmental Evaluation research frame where MUs at different stages target learning at different levels from different constituencies or perhaps from different parts of a chain of innovation (from R&D to innovation adoption).

There are four variations of Sequential configuration: Deeper Learning, Explanatory, Exploratory, and Programmatic/Accumulative. The choice of variant to use is generally informed by the researcher's resourcing/feasibility constraints, research goals, and research frame as well as by their scoping choices with respect to breadth-depth balance for learning, interesting in mapping potential causal patterns, and extensional reach. Note that Sequential MU configurations need not be restricted to just two MUs, but each additional MU will demand more time and resources.

Deeper Learning Sequential Configuration

In the Deeper Learning Sequential configuration, the goal is to extend, expand, and enhance learning achieved from one MU to the next, irrespective of the forms of data involved. Such learning may influence how subsequent stages are carried out, what questions are asked, what data sources are sampled/accessed, and what contexts/events should be observed (in short, how movement through the 'Data Triangle' in MU_2 should unfold to enhance learning). In terms of scoping, the Deeper Learning configuration seeks depth of learning through the sequencing of the two or more MUs. Two examples would be: (a) where the researcher facilitates qualitative focus groups with product consumers (MU_1) followed by qualitative interviews with specific product development managers and advertisers (MU_2), or (b) where the researcher carries out unstructured qualitative interviews with primary curriculum development specialists and academic subject matter experts (MU_1) followed by semi-structured qualitative interviews with teachers who use their materials sampled from primary schools (MU_2).

Explanatory Sequential Configuration

In the Explanatory Sequential configuration, the goal is to gather qualitative data in the second MU to help explain, qualify, or further characterise the patterns and relationships uncovered in quantitative data gathered in the first MU (see Creswell & Plano Clark, 2018, pp. 75–83). This is one of only two MU configurations that are typically mixed methods in orientation, where a different form of data is gathered in each MU. [Note, however, that this is not absolutely mandatory; the same form of data can be gathered in the sequential MUs using different data gathering strategies and/or different types of data sources, but the configuration would not technically qualify as 'mixed methods'.] Essentially, in this configuration, the different forms of data are gathered strategically to achieve a specific research goal. Here, the researcher's goal is retrospective in focus, namely MU_2 is undertaken to help flesh out/unpack/qualify/amplify/further understand the conclusions, especially anomalous, confusing, or unexpected findings from MU_1. One benefit of the Explanatory Sequential configuration is the ability to seek breadth of learning in MU_1, then depth of learning in MU_2. Two examples could include: (a) where the researcher administers a quantitative questionnaire to a stratified random sample of community members (MU_1) followed by conducting a series of qualitative interviews (MU_2) with volunteer community member participants who completed the questionnaire in MU_1 or (b) where the researcher employs systematic observations of assembly line workers to obtain quantitative measurements of work efficiency and safety in a large manufacturing firm (MU_1) followed by qualitative focus group interviews (MU_2) with groups of employees meeting to discuss general problems and issues discovered in MU_1. Equally, though, as an alternative for (b), the researcher could undertake qualitative participant observations of assembly line work groups to focus on work and safety behaviours in MU_1, followed by qualitative semi-structured interviews (MU_2) with those groups to discuss the observations made in MU_1, as a pathway toward building a grounded theory.

Exploratory Sequential Configuration

In the Exploratory Sequential configuration, the researcher's goal is to use learning from qualitative data in MU_1 to identify and inform the constructs, patterns, and relationships to be further studied in the quantitative MU_2 (see Creswell & Plano Clark, 2018, pp. 84–93). This is the second of the two MU configurations where a different form of data is typically gathered in each MU (although, as for the Explanatory configuration, this is not absolutely mandatory). Here, the researcher's goal is prospective in focus, where MU_1 is undertaken to produce learning intended to shape the focus and perhaps the content for MU_2. Two examples could include: (a) where the researcher conducts qualitative focus group interviews with consumers interested in purchasing organic and sustainably produced foods (MU_1) followed by a quantitative questionnaire, designed using input from what was learned from the focus groups in MU_1, administered to a random sample of shoppers at various supermarkets and specialist food shops in shopping malls in a metropolitan city (MU_2) or (b) where the researcher examines a sample of historical text documents providing qualitative commentary and reflections on major decision mistakes made by CEOs, politicians, and world leaders (MU_1) followed by a laboratory experiment (MU_2), conducted using postgraduate student volunteers as participants, designed to yield quantitative measurements of decision making behaviours under conditions arranged to simulate those encountered in specific historical instances explored in MU_1. Alternatively, for (a), the researcher could conduct qualitative focus group interviews with consumers interested in purchasing organic and sustainably produced foods (MU_1) followed by qualitative journaling undertaken by a purposive sample of organic shoppers, using input from what was learned from the focus groups in MU_1 to provide a partial structure for recording shopping activities in the journals.

Programmatic/Accumulative Sequential Configuration

The Programmatic/Accumulative Sequential configuration is a pattern of MUs that reflect a planned series of investigations which successively vary specific characteristics of a research context in order to further tease out, qualify, or extend meanings, patterns, and relationships that are being explored (similar to the 'multiphase design' in Creswell & Plano Clark, 2011, pp. 100–104). Here is a configuration that is purpose-built for systematically accumulating knowledge through a series of linked investigations. Under the pattern of guiding assumptions for a positivist inquiry, this configuration might comprise a series of incremental investigations where specific causal variables are systematically manipulated and tested in each MU in order to carry out more well-rounded and thorough tests of theoretical propositions/hypotheses. Under the pattern of guiding assumptions for an interpretivist/constructivist inquiry, this configuration might embody a 'grounded theory' approach, where a new theoretical perspective is built up using a series of MUs. Two examples would be: (a) where the researcher employs a general quantitative questionnaire (Survey research frame) to tease out some general theoretical propositions that can then be further tested (MU_1) followed by a more formal laboratory experiment (Explanatory research frame) designed to test those emergent propositions using undergraduate university students (MU_2) followed then by a quasi-experiment (Explanatory frame again) designed to test the theoretical propositions in a field setting (MU_3) or (b) where the researcher conducts a series of qualitative semi-structured interviews with dairy farmers about their views on innovation needs in their industry (MU_1) followed by qualitative semi-structured interviews with research and development scientists in a major farming innovation company working on a new dairy farming innovation to meet dairy farmer needs (MU_2) followed by qualitative focus groups (MU_3) with potential and actual dairy farm innovation adopters, evaluating (e.g., would it work? does it work? meets your needs? cost effective? changes needed? etc.) an innovation once it has been released into the market for testing after incorporating the feedback and learning from both MU_1 and MU_2 followed by focus groups with actual dairy farm innovation adopters once the company had further modified the innovation in response to what was learned during MU_3 (MU_4; note that this entire sequence of investigations could be a pattern of research done within the Developmental Evaluation research frame). The Programmatic/Accumulative Sequential configuration is also quite useful in the Action research frame where deliberate change planning, implementation, and evaluation for learning purposes unfold in a series of MUs; learning from each MU feeds into the next MU in the sequence.

5.3.5 Hierarchical Configurations

The basic Hierarchical MU configuration involves one MU being defined as primary, major, or dominant and other MUs as secondary, minor, supportive, or subordinate. The basic configuration was set out by Creswell and Plano Clark (2011, pp. 90–96) and its goal is to support what is to be learned/what is being learned/what has been learned in the primary MU with data gathering using a strategically chosen second (or more) MU. This helps the researcher to round out the research story they are constructing in such a way as to enhance convincingness. In Creswell and Plano Clark's (2011) intentions for this configuration, the primary MU focuses on quantitative data and the secondary MU focuses on qualitative data or vice versa. In this unifying framework, such a constraint does not exist; the configuration can be just as viable using two MUs that gather the same form of data, but using different data gathering strategies and/or focusing on different types of data sources.

Some advantages of this configuration include:

- it is resource efficient in terms of time, effort and finance, in that all data are gathered simultaneously or nearly simultaneously (i.e., the secondary MU can be implemented before, during, or after the primary MU);
- it seeks to provide triangulated stories through diversifying MUs, thus extending what might be achieved with a Simultaneous configuration; and
- it may encourage enhanced participation and participation retention through deeper involvement of participants in the data gathering processes that unfold in what amounts to a single data gathering stage.

Some disadvantages of this configuration include:

- it is not an effective configuration for drawing inferences about externalised cause and effect, since temporal precedence of cause before effects cannot be assured;
- since primary and secondary MUs differ in emphasis, the weaknesses of one cannot be fully offset by strengths in the other;
- this configuration would be challenging to implement using different patterns of guiding assumptions for each MU; and
- there may be incommensurabilities between what is learned from the dominant MU and what is learned from the subordinate MU, which may create ambiguities that are difficult to resolve.

A Hierarchical configuration may be useful in a range of research frames:

- the Survey research frame, embedding one type of questionnaire (e.g., with qualitative open-ended questions) inside of another (e.g., with quantitative rating scales) or vice-versa in order to provide supportive triangulation data;
- the Descriptive, Exploratory, or Case Study research frames, where a secondary MU is employed to tap into supplemental, often textual or other documentary data sources beyond those involved in the primary MU in order to achieve a more well-rounded contextual story;
- the Indigenous/First Nations or Othering/Emancipatory research frame, where a secondary MU is used to implement an alternative data gathering strategy to that used in the primary MU in order to better connect with certain sub-groups in a sample (e.g., children, Elders, minorities); or
- the Pan-disciplinary or Developmental Evaluation research frame, where a secondary MU uses semi-structured interviews to probe the views of a specific sub-group of participants more deeply, such as community leaders or research scientists, in support of the primary MU of undertaking structured interviews with samples from different community contexts (e.g., business, farming & agricultural, public transport).

There are two basic variations of Hierarchical configuration: Embedded and Supplemental. The choice of variant to use is generally informed by resourcing/feasibility constraints and research goals as well as by the researcher's scoping choices with respect to breadth-depth balance for learning and extensional reach. In general, with respect to resourcing issues, the Hierarchical configuration implements an intentional imbalance in resourcing: more resources to support the primary MU, fewer to support the secondary MU. The choice of secondary MU is then made in terms of achieving maximum value in supportive learning within the context of fewer allocated resources.

Embedded Configuration

The Embedded configuration is a hierarchical pattern of MUs where a secondary MU is embedded or nested within the primary MU. The secondary MU is typically implemented contiguously with the primary MU, along the lines of arguments presented by Creswell and Plano Clark (2011, p. 75). The Embedded configuration is generally oriented toward achievement of depth rather than breadth of learning, allowing the researcher to dig deeper within the context of the primary MU and is somewhat more resource-efficient relative to the Supplemental configuration because of its contiguous data gathering focus. Three examples include: (a) where the researcher employs a quantitative questionnaire (MU_1) that also contains a small number of qualitative open-ended questions (MU_2) (a common practice within the Survey research frame, which often produces a side-benefit of enhancing response rates due to inviting participants to share their own views); (b) where the researcher conducts an experiment gathering quantitative measurements (MU_1) and immediately follows up those measurements with qualitative focused interviews focusing on participants' perceptions of their experiences in the experiment (MU_2) (useful in the Explanatory research frame, for instance); or (c) where the researcher conducts a qualitative participant observation study (MU_1) in which they also conduct qualitative semi-structured interviews with specific participants (MU_2) (useful in a grounded theory investigation conducted within the Explanatory research frame, for instance).

Supplemental Configuration

The Supplemental configuration is a hierarchical pattern of MUs where a secondary MU is not embedded within the primary MU; rather it is implemented separately to it. The Supplemental configuration is generally oriented toward achievement of greater breadth rather than greater depth of learning, allowing the researcher to gather data outside the context of the primary MU. Two examples would be: (a) where the researcher administers a short quantitative questionnaire within a specific workplace (MU_2) as a prelude to selecting employees to participate in the main study involving semi-structured qualitative interviews about their workplace experiences (MU_1) (useful in the Exploratory research frame, for instance); or (b) where the researcher gathers text evidence in the form of annual reports, media releases, and other documents relevant to a Board of Directors and its decisions within a specific company (MU_2) in addition to carrying out a participant observation study of several meetings of that Board of Directors (MU_1) (useful in an Exploratory research frame, for instance).

5.3.6 Case-Based Configurations

The basic Case-based MU configuration focuses on a defined entity or event (i.e., the 'case', which may be a group, an organisation or institution, a community, or a focal event such as a natural (e.g., earthquake, flood or fire) or man-made (e.g., plane or spacecraft accident, military invasion or incursion, terrorist action) disaster, sporting or community event, or economic crisis) in its natural context. In this configuration, contextualisation of data gathering is of primary importance where the case provides the focal context and the MU focuses on data gathering within and with an understanding of that context. As Yin (2018) discussed, the focus of data gathering may reside at the level of the case as a whole, yielding a 'holistic' case study, or at the level of units of analysis contained with the case context, yielding an 'embedded' case study. Embedded units of analysis may comprise one or more different groups (e.g., different classrooms, different departments, enrolments in different courses, work teams, social groups), different levels/roles within an organisational hierarchy (e.g., senior executives, managers, teachers, students, nurses, doctors, line workers, union officials) and/or different events (e.g., team meetings, emergencies, new staff inductions/indoctrination, ritual events such as retirements, graduations). In an embedded case study, the case context forms a backdrop against which data for the different units of analysis may be gathered and interpreted. What is learned from the different units of analysis is therefore foregrounded against the background of the general case context. In a holistic case study, some aspects of data gathering may focus on contextual influences outside of the case context itself, lending a wider systems focus to the research. For example, a holistic case study of a specific school may gather data within the school that speaks to wider systemic concerns (e.g., government funding, school system regulations and policies, demographic, and economic conditions of the geographic area in which the school resides and so on).

Irrespective of whether the researcher is conducting a holistic or embedded case study, one important implication of the Case-based configuration is that the researcher is obligated to gather contextual background information about the case in parallel with or prior to implementing the MU. This contextual learning and backgrounding may be achieved through activities such as reading case-related documents (e.g., historical documents, annual reports, minutes of public meetings, media stories), accessing websites (perhaps even staff biographies) associated with the case context, and/or informal discussions with key people in the case context. In some instances, what the researcher learns from this contextual background information may help shape what they do in the MU and, perhaps, identify who important data sources might be. [Note that if the contextual backgrounding activities are planned and undertaken as more formalised and structured data gathering activities, rather than as the more informal learning activities implied here, they become their own MU, meaning that the Case-based configuration becomes hybridised with multiple MUs in play (see discussions of the Case-based Simultaneous, Case-based Sequential and the Case-based Hierarchical/Supplemental configurations below).]

Some advantages of the Case-based configuration include:

- it is explicitly oriented toward providing depth of contextualised learning;
- it facilitates the interpretation of data patterns through reflecting on those patterns in light of contextual circumstances in the case; and
- it is particularly well-suited for research guided by any pattern of assumptions that adopts a more subjectivist or critical stance as the researcher often becomes embedded in the case during their research journey (however, this does not mean the researcher cannot adopt the pattern of guiding assumptions for a positivist inquiry in the Case-based configuration; it is just less common).

Some disadvantages of the Case-based configuration include:

- it is generally a more resource-intensive and challenging form of research, especially with respect to the researcher's effort and time;
- writing the story of a case study investigation in a research outcome is generally more challenging because contextualisation needs to be thorough and sensitively utilised to make sense of data patterns and interpretations; and
- unless the researcher studies multiple cases, transportability of learning/meaning to other contexts/cases (under any subjectivist pattern of guiding assumptions) or generalisability of relational patterns to a larger population or other contexts (under a more objectivist pattern of guiding assumptions) is hard to convincingly argue (combatting this disadvantage is what gives rise to the Multiple Case-based configuration variation to be discussed below).

A Case-based configuration may be useful in a range of research frames:

- most centrally, of course, is the Case Study research frame, where the researcher's goal critically depends upon their awareness, understanding, and use of contextual nuances in the focal case;
- the Descriptive or Exploratory research frame, where the researcher may be seeking greater depth of learning in a specific case context, but with an orientation toward providing stepping off points for further research;
- the Evaluation research frame, where the researcher might be evaluating a specific program in the context of one or more selected case organisations/institutions/communities;
- the Developmental Evaluation research frame, where the researcher might seek to develop, modify, and systematically evaluate an innovation in a specific case context (e.g., in a hospital or other public or private organisation, community of farmers or an industry);
- the Indigenous/First Nations research frame, where the researcher might seek to achieve depth in understanding life within a specific Indigenous/First Nations community;
- the Cross-Cultural (cross-cultural learning emphasis) research frame, where the researcher might seek to understand life within a specific cultural context/community, which is outside their own cultural context; and
- the Explanatory research frame, where the researcher pursues case-contextualised theory building or testing (as with grounded theory, for example, or where the case is a specific event in which the applicability of a theory might be tested).

There are two basic variations of Case-based configuration: Single and Multiple. The choice of variant to use is generally informed by resourcing/feasibility constraints and research goals as well as by the researcher's intentions with respect to extensional reach. Extensional reach is really the pivotal question: does the researcher intend or hope to point to transportable meanings/social constructions and/or make generalising statements about the relevance of learning achieved within the case to or in other cases/contexts/situations? If the answer is no, then the researcher would prefer the Single Case-based configuration; if yes, then the researcher would prefer the Multiple Case-based configuration. The choice will primarily be a strategic one for the researcher, but once made, will also have resourcing implications associated with it.

Single Case-Based Configuration

The Single Case-based configuration involves a MU that focuses on a single case entity or event, where the yellow oval surrounding the MU (refer to Fig. 5.4, lower right quadrant) signifies the contextual learning and backgrounding that the researcher needs to achieve as part of their research journey. This configuration is generally more feasible, from a resourcing perspective, especially where the researcher's research time is limited (as with a postgraduate degree or time-sensitive research grant). The drawback, of course, is that any transportable conclusions or generalising statements can only be logically and speculatively argued for (very challenging but not impossible), rather than empirically demonstrated, with consequent adverse impacts on convincingness. In many instances, making such statements will not be the researcher's intention; their goal instead would be achieving a deep understanding of that one particular case. Two examples would be: (a) where the researcher administers an organisation-wide survey, measuring organisational climate and job satisfaction, to a sample of medical, nursing, technical, and support staff employed in a large public hospital (where the focus and content of the questionnaire is informed by contextual information about that hospital); or (b) where the researcher conducts qualitative semi-structured interviews of managers in a particular city council under threat of amalgamation with other city councils as a result of changes to state government policy (where the topical landscape of the interviews is informed by the researcher's contextualisation learning and backgrounding with respect to the city council's context).

Multiple Case-Based Configuration

The Multiple Case-based configuration involves a MU that focuses on multiple case entities or events, where each yellow oval (again refer to Fig. 5.4, lower right quadrant) signifies contextual learning and backgrounding that the researcher must achieve for each distinct case context as part of their research journey. The resource demands for the Multiple Case-based configuration multiply with each case considered. Not only must data be gathered and analysed within each case context, but the researcher will also need to make cross-case comparison analyses as well as carry out contextual backgrounding for each case. A single researcher can effectively manage such research activities, including managing all 'Data Triangle' activities, but the timeline for the research will be greatly lengthened and the drain on the researcher's personal resources (energy, motivation, cognition) will generally be considerable. This means that using a research team (one team member for each case) may be a more viable way of implementing the configuration, but with the added necessities of needing to manage and coordinate the efforts of each team member and the complexities of trying to integrate what is learned from the different case stories. If the researcher takes appropriate steps to ensure that the commonalities and uniquenesses of all case contexts are well-understood, they may be able to advance empirically defensible and convincing generalising statements or transportations of meanings/constructions/patterns, drawing upon those commonalities with caveats and qualifications reflected in the uniquenesses. Two examples would be: (a) where the researcher undertakes qualitative focused interviews about changes in/impacts on job demands and stress with sworn and unsworn officers sampled from two police organisations in regional New South Wales which have recently been forced to downsize due to reallocations of state and federal government resources; or (b) where the researcher undertakes participant observations of classroom teaching, focusing on classrooms embedded within three schools sampled from a particular school district in a metropolitan city.

5.3.7 Longitudinal Configurations

The basic Longitudinal MU configuration involves mapping/tracking data movements/patterns/meanings against the passage of time. The centrality of time is a key scoping consideration that leads to choosing this research configuration and is what primarily distinguishes this configuration from the Sequential configuration. In the Longitudinal configuration, the passage of time, whether relative to some intervention, event, or program or not, becomes a critical aspect of contextualisation. It allows the researcher to assemble coherent and convincing 'before-during-after'-type stories (where a focal intervention, event or program is involved) or stories about patterns of movement, change, or development over time with respect to some context or social or economic process of interest. In some instances, predictions of future occurrences beyond the time window of the research may be pursued (as in certain forms of time series research). The important thing to note here is that in the Longitudinal configuration, time itself does not play a causal role (in that the passage of time itself does not cause change), it merely provides a linear reference path and signposts against which to map trajectories and events of change. The unit of analysis for the Longitudinal configuration may be an individual economic, social, or financial indicator, a person, a group, an organisation or institution, a nation, with the basic requirement, in most cases, being that the same units are observed on each data gathering occasion.

Some advantages of the Longitudinal configuration include:

- temporal precedence of cause and effect can be ensured by observing/measuring causal events/behaviours/situations (using the Non-manipulated experienced-focused data gathering strategy) or by the researcher manipulating them to occur (using the Manipulated experienced-focused data gathering strategy) before any observations of outcomes (the effects to be attributed to the previously observed or manipulated causes) are made, an arrangement that is important if understanding and making inferences about external causation of events is an important research goal (normally associated with the pattern of guiding assumptions for a positivist or perhaps a critical realist inquiry);
- patterns of movement, changes or developmental evolution over time can be explicitly tracked/understood/observed; and/or
- the researcher can achieve greater depth of contextual immersion through repeated episodes of data gathering (may be important under any pattern of guiding assumptions that adopts a more subjectivist stance as well as for research approaches such as grounded theory or data gathering strategies such as participant observation).

Some disadvantages of the Longitudinal configuration include:

- it is, potentially, a time and resource-intensive configuration to implement, especially where many data gathering episodes are involved or where repeated contextual immersion is involved (except in instances where the researcher can access the time-based measurements or other data from a secondary database);

- where repeated contextual immersion is involved for research guided by any subjectivist stance pattern of assumptions, there is a risk that the researcher may lose their 'researcher' attitude and begin to adopt/reflect the perspectives of those they are sharing the context with (so-called 'going native'; an especially potent risk when implementing the Participant Observation data gathering strategy);
- depending upon the spacing of data gathering activities between occasions, the researcher may not be able to monitor what goes on between the episodes of data gathering (which may result in obscuring potentially important aspects of learning about the phenomena being researched, especially where the researcher is interested in external causal inferences under the pattern of guiding assumptions for a positivist or critical realist inquiry);
- in cases where the researcher wishes to pursue external causal inferences (again under the pattern of guiding assumptions for a positivist or critical realist inquiry), it may not be possible to arrange or manipulate the some or all intended causal conditions, which means that causal inferences may be more difficult to defend (this problem is what produces a quasi-experiment in the Manipulated or Non-manipulated data gathering strategies);
- where human participants are involved, there is a greater risk of participant dropout (i.e., 'mortality') from the research due to fatigue, boredom, changes in life, work, or health circumstances (the longer the timeline or the more data gathering episodes they are involved in along the timeline, the greater the risk); and/or
- in research that depends upon time-based data from secondary databases (e.g., a government, educational, financial or economic database), there may be gaps in the data for specific time periods for specific units of observation (e.g., data for one year for a company or a school might be missing), which the researcher will need to decide how best to handle the gaps.

A Longitudinal configuration may be useful in a range of research frames:

- the Explanatory research frame, where the researcher explicitly pursues an external causation story where time is either a critical independent variable or a reference point for different events (under the pattern of guiding assumptions for a positivist or critical realist inquiry, giving rise to a configuration labelled as a repeated measures or time series design) or where deeper contextual immersion is intended to yield a theoretical account for what the researcher has observed (e.g., grounded theory under pattern of guiding assumptions for an interpretivist/constructivist inquiry);
- the Action research frame, where the researcher works with a specific group or organisation through a cyclical research process to facilitate and evaluate localised contextual changes/improvements and where mapping those changes over time is critical for causally linking planned change interventions to observed improvements as well as to possible side effects;
- the Survey research frame, where the researcher gathers data from the same sample of participants on several occasions (a so-called 'panel study', which creates a longitudinal Survey research frame as distinct from a cross-sectional Survey

research frame) or where the researcher gathers data from approximately equivalent (with respect to one or more focal characteristics, such as age or school grade) samples from the same population on each data gathering occasion (a so-called 'cohort study');

- the Evaluation research frame, where the researcher sets out to evaluate the effectiveness of an intervention, treatment or program using, at a minimum, a pre-test (before undertaking the program or experiencing the intervention) and a post-test (after undertaking the program or experiencing the intervention) data gathering occasion (intermediate or follow-up data gathering occasions may also occur during or after undertaking the program or experiencing the intervention) [in psychological, educational, medical (e.g., clinical trials) and other types of social and behavioural research pursuing external causal inferences regarding the effectiveness of an intervention, treatment or program, a comparative or placebo control group (which does not receive or experience the undertaking the program or experiencing the intervention or treatment) may be used to enhance the convincingness of the causal claims];
- the Developmental Evaluation research frame, where the researcher is tracking and gathering contextualised data on the development, modification, adoption, and effectiveness of an innovation that necessarily unfolds over time and which may be linked to the seasonal or periodic occurrences of events (see, for example, discussions in Patton, 2011).

There are three basic variations of Longitudinal configuration: Time-Aligned, Event-based Time-Aligned and Developmental Time-Aligned. The term 'time-aligned' is meant to signal that data gathering episodes occur at (i.e., are aligned with) specific points along the timeline. The choice of variant to use is generally informed by the researcher's resourcing/feasibility constraints and research goals as well as by their scoping choice with respect to pursuing external causation explanations. In terms of strength of external causal inferences, the Event-based Time-Aligned configuration offers the strongest opportunities for convincingness, where the event (broadly defined) is intended as the causal factor and data are gathered, at a minimum, both before and after that event. [Note that when paired with one or more control groups and random assignment of patients to a control or treatment group, this configuration can become the gold standard randomised clinical trial (RCT) in medical research, although not without some reservations, see, for example, Kaptchuk, 2001)]. The Time-Aligned configuration is commonly employed in government (e.g., policy- or compliance-related research), institutional (e.g., tertiary education institution research), economic (e.g., labour and productivity studies) or financial (e.g., stock market and organisational profitability research) as well as in sociopolitical systems research (e.g., marketing and social policy research such as tracking the impacts of social or advertising interventions). In this configuration, the power of the story comes from many occasions (e.g., days, weeks, or years) of data gathering, however, external causal inferences are more difficult to defend and may be tied to the length of the series. The Developmental Time-Aligned configuration is less concerned with mapping external causation over time and more concerned with addressing the evolution of people, groups, institutions, or innovations over time, taking account of input and feedback from many different stakeholders at various points along the innovation development-adoption timeline. In this configuration, learning is an especially important focus and early learning can influence and shape what is done and what is learned later in time.

Time-Aligned Configuration

The Time-Aligned configuration involves the implementations of a MU over multiple, sometimes quite numerous, occasions over time. If, under the pattern of guiding assumptions for a positivist inquiry (the most common pattern of guiding assumptions underpinning this specific configuration), many time-aligned quantitative data measurements are gathered using the same MU on each occasion. This gives rise to a variation of this configuration called a 'time series' design (see, e.g., Cryer & Chan, 2008) and the spacing between data gathering occasions is intended to be approximately equal (e.g., days or years). Two examples include: (a) where the researcher uses secondary database records to track car accident occurrences attributable to mobile phone use, drink driving, drug use, or speeding over a period of many months or years to provide data to inform the creation of new government legislation for penalties; or (b) where the researcher undertakes systematic observations in a classroom over the course of a semester to record teacher-student interaction patterns over time.

Event-Based Time-Aligned Configuration

The Event-based Time-Aligned configuration involves implementations of a MU over one or more occasions in time typically before, during, and after some 'event' of interest has occurred. The 'event' may constitute a deliberate researcher-devised intervention or contextual manipulation (i.e., a training program, drug or other clinical treatment, change in environmental, social, task-related, or psychological conditions), an event of human origin that is not devised by the researcher (i.e., acquisition of an organisation by a competitor, terrorist attack, assassination of a president or prime minister, new social policy, drink driving advertising campaign), or an event of natural origin (e.g., flood, earthquake, extreme weather event, drought, sea level rise (although arguably, this could have a human origin component as well)) in the research context. Two occasions (one pre-intervention and one post-intervention) are the minimum requirement for this configuration if external causal inferences are to have some chance of being convincing and defensible. The external causal inference being pursued in this configuration is that whatever happened during the event of interest is what causes any changes the researcher observes between the pre- and post-occasions (including intermediate observations during the event may enhance the researcher's causal inference capability (often done in formative evaluations, for example), but would be more resource-intensive). A variation of this configuration could involve additional implementations of the MU within the period of an event as well. If, under the pattern of guiding assumptions for a positivist or critical realist inquiry, many time-aligned quantitative data measurements are gathered which both precede and follow an intervention of some kind, this yields a variation of this configuration called an ‘interrupted time series’ design (see, e.g., Glass et al., 2008). If only a few data gathering occasions are involved (e.g., a pre-test and a post-test), the configuration would be called a ‘repeated measures’ design. Two examples would be: (a) where the researcher employs a quantitative survey obtaining specific employee performance and attitude measurements before and after a corporate takeover of a company; and (b) where the researcher carries out focused qualitative interviews of individual group members before and after each group counselling session that they participate in with a specific group counsellor. Note, however, that a repeated measures design would also result if each occasion related to a different researcher-manipulated contextual event (meaning that time is just a positional marker for the ordering of events experienced by participants). For example, a researcher may conduct an experimental preference study where participants are provided with different experiences, which constitute the events (e.g., foods, drinks, photographs, beers, choices, scenarios, environmental conditions such as heat, cold, and/or noise, technological devices to use), on each testing occasion where event experiences are order-based not time-based.

Developmental Time-Aligned Configuration

The Developmental Time-Aligned configuration involves implementations of a MU occurring over multiple occasions in time where the goal is to track evolution in the context of incremental changes introduced to or made within the research context. In many cases, the configuration could be used in formative evaluation research within the Evaluation research frame or to track the development and modification of an innovation or program to meet the needs of a specific community, industry, or target group of users/potential adopters in the Developmental Evaluation research frame (Patton, 2011). An example would be where the researcher conducts group-based focus group interviews with dairy farmers in various dairy-producing regions of a state where groups identify innovation needs, that information is fed back to the Research & Development division of a major agricultural manufacturing company who create and provide an innovation solution to the members of the groups to meet their identified needs for a specified period of time, the focus groups then reconvene to reflect on and evaluate the innovation and suggest modifications which are then fed back to the R&D division and so on (the time taken from innovation idea to successful realisation becomes one aspect of learning important in this configuration). Note, however, that it is also possible to employ a variation of this configuration that uses different MUs on each occasion (something that Patton, 2011 envisaged). This would occur, for example, if, in the above scenario, the follow-up MUs involved semi-structured interviews with relevant R & D personnel and then a quantitative questionnaire to innovation users.

5.3.8 *Hybrid Configurations*

A hybrid configuration refers to any more complex configuration of MUs that combines features of two or more of the previous basic configurations. Hybrid configurations are becoming increasingly more commonplace in modern day research as researchers, research users, stakeholders, funders, and sponsors seek more convincing stories. While a basic MU configuration may work well in many instances, a hybrid configuration may add substantive value to a research investigation.

Some advantages of a hybrid configuration include:

- it can generally enhance convincingness relative to what any one of its component basic configurations could offer on their own;
- it can harness synergies between different configurations which may assist in offsetting the weaknesses of one basic configuration with the strengths of another configuration, including situations where different patterns of guiding assumptions are employed for each basic configuration; and/or
- it may assist in achieving both depth and breadth in research scope, where one basic configuration may be oriented toward achieving breadth of learning and another basic configuration oriented toward achieving depth of learning.

Some disadvantages of a hybrid configuration include:

- most hybrid configurations are more resource and effort intensive to plan and implement in that the research will almost always take more time to plan and complete than a basic configuration would; or
- when a single researcher is involved, as in postgraduate research, implementation could be very challenging, especially where the researcher's expertise in a particular facet of a hybrid configuration is weakest or, at the very least, requires development (this disadvantage can be offset through the use of a research team, where appropriate).

Seven different hybrid configurations are discussed below. Each offers its own advantages and disadvantages, and each can be used to achieve very different research purposes. However, it is important to note that these seven hybrid configurations should not be considered exhaustive of the possibilities; there is ample room for creativity in hybridisation depending upon the researcher's goals, research frame, contextualisations, and positionings and as well as on their available resources.

Case-Based-Simultaneous Hybrid Configuration

The Case-based-Simultaneous configuration involves the implementation of multiple data gathering strategies in the same time period within a single focal case. The purpose of using multiple data gathering strategies, multiple data forms, and/or multiple types of data sources is to achieve a deep, rich, triangulated, and well-rounded understanding of the case context and its people. However, this hybrid configuration provides more of a 'slice of life' window on the case, rather than depth through time, meaning that this configuration would not be very useful in change scenarios. In many single case studies, the minimal configuration would involve the researcher analysing documents and texts relevant to the case context to gain a deeper understanding of the case context coupled with qualitative interviews of some type with people embedded in that context, often during a single visit to the case context. A concrete example would be where the researcher administers a quantitative survey to teachers (MU_1), conducts qualitative semi-structured interviews with the principal, vice-principal and head teachers (MU_2), and observes a parent-group meeting carried out simultaneously within a selected high school during a week-long visit (MU_3). A variant of this hybrid configuration could synergistically combine the Multiple Case-based configuration with the Simultaneous configuration. A concrete example of this variant would be where the researcher prepares, pre-tests, and administers cross-culturally equivalent quantitative questionnaire to workers (MU_1) combined with semi-structured qualitative interviews of managers or executives in two culturally different contexts (MU_2) (e.g., two schools, organisations, or communities in different countries).

This hybrid configuration may be useful in the following research frames:

- The Case Study research frame is the most obvious frame for this configuration, where the researcher employs different data gathering strategies (e.g., interviews, questionnaires, document analysis, participant observation) to obtain information (qualitative, quantitative, or merged) from different data sources within a single case study context;
- The Descriptive research frame, where the researcher seeks to develop a more well-rounded descriptive picture of a situation, event or sample, including contextual case information;
- The Exploratory research frame, where the researcher seeks to undertake a thorough exploration of a new or unfamiliar case context through gathering data from different sources, perhaps as a prelude to further research;
- The Cross-Cultural (comparative emphasis) research frame, where the researcher implements a Multiple Case-based Simultaneous configuration within two culturally different case contexts as described above;
- The Pan-disciplinary research frame, where the researcher connects with various stakeholders (e.g., community members, scientists, relevant local government and state politicians, business owners, agricultural producers) using various data gathering strategies, in order to establish strategies for managing policy or natural resource impacts in an applied learning case context (e.g., a local farming community); or
- The Indigenous/First Nations or Othering/Emancipatory research frames, where the researcher connects with a variety of data sources, some of whom might be gatekeepers (e.g., tribal Elders, senior management) for gaining access to other people in the case context (e.g., an Indigenous/First Nations community, a public service organisation such as the police, a school) perhaps using different data

gathering strategies, in order to construct a more complete perspective or learn as much as possible about relationships and issues.

Case-Based-Sequential Hybrid Configuration

The Case-based—Sequential hybrid configuration involves multiple data gathering strategies that are sequentially implemented within a focal case with subsequent MUs building on what is learned from previous MUs. This hybrid configuration uses multiple data gathering strategies, multiple data forms, and/or multiple types of data sources to build up a deep, rich, increasingly focused (through the sequencing of MUs) and well-rounded understanding of the case context and its people. An example would be where the researcher implements an organisational document analysis (MU_1) to learn about the case context and the recent background history of a failed takeover bid by a larger competitor, then conducts a focus group interview (MU_2) with departmental managers in the case context, discussing the failed takeover bid and its impacts on the case organisation, then using what has been learned from the focus group, constructs and administers a quantitative questionnaire focusing on job security perceptions to a representative sample of employees (MU_3). A variant of this hybrid configuration could synergistically combine the Multiple Case-based configuration with the Sequential configuration. This could be illustrated in the above example if the researcher adds two more case study organisations, each surviving a recent takeover bid in order to build up a more transportable (i.e., cross-case) set of theoretical arguments to explain what happened in each case.

The Case-based-Sequential hybrid configuration may be useful in the following research frames:

- The Case Study research frame, illustrated in the failed takeover bid example presented above;
- The Action research frame, where the researcher effectively becomes a joint participant,, in conjunction with relevant groups in a specific case context (e.g., a school, community group or business), in action learning cycles intended to improve some process or outcome, where there is a MU focused on specific stages in the cycle: diagnosing the problem/formulating a solution and implementing/evaluating/reflecting on that solution;
- The Explanatory research frame, where the researcher wishes to implement an explanatory sequential configuration within one or more case contexts; or
- The Indigenous/First Nations or Othering/Emancipatory research frames, where the researcher might need to undertake a small-scale research investigation in order to build trust and good will to pave the way for gaining deeper contextual access for further data gathering within a specific community or institution.

Case-Based Hierarchical/Supplemental Configuration

The Case-based Hierarchical/Supplemental hybrid configuration where the supplemental MU_2 focuses on data gathering from larger contexts surrounding and perhaps influencing/impacting the case context that provides the primary data gathering focus for the primary MU_1. This hybrid configuration employs multiple data gathering strategies, multiple data forms, and/or multiple types of data sources, but augments the research in such a way as to provide a larger, more systemic and contextualised understanding of the case, perhaps connecting with relevant stakeholders outside of the case context but who have an interest in what happens within the case context. An example would be where the researcher conducts semi-structured interviews with managers within a specific case context focusing on new government regulatory requirements (MU_1) as well as accessing and analysing external documentation focusing on the new regulatory requirements and what their implications are for the case context (MU_2). One variant of this hybrid configuration could synergistically combine the Multiple Case-based configuration with the Hierarchical/Supplemental configuration. This would be illustrated in the above example if the researcher added another case study organisation, conducting semi-structured interviews in each as well as accessing external case-related documents. Such a configuration would facilitate making cross-case comparisons/contrasts with respect to the implications of the new regulatory requirements for each case. A second variant could blend the Single or Multiple Case-based-Simultaneous or Sequential configuration with one or more Hierarchical MU elements, yielding richer data both within the case context as well as from the larger contexts in which the case is embedded.

The Case-based Hierarchical/Supplemental hybrid configuration may be useful in the following research frames:

- The Case Study research frame, illustrated in the new government regulatory requirements example previously presented;
- The Pan-disciplinary research frame, where the researcher conducts participative action research within a case context, but also connects with key stakeholders outside the case context including those who may be in a position of power to influence events and perceptions within the case context (here is where the second variant described above could be useful); or
- The Exploratory research frame, where the researcher wishes to round out as well as deepen the exploratory learning from a case study, by seeking out additional contextual information outside the case.

Longitudinal-Simultaneous Hybrid Configuration

The Longitudinal-Simultaneous hybrid configuration emerges when simultaneous configurations of MUs are repeatedly implemented over time, perhaps to track the impact of an event or change occurring between or during implementation occasions. This hybrid configuration can be used for panel studies where a sample of people is followed through time to track development or change, but with the additional advantage of using multiple data gathering strategies and/or multiple forms of data on each occasion to enhance triangulation. An example would be where the researcher administers a quantitative questionnaire to employees in a company and conducts qualitative interviews with managers in the company at four different times of the year (each after quarterly company reports have been submitted), to track changes in organisational climate and leadership perceptions following the appointment of a new CEO.

The Longitudinal-Simultaneous hybrid configuration may be useful in the following research frames:

- The Action research frame, where the researcher participates in several cycles of action learning with one or more groups, each time using MUs that focus on different stages of the cycle (e.g., focus groups for the diagnosis and planning stages; participant observation for the implementation and evaluation stages), and where tracking how the change or solution evolves over time is important;
- The Evaluation research frame, where the researcher is interested in conducting a formative evaluation of a program over time (while the program is being implemented), using different data gathering strategies and tapping into different data sources;
- The Cross-Cultural (comparative emphasis) research frame, where the researcher implements MUs, focusing on different cultural groups as data sources on several occasions, perhaps as a way of studying the impact of a major historical event (e.g., such as a change in global economic conditions) over time on the different groups; or
- The Developmental Evaluation research frame, where the researcher taps into several different types of data sources (e.g., research and development scientists, community members and consumers, workers/producers) on multiple occasions along the evolutionary timeline for a specific product or social innovation.

Longitudinal—Case-Based Hybrid Configuration

The Longitudinal—Case-based hybrid configuration involves a MU being repeatedly implemented (perhaps even continuously implemented) in the context of a specific case. For example, the researcher might undertake specific short periods of participant observation of a local city council over several occasions as the council works toward amalgamation with another local city council. It is quite straightforward to envisage a more complex form of this hybrid configuration where multiple simultaneous MUs are implemented. For instance, a simultaneous multiple MU variant applied to the local city council context example above might include some focused qualitative interviews with senior council managers as well as the amassing of documents and field notes on each occasion as separate sources of data.

The Longitudinal—Case-based hybrid configuration may be useful in the following research frames:

- The Case Study research frame, where the researcher wishes to track/understand changes over time within one or more specific cased contexts;
- The Action research frame, where the researcher participates in several cycles of action learning with a group in a specific case context, each time using MUs that focus on different stages of the cycle (e.g., focus groups for the diagnosis and planning stages; participant observation for the implementation and evaluation

stages), and where tracking how the change or solution evolves over time is important;

- The Developmental Evaluation research or Pan-disciplinary research frame, where the researcher wishes to focus data gathering strategies on processes and stakeholders implicated in the development, evolution, adoption, and use of one or more innovations within a specific case context (e.g., within a specific industry or community);
- The Exploratory research frame, where the researcher wishes to explore a new case context and to achieve greater depth of understanding by following the case over time, perhaps as a way of exploring the evolution and dynamics of the case context;
- The Explanatory research frame, where the researcher plans to build up and/or test a set of theoretical accounts or explanations for how a case context develops/changes over time, perhaps because of or in parallel with events/changes in within the case context itself or in one or more larger context(s) within which the case is embedded; or
- The Indigenous/First Nations or Othering/Emancipatory research frames, where the researcher wishes to examine the dynamics over time in an Indigenous/First Nations case context, perhaps in concert with an increase in depth focus over time, or in dynamics over time within that case context, but under the pattern of guiding assumptions for a critical or systemic othering inquiry.

Longitudinal-Hierarchical Hybrid Configuration

The Longitudinal-Hierarchical hybrid configuration involves embedded (or supplemental) hierarchical MUs being repeatedly implemented across time. The advantage of this specific configuration is that it allows the researcher to primarily map out/understand changes over time or to tease out cause-effect relationships (by controlling for temporal priority in obtaining cause and effect information) while, at the same time, providing them with additional participant-focused contextual information to help round out or qualify the emerging story. For example, the researcher might undertake a panel study using quantitative questionnaires for measuring attitudes and preferences for products and product attributes that also contain open-ended questions focusing on personal views about products and their attributes, administered on several occasions, to the same sample of consumers, with the intention of testing specific theoretical propositions by ensuring that potential causal factors are measured before any measurements of effects are obtained.

The Longitudinal-Hierarchical hybrid configuration may be useful in the following research frames:

- The Survey research frame, where the researcher undertakes a panel study as described above (a popular research approach taken by marketing and organisational behaviour researchers, for example);

- The Evaluation research frame, where the researcher is interested in conducting a formative evaluation of a program over time (while the program is being implemented) using a primary data gathering strategy (with a specific associated data form, quantitative, qualitative, or hybrid/merged) and a secondary data gathering strategy (often in conjunction with the other data form, qualitative, quantitative, or hybrid/merged); or
- The Developmental Evaluation research frame, where the researcher taps into several types of data sources (e.g., research and development scientists, community members and consumers, workers/producers) on multiple occasions along the evolutionary timeline for a specific product or social innovation, but where some data gathering strategies and/or types of data sources have greater priority of focus for the researcher.

Longitudinal–Sequential Hybrid Configuration

The Longitudinal–Sequential hybrid configuration involves different MU implementations at specific points along a timeline, perhaps to track the evolution of a change process with increasing depth or breadth of focus. The advantage of this hybrid configuration is that it facilitates the progressive inward (or outward) focusing on some phenomenon of interest, using what is learned from previous stages to influence what is sought after or undertaken in subsequent stages. Two examples could be (a) where the researcher carries out a quantitative choice experiment with a sample of consumers to determine the most important features to include in a new product that a manufacturing company plans to develop (MU_1) followed by conducting focused interviews with R&D scientists who then develop that product (after receiving feedback on what was learned from the choice experiment) (MU_2) followed by administering a short quantitative questionnaire to a large sample of consumers to assess their satisfaction with the product and its features after its release (MU_3); or (b) where the researcher carries out general participant observations within a specific organisation (MU_1) followed by semi-structured interviews with key people within that organisation (where the focus/content of each interview is suggested by analyses of the earlier participant observations) (MU_2) followed by more focused observations of specific events and interpersonal interactions within the company (MU_3) (a common type of research trajectory for grounded theory research, guided by an interpretivist/constructivist pattern of assumptions).

The Longitudinal–Sequential hybrid configuration may be useful in the following research frames:

- The Action research frame, where the researcher, in conjunction with a focal group or groups, employs a distinct MU for every stage of the action learning cycle to diagnose (e.g., focus group discussion), plan and implement (e.g., participant observations), evaluate (e.g., quasi-experiment) and learn (focus group discussion) about a change process or problem solution;
- The Evaluation research frame, where the researcher carries out a combined formative/summative evaluation of a new training program, employing different

MUs at different stages of the evaluation process (e.g., focused interviews with trainers and potential trainees prior to final program design and implementation; quantitative measurement of key desired outcomes for trainees from the program immediately prior to its commencement (pre-testing); measurement of milestone desired outcomes for trainees midway through the program (intermediate testing); measurement of final desired outcomes at conclusion of the program (post-testing); focused interviews with trainers and trainees reflecting on the program after its conclusion; and measurement of desired outcomes six months after program completion (follow-up testing to assess program impact duration);
- The Developmental Evaluation research frame, where the researcher undertakes an innovation- or product-focused research trajectory like example (a) above; or
- The Cross-Cultural (cross-cultural learning emphasis) research frame, where the researcher employs several different but linked MUs intended to facilitate a deeper understanding of a specific cultural context through a period of change and, with appropriate feedforward from earlier MUs, provide a foundation for one or more subsequent MUs (e.g., conducting qualitative semi-structured interviews with key people in the focal cultural group early in the change process where what is learned influences the nature and content of a quantitative questionnaire intended for a larger sample from that cultural context focusing on the change process itself and where what is learned from the questionnaires helps to identify which participants to interview and what to focus on in those interviews after completion of the change process). [Note that a similar configuration logic could be applied within the Indigenous/First Nations or Othering/Emancipatory research frames.]

5.3.9 Evolutionary Configurations

An evolutionary configuration is more complex and open-ended configuration that employs a planned initial MU during which, through which, or as a consequence of, a critical leverage/choice point emerges for undertaking one or more additional unplanned MUs (Creswell & Plano-Clark, 2018, Chap. 3, refer to such configurations as 'emergent'). The term 'evolutionary' is intended to suggest that the research configuration evolves as circumstances and complexities dictate, meaning that the researcher maintains maximal flexibility and adaptability in tailoring the configuration to emergent research needs, constraints, and opportunities.

Some advantages of an evolutionary configuration include:

- it provides maximal research flexibility and agility in that signals and new directions that emerge from the first planned stage can be followed as appropriate in subsequent stages;
- it can take advantage of emerging opportunities such as tapping into new types of data sources that become available during the course of the research;

- it can adapt to emerging constraints or changing circumstances, such as an organisation or group withdrawing its participation part through the research; and/or
- it is responsive to/dependent upon learning in that what has been learned during the first MU helps the researcher to decide what to implement as the next MU.

Some disadvantages of an evolutionary configuration include:

- it presents greater challenges and uncertainties associated with planning and coordinating the implementation of the MU components, since only the first MU is scoped and planned at the outset;
- it may create a greater sense of uncertainty for the researcher, knowing that their entire research journey cannot be fully planned;
- it creates a greater dependence upon pluralist systemic thinking in order to more effectively conceptualise and capitalise upon a research pathway that will maximise the researcher's learning and chances of achieving a convincing outcome; and/or
- it may be more challenging to integrate learning holistically across the evolutionary configuration, because learning from the first MU could overly narrow or constrain the focus for planning the next MU (i.e., premature convergence).

There are two types of evolutionary configuration discussed below: the Adaptive configuration and the Developmental configuration. What influences the choice of configuration is the source of pressures that create the leverage/choice point, namely whether the choice point emerges predominantly as a result of external influences on the research journey or predominantly from learning from previous stages of the journey (i.e., internal influences). The former situation is reflected in the Adaptive configuration and the latter situation is reflected in the Developmental configuration.

Adaptive Configuration

In the Adaptive configuration, a MU leverage/choice point is created as a consequence of an unanticipated opportunity, constraint, situation, event, or new data source that the researcher becomes aware of. Here, predominantly external forces create the leverage/choice point which offers multiple optional pathways to travel down, and the researcher navigates that choice point by considering what would best fit their research needs and resource capabilities. Such flexibility is consistent with the concept of bricolage (see, e.g., Kincheloe, 2005) and embeds a pluralist systemic complexity perspective. Three examples would be: (a) where the researcher administers a quantitative questionnaire focusing on organisational climate to a sample of employees (MU_1), but then an opportunity arises where they can access and interview some ex-employees about their perceptions of the climate of the organisation they have just left (MU_2), which strikes the researcher as more important than interviewing senior managers within the organisation (MU_3, not implemented); (b) where the researcher is conducting climate change management research within a community suffering from drought and after conducting initial focused interviews with community members, farmers, local government managers, and research scientists (MU_1), it becomes clear that the researcher should engage in data gathering from farmers and their families, who are claiming that climate change is affecting their access to water, because their needs are more immediate and urgent to address (MU_2) than other types of data sources (MU_3, not implemented); or (c) where the researcher, while conducting preliminary qualitative interviews with staff in a school (MU_1), becomes aware of two upcoming events in that school, a staff meeting and a student awards presentation evening, and decides to conduct participant observations of the awards presentation evening (MU_3), because it could potentially lead to a wider range of learning opportunities from teachers, students and parents than gathering data at a staff meeting (MU_2, not implemented).

The Adaptive evolutionary configuration may be useful in a several research frames:

- The Exploratory research frame, where the researcher, going into an unknown or little known area of research or situation, carries out an initial planned stage of data gathering, following which unanticipated opportunities emerge which create a critical leverage/choice point offering several possible pathways for the researcher to take and explore next (here the researcher must consider which option offers the greatest promise for expanding/extending the preliminary knowledge base being assembled as well as meeting stakeholder needs and try to avoid those pathways likely to be blind alleys, tangential, or of minimal benefit);
- The Developmental Evaluation research frame, where the researcher implements an initial planned stage of data gathering, focusing on innovation in complex contexts and circumstances, and the resulting learning coupled with larger contextual dynamics (reflecting the confluence of internal and external forces) creates a complex leverage/choice point offering several pathways for where, how and on whom to focus next (this configuration is largely consistent with Patton's, 2011, conceptualisation of developmental evaluation and is often the way things play out in practice, but this configuration is accompanied by a greater sense of uncertainty in terms of which pathway would be most productive to pursue and with whose interests such pursuit would be most congruent);
- The Pan-disciplinary research frame, where the researcher implements an initial planned stage of data gathering with key stakeholders in a specific issue or problem and the complex nexus of circumstances and competing stakeholder expectations creates a critical leverage/choice point offering multiple possible pathways to follow with subsequent research activities (here the researcher must attend to many voices and sources of input, while avoiding disciplinary constraints/expectations) in order to choose the most productive pathway to follow; or
- The Indigenous/First Nations research frame, where the researcher implements a participatory focus group-type of data gathering process with community Elders to learn about the important traditions, issues, and concerns of the community and then negotiates with those Elders to gain access to and carry out either semi-structured interviews or participant observations within the community to see how those traditions, issues, and concerns play out over time, under the condition that a

tribal Elder must accompany the researcher at all times while in their community gathering data.

Developmental Configuration

In the Developmental configuration, an MU leverage/choice point emerges as a consequence of what was learned from MU_1. Here, predominantly internal forces create the leverage/choice point, and the researcher navigates that choice point by considering what would best suit the evolving trajectory of their research needs while providing the best opportunities for further learning. An example would be where the researcher commences their investigation with participant observations in a community and, following on from their preliminary analyses and interpretations of those observations, decides to pursue a set of semi-structured qualitative interviews with key people, identified during the participant observation phase, where the intention of the interviews is (a) to test out some of the preliminary interpretations from the participant observation MU and (b) probe for additional information that would help to flesh out and extend those interpretations and provide guidance for subsequent data gathering activities (a grounded theory approach).

The Developmental evolutionary configuration may be useful in a several research frames:

- The Developmental Evaluation research frame, where the researcher implements an initial planned stage of data gathering, focusing on innovation in complex contexts and circumstances, and the resulting learning creates a critical leverage point for where, how and on whom to most productively focus on next (this configuration is most consistent with Patton's, 2011, conceptualisation of developmental evaluation where downstream MU choices depend upon earlier MU choices and what they have yielded and in what contexts—building up a coherent chain of logic);
- The Explanatory research frame, where the researcher implements an initial planned MU with the intention of building up an explanatory theoretical account (i.e., a grounded theory approach) but refrains from planning subsequent research activities until preliminary interpretations from this initial MU have been achieved, whereupon they make a choice as to how best to leverage opportunities to build upon and extend that learning in terms of theoretical sampling and/or more selective or focused data gathering and context choices (Charmaz, 2014); or
- The Action research frame, where what the researcher has learned during the initial planned action learning cycle creates a critical leverage point that motivates changes in direction, strategies, and focus of the research for another action learning cycle (navigating this critical leverage point usually involves a group decision-making process).

5.4 Implementation Considerations for Navigating the 'Data Triangle'

Recall that Fig. 5.2 depicted a more elaborated view the relationships between scoping, configuring, and implementing activities, where each vertex of the 'Data Triangle' had a number of considerations associated with implementing choices for that vertex. We explore those considerations and their implications in more depth here.

5.4.1 Implementation Considerations for Data Source Strategies

There are several critical considerations associated with data source choice strategies that researchers need to effectively manage. These considerations are generally relevant regardless of the particular data source choice strategy being implemented. However, they take more specific and locally relevant shape once a particular data source choice strategy (or strategies) has been adopted by the researcher. The concept map presented in Fig. 5.5 unpacks these considerations and serves as a stepping off point for further exploring their implications for the researcher's journey. Note that, in Fig. 5.5, the researcher's contextualisations for the research, including importantly, pattern(s) of guiding assumptions, frames, and research questions/propositions/hypotheses, inform the implementation processes for data source strategies. As well, the success/usefulness of realised data source choices is judged by reflecting back on those contextualisations. They do this by influencing the criteria the researcher uses to make decisions about which specific data source strategies to employ and those decisions then signal what the researcher needs to do in terms of implementation processes. Those implementation processes include strategies for accessing and engaging with data sources, managing interactions with data sources, and ensuring ethical safeguards, each of which shares dynamic relationships with aspects of the others. There are a number of potential choice criteria that researchers can employ to influence the nature of the data source choice strategy they employ while simultaneously addressing the question of what counts as a 'good' sample. Note here that 'good' is always defined relative to the criterion or criteria used to drive the data source choice process and how and why specific data sources are chosen will always be a precursor to convincingness.

Criteria for Population-Oriented Data Source Choice Strategies

Statistical Efficiency = Precision

This choice criterion focuses on achieving a sample of a size sufficient to provide statistical information about a population within a specified margin of error. Statisticians and polling organisations tend to focus on the precision of their estimates in

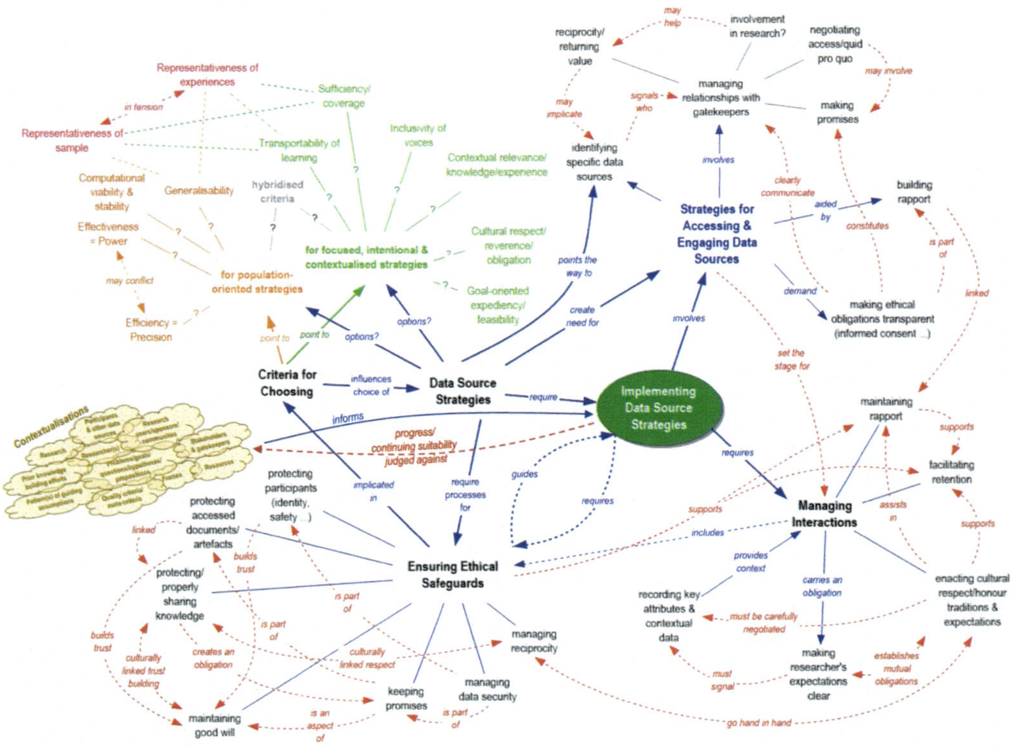

Fig. 5.5 Issues, considerations, and dynamics associated with implementing strategies for choosing/connecting/engaging with data sources

their sampling logic. The focus on precision involves generalising statistical information from sample to a population. In terms of planning research, this translates into a decision about the size and nature of the sample required. The basic question is how to achieve the desired precision most efficiently. [Note that it is important to differentiate between the population of interest, which is the ultimate target of a researcher's generalisations (e.g., all employees who have ever worked in a specific industry) and a sampling frame (see Schaeffer et al., 2012), which is that subset of a population that the researcher has practical access to for drawing their sample (e.g., all employees currently working in a specific industry in a specific city). Technically, one samples from a sampling frame, not from the population at large.]

The converse of precision in statistical estimation is error (i.e., more precision equals less error and vice-versa) and to increase precision, we focus on reducing errors. Greater statistical precision is generally easier to achieve if a random sampling process is used to select the data sources from which data are gathered. Random sampling enhances precision by minimising systematic (i.e., potentially explainable and controllable) errors and ensures that only the rules of statistical probability influence who ends up in the sample. Thus, random sampling is a form of control procedure that has the goal of trying to ensure that any characteristics of data sources and any researcher biases cannot influence the sampling process. The outcome is that statistical estimates are more precise. Computational procedures, often with the aid of a computer support system, are used to estimate the size of sample required to achieve a certain level of precision with a specific level of confidence, but their use often requires some intelligent simplifying guesses about the variability in measurements. To simplify things, some authors (e.g., Bartlett et al., 2001; Cohen, Manion & Morrison, 2011, p. 147; Gill & Johnson, 2010, p. 130; Saunders, Lewis, & Thornhill, 2012, p. 266) have produced tables and simplified formulas to help with selecting a desired sample size.

Statistical Effectiveness = Power

Choosing data sources in pursuit of precision is most appropriate if the researcher's interest is in the precision of point estimates (statistic to parameter, especially important in polls). However, if the researcher's intention is to demonstrate or test for the existence of a relationship (e.g., group differences, correlation, prediction; generically termed an effect), then the precision criterion is generally not adequate, especially for stakeholders who may be interested in or wish to use/apply the research findings (such as doctors interested in the outcomes of a clinical drug trial). When demonstrating or testing for a relationship, the researcher is more interested in having a high likelihood of finding evidence for their hypothesised relationship, using sample information, if that relationship truly exists in the population. This is the question of power, and it changes the focus from efficiency in estimation (has one achieved a precise enough estimate?) to effectiveness in estimation (has one found a genuine relationship?). Behavioural and medical researchers tend to focus on power in their sampling logic.

The power focus is still predicated on the use of random sampling, but we now look at setting sample size at a level that will provide enough power to detect an effect of a desired size. A useful metaphor for statistical power is a microscope. The researcher will not be able to see details in an object of a certain size unless their microscope has enough magnification (i.e., enough power) for resolving details of that size. Generally speaking, progressively larger samples are required to have enough power to detect progressively smaller effect sizes. For greater depth of discussion of statistical power and its implications for research gathering quantitative data, see Cohen (1988; considered the classic text on the subject). Computational procedures, often with the aid of a computer support system, are used to make power and sample size calculations (e.g., *G*power*; *SYSTAT*; *SPSS SamplePower*; *PASS*; *Stata*; for online power and sample size calculators, see http://powerandsamplesize.com/). The use of such systems generally requires an intelligent guess about the effect size one wants to detect. A reasonable guess may be gleaned from prior research, stakeholder interests, and/or researcher hunches/expectations. It has now become a routine expectation that researchers use power analysis to make informed choices about sample sizes in their proposals to funding organisations and research ethics committees, especially for health and medical research. Some authors (e.g., Cohen, 1988, 1992; Smithson, 2000, p. 207) have produced simplified tables for roughly establishing the desirable random sample size assuming a specific confidence level and size of effect to be detected.

One interesting dilemma created by the power criterion is that it is possible to have too much power conferred by very large sample sizes. What this means is that with a large enough sample, even a miniscule effect will be detected as significant, even though, in practical terms, the effect is virtually meaningless if someone plans to make policy or program changes, invest money, or adopt an innovation based on the research outcomes. The dilemma hinges on the distinction between statistical significance (i.e., inferring that a difference or relationship exists in the population from which the sample was drawn) and practical utility (i.e., inferring that the observed difference or relationship is large enough to do something useful with). When samples get very large, statistical significance wins out over practical utility and this can have downstream adverse implications (including increased costs or wasted resources) for certain stakeholders who might wish to make social policy changes, invest in or advocate adoption of an innovation based on what the researcher has found. Statistical significance does not equate with utility. One can have a statistically significant finding that is either practically useful (best case scenario, showing a reasonable effect size) or practically useless (worst case scenario, showing a very small effect size). Conversely, however, one cannot have a practically useful finding that is statistically insignificant. If the practical utility of one's research findings using quantitative data is important, then the power criterion is to be preferred over the precision criterion in sampling.

Statistical and Computational Viability and Stability

For more complex research configurations involving the quantitative measurement and evaluation of relationships between multiple constructs (typically referred to as multivariate research), social and behavioural researchers typically need to shift their data source choice criterion from precision or power toward a criterion oriented toward maximising the viability and stability of statistical estimates. Accordingly, the size of a random sample is set at a level that will more likely yield stable statistical estimates and viable solutions for the models being tested. In some instances, if the sample size is too small, the analytical procedure being used may not be able to achieve a viable or interpretable solution. This means that sample size considerations are driven equally by power or precision concerns (which still play a part in the larger scheme of things) as well as by rules of thumb or heuristics that signal a desirable ratio of cases/participants to variables/predictors in the model being tested (e.g., 5 to 1, 10 to 1 or 20 to 1). Hair et al. (2010) suggested a heuristic minimum of a 10 to 1 ratio of cases/participants to variables for many types of multivariate analyses (including multiple regression, factor analysis, and structural equation modelling) and a preferred ratio of at least 20 to 1. van Voorhis and Morgan (2007) suggested that for group-based comparisons, the minimum sample size for each group should be 30; for regression and correlation analyses, sample size should be based on a 10 to 1 ratio of cases/participants to predictors; and for factor analyses, a sample size of at least 300 would be considered "good". In general, the safest heuristic to use is a minimum ratio of 20 to 1 for any research involving modelling with multiple constructs.

Representativeness of Sample

This criterion shifts the emphasis from sample size to sample composition. A sample can be considered representative of some population of interest if it is statistically comparable to that population in terms of the distribution of all relevant characteristics of participants. What counts as a relevant characteristic is one that (1) can potentially influence the patterns of measurements that are obtained, over and above those variables that the researcher explicitly wishes to examine, or (2) imposes constraints on the extent of permissible generalisations, which relates to the external validity quality criterion associated with the pattern of guiding assumptions for a positivist inquiry. In other words, seeking a representative sample constitutes a form of procedural control, where appropriate sample composition helps to control for the influence of certain alternative plausible explanations for observed patterns of results or for certain limits on generalisation. Interestingly, while random sampling is ideally intended to produce a representative sample, it does not always do so. By chance alone, an unrepresentative sample can result from a random sampling process. This is the primary reason why alternative sampling schemes such as stratified random sampling or the quota strategy have been developed. They are intended to enhance sample representativeness by requiring the researcher to sample separately within each relevant category of population member. The more representative the sample, the stronger the case for generalisability. It is important to note that this criterion may also be useful/relevant for focused, intentional, and contextualised data source choice strategies (see also the sufficiency/coverage and inclusivity of voices criteria discussed below). However, where qualitative data are gathered, representativeness of sample will be decided on contextual grounds (such as achieving a sample composition that will provide for meaningful compare and contrast discussions or ensuring that minority or disenfranchised voices are included) rather than statistical grounds, a logic related to the sufficiency quality criterion associated with the pattern of guiding assumptions for an interpretivist/constructivist inquiry.

Representativeness of Experiences

This consideration is specific to research involving human participants and to data gathering strategies that either provide researcher-manipulated experiences for people or focus on their naturally-occurring experiences in some context of interest. It shifts the representativeness focus from the sampling of participants to the sampling of experiences (i.e., tasks, conditions, or situations) that sample members will experience during their participation in the research, which relates to the external validity quality criterion associated with the pattern of guiding assumptions for a positivist inquiry. This is an especially important issue to consider in laboratory-based research and in research intended to impose control over context in field settings. Brunswik (1952, 1956; see also Hammond & Wascoe, 1980, for discussion and examples) persuasively argued that researchers, who offer structured experiences for people during the course of their research, should design the tasks, conditions, and situations under which research participants will be measured to be representative of (i.e. similar to, in all relevant features) those tasks, conditions, and situations the participants would likely encounter in their normal life or work circumstances. Such tasks/conditions could include: decisions to make (e.g., which job applicants to make redundant, hire, or promote), problems to solve (e.g., diagnosing medical illnesses; assessing the threat level of an unidentified aircraft or ship), evaluations, judgments or forecasts to make (e.g., grading students' homework, forecasting severe hail or wind shear at airports), preferences to rate (e.g., preferred products to purchase or apartments to rent), work tasks to complete (e.g., performance appraisals), tasks conducted under stressful conditions (e.g., flying an aircraft through severe weather or during an emergency event; identifying enemy aircraft by radar under battle conditions) or tasks with deadlines or with conflicting, inaccurate or missing information (e.g., deciding upon guilt or innocence of a defendant, where testimonies from witnesses conflict).

Egon Brunswik called this principle 'representative design' and argued that it was as important—and in many cases, more important—for researchers to focus on achieving this than a representative sample. This is because generalisations from highly-controlled tasks, undertaken under artificially manipulated conditions (as in a laboratory experiment, undertaken where the Manipulated experience-focused data gathering strategy was employed), to the 'real world', with all its attendant complexities and messiness, in which participants must live and work are very hard to defend. This means that representativeness of samples and representativeness of experiences operate in tension with each other, where the relevant balance between them must be judged according to local contextual circumstances. The central philosophy behind representative design is understanding how each individual person behaves under conditions and in situations that approximate what they would encounter in their normal life space (giving rise to what Brunswik, 1952 termed the 'idiographic' perspective) before attempting to generalise behavioural patterns to other individuals (what Brunswik, 1952 identified as the 'nomothetic' perspective). Representative design thus focuses on the meaningfulness and relevance of the experiences that sample members undergo or encounter during their participation in the research. There are two ways to implement the principles of representative design: directly sample the sorts of tasks participants confront in their lives to use in the research (a holistic strategy) or create one or more simulated tasks that exhibit all the formal characteristics of tasks that participants confront in their lives (a decompositional strategy). It is important to note that this criterion may also be useful/relevant for focused, intentional, and contextualised data source choice strategies (see also the sufficiency/coverage and contextual relevance/knowledge/experience criteria discussed below). However, where qualitative data are gathered, representativeness of experiences will be decided on contextual relevance grounds (e.g., choosing from experiences that participants naturally encounter in the research context of interest) a logic that is related to both the authenticity and sufficiency quality criteria associated with the pattern of guiding assumptions for an interpretivist/constructivist inquiry. In this situation, representativeness of experiences could be enhanced by attending to the range of experiences that participants have in some context and making choices for data gathering that would maximise exposure to the widest range of those experiences.

Criteria for Focused, Intentional and Contextualised Data Source Choice Strategies

Sufficiency/Coverage

Under certain patterns of guiding assumptions where a more subjectivist stance is adopted, representative samples and representative design do not carry the same meaning or implications. This is partly because generalisations to a population are not typically sought in such research, and partly because such research typically does not employ the deliberate manipulation of tasks and conditions as a strategy for data gathering. Instead, the most desirable sample is one that affords the researcher access to sufficient data upon which to base their interpretations and conclusions. The most desirable observations are those that are made under natural environmental and social circumstances, with all the messiness that may imply. However, it is important to realise that participants are not the only type of sampled entities relevant to sufficiency. Sufficiency/coverage asks whether the researcher has sampled enough participants, handiworks, artefacts, documents, experiences, times, events, and contexts to develop, anchor, and sustain the interpretations and conclusions that emerge from the data. Longer immersion in a naturalistic context will generally yield samples affording greater sufficiency/coverage.

Inclusivity of Voices

This criterion explicitly focuses on sample composition by attempting to ensure that the widest possible variety of voices is engaged with by the researcher. This is a criterion explicitly focus on achieving diversity in data sources. Enhancing inclusivity requires the researcher to remain open as to what they may encounter and learn from data sources, even if what emerges or is expected to emerge runs counter to or is harshly critical of 'mainstream' or majority perspectives (i.e., seeking alternative, marginalised, ignored, suppressed, imprisoned, or silenced voices as well as 'mainstream' or majority voices). Employing the inclusivity of voices data source choice criterion can help the researcher to achieve a more fully representative sample. However, it also implies the researcher knows, at the outset, that they may not be able to avoid having to deal with political, legal, or cultural fallout that may ensue from their inclusivity choices. In short, employing the inclusivity of voices choice criterion deliberately overlays an 'fairness' or 'equitable hearing' value-position on the research, which some stakeholders may find objectionable.

Contextual Relevance/Knowledge/Experience

This choice criterion focuses on sample composition, where the goal becomes ensuring that the most appropriate people and other data sources (i.e., those with, or offering information pertaining to, contextually relevant knowledge/experience/roles/situations) are included in the researcher's sample. The goal of this choice criterion is to maximise the researcher's contextualised learning. It may also help enhance or facilitate access to other desired data sources and contexts, as they become known to the researcher. What may count as contextual relevant knowledge and/or experiences may be identified by the researcher or by other stakeholders in the research.

Transportability of Learning

Transportability of learning refers to whether the interpretations and conclusions that have emerged from the researcher's current study, in its context with the data sources that they have sampled, might have meaning for or be relevant to other contexts. It is especially of potential relevance to research that has a case-oriented component (i.e., a deeper focus on one or more specific contextual places/instances/events/occasions). While it is easy to confuse transportability with the concept of generalisation (which focuses on generalising from a sample to a relevant population), it is not the same thing. Note, however, that both sample and experience representativeness do contribute to generalisability as well as transportability considerations, just in different ways. Whereas both sample and experience representativeness are more determinative for generalisability, they simply provide contextual background for transportability. Transportability asks 'does what I learn in one research context have meaning for any other contexts?' Not every study that has a case-oriented component has transportability as a goal, but if it is a goal, data sources choices must be such that they would facilitate the convincing achievement of cross-contextual meaning and relevance. Cross-site participant observation studies, for example, achieve transportability through sampling in each context and then cross-comparing the interpretations that emerge. One prerequisite for transportability is a deep focus on local contextual features and nuances, as these will either facilitate or inhibit cross-context transportation of meanings.

Goal-oriented Expediency/Feasibility

This criterion shifts the focus from statistical efficiency/effectiveness with respect to sample size and/or representativeness with respect to sample composition to what is expedient/feasible for the researcher to access in building up their sample in order to achieve their research goals. The need to use this criterion is often driven by the researcher having to achieve their goals within severe resource, geographical, or data source access constraints. It may also provide the most feasible pathway for gaining access to willing participants (online research participant pools may provide one such pathway). It is important to understand that the expediency/feasibility criterion generally means that the researcher deliberately sacrifices the potential to generalise what they learn to some population of interest. It may, in fact, be the result of a trade-off decision. It may also adversely impact on transportability of learning. In short, the expediency/feasibility criterion is generally not a desirable first option for selecting a data source choice strategy but might emerge as the only viable option under highly constrained circumstances.

Cultural Respect/Reverence/Obligation

For certain research projects and contexts (e.g., cross-cultural or Indigenous/First Nations contexts), there may be cultural expectations regarding who the researcher should connect or engage with, whether in the role of a specific data source or for obtaining more contextually-sensitive knowledge and, perhaps, even access to other data sources. The emphasis on the cultural dimension distinguishes this criterion from the more general contextual relevance/knowledge/experience criterion. In such research contexts, it would be appropriate for the researcher to explicitly display respect and reverence for cultural expectations by specifically choosing to connect/engage with certain individuals or groups (e.g., tribal or community Elders/leaders or councils, political leaders, religious leaders). In some cases, this may be imposed as an obligation, on the part of the researcher, for gaining access to culturally-privileged land/country, knowledge, and experiences. Here, we can see that, in many instances, this choice criterion operates hierarchically, constituting a primary condition of cultural access that should or must be met before other choice criteria can be imposed for selecting data sources. It should also be noted that in certain circumstances where a government acts in a cultural guardianship role, the focus on cultural respect/reverence/obligation may be largely or wholly inseparable from political obligations.

Hybridised Choice Criteria

Data source choice criteria can be creatively hybridised to gain additional advantages over what reliance on a single choice criterion could achieve or to overcome or mitigate the disadvantages of one criterion using the strengths of another criterion. For example, one could focus on precision or power while, at the same time, focusing on representativeness of sample in political polling research. One could focus, at a localised level, on statistical efficiency or effectiveness combined with a focus on expediency in field research. A focus on representativeness of sample could be combined with representativeness of experiences to provide more fully generalisable learning from laboratory-based research. A focus on representativeness of experiences could be combined with a focus on contextual relevance/knowledge/experience and inclusivity of voices in qualitative interview research. The contextual relevance/knowledge/experience criterion could be synergistically combined with the cultural respect/reverence/obligation criterion in Indigenous/First Nations research.

Strategies for Accessing and Engaging Data Sources

Once one or more data source strategies have been selected by the researcher, implementation requires that concrete strategies for accessing and engaging data sources be developed. No matter which data source choice strategy is implemented, the researcher must take steps to ensure that all relevant gatekeepers are identified, and processes put in place to effectively and authentically (i.e., with no deception or false promises) manage the researcher-gatekeeper relationship. A gatekeeper may occupy a social or cultural role (e.g., parents, community leaders, Elders in an Indigenous or First Nations community), an institutional/organisational role (e.g., supervisors, principal of a school, CEO or Manager of Human Resources, manager of historical archives), or a regulatory role (e.g., government department or Minister, ethics review committee). From the researcher's perspective, they want access to the data sources (whether human or non-human) over which the gatekeeper has 'control'; from the gatekeeper's perspective, they want to ensure that, if they provide such access, there will be no blowback, harm, or other adverse outcomes for themselves and/or the data sources or for the larger context within which the gatekeeper operates. If data sources are non-human (e.g., documents, emails, minutes of meetings, spreadsheets/databases, websites) and not publicly accessible, then they can be accessed only with the permission of a relevant gatekeeper. If data sources are human, then there could be one or more layers of gatekeepers to pass. In some instances, the researcher may directly approach people, either face-to-face or via some medium or technological pathway (such as online), to enlist their participation. In other instances, one or more gatekeepers may constitute hurdles the researcher must pass before individuals are approached. In short, approaching individuals about their potential participation will always be the final step in the process of gaining access.

Regardless of whether data sources are human or non-human, effective management of interactions with gatekeepers is the key to gaining access. The researcher's interactions with gatekeepers must be genuine and transparent, with respect to the researcher's goals and intentions. The relationship with each gatekeeper may take a bit of time to build sufficient rapport and trust before a request for access can be made. The researcher may be asked by a relevant gatekeeper to consider pathways for returning value/ensuring reciprocity to the gatekeeper's group, community, or organisation, something that is critically important where Indigenous/First Nations people are concerned. At a gatekeeper's request, reciprocity may commence by involving the gatekeeper or other stakeholders in the research itself. A gatekeeper may ask for a quid pro quo in exchange for granting access. This is not the same thing as reciprocity, which creates an expectation that the researcher will provide some benefit from the research back to the group, community, or organisation that the gatekeeper represents or is a part of. A negotiated quid pro quo often has a more personal benefit to the gatekeeper as its focus and some requests may create ethical dilemmas for the researcher, if those requests potentially threaten the rights of individual participants. No matter what avenue of access is pursued by the researcher, the result will effectively be a promise made to each gatekeeper to honour any agreement reached for granting access. In making such promises, the researcher is obligated to make clear what ethical safeguards they will be providing to individual participants.

Some gatekeepers may adopt a simple protective stance with respect to the data sources in their charge (call them 'protectionists'), in which case, the researcher's main objective is to provide clear information about why their access to data sources is required for their research, signal willingness to comply with any constraints or restrictions the gatekeeper might wish to impose conditional upon providing access, and to offer a set of guarantees/promises regarding actions they will take to ensure data source protection (which may need to extend outward into the community in which they live or belong to) and freedom from adverse outcomes. Other gatekeepers may wish to get something they want in return for their provision of access, thereby creating a political scenario (call them 'instrumentalists'), rather than a purely protectionist focus. In these circumstances, in addition to implementing the same actions as for protectionist gatekeepers, the researcher must consider what they are willing to agree to with respect to what the gatekeeper wants in return for providing access. For example, a gatekeeper in an organisation may ask for a final report on the research to be presented to the organisation, may ask to be told the identities of data sources they provide access to and what information they divulge, or may request that a fee or some other consideration be paid in exchange for providing access. The researcher must consider carefully and in full light of their ethical obligations what promises they make to an instrumentalist gatekeeper; backing out of a promise downstream is generally a destructive option. The researcher's early interactions with a gatekeeper will usually signal whether they are protectionist or instrumentalist in orientation and can adapt their management strategies accordingly.

A unique situation arises where human participants are desired as data sources. While they may be subject to the say so of specific gatekeepers within their organisation, institution, community, or culture/society, every human above the age of consent serves as their own gatekeeper (parents or guardians become gatekeepers for their children or adolescents). This later situation arises because of the researcher's ethical obligation to obtain informed consent from each participant (which encompasses the right to withdraw their participation at any time, even after giving initial consent). The researcher must spend time, effort, and, perhaps, monetary resources to gain their individual agreement to participate. However, the protectionist ('will I be safe?; could I lose my job?) and instrumentalist ('what do I get in return for my participation?) distinctions remain relevant, becoming the drivers for such strategies as promising anonymity/confidentiality, offering payment or some other valued outcome in exchange for participation, or getting access, at a later date, to something that the research shows to be beneficial (e.g., a specific treatment or training regime). Furthermore, in order to avoid the situation where participants who consent later withdraw their consent, the researcher must be cognisant of how participants' experiences during the research might inadvertently stimulate a downstream need for participants to withdraw and take steps to minimise this risk. This may, in turn, create the need for the researcher to invest time and energy in establishing some level of rapport with participants to enhance their commitment to stay the course with their involvement in the research. The researcher needs to ensure that whatever they do or say to participants does not cause the withdrawal of their participation, which means that interactions with participants, whether a formal part of research activities or not, must be managed with sensitivity to their circumstances and respect for their rights. The bottom line here is that, because human participants are their own last gatekeeper, they effectively have control over the nature, shape, and implementation of the researcher's data source choice strategy. The right to informed consent and the right to withdraw at any time supersedes any demands that the researcher's specific data source choice strategy might impose (e.g., may make a random sampling process non-random; may render a purposive strategy incomplete; may inadvertently create a negative case) and may therefore work counter to what a specific choice criterion might be seeking (e.g., representativeness of sample, power, inclusivity of voices, sufficiency/coverage).

Managing Interactions

Managing interactions, beyond the boundaries of data gathering activities (such as interviews, laboratory tasks, questionnaires, or activities where handiworks are produced by participants), during the implementation of research activities means keeping the lines of communication open with participants. This allows the researcher to gain some insights into their thoughts and experiences while participating. Even in highly controlled experimental research, under the pattern of guiding assumptions for a positivist inquiry, the researcher should be alive to the perceptions and reactions of participants with respect to what is happening to/with them during the research. Through such communications, the researcher may become aware of adverse impacts of ongoing research participation or of participants' inability to carry out research tasks in the way they were intended and can then make adaptive changes or trade-offs to ameliorate or remove those impacts. Here, pluralist systemic thinking can play an important role in ensuring that the widest possible array of potential adaptations is considered, whilst keeping convincingness and feasibility in mind. Managing interactions also allows the researcher to maintain rapport with participants, which builds a foundation of trust. Where appropriate, researcher-participant interactions may serve as a platform for showing cultural respect and honouring traditions and cultural expectations, especially when it comes to Indigenous/First Nations research. Maintaining trust, rapport, and respect for culture and traditions facilitates the participants' continued willing participation in the research, thereby reducing attrition rates and minimising problems with missing or misleading data.

In all communications with participants, the researcher has the obligation to make their expectations and ethical protections clear and to consistently demonstrate that their obligations are being met. Where different cultures are concerned (i.e., that of the researcher and those of the participants), managing interactions is one way to ensure that expectations are being met from the points of view of all cultures involved.

In some instances, interactions with gatekeepers may also be warranted to ensure that expectations are being met. Where non-human data sources are concerned, the important interactions to manage are those with the gatekeeper(s) for those sources. Gatekeepers may need periodic assurances that the information to which they have granted access is not being misused or inappropriately shared.

One important aspect of managing interactions with or about data sources is ensuring that the researcher clearly indicates what potentially identifying attributes about participants and what contextual data they will be recording in their data gathering activities as well as what protections they will place on whether, how, and in what form such data are shared. There may be cultural expectations that need negotiating as well, especially from an Indigenous/First Nations perspective, where culturally-privileged knowledge is potentially being shared. These become important considerations both from a data analysis and reporting perspective (e.g., using pseudonyms to identify sources of quotes or specific data patterns such as outliers, summarising participant attributes in an aggregate format, redactions where privileged data are blocked from access) as well as from a meta-data (i.e., descriptors and formatting of data content, descriptions of contexts of gathering, and any geo-spatial identifiers) perspective, if the researcher intends to ultimately share their databases/files with other researchers or with a data warehouse of some description. Such data can also help the researcher to contextualise their interactions with participants and gatekeepers in records such as their research journal.

Ensuring Ethical Safeguards

Ensuring ethical safeguards is mandatory requirement of any research involving human participants and/or artifacts produced by human behaviours. In many cases, where researchers are undertaking their research journeys as part of their roles in an employing organisation (e.g., university, research organisation, government office), those institutions will typically have processes (e.g., review committees, document templates, pro formas, policies, and guidelines) in place for vetting and approving the ethical safeguards the researcher intends to offer. However, ethical safeguards may also need to be vetted/approved by relevant gatekeepers in communities and organisations that will serve as or have interests in contexts for the research and those gatekeepers may be bound by their own policies as well as, perhaps, cultural expectations. Ensuring ethical safeguards creates concomitant obligations to clearly communicate those safeguards (i.e., protecting identities and privacy, guaranteeing freedom from harm (physical, work-related (e.g., employment and relationships), psychological, and cultural), protecting/controlling access to data provided and knowledge shared, and gaining informed consent and reinforcing the right to withdraw) and their implications to gatekeepers and potential participants and to guarantee/promise that such safeguards will not/cannot be violated. Such communication is essential processes for building trust as well as for supporting the establishment and maintenance of rapport between the researcher and participants and retention of participants for the duration of the research (i.e., avoidance of premature withdrawals from participating). Thus, when it comes time to implement research activities around the 'Data Triangle', the researcher must create, gain approval for, and enact all promised safeguards, develop all necessary documentation and pathways for obtaining informed consent and passing gatekeepers, and signal a clear process for handling any complaints that gatekeepers or participants might have about the researcher or research processes or for seeking assistance if something unforeseen occurs or if a safeguard breaks down in some way. Maintaining good will is a function of trust, aided by ensuring ethical safeguards, and is an emergent obligation of the researcher. It offers a dual assurance that the researcher will do everything in their power to avoid actions that could (1) cause adverse impacts in research contexts in ways that could disrupt the lives of participants and/or disturb the contexts they inhabit or (2) spoil access to such participants/contexts for future researchers. However, maintaining good will also involves transparently acknowledging that the researcher cannot foresee all possible outcomes or unintended side effects and therefore puts processes in place for handling such occurrences. If the research is to involve research assistants in data gathering activities, participants need to be formally notified of this fact, advised that these assistants will be under the direct supervision of the researcher, and assured that these assistants are accountable for upholding the same ethical responsibilities and obligations as the researcher.

Where human participants are involved, forms for obtaining informed consent from participants must be developed (universities often offer templates and guides for such documents) and those documents will need to set out exactly what the researcher expects of participants (or what will happen to them, as appropriate), invite participation (including spelling out any rewards the person could achieve by agreeing to participate, i.e., incentives), spell out all ethical safeguards the researcher promises to implement, provide contact details for the researcher, and spell out pathways for filing grievances or obtaining counselling should promised safeguards fail or unforeseen circumstances eventuate. Such documents only address the obtaining of informed consent at the individual level. The researcher may need to develop other documentation for passing gatekeepers in order to gain the opportunity to seek such individual permissions to participate.

Where non-human data sources are concerned, establishing, ensuring, and communicating ethical safeguards for appropriately accessing and acknowledging such data sources, appropriately protecting what the researcher learns from such data sources, and maintaining goodwill with the gatekeepers who provided access are essential for trust building. In such cases, documents seeking permission to access from relevant gatekeepers become critical to develop. Where data sources (human and/or non-human) are associated with or part of Indigenous/First Nations cultures, establishing, ensuring, and communicating ethical safeguards becomes part of the processes for respecting/honouring cultural traditions and knowledge, appropriately accessing, sharing (including considerations of if, how, when, by whom, and with whom), and safeguarding culturally-privileged knowledge, and managing reciprocity/returning value from the research back to the communities that participated. As part of this process, the researcher must identify the right gatekeepers to engage with, how, when, and where. In all cases, the expectations of the researcher vis-à-vis their research agenda must be clearly communicated (in some cases, negotiated) and any promises made by the researcher must be honoured in a timely fashion. It is important to remember that, with respect to Indigenous/First Nations research, formal written documentation seeking access and clarifying the researcher's goals and ethical assurances may be secondary to undertaking in-person verbal discussions with Elders and with communities, possibly over a series of conversations to allow trust relationships to develop.

Managing Data Security

Once data sources have been enlisted or accessed, the researcher must attend to keeping whatever information is revealed by, or contained within, a data source secure and protected. Nowadays, this is an even more critical concern with the push toward making social and behavioural research data openly available to other researchers (e.g., via open access research databases). Increasingly, data archiving and sharing are becoming requirements of research policy at many institutions, a policy position that remains contentious with many academic and professional researchers, especially in the social and behavioural sciences and particularly with respect to qualitative research. Data security means that the researcher has put in place steps to control access to the data they have gathered and, as a corollary to that process, taken steps to ensure that the identity of any human data sources, as well as their connection to any information they have provided, is protected or anonymised/de-identified (see, for example, discussions in Ross et al., 2018; Mauthner & Parry, 2013). Controlling access may require physical measures (e.g., a locked file cabinet in a locked room where only a few known people have access) and/or virtual measures (e.g., password-, facial recognition-, or biometrically-protected data files on a personal computer or secure server; copy protection; de-identification and, perhaps even encryption, of data). No matter how data security is ensured, human participants must be advised of these safeguards during the process of obtaining informed consent. If data are ultimately to be openly shared via some sort of data warehouse or data sharing platform, participants should be advised of this fact and should be asked to give their informed consent for this to happen. In addition, if the data are to be shared or warehoused, the researcher must carefully attend to assembling meta-data to accompany the dataset, in order to fully contextualise the data and avoid potential misinterpretation, misreading, or misuse. If data sources are non-human and not publicly available (e.g., documents, artefacts, company emails), then the appropriate gatekeeper(s) for those data sources should be provided with the assurances described above. Managing data security extends to more general social contract issues as well, so that future opportunities for social and behavioural research are not threatened or damaged. The researcher should ensure that they maintain positive and effective relationships, faith, and goodwill with gatekeepers (as well as the institutions/organisations/communities they serve) through authentic actions, so that the potential opportunities that other researchers may wish to pursue with the gatekeepers are not adversely impacted. If the research concerns Indigenous or First Nations people or communities, then the researcher must undertake to follow cultural expectations and traditions regarding the proper sharing of their knowledge and fulfilment of reciprocity obligations within the community.

5.4.2 Implementation Considerations for Data Gathering Strategies

There are a range of issues, considerations, and dynamics that emerge with each choice of data gathering strategy and these are depicted in the concept map in Fig. 5.6. Procedures for data gathering activities must be developed and formalised, decisions must be made about what and how best to organise the financial, physical, technological, logistical, and human resources needed to implement the strategy, data recording procedures must be set up, measurements and instruments, where appropriate, must be developed or adopted and validated, and cultural and other contextual adaptations may need to be made to the implementation process. All these procedures and choices should be pilot tested or trialled before actual implementation commences. While these might seem like discrete activities, they are all interconnected, and choices made in one area may have implications for what is done in other areas. Systemic thinking is useful for identifying many of these interconnections so that the researcher may take appropriate actions to address the implications. Of course, the research contextualisations will strongly inform and influence the nature and content of many of these procedural choices and their interconnections.

Procedures

Every data gathering strategy requires detailed procedures for implementation purposes. Procedural requirements will depend heavily upon the research contextualisations, including patterns of guiding assumptions and framing. The researcher will need to develop procedural documentation specific to each data gathering strategy. Some procedures will take the form of concrete instructions (delivered verbally, online, or in writing, as appropriate) to participants with respect to what they should do or are expected to do when they are, for example, completing a questionnaire, participating in an interview, keeping a diary or journal, producing a drawing, cause map, or other creative work, participating in an interview or an experiment (e.g., completing experimental tasks). Other procedures will focus on what the researcher needs to do when implementing a data gathering strategy, for example (note that when 'where appropriate' is stated, this means 'appropriate for the chosen strategy, pattern of guiding assumptions, research frame, ethical requirements, and consistent with expectations for maintaining research quality'):

- setting out, where appropriate, the protocols for conducting an experiment or quasi-experiment (e.g., manipulating conditions (e.g., what, how, when, and where to change experimental conditions and for whom), resources to be allocated);
- spelling out, where appropriate, procedures for generating artificial data (e.g., software platforms to use, assumptions made, models to implement, programming to be done, resources to be allocated);
- instituting, where appropriate, control procedures (e.g., resources to be allocated, control by measurement processes, control by design elaborations such as

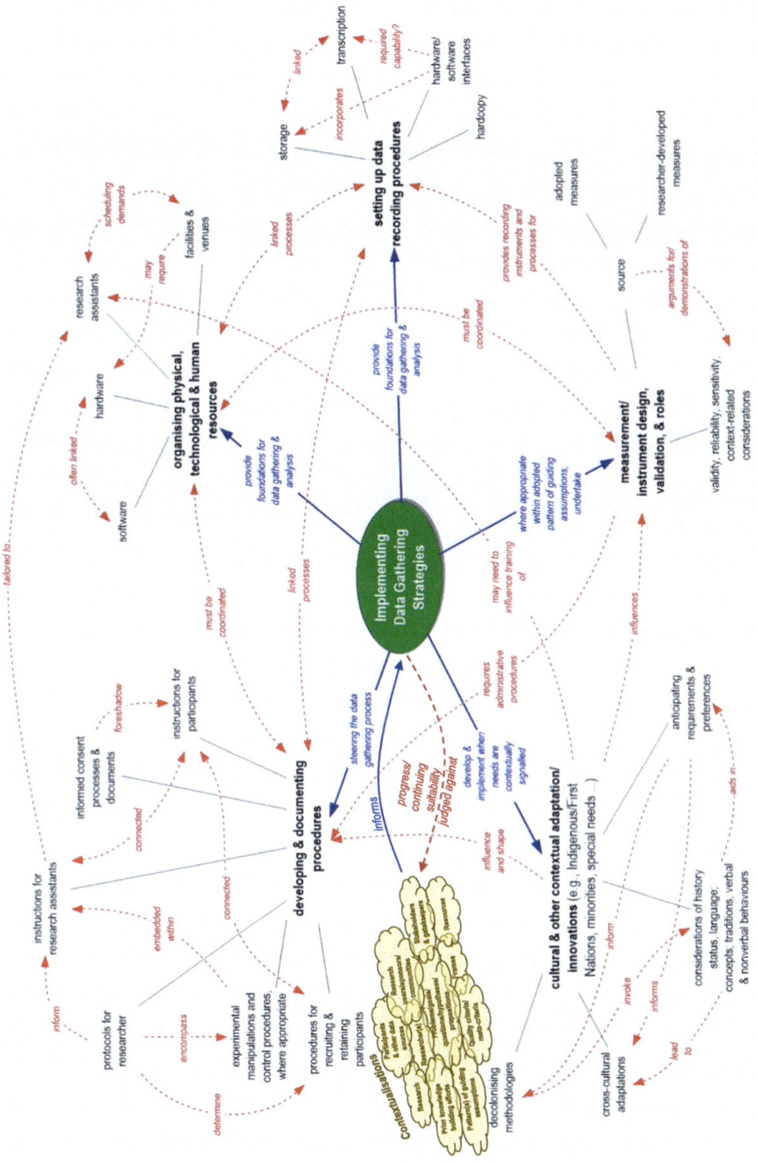

Fig. 5.6 Issues, considerations, and dynamics associated with implementing data gathering strategies

control groups and repeated measurements, procedural control through instructions (including blind and placebo conditions), control through model specification, control through statistical analysis; see more elaborated discussions in Cooksey, 2020, Fundamental Concept VII),
- setting out, where appropriate, protocols for scheduling and conducting/recording of testing sessions (e.g., what measurements to gather, when, where, and how to gather them, resources to be allocated);
- detailing, where appropriate, protocols for scheduling and conducting/recording observations (e.g., sites, occasions, recording instruments/forms, procedures for recording field notes, technological support to be used, resources to be allocated);
- detailing, where appropriate, procedures for dealing with documents or other handiworks (e.g., accessing, coding, recording/storing, using technological support, resources to be allocated); and
- setting out, as appropriate, protocols for scheduling and conducting/recording interviews, (e.g., when, where, how, physical arrangement of interview sites, using what technological support, development of scripts, guidelines for transcription, and/or topical landscapes, resources to be allocated).

If the research is to involve research assistants in data gathering activities, protocols need to be established for training those assistants, reinforcing the standards they must meet, and allocating and managing their responsibilities and activities. Instructions for research assistants need to be tailored to meet their level of experience and expertise. The use of research assistants creates additional organisational and scheduling demands for the researcher and will draw heavily upon the researcher's capacities to interact and supervise others. In cross-cultural research, research assistants may require additional training with respect to verbal and non-verbal interactions if the data gathering activity crosses cultural boundaries. For Indigenous/First Nations research, the researcher, especially if they are a member of the colonising culture, may be working with key members of the Indigenous/First Nations communities as research partners and protocols need to be established for effectively managing and respecting such relationships.

Organising Physical, Technological, and Human Resources

In addition to procedural developments, the researcher must organise all the physical, technological, and human resources needed to implement each data gathering strategy. On the physical side, the researcher may need to organise facilities required for data gathering (e.g., laboratory space and time, equipment) and, where appropriate, the venues and equipment required for data gathering events (e.g., interview rooms, furniture, equipment, and layout). For research conducted on-site in field settings, the researcher may need to negotiate with relevant gatekeepers about access to physical space and other organisational resources to support data gathering activities. Such negotiations may extend to how space is furnished, what equipment it needs to contain or have access to, how isolated it needs to be from noises and other potential distractions, and what times of day it may be accessed. On the technological side, the researcher needs to make decisions about what software and hardware they need to support their data gathering efforts; in many cases, software and computer hardware go hand in hand. This might include software to support data reading, recording, and storage (e.g., PDF readers, transcription software, spreadsheets, programs for designing online data gathering environments) and/or hardware for supporting participants in the production and storage of data (e.g., microphones, wearable data recorders, voice recorders, flipcharts, whiteboards, sticky notes, cameras) or supporting the researcher (e.g., laptop or tablet equipped with appropriate software, photocopier).

Where research assistants are to be employed to assist with data gathering activities, the researcher must organise all relevant human resource-related paperwork with their institution, schedule sessions for training, data gathering, and research team meetings (for Indigenous/First Nations research, these processes would be conducted in partnership with the relevant community/people). In a team-based research environment, the researcher frequently adopts a management role and must behave accordingly, drawing upon their social and leadership skills. If the researcher is also in a mentoring or postgraduate supervision role for a research assistant, there will need to be clear boundaries between what is expected with respect to the research project and what is more generally expected as part of the mentoring or supervisory role. Additionally, the researcher may need to be alive to the potential need to appropriately pair research assistants with the participants from which they gather data (e.g., via interviews or observations) and those pairings might need to be made according to cultural or other relevant considerations.

Data Recording

Part of the organisational and preparation necessities for implementing data gathering strategies includes setting up the procedures, platforms, and structures and acquiring resources for recording data. Data will often be gathered and stored in as close to their original form as possible, with minimal preprocessing. Such data are referred to as raw data. The researcher may then transform the raw data according to their needs downstream, but the original raw form of the data is typically preserved. However, depending on circumstances and pattern of guiding assumptions, data may sometimes be transformed prior to recording and storage (reflecting a Transformative data gathering strategy). In such cases, the raw form of the data may be lost once the transformations have been completed.

Data recording may be carried out via hardcopy instruments (such as printed questionnaires, drawings/maps on paper, hand-written field notes or journals/diaries), in which case, the raw data have a physical form that must be stored. For many types of hardcopy data gathering, participants directly produce the raw data by ticking boxes, circling numbers, writing answers to questions, making sketches, maps, or drawings, telling stories, making journal entries, and so on. The researcher's task is to make this data production process as transparent, unthreatening, and simple as possible. If printed questionnaires are employed, the researcher automatically enacts a Measurement data gathering strategy and must attend to instructions, layout, formatting, question complexity, question ordering, wording and readability, response scaling, and minimising the chances of response biases (e.g., social desirability, where participants make responses that put them in what they think is a 'good light' with respect to how society might perceive them; acquiescence, where participants tend to agree with every statement put to them, even when they don't actually agree; and counter-productive patterned responding, where participants respond according to some pattern facilitated by the design of the instrument). If maps, drawings, or other types of artwork are to be produced, the researcher must attend to instructions and provide the necessary materials. If participants tell stories orally, as is common in Indigenous/First Nations research, the researcher must negotiate with the appropriate community gatekeepers to arrive at what the most permissible and least intrusive data recording process might be. Here, the data recording processes become a collaborative effort. For data gathering strategies that rely on documents produced by others for other purposes (e.g., meta-analysis, meta-synthesis), the researcher may gather hardcopy forms of documents, in which case, filing structures and storage methods must be set up and copyright issues managed. For other types of hardcopy data gathering, the researcher produces the raw data themselves (as when recording field notes, making entries in their research journal, coding and recording observations, and so on). Here, the researcher must attend to assembling the necessary forms and materials as well as ensuring they have consistent processes in place. No matter who produces the raw data, hardcopy forms of raw data are then typically transformed (thereby enacting a Transformative data gathering strategy) via transcription, data entry into a spreadsheet, or photograph in order to facilitate electronic storage and analysis. Note that error checking is an essential process to engage in whenever raw data are transformed and stored.

Research data may also be directly recorded using some type of technological or hardware/software interface and stored electronically. As with hardcopy forms of data, we can distinguish between data produced directly by participants, data produced by the researcher, or data produced by others for other purposes. When participants produce data electronically, they typically do so via some sort of hardware/software platform (e.g., electronic diary/journal, online questionnaires and experiments, mapping or drawing software) and/or a technological device (such as a voice recorder, smart phone, tablet, camera, wearable data recorder, or custom-built device). In such cases, the researcher must attend not only to the production of the data gathering interface itself (e.g., an online questionnaire or an online experiment, usually via a formal software platform (e.g., *Qualtrics*, *SurveyMonkey*, *PsyToolkit*, *Q* (for online choice modelling experiments), *SurveySparrow*) but also to any requisite training for participants, the minimisation of errors, and ease of use. Participants may also use platforms (e.g., *X (Twitter), Facebook, Instagram, SnapChat, YouTube, TikTok ...)* to produce raw data in the form of social media posts, blogs, video segments, and so on. In these cases, the researcher will need to attend to any necessary instructions for production. When researchers produce electronic data, they may use technological devices such as voice recorders, cameras, smartphones, eye tracking monitors, or software such as online journaling programs (e.g., for recording field notes and observations) or data transcription programs (e.g., for transcribing recorded interviews) and they will need to systematise and organise how they produce such data. It is also prudent for the researcher to have backup systems in place (e.g., extra batteries, additional flashdrives, backup camera, or smartphone) in case of failure as well as double check the devices to ensure they are in working order before every data gathering event. When using electronic data are produced by others for other purposes, the researcher will need to attend to file structure and format, downloading considerations, storage medium, and potential copyright issues. In all cases, electronic data may be stored locally on the researcher's computer, stored centrally on their organisation's computer system, or stored virtually in the cloud using services offered by Google, Microsoft, and other providers. Data security must be managed for all these modes of storage so that access to the data can be controlled, as much as possible, by the researcher.

Measurement/Instrument Design, Validation, and Roles

Measurement processes are enacted within the Measurement data-shaping strategy and invoke specific data recording considerations. Measurement processes are specifically focused on producing quantitative data, using numbers to represent some observational or behavioural phenomenon, characteristic, or event, and are typically embedded in an instrument of some kind (e.g., a questionnaire, a physical device like a ruler, or a technological device, such as a meter or gauge). If the researcher intends to gather quantitative data as part of their research journey, measurement considerations must be addressed, and these would include:

- *Operational definition of constructs*: The process of building a defensible measurement system, especially for something that we cannot directly observe, commences with a clear identification of the construct to be measured. A construct is basically a concept or property theorised to describe specific differences between people. A theory places a construct (sometimes referred to as a hypothetical construct) in a larger context by showing how it is hypothesised to relate to other constructs of interest. Testing a theory therefore requires, as a first step, the design of a measurement system for every construct embedded within it. The process of translating a construct into a measurement system is called operational definition. Technically, an operational definition is defined as "spelling out, very clearly, the steps the researcher will go through to obtain concrete measures of that construct" (Cooksey, 2020, p. 24). Operational definition thus describes how observations of manifestations of the construct should be translated into numerical values.
- *Processes for assigning numbers to phenomena*: This concerns scaling decisions, which range from the simplest nominal scale (numbers relate only to categories or classifications; some consider this a qualitative representation) to an ordinal/ranking scale (where numbers reflecting an ordering against some construct of interest) to an interval rating scale (having no definable zero point, but where the intervals between numbers have meaning) to the most precise ratio scale (having a definable zero point). These are the processes that yield the actual numerical data to be recorded.

- *Measurement validity*: Validity is characteristic of measurement quality that concerns the inferences that may properly/defensibly be made from a measurement scale. The most fundamental type of validity is *construct validity*, which reflects the quality of the operational definition process. It goes to the question of whether the measurements obtained reflect the phenomenon the researcher claims they are. Construct validity may be demonstrated statistically (via analyses), empirically (via group comparisons), conceptually (using logical argument), or by reference to prior research (i.e., evidence provided by others). *Predictive* or *criterion-related validity* becomes relevant when measurements are to be used in decision-making contexts. Such validity relates two measurement systems applied to the same objects or individuals, one system that serves as predictor and one system that serves as the criterion of interest, i.e., that which is to be predicted. The decision-making emphasis emerges when the predictor measurement is used to make anticipatory statements about the criterion and validity is high when the two systems are highly statistically related. *Content validity* addresses the question of whether the measure adequately samples some learning, task, or skill domain of interest. This form of validity is most useful for measures of ability and achievement to possess and is usually achieved by effective content-sampling. *Face validity* addresses the question of whether an instrument 'looks/feels' like it measures its intended construct, generally from the viewpoint of participants providing the measurements. This form of validity is most useful for self-report measures to possess.
- *Measurement reliability*: Reliability is a characteristic of measurement quality that concerns how consistently the measurement process produces outcomes over time and conditions. *Internal consistency reliability* addresses the question of how consistently a set of items or measurements work together to measure a common theoretical construct, typically assessed using a statistical index called Cronbach's Alpha. This form of reliability is important to demonstrate if the researcher is creating a composite measure from specific individual measurements. *Test–retest reliability* addresses the question of how stable construct measurements are over different measurement occasions. This form of reliability is important to demonstrate if measurements are to be repeated over time with the same participants (assessed by correlating measurements on two occasions for the same sample of participants). *Inter-observer/inter-rater reliability* addresses the question of how consistently two or more observers or raters independently record the same observations on the same entities at the same time. This type of reliability is important to demonstrate in systematic observational research (e.g., where the researcher is looking to measure specific aspects of human behaviour in some context).
- *Measurement sensitivity*: Sensitivity is a characteristic of measurement that concerns the ability of a measurement scale to reflect the magnitude of changes or degree of differentiation the researcher is looking for. The key feature of a scale that impacts on its sensitivity is the number of discrete rating or measuring points it offers. More sensitive measures can reflect finer differences in the amount of what is being measured. If the researcher seeks to measure change over time or

measure differences between groups, then measurement sensitivity becomes an important issue to address.

- *Contextual relevance*: This aspect of measurement quality concerns the context(s) in which/for which the measurement can be most defensibly applied. This is especially important to understand when a measurement process/instrument is applied in several cultural contexts. A measurement instrument is most contextually relevant to the culture in which it was developed (although the researcher must remain aware of potential sub-cultural issues). If that instrument is then applied in other cultures, as in the cross-cultural research frame, contextual relevance must either be argued for or systematically developed. There is, however, another angle to the question of contextual relevance. Under patterns of guiding assumptions that examine externalised causality and for experience-focused and Measurement data-shaping strategies, contextualisations (e.g., prior research and theorising by others, research propositions and/or hypotheses, research contexts, positioning of participants) help the researcher to appropriately allocate independent, dependent, moderating, mediating, and extraneous variable roles to different quantitative constructs and measurements for data gathering (as well as for analysis and model building/testing) purposes.

What complicates things for social and behavioural researchers is that many of the characteristics of people that they seek to measure are not directly observable (e.g., intellectual or cognitive capabilities, attitudes, values, beliefs, emotions, perceptions), therefore, they must infer their existence or strength on the basis of indirect measurement processes. Furthermore, many types of measurements in social and behavioural research vary in quality and degree of error depending upon the individual generating the measurements. The challenge for the researcher is to design measurement systems that simultaneously: (1) measure their intended construct in a consistent manner; (2) keep measurement errors/biases to a minimum; and (3) are contextually relevant. These challenges differ depending upon whether the researcher adopts measurement systems/instruments created by other researchers or the researcher builds their own measurement systems/instruments.

Where the researcher adopts the measurement systems/instruments created by others, the researcher must amass evidence for validity, reliability, and cultural relevance from the prior research they review. Measurement design issues are largely out of the researcher's hands. Cultural relevance may become a critical issue here if the researcher intends to employ a measurement system/instrument in a culture or sub-culture other than the one in which it was originally developed. Wording of questions, concepts invoked by questions, subject matter of questions can all become stumbling blocks when crossing cultural boundaries (even more closely-related culturally boundaries such as between the USA and UK or the USA and Australia) and the researcher has an obligation to argue or empirically demonstrate that such matters have been suitably addressed before claiming that the measurement systems/instruments are equivalent across contexts.

Where the researcher creates/develops their own measurement systems/instruments, they must attend to/contend with the full array of considerations that are associated with the design of measurement systems/instruments, including tailoring measurements/instruments to meet needs and expectations of the researcher as well as the target participants. Furthermore, the researcher has an obligation to empirically demonstrate appropriate kinds of validity and reliability, whether in a pilot study or in a distinct phase of their research (which immediately creates a sequential MU configuration). If the researcher adopts another researcher's measurement system/instrument but it is found to require significant modifications to make it relevant to the researcher's intended context(s), then the modified measurement system/instrument should be treated as if it was researcher-designed.

Hand (2004) provides a very useful overview of different kinds of measurement systems one might use in social and behavioural research. Keeves (1997, section III) presents a range of contributions covering many facets of educational measurement, a number of which are relevant to more general kinds of social and behavioural research. Shum et al. (2017), Reynolds et al. (2010) and Athanasou (1997) are three general texts that provide full discussions of psychological and educational measurement, testing and assessment. Lane, Raymond and Haladyna (2015) provide a thorough discussion of issues surrounding test development, whereas Haladyna (2004) discusses the development of multiple-choice tests in particular. Furr (2021) covers a range of issues and processes associated with psychometrics. Bond and Fox (2001) discuss the mechanics and applications of the Rasch model to measurement problems. Thornton and Kedharnath (2013) discusses issues surrounding the development and use of work sample tests in organisational settings. Nunnally (1978, Chap. 14, focusing on personality measures, and Chap. 15, focusing on measures of sentiments) and de Vellis (2016) discuss self-report scale development. The Boyle et al. (2015, Section I) handbook provides a useful exploration of issues associated with using self-report measures. Carmichael (2016) provides a thorough coverage of concepts, measures, and methods for demographic analysis. Yamarone's (2017) handbook focuses on economic indicators and Frumkin (2015) explores how economic and financial indices are constructed.

Cultural & Other Contextual Adaptations/Innovations

Depending upon pattern of guiding assumptions, frame, and research questions/propositions/hypotheses, implementing certain data gathering strategies may require modification, adaptation, augmentation, or reconceptualisation. Such changes in the context of cross-cultural or Indigenous/First Nations research frames, in particular, pose a set of unique challenges for the researcher. In a cross-cultural comparative research frame, the researcher must be able to argue/demonstrate that the data arising from each cultural group involved in the comparison are conceptually equivalent with respect to meaning and implications and free from any influence of culturally-specific factors. In a cross-cultural learning frame, the researcher must be able to gain insights and understanding of one or more cultural groups of interest, a goal made much more difficult if the researcher is not a member of any of those culture groups. In either type of cross-cultural frame, the researcher will not be able to employ what might be termed 'standard' or 'normal' approaches for mono-cultural studies or for studies where the researcher and the culture of interest are shared.

Prior to implementing data gathering, the researcher may need to engage in one or more additional processes, each of which may add some complexity to their research journey. Such engagement will help the researcher to anticipate requirements and preferences for their data gathering activities. If a questionnaire, structured, focused, or semi-structured interview is to be implemented, the researcher needs to make decisions about the language to be used in those instruments or interactions. If a common language, say English, is to be used, then competency in the language becomes an issue for participants from non-English speaking or English as second language backgrounds. Conceptual equivalence (between the common language and native language) of words, sentences, concepts, and instructions must be established, and use of idioms, jargon, and localised expressions must be avoided, so that when the common language is used, the researcher can be relatively certain that participants will be making the same interpretations and doing the same things. To facilitate this process, the researcher could rely upon experts in the native language or work with relevant cultural communities. For questionnaires or structured interview scripts, a process of backtranslation can be implemented which produces the original instrument/script in the intended common language, then a set of experts translates them into the required native language, a different set of experts backtranslates the native language version back into the common language, then the original and resulting back-translated versions are compared for equivalence (see, for example, Gjersing et al., 2010; Halder et al., 2016). Conceptual equivalence can be argued if the two versions match; if they don't match, further discussion or refinement of the process will be required. Alternatively, the researcher may decide to use the native language of each cultural group of interest and rely on experts in each culture to argue for equivalence. This is a more challenging and less convincing pathway to follow. For data gathering processes involving participant interactions with the researcher or a researcher assistant in cross-cultural contexts, verbal as well as non-verbal (e.g., gestures, eye contact, social proximity, mode of dress) behaviours must be attended to in order to avoid giving offense or creating confusion, the gender, religious and/or cultural background of the interviewer, vis-à-vis the interviewee, may also need to be carefully considered. If the researcher has the time and resources to obtain some level of cultural and context-specific learning and knowledge (either through participant observation or discussions with experts or community members), this can facilitate the achievement of culturally-equivalent comparisons. Broesch et al. (2020) provide a very insightful discussion of issues and processes to consider when implementing a cross-cultural data gathering process, including thinking carefully about site selection for the research within each cultural group, involving communities from each cultural group in the production and configuring of data gathering processes, making localised accommodations for implementing data gathering processes, and working in a research team where at least one member has cross-cultural research experience with and/or contacts in the cultures/communities of interest. In all cases, pilot-testing/trialling is essential, and it would be prudent to incorporate some processes for participants to discuss their participation experiences.

For research in Indigenous/First Nations contexts, the above cross-cultural considerations will typically still be relevant. However, because Indigenous or First Nations communities/people are involved, there is a whole new layer of considerations that arise because of the political, economic, spiritual, environmental, and social ways such people have historically been, as well as currently are, colonised, marginalised, culturally denuded, knowledge stripped, oppressed, and, in many cases, subjected to programs of eradication or wholescale displacement by more dominant, typically western cultures. This means that such histories cannot be ignored or diminished and must be actively acknowledged throughout the research process. This can be accomplished by through a process of decolonisation (see, for example, Archibald et al., 2022; Chilisa, 2020, especially Chaps. 1, 6 and 7; Evans et al., 2014; Kovach, 2021, especially Chap. 2 and Conclusion; Krusz et al., 2020; Mataira et al., 2005; Smith, 2021; Windchief et al., 2024; Zavala, 2013). In clarifying the intent of decolonisation, Mataira et al. (2005) stated that:

> [t]o a certain extent, the decolonization of methodologies (Smith, 1999) requires abandoning established epistemologies and perspectives and creating new methods from the inside out. That is, we must accept the relative nature of the culture-related pathology and well-being and create and align methods with elements of the experience to best represent them. These representations must be woven together as a way to restore the integrity of indigenous cultures and histories and counter the systematic fragmentation that has served to dehumanize and disempower indigenous peoples. (p. 43)

There are multiple approaches to the process of decolonisation. One approach involves modifying, adapting, and critically repurposing western methods of data gathering and knowledge building to expressly prioritise and acknowledge Indigenous/First Nations histories, perspectives, knowledges, and traditions (see, for example, arguments offered by Krusz et al., 2020). This approach typically leads to more participatory involvement of Indigenous/First Nations communities in the research process. Another approach advocated by authors such as Archibald et al. (2022), Evans et al. (2014), and Smith (2021) is to reject western knowledge building approaches in favour of uniquely Indigenous/First Nations methodologies. Here, the methodological landscape and associated 'Data Triangle' is creatively expanded to incorporate innovative approaches to data gathering and analysis, such as talking or yarning circles, lilyology, Indigenous storywork, and postcolonial Indigenous interviews. Such approaches will be wholly Indigenous and much better suited for insider rather than outsider researcher positioning. Yet another approach, better suited for an outsider researcher positioning, asks that the researcher engage in critical self-reflexion in order to commence 'decolonization of the self' (see, for example discussions in Krusz et al., 2020, pp. 205–208). Decolonisation of the self involves the researcher taking ownership for dominant culture privilege and power, learning to question notions and assumptions of the superiority of western knowledge production, subordinating one's own history and cultural views in favour of the Indigenous or First Nations perspective, and cultivating trust through honesty and humility within oneself as well as with others. Decolonisation foregrounds Indigenous/First Nations traditions, knowledge, and culture and any data gathering approach needs to fully and authentically enact such foregrounding.

In practical terms, decolonisation may involve such considerations such as orienting interviews toward inviting story telling/sharing instead of engaging in alternating question–answer interactions; preferencing interviewing approaches over questionnaire approaches; ensuring that the proper pathways for seeking access to communities and individuals are followed; and ensuring that issues such as using photography, voice recording, and other technologies, and who, when, where, with whom, and on who's turf/land data gathering should occur are all appropriately discussed/negotiated. With respect to interviewing, without decolonisation, a westernised relational mindset/stance between the researcher and those who are participating in the interview exists. The westernised relational mindset/stance emphasises (1) the importance of the individual rather than the collective or community when gathering data, (2) the expectation that the researcher should have access to the worldviews of the participants as a social science investigative right; (3) that the knowledge possessed by the participants is not privileged nor has primacy (including historical primacy) over knowledge associated with the researcher and their culture; and (4) the only relationship that needs to be safeguarded is that between the researcher and the researched (the participants) in their individual interactions, ignoring the fact that, historically, the westernised stance has dominated Indigenous/First Nations stances and lives through the processes of colonisation (see, for example, the various discussions in Chilisa, 2020; Chilisa & Tsheko, 2014; Kovach, 2021; Smith, 2021). From an Indigenous/First Nations perspective and in the context of the Indigenous/First Nations research frame, these emphases are seen to be inappropriate, oppressive, and culturally-insensitive, which is why certain aspects of the pattern of guiding assumptions for an Indigenous/First Nations inquiry mirror aspects of the pattern of guiding assumptions for a critical or systemic othering inquiry.

Adopting a postcolonial stance works to counteract the view that the researcher is the 'coloniser' and participants are the 'colonised' by decolonising these roles through how interactions for gathering data, and the knowledge they grant access to, unfold. This gives rise to the process of engaging in postcolonial Indigenous interviews (Chilisa, 2020, Chap. 11), where the westernised stance is rejected through adoption of the inverse of every emphasis listed in the previous paragraph. This repositions participants as 'cultural insiders/knowers' and the researcher as a 'cultural outsider/learner', especially when and where the researcher is of non-Indigenous/non-First Nations background. It also inverts the typical power relationship between researcher and participant by prioritising the participant and their positioning as a member of an Indigenous/First Nations community/culture. Virtually every interviewing strategy can be recast in a postcolonial mode. For example, focused, semi-structured, or unstructured interviews can be recast as relational interviews or conversations that emphasise relational ways of knowing. This gives primacy to the worldviews, knowledge, and relationships that are part of the life spaces, histories and environments/ecologies relevant to the participants (i.e., culturally-privileged knowledge building). Such interviews may not work well as one-on-one interactions, needing instead to incorporate voices and value the knowledge shared by others and with each other at the same time, becoming a form of group-based conversation. This comes about because, in general, no one Indigenous/First Nations person holds or has access to all relevant knowledge or relationships. As part of the interview process, cultural objects and symbols may serve as important focal stimuli for the interactions and these must be respected as such by the researcher. As another example, focus group interviews can be recast as talking circles where every member of the group enjoys equal voice, time, and space to contribute. An important process for gaining access to conduct interviews with Indigenous/First Nations people (whether on their land/country or not) may be to interview tribal Elders/leaders first to gain insights and build upon their wisdom, extensive life history, and experience (what Chilisa, 2020, pp. 254–255, refers to as 'philosophical sagacity'). This may also serve as a springboard to bringing others into the interview process where they have relevant knowledge. Kovach (2021) indicates that there may be cultural protocols to be followed (such as the offering of a gift) before access may be given for interviews (an ethical consideration) and reinforces the importance of negotiating all aspects of the data gathering process with relevant tribal representatives (e.g., whether conversations can or should be recorded or transcripts produced, whether pictures or videos can be taken, what can and cannot be shared beyond the boundaries of the research). This consideration of cultural traditions and protocols is an important sign of respect for an Indigenous/First Nations community and is not something that can or should be taken lightly.

5.4.3 Implementation Considerations for Data Analysis Strategies

There are a range of issues, considerations, and dynamics that emerge with each choice of data analysis strategy, informed by contextualisations, and these are depicted in the concept map in Fig. 5.7. Data must be organised, stored, and protected against unauthorised access. Raw data must be prepared for analysis, usually through some transformative strategy, and then screened for accuracy/errors. Software support systems for analysis, where appropriate, must be acquired and competencies in their use achieved through training, practice, and/or use of an expert. There is a strong interconnection between data preparation/screening and the use of software support systems. Analyses themselves are typically implemented by following specific pathways and those pathways are highly dependent upon research contextualisations, particularly the research questions/propositions/hypotheses, pattern(s) of guiding assumptions, and research frame(s). Where quantitative data analyses are concerned, measurement variables will take on specific roles (e.g., independent, dependent, moderator, mediator, extraneous) in model building and testing activities and analysis pathways are typically sequentially organised with prescribed steps, which statistical software systems are designed to facilitate. Where qualitative analyses are concerned, analysis pathways are much more open and typically iterative; the software supporting such pathways facilitate the memoing, recording and display of data coding and analysis while maintaining live links between the data and the displays they inform. Where multiple data forms are involved, the researcher may need to build creative bridges between the forms (e.g., assigning numerical codes to

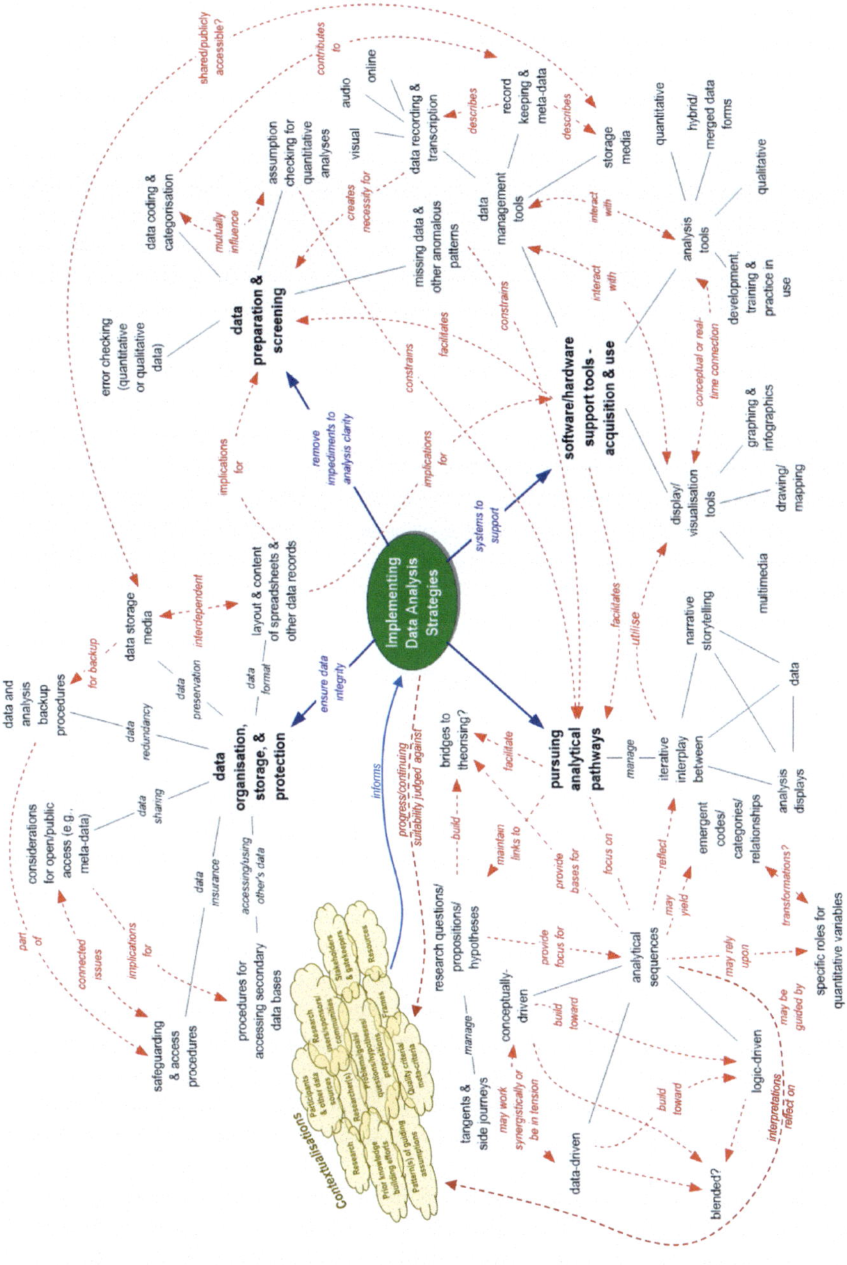

Fig. 5.7 Issues, considerations, and dynamics associated with implementing data analysis strategies

emergent qualitative categories or categories of participants, constructing/connecting qualitative characterisations/qualifications to quantitative patterns). As well, hybrid/merged data forms may require more specialised and innovative analysis processes.

Data Organisation, Storage, and Protection

Data analysis should not proceed until the researcher has carefully planned for how they will manage the organisation, recording, and storage of their data and how they will prepare those data for analysis. One aspect of this process involves planning for recording the data in as close to their original form as possible, i.e., with minimal interpretive work. As data analysis activities progress, the researcher should ensure that their reasoning behind undertaking each activity is recorded in their research journal. Access to the data while they are being gathered must be fully controlled by the researcher. If the data will ultimately be shared in some type of data warehouse or open access platform, care must be taken to ensure that, if anonymity/confidentiality has been promised to participants or gatekeepers, the researcher removes all potential indicators of individual or organisational identity. This is a process that should commence at the outset of data gathering, not as an afterthought, when things can be missed or forgotten. This also means the organisation of contextual information and meta-data to accompany the data needs to commence at the same time. Arrangements should be made for making regular back-ups of data files as well as analysis files and notes about the organisation of data and analysis files should be recorded in the research journal (where such notes can then be used as a source of contextual information as well as serving as one basis for the meta-data). If data are obtained from third-party sources, such as secondary databases, the researcher must follow all protocols for access and identity protection.

When qualitative data are gathered, the researcher should ensure that interviews are either voice- or video-recorded, documents are obtained in their original form and format (e.g., PDF, text, image, video or audio file), and field notes are recorded in such a way as to ensure that the researcher's personal thoughts, emotions, impressions, and interpretations are separated as far as is possible from the narrative recording of experiences, observations, interactions, and conversations. To facilitate recording qualitative data in as close to original form as possible, technological support, such as recording and storage devices, will often be necessary. For analysis purposes, such data will typically then be transformed into a format that is more amenable to/suitable for the types of analyses the researcher undertakes (this, of course, means that the researcher will need to know, when commencing, exactly which software support system(s) they intend to use and what their requirements are for data files). Thus, any data gathering strategy that produces or yields qualitative data will automatically be paired with the Transformative data-shaping strategy. The researcher should also ensure that the original/raw data are never destroyed or corrupted during the processes of transformation. Many of the software packages used for qualitative analysis of data automatically ensure that only copies of the original datafiles are used for analysis purposes, but each may also have esoteric requirements as to the formats of data they will accept.

When quantitative data are gathered, the data should be recorded in as raw a form as possible, with minimal preprocessing and the researcher needs to organise the structure and format (e.g., spreadsheet, database) of the files that will contain such data. Quantitative data may be electronically recorded (e.g., via a software process or hardware device) directly into a spreadsheet or database, in which case the researcher must decide on exactly what and how each measurement is stored. Backups are essential to maintain. Quantitative data may also be obtained via hardcopy (e.g., via written questionnaire) in which case the researcher will typically need to arrange for manually entering the data into an electronic storage file (including specifying all the rules that will guide the data entry process and spelling out how missing data are to be handled, thereby enacting a Transformative data-shaping strategy. However, in some cases, the hardcopy may be directly scanned for electronic storage. Once the raw data have been recorded, the researcher will need to decide if and how the data will be transformed from their original recorded form into a form that they can use for data analysis purposes (again, pairing with the Transformative data-shaping strategy) and that form may be determined by the specific quantitative analysis program/package they intend to use for the analyses. If multiple quantitative analysis programs/packages are to be employed for different types of analyses, there may be different data formatting requirements for each program that the researcher will need to adapt to. If the researcher is accessing one or more secondary quantitative databases, they will need to devise rules/processes for how to sample, extract and store just those cases, variables, and measurements they need into their own database for analysis as well as access relevant meta-data.

Data Preparation and Screening

Prior to commencing analysis activities and irrespective of the type of data being dealt with, the researcher needs to ensure that the data are error-free. Thus, it is worth the time and resources to invest in data checking processes. Any manual data entry process certainly runs a high risk of data entry errors, and it is worth checking every data file or, in the case of a spreadsheet, every line, for typos or data entry confusions. However, any electronic recording process also runs the risk of including data entry errors (e.g., batteries run out partway through recording an interview, extraneous signals interfere with recording some aspect of the data, a disk is full, an unexpected response that slips through any error trapping is made), so error-checking is important in these cases as well. Where quantitative data are concerned, there are two additional interconnected requirements to consider before analyses proceed (these requirements may also be revisited once preliminary analyses have been undertaken). One requirement is to decide how to encode quantitative information to best suit analytical choices. This includes considering whether to collapse multi-categorical variables into a smaller number of categories (so that each category has a larger sample size), how to represent/utilise information about missing data, whether to reverse score certain variables, whether to mathematically transform or combine certain variables, and what numbers to best use to code categories for analysis (e.g., binary variables are best coded as 0 or 1 rather than 1 or 2 for building linear models; see, e.g., discussions and illustrations in Cooksey, 2020, Fundamental Concept VI). The second requirement employs certain analytical and data display processes to check whether specific variables/collections of variables satisfy the assumptions (e.g., normal distribution, homoscedasticity, linearity, absence of multicollinearity between variables, independence of errors; see, e.g., discussions and illustrations in Cooksey, 2020, Fundamental Concept V and pp. 380–383) required by an analytical procedure the researcher intends to use and for checking for missing data biases and anomalous data patterns (see, e.g., discussions and illustrations in Cooksey, 2020, Procedure 8.2). All these considerations form part of data management and these processes interact with the requirements and capabilities of analysis tools and procedures.

Where qualitative (and in many cases, hybrid/merged) data are concerned, the major consideration the researcher needs to address is what the best format is for the qualitative or mixed data analyses they envisage. For recorded interviews, this generally means that the original recorded interview needs to be transcribed into a text file, prior to analysis. However, the researcher needs to attend very closely to managing the transcription process to ensure consistency. This means that the researcher must define the rules for transcription, for what information in each recording (e.g., contextual interjections/disruptions, pauses, length of time of pauses, unclear utterances, over-speaks where interview and participant talk overlaps, utterances that cannot be heard because of recording failure or background noise; for group interviews, identifying who is speaking) is to be transcribed and how it will be represented in each transcript. This all becomes material to record in one's research journal. Once transcripts have been produced, they need to be checked for accuracy, errors, and consistency with the rules of transcription. This may require listening to the recorded material while reading through the transcript. For non-textual data, the researcher needs to decide whether and how it will be used in the intended computer analysis support system. Many modern software packages can deal with a variety of formats but capabilities for coding such data may vary. For data from video recordings or social media or blogging websites, the researcher needs to decide whether only the text data is of interest, or texts plus images/videos.

Software/Hardware Support Tools—Acquisition and Use

Where the researcher intends to use one or more computer software support systems, programs, or packages (e.g., *SPSS, SYSTAT, NCSS, Stata, eViews, SAS, Excel* for quantitative data and/or *MAXQDA, NVivo, dedoose, Decision Explorer, ATLAS.ti, The Ethnograph* for qualitative or hybrid/merged data), they must become familiar with its requirements and capabilities for data entry, formatting, storage, and analyses, which creates a training/practice burden for the researcher to deal with. That knowledge will constrain how they manage and analyse their data, and they will need a system of procedures to help them manage their data and associated analyses within those constraints. If multiple types of data are to be analysed, the researcher needs to check the capabilities of their intended software support systems to handle such complexities in ways that can help to assemble a convincing story. Different software systems and packages will have different capabilities in terms of analysis pathways offered, options for customisation available, and data summarisations and visualisations they can produce. There will typically be a close relationship between the data management capabilities of the package and the types of analyses and data displays it can produce. More comprehensive statistical and qualitative analysis packages will offer a wider range of data management tools, analysis options, and data display options, but perhaps at the expense of not offering certain types of analyses and/or data displays, which may be more purpose-built. For example, some general data analysis software packages may not have adequate capabilities for drawing cause maps, infographics, or mind maps. If the researcher requires such representations as part of their data displays for analysis, they may need to supplement their data analysis activities with other types of software (such as *Decision Explorer*, *Inspiration, Venngage, Canva, Piktochart, Visme*, or even *Adobe Illustrator, PowerPoint, Visio* or *SmartDraw*). The advantage of using such supplemental software packages is that the researcher can access extensive capabilities for producing highly customised and specialised diagrams, maps, and images. The drawback is that when one moves outside of a general analysis support package, the capacity for maintaining live links between the data and the data display being produced is typically lost, which makes updating and making changes, upon encountering new data, a challenge.

Pursuing Analytical Pathways

Analytical pathway choices can be influenced by contextualisations, pattern(s) of guiding assumptions, and researcher frame(s) as well as by the researcher's questions, propositions, and/or hypotheses. Data analysis requires close connections between the data being analysed and the displays of those analyses. For qualitative analysis, the relationship between data and data display is typically iterative and evolutionary. Data displays may range from pictures to maps/diagrams to specific types of tabular and symbolic displays and various combinations, as appropriate (see, e.g., the multitude of examples in Miles, Huberman & Saldaña, 2020). For quantitative analysis, the relationship between data and data display often centres on identification of characteristics and patterns in the data and localisation of certain patterns to certain segments or individuals in the data set (e.g., outliers or extreme values, the specific variables included in the analysis, the nature of the variables included in the analysis, and so on). Certain analysis outcomes may suggest certain data transformations to implement and/or additional analyses to undertake. Data displays may range from individual indicators and statistics to tabulated statistics to graphs and diagrams and various combinations, as appropriate (see, e.g., the multitude of examples in Cooksey, 2020). As a general rule, regardless of which computer software support system or package is employed, the researcher would benefit from making use of all of the labelling, memoing, and tagging facilities it offers. Such capabilities make it much easier for the researcher to manage and revisit analyses over a long stretch of time.

Typically, narrative storytelling builds the bridge between the data and a data display adding to convincingness by making the connections clear. Analyses are guided by but not generally fully determined by research questions, propositions, or hypotheses. A narrow or wide set of possibilities may exist, and the researcher will need to reason their way through toward the most appropriate choice for the story they want/need to convey. In some cases, the analyses are intended to build bridges toward some kind of theory or account for what the researcher learned via their navigation of the 'Data Triangle'. In other cases, they may simply inform a descriptive account. Any analytical tangents, side journeys, or supplemental analyses pursued should be carefully managed so as not to detract from the main stories where convincingness resides. However, in some instances, such analyses may help to support the main stories the researcher wants to convey, in which case they are worth pursuing. In short, sound and logical judgment is required here. The researcher should rely on their research journal to record all their thoughts about their data management and analysis processes and to record all the decisions/choices they made along the way.

If there are any specific criteria being used by the researcher to make analytical choices and draw analytical conclusions and interpretations, they must spell these out and apply them consistently. For quantitative analyses, this could include discussions of level of significance for statistical tests, how effect sizes will be characterised, choice of statistical indicator or index to report. For qualitative data, this could include discussions about what counted as an instance for specific coding categories, how specific quote materials or focal participants for providing exemplar material were selected, how a first cycle coding system was dealt with during second-cycle coding (see discussions in Saldaña, 2021), how choices for consideration and follow-up in grounded theory work were made, and where mixed data or hybrid/merged data forms are involved, which aspects of the various types of data were brought together to inform an integrated story and why.

Many analyses, especially but not exclusively for quantitative data, unfold in a logical sequence and making this sequence clear and adhering to it help to add to convincingness (i.e., they are logic-driven sequences). For example, in many multivariate analyses, there are specific pathways for analyses to support different research goals (e.g., exploratory, confirmatory, cross-validation, typology discovering and testing, see discussions in Appendix B of Cooksey, 2020). Each pathway follows a particular sequence of analyses that will help the researcher to build the story they want to convey. Qualitative analyses tend to be much less sequential and formulaic, instead tending to be highly iterative, emergent, and evolutionary. Often, the researcher does not know what they are looking for or what they will find until they commence analysis, and such learning keeps getting revisited and refined as more and more data are encountered. Such an approach would be characterised as data-driven or bottom-up. Alternatively, the researcher could start with a framework or a set of pre-existing categories and then work to map the data onto that framework or those categories. Such an approach would be characterised as conceptually-driven or top-down. If the commencing framework or set of categories is not allowed to evolve, then analysis provides a test for that pre-existing system. If, however, the commencing framework or set of categories is allowed to evolve and change, then a synergistic dynamic is established between data-driven and conceptually-driven approaches. Various blends of logic-driven, data-driven, and conceptually driven approaches can be envisaged, especially where multiple types of data are involved (as in mixed methods research). It is the researcher who builds up their analytical sequences in such a way as to best support a conveying convincing story from their data. How they arrived at their attack on analysis problems is something they must convey/defend as well. Convincingness may rest as much on how the researcher arrived at what they learned as on what they, in fact, did learn; how well the former is unfolded will influence how well the latter is received. When interpreting what emerges from analytical processes and pathways and drawing convincing conclusions, the researcher can employ pluralist systemic thinking to reflect on contextualisations to help shape as well as make sense of analysis outcomes. Such thinking about contextualisation can feed into interpretations in various ways: more fully characterising what has been learned; qualifying/delimiting the meaning and extent of what has been learned; augmenting/explaining what has been learned; and/or building up plausible accounts for unexpected or anomalous results (where abductive reasoning can play a role). This is the case irrespective of the form of data being analysed; contextualisations add richness, meanings, nuances, texture, and boundaries to otherwise more sterile analytical results.

5.4.4 Piloting/Trialling

No matter which data gathering strategies are implemented, the researcher must go into the research knowing that interview processes, measurement instruments, data source choice and access processes may not work as they intended and that problems they cannot foresee at the outset may emerge. One important but unfortunately often neglected strategy for reducing such uncertainties is to build a pilot testing or trialling stage into the research journey (note that the use of trialling here does not have the same meaning as a clinical trial which describes a concrete methodological system for navigating the 'Data Triangle', usually in medical or health contexts). We say unfortunate because, in many cases, researchers avoid piloting or trialling their intended approaches to navigating the 'Data Triangle' because of the additional resource and time demands it creates. However, the potential for learning and improving one's planned approaches far outweighs the additional resource and time burden. It should be noted that inclusion of a piloting or trialling stage in one's research journey does not necessarily constitute the use of a sequential MU configuration. In most cases, it is better to think of piloting or trialling as a preliminary step in shaping the research journey for success. However, in certain instances, pilot testing/trialling may be of such a scale that it does constitute the initial MU in a sequential MU configuration (e.g., when a new measurement instrument is developed *and validated* during the pilot phase or when a series of initial interviews forms the basis for creating the topical landscape for a semi-structured interview study).

When a researcher pilot tests or trials their intended processes for navigating the 'Data Triangle', they deliberately set up circumstances for obtaining feedback on how feasibly and how well those processes work, thereby providing important feedforward information for modifications and trade-off decisions. Pilot testing is usually associated with the pretesting of a newly developed or adapted measurement instrument and associated administration processes. Trialling is a more general process where the researcher engages in small-scale 'simulations' of their navigation of the 'Data Triangle'. In most cases, the information gathered during pilot testing/

trialling is not analysed in the same ways as data will be when actually navigating the 'Data Triangle'. Analysis focuses on observing/learning what does and doesn't work. However, as flagged above, some researchers will use the pilot testing of a measurement instrument to gather some preliminary validation data. In a pilot test or trial, one can evaluate/test/debug:

- data source choice procedures, using smaller-scale simulations of those processes (focusing more on administrative and access aspects than full-blown implementation as well as on the processes the researcher uses to recruit participants and/or the processes the researcher uses to obtain, read, and store documents or deal with other non-human data sources, where appropriate);
- interview and observational procedures/protocols and associated data recording processes using one or more instances of implementation (focusing on things such as feelings of comfort and safety, existence of rapport, appropriateness of the behaviours of interviewer/observer, appropriateness of any technology used, comments about the context(s) where they were interviewed/observed, as appropriate), often in a different context from that intended for the main research effort (to avoid depletion of potential data choice options and potential contamination of the context itself);
- participant-produced data gathering strategies, such as keeping a diary/journal, creating cause map, rich picture, or other creative artwork (focusing on things such as on difficulty of the task, availability/useability of materials and technology, impacts on other aspects of their lives, as appropriate);
- instruments and measurements (may incorporate a validation exercise) and associated data recording processes (focusing on things such as structure, wording, complexity, readability or useability, formatting, ease of responding, use of technology, meaning of concepts, cultural adaptations, areas of sensitivity or offense, as appropriate); and/or
- data preparation and preliminary analysis procedures (focusing more on training, familiarisation, and practice runs (perhaps using data from another completed project) than actual full implementation as intended in the main study).

An important secondary purpose for pilot testing/trialling is that it provides opportunities for the researcher to hone their skills (e.g., interviewing and/or observing, data recording and management, measurement instrument design, social interaction and analysis skills in a research context). Additionally, if research assistants are to be involved in navigating the 'Data Triangle', pilot testing/trialling provides a context for training and evaluating readiness. Thus, piloting/trialling is a pathway for not only improving 'Data Triangle' navigation from participant and data quality perspectives, but also a pathway for researcher development/self-improvement in the context of the intended research journey.

While pilot testing/trialling simulates navigation of the 'Data Triangle' processes, the researcher should also include data gathering processes specific to the pilot testing/trialling effort. A common strategy, when pilot testing a new or culturally-adapted measurement instrument, is to interview or seek written feedback from the participants about their experiences with said instrument including seeking comments on its face validity, ease of interpretation, structure, formatting, ease with which responses can be made, and content. When pilot testing an observational protocol, observers/raters could be interviewed/asked for their feedback on the content and ease of use of the observational protocol. When trialling interviewing strategies, a common practice is to augment the interview by allocating some time at the end to discuss the participant's experiences with the interview, the interviewer, the interview content and focus, and the interview context. When data gathering strategies are being evaluated, a researcher may elect to employ a small purposive, convenience, or volunteer sample of participants solely for the pilot testing/trialling effort (which may or may not mirror how participants will be chosen for the main study). The researcher and any involved research assistants should also engage in reflective and reflexive efforts (at an individual or group-level, as appropriate) in order to critically assess what has and hasn't worked. Wishful thinking or uncritical optimism is not a viable pathway for moving forward with the main study. If feedback demands change, then it must stimulate the creation of feedforward actions to produce the necessary change. There is no limit to the potential value that piloting/trialling can add to the achievement of convincing research outcomes.

When considering feedback obtained during pilot testing/trialling, the researcher primarily looks for information that clearly shows what works and what doesn't work. Feedforward actions taken will focus on changing what doesn't work and those changes may require one or more trade-offs to be made. Such changes might include modifications or amendments, removal of one or more processes, or inclusion of one or more new processes. Learning from pilot testing/trialling might also provide signals about potential problems that could emerge which would help the researcher devise a Plan B (for alternative data source strategies, data gathering strategies, and/or data analysis strategies) as a backup pathway for proceeding in case those problems do occur. If a Plan B is devised, the researcher should critically evaluate the pros and cons associated with it so that, if they are forced to adopt the plan, they know what they stand to gain and lose.

In all cases, the researcher should harness pluralist systemic thinking in order to successfully translate feedback from pilot testing/trialling into feedforward actions to take to improve the chances that researcher will achieve convincing outcomes. Thus, the researcher's goal is to simultaneously achieve/maintain feasibility in processes for navigating the 'Data Triangle' and convincingness in terms of what can be learned when the main study is undertaken. For MU configurations that are highly complex, multiple piloting/trialling efforts may be required. One could argue that this is a tangible risk that arises from pluralist systemic thinking, and, in one sense, this is correct. However, it is more productive to think of this as the cost of achieving more convincing research whilst keeping things feasible and viable. There are no free lunches in research. The role of (and commitment of resources to) effective piloting/trialling is to improve the chances that convincingness will be achieved.

5.5 Integrative Foreshadowing

All the work that goes into contextualising the research journey (including considerations of pattern(s) of guiding assumptions, research frames, and the researcher's questions, propositions, and/or hypotheses) continues to have impact when final choices are made for navigating the 'Data Triangle' and research activities are configured and implemented. Feasibility and convincingness become dual criteria for making such choices and contextualisations can be used to judge their progress and continuing suitability. Where choices become impractical, infeasible, or even irrelevant or when a new opportunity emerges, trade-off choices may be required. In some cases, issues may arise that were not detected or even looked for during pilot testing/trialling. Some examples (in part, inspired by the experiences of Alhosani, 2010; Hunter, 2004; Robson, 2011; Sandall, 2006; Wolodko, 2017) include:

- a researcher may restart their qualitative analyses under the pattern of guiding assumptions for an interpretivist/constructivist inquiry because they feel they are not sufficiently withholding their preconceptions from their coding activities;
- a researcher may discover that some participants are unable to cope with using a cause mapping participant-centred strategy, so they adopt the structured experience-focused strategy of gathering think-aloud protocols for these participants and construct their cause maps based on those think-aloud talks;
- part-way through a case study investigation of the culture of a not-for-profit organisation using internal documents and interviews as data gathering strategies, a media article is published in a national newspaper about the case study organisation and the researcher decides to further pursue that outside contextual information by interviewing the author of the article;
- part-way through an investigation that employs the focused interview interaction-based strategy, a researcher may move from conducting on-site (at workplace) interviews in favour of off-site interviews because some participants are worried about management potentially learning about what they are saying; or
- a researcher may, after receiving feedback about a questionnaire from early participants from a particular religious background, rework that questionnaire because some of the concepts in the certain questions struck those participants as meaningless, confusing, and potentially offensive (has implications for the data source choice strategy as well).

There are no 'free lunches' in social and behavioural research, every strategy choice (whether for data sources, data gathering, or data analysis) around the 'Data Triangle' provides potential benefits (e.g., to the researcher, to participants, to stakeholders and gatekeepers, to communities) and incurs costs (e.g., ethical requirements, resource and time requirements, falling short of allowing the researcher to achieve their goals). As the research journey unfolds in real time and space, the researcher needs to be alive to feedback that signals what is working and what is not. Pluralist systemic thinking continues to be valuable here, in terms of helping the researcher translate feedback into concrete feedforward actions (such translation may involve minor tweaks to research activities or major trade-offs). Scoping helps the researcher fine-tune and finalise their choices and formulate and configure MUs, each of which combines one data source strategy, one data gathering strategy, and one form of data. Implementation fleshes out the researcher's choices with detailed processes and necessities. In order to judge progress and continuing suitability in terms of what is working and what is not, the researcher should continually reflect on their contextualisations, goals, and intentions. Such reflection will also prove very useful when interpreting results from data analyses and drawing conclusions.

Once navigation of the 'Data Triangle' is complete or sufficiently progressed, the researcher moves on to create/produce, and share/communicate research outcomes. As we will see, this is a process of storytelling, where the story (which contains a myriad of smaller interconnected/interwoven sub-stories) concerns some or all aspects of the research journey, oriented toward one or more audiences, often with the hope of producing some kind of impact beyond the researcher's own contexts. The choice of media (e.g., written text, oral presentation, community discussion/interaction, interactions with research sponsors, stakeholders and gatekeepers), shaping, and formatting of this story is different for different types of research outcomes and for different audiences, but will typically include elements of contextualisation, culminating in the researcher's goals, questions, propositions and/or hypotheses and elements that describe how the 'Data Triangle' was navigated, what was learned, and for whom that learning might be relevant. It will be a linearised story of a nonlinear journey and the researcher must decide the order in which events and activities are described/reported as well as their interconnections and importance to the overall story. Contextualisations will certainly influence not only the nature of any research outcome produced but also its content. The researcher's journal will prove to be an invaluable asset, no matter what research outcome is envisaged, as it essentially chronicles the research journey, its contextualisations, the choices and trade-offs made, the learnings achieved, as well as the researcher's thoughts, feelings, and reactions experienced along the way. The meta-criteria will also prove useful in crafting research outcomes as they connect directly to convincingness, and it is convincingness that carries the potential for impact with an audience. In producing a research outcome, the researcher will need to be conscious of audience(s) expectations and willing to critically reflect on and be reflexive with respect to what they have done, what they have learned and its limits, what else might need to be done, and what impacts their work might have for downstream users/innovation adopters, other researchers, and where appropriate, communities and cultures.

References

Alhosani, A. (2010). *Exploring organisational citizenship behaviour in the federal hospitals in the United Arab Emirates: A cross-cultural research study*. Unpublished PhD thesis, UNE Business School, University of New England, Armidale, NSW.

Archibald, J., Xiiem, Q., Lee-Morgan, J., & Santolo, J. (Eds.). (2022). *Decolonizing research: Indigenous storywork as methodology.* Bloomsbury.
Anderson, B., Deane, D., Hammond, K., McClelland, G., & Shanteau, J. (1981). *Concepts in judgement and decision making: Definitions, sources, interrelations, comments.* Praeger Publishers.
Athanasou, J. (1997). *Introduction to educational testing.* Social Science Press.
Azorín, J. M., & Cameron, R. (2010). The application of mixed methods in organisational research: A literature review. *Electronic Journal of Business Research Methods, 8*(2), 95–105.
Bartlett, J., Kotrlik, J., & Higgins, C. (2001). Organizational research: Determining appropriate sample size in survey research. *Information Technology, Learning and Performance, 19*(1), 43–50.
Bond, T., & Fox, C. (2001). *Applying the Rasch model: Fundamental measurement in the human sciences.* Lawrence Erlbaum Associates.
Boyle, G., Saklofske, D., & Matthews, G. (Eds.). (2015). *Measures of personality and social psychological constructs.* Academic Press.
Broesch, T., Crittenden, A., Beheim, B., Blackwell, A., Bunce, J., Colleran, H., Hagel, K., Kline, M., McElreath, R., Neson, R., Pisor, A., Prall, S., Pretelli, I., Purzycki, B., Quinn, E., Ross, C., Scelza, B., Starkweather, K., Stieglitz, J., & Mulder, M. (2020). Navigating cross-cultural research: Methodological and ethical considerations. *Proceedings of the Royal Society B: Biological Sciences, 287*(1935), 20201245.
Brunswik, E. (1952). *The conceptual framework of psychology.* University of Chicago Press.
Brunswik, E. (1956). *Perception and the representative design of psychological experiments* (2nd ed.). University of California Press.
Cameron, R. (2023). A mixed methods research notation system for complex designs. In R. Cameron & X. Golenko (Eds.), *Handbook of mixed methods research in business and management* (pp. 348–364). Edward Elgar Publishing.
Carmichael, G. (2016). *Fundamentals of demographic analysis: Concepts, measures and methods.* Springer.
Charmaz, K. (2014). *Constructing grounded theory* (2nd ed.). Sage Publications.
Chilisa, B. (2020). *Indigenous research methodologies* (2nd ed.). Sage Publications.
Chilisa, B., & Tsheko, G. (2014). Mixed methods in Indigenous research: Building relationships for sustainable intervention outcomes. *Journal of Mixed Methods Research, 8*(3), 222–233.
Cohen, L., Manion, L. & Morrison, K. (2011). *Research methods in education* (7th ed.). Routledge.
Cohen, J. (1988). *Statistical power analysis for the behavioral sciences* (2nd ed.). Lawrence Erlbaum Associates.
Cohen, J. (1992). A power primer. *Psychological Bulletin, 112*(1), 155–159.
Cooksey, R. (2011). Yours, mine or ours: What counts as innovation? *Journal of Agricultural Education and Extension, 17*(3), 283–295.
Cooksey, R. (2020). *Illustrating statistical procedures: Finding meaning in quantitative data* (3rd ed.). Springer Nature.
Cooksey, R., & McDonald, G. (2019). *Surviving and thriving in postgraduate research* (2nd ed.). Springer Nature.
Creswell, J. & Plano Clark, V. (2011). *Designing and conducting mixed methods research* (2nd ed.). Sage Publications.
Creswell, J., & Plano Clark, V. (2018). *Designing and conducting mixed methods research* (3rd ed.). Sage Publications.
Cryer, J., & Chan, K.-S. (2008). *Time series analysis: With applications in R.* Springer.
de Vellis, R. (2016). *Scale development: Theory and applications.* Sage Publications.
Evans, M., Miller, A., Hutchinson, P., & Dingwall, C. (2014). Decolonizing research practice: Indigenous methodologies, Aboriginal methods, and knowledge/knowing. In P. Leavy (ed.), *The Oxford handbook of qualitative research* (2nd ed., pp. 263–281). Oxford Univesity Press.
Flick, U. (2018). Triangulation. In N. K. Denzin & Y. S. Lincoln (Eds.), *The Sage handbook of qualitative research* (5th ed., pp. 444–461). Sage Publications.

Frumkin, N. (2015). *Guide to economic indicators* (4th ed.). Routledge.

Furr, R. (2021). *Psychometrics: An introduction* (4th ed.). Sage Publications.

Gill, J., & Johnson, P. (2010). *Research methods for managers* (4th ed.). Sage Publications.

Gjersing, L., Caplehorn, J., & Clausen, T. (2010). Cross-cultural adaptation of research instruments: Language, setting, time and statistical considerations. *BMC Medical Research Methodology, 10*, 1–10.

Glass, G., Willson, V., & Gottman, J. (2008). *Design and analysis of time-series experiments.* Information Age Publishing.

Gobo, G., Fielding, N., La Rocca, G., & van der Vaart, W. (2022). *Merged methods: A rationale for full integration.* Sage Publications.

Hair, J., Black, B., Babin, B., & Anderson, R. (2010). *Multivariate data analysis: A global perspective* (7th ed.). Prentice Hall.

Haladyna, T. (2004). *Developing and validating multiple-choice test items* (3rd ed.). Routledge.

Halder, M., Binder, J., Stiller, J., & Gregson, M. (2016). An overview of the challenges faced during cross-cultural research. *Enquire, 8*, 1–18.

Hall, B., & Howard, K. (2008). A synergistic approach: Conducting mixed methods research with typological and systemic design considerations. *Journal of Mixed Methods Research, 2*(3), 248–269.

Hammond, K., & Wascoe, N. (Eds.). (1980). *Realizations of Brunswik's representative design.* Jossey-Bass.

Hand, D. (2004). *Measurement theory and practice: The world through quantification.* Wiley.

Hunter, J. (2004). *Investigating the complexity of outsourcing decisions: An interpretive approach.* Unpublished PhD thesis, New England Business School, University of New England, Armidale, NSW.

Janis, I., & Mann, L. (1977). *Decision making: A psychological analysis of conflict, choice, and commitment.* Free Press.

Jick, T. (1979). Mixing qualitative and quantitative methods: Triangulation in action. *Administrative Science Quarterly, 24*(4), 602–611.

Johnson, R. B., & Onwuegbuzie, A. (2004). Mixed methods research: A research paradigm whose time has come. *Educational Researcher, 33*(7), 14–26.

Kaptchuk, T. (2001). The double-blind, randomized, placebo-controlled trial: Gold standard or golden calf? *Journal of Clinical Epidemiology, 54*(6), 541–549.

Keeves, J. (1997). *Educational research, methodology, and measurement: An international handbook* (2nd ed.). Pergamon

Kincheloe, J. (2005). On to the next level: Continuing the conceptualization of the bricolage. *Qualitative Inquiry, 11*(3), 323–350.

Kovach, M. (2021). *Indigenous methodologies: Characteristics, conversations, and contexts* (2nd ed.). University of Toronto Press.

Krusz, E., Davey, T., Wigginton, B., & Hall, N. (2020). What contributions, if any, can non-Indigenous researchers offer toward decolonizing health research? *Qualitative Health Research, 30*(2), 205–216.

Lane, S., Raymond, M. & Haladyna, T. (2015). *Handbook of test development* (2nd ed.). Routledge.

Leech, N., & Onwuegbuzie, A. (2009). A typology of mixed methods research designs. *Quality & Quantity, 43*, 265–275.

Loomis, D., & Maxwell, J. (2003). Mixed methods design: An alternative approach. In A. Tashakkori & C. Teddlie (Eds.), *Sage handbook of mixed methods in social & behavioral research* (pp. 241–271). Sage Publications.

Mataira, P., Matsuoka, J., & Morelli, P. (2005). Issues and processes in Indigenous research. *Hülili: Multidisciplinary Research on Hawaiian Well-Being, 2*(1), 35–45.

Mauthner, N., & Parry, O. (2013). Open access digital data sharing: Principles, policies, and practices. *Social Epistemology, 27*(1), 47–67.

Miles, M., Huberman, A., & Saldaña, J. (2020). *Qualitative data analysis: A methods sourcebook* (4th ed.). Sage Publications.

Nastasi, B., Hitchcock, J., & Brown, L. (2010). An inclusive framework for conceptualizing mixed methods design typologies: Moving toward fully integrated synergistic research models. In A. Tashakkori & C. Teddlie (Eds.), *Handbook of mixed methods in social & behavioral research* (2nd ed., pp. 305–338). Sage Publications.

Nunnally, J. (1978). *Psychometric theory* (2nd ed.). McGraw-Hill.

Patton, M. (2011). *Developmental evaluation: Applying complexity concepts to enhance innovation and use*. The Guilford Press.

Plano Clark, V., & Ivankova, N. (2016). *Mixed methods research: A guide to the field*. Sage Publications.

Poth, C. (2018). *Innovation in mixed methods research: A practical guide to integrative thinking with complexity*. Sage Publications.

Poth, C. (ed.). (2023). *Sage handbook of mixed methods design*. Sage Publications.

Reynolds, C., Livingston, R., Willson, V., & Willson, V. (2010). *Measurement and assessment in education* (2nd ed.). Pearson Education International.

Robson, M. (2011). *The use and disclosure of intuition(s) by leaders in Australian organisations: A grounded theory*. Unpublished PhD thesis, UNE Business School, University of New England, Armidale, NSW.

Ross, E. (1999). *Academics' perceptions of university culture and the factors that facilitate or inhibit commitment to it*. Unpublished PhD thesis, Department of Marketing and Management, University of New England, Armidale, NSW.

Ross, M., Iguchi, M., & Panicker, S. (2018). Ethical aspects of data sharing and research participant protections. *American Psychologist, 73*(2), 138.

Saldaña, J. (2021). *The coding manual for qualitative researchers* (4th ed.). Sage Publications.

Sandall, J. (2006). *Navigating pathways through complex systems of interacting problems: Strategic management of native vegetation policy*. Unpublished PhD thesis, New England Business School, University of New England, Armidale, NSW.

Saunders, M., Lewis, P., & Thornhill, A. (2012). *Research methods for business students* (6th ed.). Pearson Education.

Scheaffer, R., Mendenhall III, W., Ott, R., & Gerow, K. (2012). *Elementary survey sampling* (7th ed.). Brooks Cole/Cengage Learning.

Shum, D., O'Gorman, J., Creed, P., & Myors, B. (2017). *Psychological testing and assessment ebook* (3rd ed.). Oxford University Press.

Smith, L. (2021). *Decolonizing methods: Research and Indigenous peoples* (3rd ed.). Zed Books.

Smith-Ruig, T., & Sheridan, A. (2009). Through my eyes: Conducting research as a vision-impaired researcher. In V. Minichiello & J. Kottler (Eds.), *Qualitative journeys: Student and mentor experiences with research* (pp. 157–175). Sage Publications.

Smithson, M. (2000). *Statistics with confidence*. Sage Publications.

Spector, P. (2006). Method variance in organizational research truth or urban legend? *Organizational Research Methods, 9*(2), 221–232.

Tashakkori, A., & Teddlie, C. (Eds.). (2010). *Sage handbook of mixed methods in social & behavioral research* (2nd ed.). Sage Publications.

Teddlie, C., & Tashakkori, A. (2011). Mixed methods research: Contemporary issues in an emerging field. In N. Denzin & Y. Lincoln (Eds.), *The Sage handbook of qualitative research* (4th ed.). Sage Publications, pp. 285–299.

Thornton, G., III., & Kedharnath, U., et al. (2013). Work sample tests. In K. Geisinger (Ed.), *APA handbook of testing and assessment in psychology* (Vol. 1, pp. 533–550). American Psychological Association.

van Voorhis, C., & Morgan, B. (2007). Understanding power and rules of thumb for determining samples. *Tutorials in Quantitative Methods for Psychology, 3*(2), 43–50.

Windchief, S., San Pedro, T., & Kovach. M. (2024). Why we do Indigenous methodologies: Contemplations on Indigenous protocol, theory, and method. In N. Denzin, Y. Lincoln, M. Giardina, & G. Cannella (Eds.). *The Sage handbook of qualitative research* (6th ed., pp. 241–258).

Wolodko, K. (2017). *The emergence of group dynamics from contextualised social processes: A complexity-oriented grounded-theory approach*. Unpublished PhD thesis, UNE Business School, University of New England, Armidale, NSW.

Yamarone, R. (2017). *The economic indicator handbook: How to evaluate economic trends to maximize profits and minimize losses*. Wiley.

Yin, R. (2018). *Case study research and applications: Design and methods* (6th ed.). Sage Publications.

Zavala, M. (2013). What do we mean by decolonizing research strategies? Lessons from decolonizing Indigenous research projects in New Zealand and Latin America. *Decolonization: Indigeneity, Education, & Society, 2*(1), 55–71.

Chapter 6
Making Convincing Choices: Sharing/Communicating Research Outcomes

Abstract This chapter explores the 'downstream' end of the research journey, where research learning, knowledge, and outcomes are shared with/communicated to relevant audiences and interested stakeholders. The researcher shapes the stories they wish to convey about their research, translating their choices made throughout their journey into forms that will resonate with intended audiences. Communication may be largely unidirectional (researcher to audience). In other cases, research sharing may be multi-directional, participative, and reciprocal, involving the researcher and interested and/or potentially impacted people/communities, and may also encompass direct or indirect actions or impacts. The researcher's stories/actions should be influenced by the choices made throughout their journey, critically reflect on relevant contextualisations, acknowledge the influences/impacts of relevant stakeholders/gatekeepers, where appropriate, and be guided by meta-criteria for research quality. A well-maintained research journal will prove immensely helpful in shaping those stories/actions. If the research produces a concrete innovation as an outcome, then at least some of the research sub-stories will need to address efforts to evaluate, share, facilitate, and spread adoption of that innovation to interested potential users and communities. If the research recommends specific actions to be undertaken, the arguments, contexts, and ultimate goals for such actions need to be clearly spelled out.

6.1 Overview

When we talk about research outcomes, we are referring to any attempt to communicate/share research learning/knowledge and/or to promote/undertake actions to create impact and change. Just as there is diversity in approaches to social and behavioural research in pursuit of the unity of knowledge, there is a diversity of potential outcomes that could emerge from a research journey to display/convey that knowledge. To illustrate, consider the following (non-exhaustive) diverse array of potential research outcomes, most being retrospective and reflective/reflexive (intended to convey, reflect upon, and critically consider what has been learned

R. W. Cooksey, *Unity from Diversity*, https://doi.org/10.1007/978-981-97-3462-7_6

through a research journey as well as its impact and meaning) and others being prospective and anticipatory (i.e., proposals intended to garner support for a future research journey):

- an autoethnography, conveying the story of a research journey using one's research journal and own research experiences as important resources and touchstones (see, for example, Jones & Adams, 2024);
- an unpublished report/thesis/dissertation/portfolio, typically produced to meet requirements of an Honours or postgraduate research degree (see, for example, Cooksey & McDonald, 2019, Chap. 22);
- a formal interim or final report to a governing body, organisation, or institution that solicited/ commissioned/paid for the research (typical of consultant reports, for example);
- a professional article, which may be published in a trade journal or as a working paper (thereby constituting a contribution to grey literature);
- an academic, professional, or industry lecture, seminar, or workshop, provided in an institutional, organisational, or public forum for presenting research findings, intended to return value to academic scholarship, a profession, organisation, or community (perhaps promised as part of a quid pro quo arrangement with gatekeepers);
- an academic conference presentation/publication, usually done in advance of more formal publication in another outlet, intended to accomplish both sharing of knowledge as well as development/maintenance of professional networks and relationships;
- a poster paper for a conference, which is a stand-alone concise research synopsis of a research journey or some aspect of that journey;
- trade conference presentation/talk, intended to focus on practices and working conditions within a profession or industry, often with the intent to stimulate change;
- a published book chapter, often solicited by other researchers or practitioners to produce a body of collected knowledge with respect to some topic of interest;
- a published book, often the culmination of an extended and in-depth research journey;
- a published article in an academic journal, the ultimate outcome sought by university researchers as well as by the institutions they work for;
- an informal public or community forum, where the researcher participates in a dialog about their research with interested members of the public or community, perhaps as an act of reciprocity for research participation and knowledge sharing by the community (see, for example, discussions in Fear et al., 2006; Kovach, 2021; Smith, 2021) or as a way for co-produced research to be shared (see, for example, Smith & McGannon, 2024);
- one or more postings on a social media platform, for more rapid sharing of knowledge, feedback, and commentary (see, for example, Hasim & De Luca, 2011; Koranteng & Wiafe, 2019);

- a discussion, talking, or yarning circle, or other context for oral story sharing and storytelling, usually associated with Indigenous/First Nations research and undertaken in the spirit of reciprocity (see, for example, discussions in Chilisa, 2020);
- a critique, which could take a written or performative form, of current policy, living, or working conditions for society or for certain sub-groups within society, with the intention of highlighting inequities and injustice, and stimulating discussion and perhaps policy or other type of social change (see, for example, discussions in Alveesson & Deetz, 2021, Chap. 5, in particular);
- a series of recommended/planned or enacted actions intended to create research impact and produce change, where the researcher engages more direct actions, often in participation with relevant others, to create change and impact (see, for example, Greenwood & Levin, 2007, Chap. 17; McIntyre, 2008, Chap. 3);
- a technological or social innovation developed/evaluated as part of the research journey, usually paired with other forms of research outcomes to share the research learning underpinning the innovation (see discussions and examples in Cooksey & McDonald, 2019, especially Chap. 22; Patton, 2011; Patton et al., 2015);
- a creative performance or artwork (e.g., in the form of theatre, ethnodrama, visual displays, dance, film/video/multimedia, song, poetry, painting, sculpture), often emerging from arts-based research (see, e.g., Chilton & Leavy, 2020; Greenwood, 2019; Siegesmund, 2024), intended to enhance interest in the research as well as inclusiveness and social impact through the sharing of the artform (see, for example, discussions in Alveesson & Deetz, 2021; Archibald et al., 2022; Denzin, 2024; Poth, 2018, Chap. 10; Saldaña, 2024);
- a research proposal, which is an anticipatory outcome requesting organisational, institutional, government, or community support for a research journey, typically produced in postgraduate contexts (see, for example, Punch, 2016);
- a grant proposal, which is another anticipatory outcome requesting financial support for undertaking a research journey, submitted by an individual researcher or a research team (see, for example, Locke et al., 2013).

Figure 6.1 displays a concept map of issues, considerations, and implications that emerge when producing or co-producing a research outcome of any form. Any research outcome is a joint product of the researcher's journey, the contextualisations that emerged during that journey (including, perhaps, the contiguous/parallel journeys undertaken by others), the pathways taken to navigate the 'Data Triangle', and the researcher's goal is to create or co-create an outcome that is convincing. Convincingness requires that the researcher knows which audience(s) they wish to speak to or work with. Given the diversity of possible forms for research outcomes (as well as the potential for innovating new forms), researchers should rely on pluralist systemic thinking to arrive at a form of outcome that will serve their own purposes and intentions as well as meet the needs of interested/potentially impacted audiences. Such a choice should be made in an organic rather than deterministic fashion, but certain circumstances and contextualisations may alter this dynamic toward a more deterministic form. For example, in research guided by the pattern of guiding assumptions for a positivist inquiry, entirely reliant on quantitative data and statistical analysis, and embedded within an explanatory research frame, the form, shape, language and content of a research outcome is determined by the strictures involved with reporting statistical analyses, a need best met by some kind of formal report. In research guided by the pattern of inquiry for a postmodern inquiry, the array of possible research outcomes is limited only by the researcher's imagination/creativity coupled with awareness of what it might take to impress or convey a viewpoint to a specific audience. In research guided by the pattern of guiding assumptions for an Indigenous/First Nations inquiry, embedded within an Indigenous/First Nations research frame, the choice of possible outcomes may be informed by considerations of reciprocity, participation in production of the outcome, community needs, and respect for Indigenous/First Nations knowledge, something that formal or even informal written reports are unlikely to accomplish, but oral storytelling/story sharing or a performative outcome likely would.

Research outcomes will typically convey multi-layered stories about the researcher's journey, sometimes culminating in proposed or enacted actions intended to produce change in some way, and the different forms of outcomes may allow the researcher to achieve these goals in ways that are most sensitised to the needs of interested parties. Such stories and actions may be conveyed within one or more specific settings. We differentiate research outcome creation/production from co-creation/co-production because, depending upon contextualisations, and particularly the researcher's adopted pattern(s) of guiding assumptions and research frame(s), the researcher may produce an outcome on their own or work in concert with others (e.g., other researchers, participants, or other key stakeholders) to jointly produce the outcome. Employing pluralist systemic thinking will allow the researcher to consider the widest set of potential research outcome possibilities and assess their potential using the convincingness and associated meta-criteria. By doing this, the researcher will consciously reason their way to a specific choice of research outcome, in essence, contextualising that outcome. In some cases, more than one outcome will be required to meet the needs of more diverse audiences.

6.2 Trajectories for Choosing Research Outcomes

Figure 6.1 allows one to trace numerous potential trajectories that can be taken for choosing a research outcome. Contextualisations and actions taken/knowledge gained/learning achieved while navigating the 'Data Triangle" provide the background against which the considerations and implications of possible research outcomes can be critically evaluated. The primary driver for choosing a research outcome to produce will typically be the researcher's intentions, although for some kinds of research, stakeholder/audience expectations may dictate those intentions (as when research is publicly funded or commissioned by an organisation or community or a thesis/dissertation/portfolio is produced to meet postgraduate degree requirements in a university). The simplest non-pluralist trajectory would be taken if a

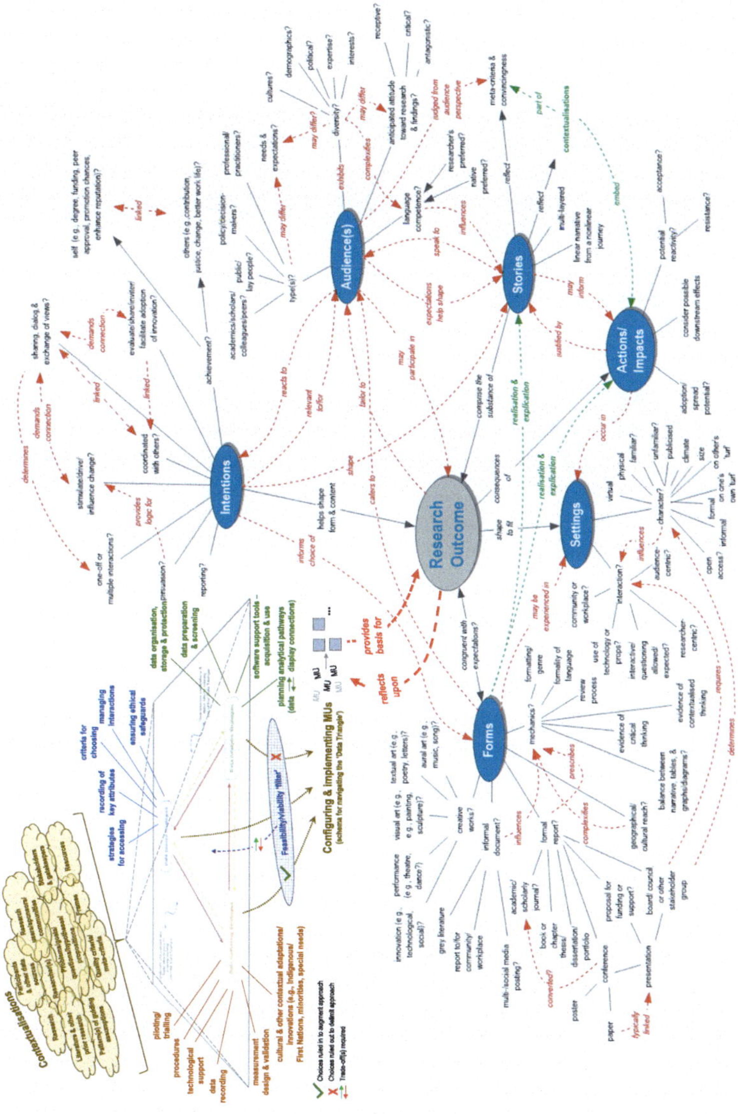

Fig. 6.1 Issues, considerations, and implications for shaping research outcomes after contextualisation and navigation of the 'Data Triangle'

researcher undertook their research journey with the sole intention of achieving a published journal article (perhaps to fulfil promotion or supervisory aspirations). In such a case, the form, setting, and audience is pre-determined as are the rules and expectations for producing and shaping the story and format of the outcome for submission to a specific journal. The initial arbiters of convincingness will be the journal editor and the reviewers they appoint. However, under a wide variety of other circumstances, researcher and audience (key stakeholders are included here) needs and expectations may be more complexly interconnected, and the researcher will want to think carefully about what kind of outcome would be most feasible for the researcher to achieve as well as most convincing to intended audiences. Here, researcher intentions and audience considerations intertwine to different extents, potentially altering the ultimate motivation and purpose of a research outcome. Where the researcher employs multiple types of data (e.g., mixed methods research) or intends their research to speak not only to academics/scholars, but also to practitioners/professionals or potential adopters of an innovation, multiple audiences are implicated, and the researcher will need to decide if a single research outcome will accomplish that goal or they need to produce multiple outcomes. If pursuing a single research outcome (as, perhaps, with a professional doctorate portfolio), the writing task comes with a complex array of additional challenges (in terms of choice of language, the balance between different aspects of the research story, and the conclusions and implications drawn for the different audiences). If the latter, the challenge becomes one of coordinating the different stories that different audiences expect to experience, judging what to include/play up and exclude/play down for each outcome (i.e., tailoring each outcome to a specific audience). Since audiences are arbiters of convincingness, different audiences may require different types of stories in order to be convinced.

Settings are physical or virtual spaces where audience(s) encounter or experience research outcomes. As such, they are an important consideration for shaping the stories and actions implicated in the research outcome. For certain kinds of research (e.g., Indigenous/First Nations or cross-cultural), the setting where audiences/stakeholders experience a research outcome is at least as important as the stories and actions themselves. For example, discussing or yarning about research carried out in an Australian Indigenous community would take on completely different meanings if experienced on their country or away from their country. If done away from country, decolonisation, as a necessary part of the reciprocity process, will likely be counterproductively negated and erode trust, which would completely taint or put at risk the knowledges being shared. If yarning is done on country, decolonisation can be effectively maintained. If the researcher produces an innovation as part of their journey, then the settings where they share their research with potential adopters becomes a critical consideration. For example, if a new medical or agricultural innovation has been produced/evaluated by the researcher under the Developmental Evaluation research frame, the potential impact of the research on adoption rates may be diluted if the experiences with this audience occur in a neutral common area, such as a conference centre or university lecture hall, rather than in hospitals or on farms where connections between meaning and context can be readily and convincingly made.

Since the potential forms of research outcome involve complex mixtures of medium and message (in potentially different settings), various aspects of research contextualisation (such as researcher contextualisation, research frame, pattern of guiding assumptions, mode of knowledge building, stakeholder and gatekeeper expectations) can influence the choice. Patterns of guiding assumptions for a post-modern or critical inquiry can move the researcher toward producing a creative work. Commissioned research within an organisation may trigger the need for a report to the governing board. Research for a postgraduate degree will point to a thesis or dissertation that speaks to a predominantly academic audience, but, for a professional doctorate, may point to a portfolio outcome that speaks to multiple audiences (i.e., academics and practitioners, policy makers, industry professionals, and/or communities of potential or actual innovation adopters). Research in an Indigenous/First Nations community may create the need for oral dialog and story sharing with the community (on their own country) with which the research was carried out. Research guided by patterns of assumptions for a critical or systemic othering inquiry may move the researcher toward a series of research-related postings and recommendations for action and change on a social media platform. Research within an Action research frame may demand that the research outcome take the form of negotiated actions, for implementation in pursuit of change and improvement of localised working practices, between researcher and participants/co-researchers.

It can pay for the researcher to be mindful of at least some of their preferred research outcomes early on in their journey because aspects of the journey may need to be shaped so as to better facilitate the achievement of such outcomes. This is an instance where downstream-oriented thinking about what the researcher wishes to achieve with their research can actively influence the thinking that goes into the choices made for navigating the 'Data Triangle'. In effect, such thinking forms a part of research contextualisation. However, it is important to realise that such downstream-oriented thinking cannot ignore the expectations of potential audiences. No matter what research outcomes are produced or co-produced by the researcher, the shaping and substance of each outcome should draw heavily upon the researcher's journal as a continuous record of the journey. A well- and regularly-maintained research journal supports an autoethnographic account of the researcher's journey, yielding vital prompts to memory, evidence of reasoning chains for the choices and trade-offs the researcher made during their journey, and initial thoughts, impressions, ideas, interpretations, and emotional as well as cognitive, reflective, and reflexive perspectives on the researcher's experiences.

For research where critique, change, improvement, or emancipation is a goal, research outcomes that involve only stories (whether in written or multimedia form or orally delivered) will likely not have the desired audience impact. In such cases, impact is carried, not necessarily in the stories, but in the concrete recommendations or actions the stories point to. Here, the researcher takes on the role of change agent or, at the very least, change provocateur, and their suggested or realised actions may target, for example, social policy, systemic/institutionalised societal conditions, social justice and fairness, improvement of processes, or correction (after illumination) of marginalising or discriminating processes. This kind of research is active, not passive, and comes with a specifically espoused value positioning that must be clearly communicated with the target audience(s). The task of the researcher is to correctly balance story and action in ways that will enhance impact in the desired directions. It is the actions that carry the impact of the research.

6.3 Shaping Research Outcomes

Communicating and sharing what a researcher has learned through their research journey is basically a process of storytelling. The researcher creates or co-creates (in conjunction with other researchers, participants, and/or stakeholders) the stories that depict, unpack, and reflect on their journey. While the final story needs to have a unified appearance and feel in order to have impact, it will usually comprise many layered and interconnecting sub-stories brought together in such a way that the researcher's intentions can be achieved and audience(s) convinced, while, at the same time, meeting the needs and constraints imposed by the chosen story form, setting, and audience(s). Depending upon circumstances, research stories may be associated with, promote, or directly enact actions for innovation and change that the researcher argues are necessary to meet the implications of what they have learned. Successful, convincing, and impactful research storytelling cannot occur without an understanding of the audience(s) that the researcher wishes to speak to or share with in terms of their needs, competencies, hopes, fears, and expectations. Shaping a research story thus requires not only researcher effort to produce, but also adaptive responses to feedback (either anticipated or realised). This means that the researcher is not the sole shaper of their story. They are, of course, a major actor within their story, but other actors play roles as well, some of whom may have actually participated in the researcher's journey whereas others observed or were interested/implicated in or affected by the undertaking of that journey.

6.3.1 Reflexivity in Crafting Research Outcomes

Part of pluralist systemic thinking involves reflexivity, the researcher's capacity to critically and continually re-examine their own positioning with respect to other positionings and perspectives they encounter (see, for example, Bettez, 2015; Kleinsasser, 2000; Smith, 2006). It is a critical form of self-questioning. Flexibility and adaptability flow naturally as reflexive considerations move the researcher toward change. Being reflexive means being constantly on the lookout for feedback that signals important disparities and incongruencies that the researcher needs to address. Reflexivity reminds audiences that the researcher acknowledges and leverages their own humanity and subjectivity. While most discussion of reflexivity tends to occur in the context of qualitative research, an argument can be made that it is equally important for research guided by an objectivist stance and a focus on external causation (e.g., for positivist or critical realist inquiries). The key difference between reflexivity in subjectivist research and objectivist research is when it comes into play: for subjectivist research, reflexivity is built into the research process from the beginning; for objectivist research, reflexivity becomes important (1) when the researcher is compelled by emergent circumstances to consider trade-offs, and (2) when the researcher looks back on their research in light of contextualisations, when considering interpretations, conclusions, and implications. No matter the commencing stance, crafting a research outcome allows the researcher to think reflexively about what they have learned, their role in that learning, problems, issues, and surprises they had to confront during their journey, and impacts they have had on others and that others have had on them. Making such thinking transparent to relevant audiences carries a dual impact: (1) it conveys a genuineness to the audience, that the researcher is willing to change how they think based on what they have learned; and (2) it encourages the audience to consider their own positioning in a reflexive manner. The contextualisation meta-criteria for judging research quality facilitate reflexive thinking, as they ask the researcher to consider their own positioning juxtaposed with other research, with the positioning of participants and other data sources, and with sensitivity to the contexts from which their learning is emerging and in and for which what they are learning is relevant. The realisation meta-criteria facilitate reflexive thinking through critical examination of the congruence between the researcher's adopted pattern(s) of guiding assumptions, frame(s), mode of knowledge building, and choices for navigating the 'Data Triangle'. Congruence enhances research quality and convincingness; incongruence or inconsistencies erode research quality and detract from convincingness. Reflexivity provides a pathway for the researcher to critically examine any emergent incongruencies or inconsistencies. The explication meta-criteria also facilitate reflexive thinking and how best to display it to relevant audiences (through critical examination of unexpected outcomes, acknowledgement of imitations, unpacking value for learning, change and impact, and fertilisation of ideas and actions). The presentational character meta-criterion asks the researcher to explicitly consider audience roles, needs, and expectations as part of their reflexive work.

6.3.2 Crafting Research Outcomes to Communicate Researcher Intentions

When crafting a research outcome, the researcher needs to clearly communicate their intentions, and this requires knowledge of their intended audiences. Intentions will be complexly interconnected with considerations such as whether a research team was involved in the research journey (meaning that inter-researcher dynamics and multiple positionings may influence intentions), whether stakeholders, gatekeepers, and/or participants play a role in determining intentions, the form the research outcome is taking, the setting(s) in which the research outcome is experienced, and/or whether the researcher intends to recommend or undertake specific actions informed by what they learn. Certain patterns of guiding assumptions (e.g., for Indigenous/ First Nations or systemic othering inquires) may influence intentions by signalling an explicit value positioning for the researcher (e.g., decolonisation, illumination of injustice or inequity, identification with a marginalised or oppressed group) If relevant audiences remain unaware of researcher intentions, uncertainty and doubt, flawed inferences about convincingness or unbiasedness, and potentially even suspicion/mistrust may be fostered. If the audience feels or assumes that the researcher's true intentions are hidden, the researcher risks running afoul of ethical expectations about informed consent and lack of deception in research. Countering these risks means that reasoning about researcher intentions, their sources, and implications, must be made transparent. It is one important pathway toward building a connection between the researcher and the audiences they wish to engage with.

6.3.3 Crafting Layered and Interconnected Research Sub-stories

Although research stories can take many different forms, be experienced in different settings by one or more audience(s), and be motivated by varying intentions, there are a number of specific sub-stories that will typically be expected to be embedded, reflected, and integrated, in some way, however briefly, in the final story. There is a typical logical ordering for these sub-stories, but that ordering may be flexible depending upon the setting and the expectations of one or more audiences. These sub-stories may be weighted very differently depending upon intentions, audience(s), form, and setting and such differential weighting may influence the content, depth, and the ordering of the sub-stories. Depending upon form, setting, and audience, some sub-stories may be given almost no emphasis, which means they will not appear except via brief references within other sub-stories. For research outcomes that take the form of a creative work, other than a substantive innovation, the various sub-stories operate behind the scenes to shape, steer, focus, and inform the work. Also, the sub-stories to be discussed in more detail below do not constitute all the possible sub-stories the researcher may wish or need to convey. Contextualisations, unusual or complex research contexts, recommended or enacted actions, development of an innovation, unexpected constraints, or opportunities may point to new sub-stories that need to be conveyed, in terms of setting the stage for and conceptualising the research journey as well as helping to understand/characterise what was learned from any data that were gathered.

Table 6.1 provides a symbolic tabular depiction of how patterns of emphasis for expected sub-stories might play out in a sampling of different forms of research outcome. The bars in the table provide a qualitative impression of the relative importance of and amount of effort (or word space or verbal/thematic content) devoted to developing each sub-story. Where the bars are dashed, the sub-stories operate behind the scenes, to be experienced vicariously and indirectly by the audience through interpretations offered in the research outcome, rather than being directly experienced through reading or verbal communication. Indications of potential audiences and settings are also provided in Table 6.1 as are some general comments about the particular form of research outcome. It is important to remember that this table only focuses on the typical expected sub-stories for research outcomes. Any other sub-stories needed to help create a convincing research outcome would need to be considered as well. For some outcomes, these emergent tailored or specialised stories may be quite impactful, thus requiring serious attention. This is more likely to be the case for research frames and patterns of guiding assumptions that have a decolonising, critical, emancipatory, or participatory intent.

Every sub-story can benefit from considerations of relevant meta-criteria to enhance the prospects that it will contribute to rather than detract from convincingness. Achieving a balance between meeting the demands of the contextualisation, realisation, and explication meta-criteria is vital. From the perspective of creating a convincing research outcome, the explication domain provides the primary meta-criteria for judging convincingness, supported by appropriate reflective/reflexive consideration of the contextualisation and realisation meta-criteria domains (researcher positioning is especially relevant to reflexive intent). In the discussions that follow, we will emphasise the role that each meta-criterion can or should play in the crafting of specific sub-stories. The dynamic nature and interconnectedness of the meta-criteria domains (recall Fig. 4.4) ensures that no sub-story stands in isolation from other sub-stories. Any research outcome, especially one that is in written or oral form, typically embodies a linearised distillation and representation of a research journey, simply because that is how storytelling unfolds. However, the connective tissues gluing the outcome together typically emerge from nonlinear linkages between different aspects of the journey. For creative research outcomes, such as performance and art forms, the nonlinearity of the research journey may be more directly or indirectly captured and represented (see, for example, discussions in Alvesson & Sandberg, 2021, Chap. 8; Saldaña, 2024). In any case, the meta-criteria allow one to evaluate sub-stories by focusing attention not only on the sub-story itself but also its connective tissue shared with other sub-stories. Table 6.2 presents an impressionistic representation of where each meta-criterion might be useful, for which sub-stories, and with what emphasis. In the discussions to follow, it will be useful to continually refer to Tables 6.1 and 6.2 as the discussions will expand/clarify the symbolic stories conveyed in the two tables.

'Precis' Sub-story

A 'precis' sub-story, often referred to as an abstract, provides a detail-light helicopter overview of the research journey being conveyed and gives foreshadowing glimpses into some of the other sub-stories that audience(s) will encounter. This sub-story invites audience(s) to invest in following the holistic and unified final story. The

Table 6.1 Qualitative impressions of patterns of sub-story emphasis for different forms of research outcome

Sub-Story	Academic Journal Article	Doctoral Thesis	Doctoral Portfolio	Conference Presentation	Presentation to a Board	Community Meeting	Book Chapter	Creative Work	Innovation
Precis									
Introductory									
Prior research/ knowledge building									
Conceptualisa- tion									
Navigating the 'Data Triangle'									
Learning from/meaning of data									
Pulling it all together									
Potential audiences	Primarily academic researchers & scholars	Examiners, then primarily academic researchers;	Examiners, then academics as well as managers; industry professionals; practitioners	Academic researchers as well as some professionals	Executives, managers, professionals	Community leaders, Elders, and members; lay people; interested members of public	Academics and/or practitioners, depending upon focus of the research journey	Academics; professional;, communities; members of the public	Practitioners; potential adopters; industry professionals
Potential settings	Print or online; internationally or locally	Primarily print; may have an associated oral presentation	Primarily print; may have an associated innovation or set of change actions	Conference venue meeting rooms (may later be published in proceedings)	Meeting room in board's organisation; accompanied by written report	Town hall or other public meeting space; meeting place on Indigenous/ First Nations country/land	Primarily print, but online access also possible	Varied media possible - theatre stage, online, film, print, recording	Settings where innovations will be deployed; commerciali- sation settings
General comments	Word limits and reviewer comments typically constrain this outcome.	This form provides optimal opportunity to fully develop all sub-stories	This form provides optimal opportunity to fully develop all sub-stories	This form is constrained by time limits, highly focused; invites Q & A	This form typically commissioned as a consultant report; invites Q & A	This form invites Q & A interactions or more open discussions	Word limits and reviewer comments typically constrain this outcome.	Translating research learning/ knowledge into the form can be challenging	An innovation will typically have a developmental & evaluative story accom- panying it

Table 6.2 Impressionistic representation of meta-criterion emphasis relevant to each sub-story

Domain	Meta-criterion	Sub-story						
		'Precis'	'Introduc-tory'	'Prior research/ knowledge building'	'Conceptual isation'	'Navigating the 'Data Triangle''	'Learning from/ meaning of data'	'Pulling it all together'
Contextualisa-tion	Juxtaposition-ing with Other Research		◉	●	◎	◌		◎
	Researcher Positioning		◉	◌	●	●	●	◉
	Positioning of Participants & Other Data Sources		◌	◎	◌	●	●	◉
	Contextual Sensitivity	◌	◉	◉	◎	◎	◉	◉
Realisation	Internal Coherence			◎		●	●	◌
	Analytical Integrity			◎		●	●	◎
	Extensional Reasoning			◎	◌	●	◎	◎
Explication	Value for Learning, Impact & Change	◌	◉	◎				●
	Fertilisation of Ideas & Actions			◎	◎			◎
	Handling of Unexpected Outcomes						●	◉
	Acknowledge-ment of Limitations					◎		◉
	Presentational Character	●		◉		◉	●	●

● relevant for entire story

◎ relevant for most aspects of the story

◌ relevant for several aspects of the story

○ relevant for one or two aspects of the story

'precis' sub-story thus has significant power to influence audience choice whether to pursue or ignore further understanding of the researcher's journey and the knowledge it has built. In short, it is an invitation to the audience to 'turn the page' to see what happens. It is difficult for research to be convincing if the 'precis' sub-story cannot entice further experiencing of the researcher's story. In creating this sub-story, the researcher needs to think concisely and clearly about what to emphasise. There will likely be limited space to detail sub-stories beyond a single sentence summary statement for each.

The'precis' sub-story should be gauged against the presentational character meta-criterion, as the way the sub-story is crafted is determined by relevant audience considerations coupled with constraints imposed by the particular form of research outcome being created (e.g., word limit, expected content, composition). Thus, the 'precis' sub-story should speak directly to intended audiences, avoiding jargon and esoteric language, and excessive detail. Aspects of the 'precis' sub-story should reflect contextual sensitivity, so the audience gets a feel for the context(s) the research is conducted in, as well as value for learning, change and impact, so the audience gets a sense of what the researcher has learned and its implications.

'Introductory' Sub-story

An 'introductory' sub-story sets the stage for the research and provides motivational insights into the purposes of the research, unpacks some of the contextualisation of the research (e.g., signalling research contexts, participant contexts, and stakeholder contexts), and makes clear the positioning and intentions of the researcher. It is an extension of the precis, providing a sub-story which invites the audience to commit to experiencing the remaining sub-stories as they are presented and integrated into the holistic and unified final research outcome. To achieve this, the 'introductory' sub-story should show, in transparent terms, where the researcher is coming from (i.e., contextual considerations) and where they want to get to (i.e., their goals and intentions), encouraging the audience to come along for the ride. Thus, the main goal of the 'introductory' sub-story is to persuade relevant audience(s) that the researcher's journey is one worth following and this is best done by providing foreshadowing statements and brief descriptions, rather than detailed argumentation and logical defence (which is accomplished via other relevant sub-stories).

The 'introductory' sub-story should be evaluated against all four contextualisation meta-criteria. This sub-story is meant to draw in and engage relevant audiences, inviting them to commit to complete their experience with the research outcome. Including some important aspects of prior research (juxtapositioning with other research to create a bridge between what the researcher intends and what has been learned before), clarifying aspects of researcher positioning (e.g., researcher's motivation and intentions), and pointing out the context(s) in which the research will occur (linking and tailoring intentions sensitively to context) all help the audience to see where the researcher is coming from and where they want to get to and why. Some insight into the positioning of participants and other data sources enhances engagement, particularly if participants belong to relevant audiences. Finally, engagement may be enhanced if the researcher briefly foreshadows some of their major findings and sensitises those findings to the context in which they emerge (demonstrating value for learning, change and impact).

Prior Research/Knowledge Building' Sub-story

A 'prior research/knowledge building' sub-story, often but restrictively referred to as a literature review, explores previous research done and knowledge built by others (including, perhaps, the researcher's own previous and relevant work), critically reflects on that prior research and knowledge, and leverages what is sensible, meaningful, and relevant for the researcher's purposes when setting out their current research questions/propositions/ hypotheses. Here, the research stories from others are experienced by the researcher during their journey and the researcher renders critical judgments about their convincingness when evaluating the take-away messages from each story. Pluralist systemic thinking is important to showcase in this sub-story as it influences which/whose prior research/knowledge is explored, under which patterns of guiding assumptions, in which contexts and research frames, reflecting which mode of knowledge building, and for what purposes. Without effective use of pluralist systemic thinking, the more likely it will be that the fields of prior research/knowledge explored and critically evaluated will be curtailed/restricted, which will overly constrain/limit contextual fit as well as downstream strategic choices for navigating the 'Data Triangle'. This is especially the case when disciplinary, institutional, or paradigm norms are given too much power to direct the researcher's attention. It is much harder for research to be convincing if the basis for its genesis is overly constrained and too selective or one-eyed.

The 'prior research/knowledge building' sub-story should be logically ordered and argue progressively for the research context(s), problem, and questions/propositions/hypotheses. In order to do this, the researcher may need to build interconnecting linkages with a conceptualisation sub-story, where appropriate, to establish a framework within which the research questions/propositions/hypotheses are set out. Here, 'where appropriate' might be judged against the researcher's adopted pattern(s) of guiding assumptions, since not all research journeys will commence with a preformed conceptual framework. The researcher should be critical but fair in their handling of prior research and knowledge building efforts. Pluralist systemic thinking will seek to achieve a balanced view by considering prior research that adopts different perspectives, even if they yielded contradictory results (which, interestingly, automatically creates a gap which the researcher can leverage to focus research intentions that could potentially resolve the contradictions). Where gaps have been observed in prior knowledge building efforts, the researcher should clearly identify the gaps and provide logical arguments for why they see them as gaps. This can help the researcher to set the stage for their own research later in the sub-story. Part of what makes the 'prior research/knowledge building' sub-story convincing to an audience are judicious choices of the most relevant prior work to review and fair-minded critique of what was learned from such prior efforts that could be used to inform the researcher's current journey.

The'prior research/knowledge building' sub-story should primarily be evaluated against the juxtapositioning with other research meta-criterion as this entire sub-story concerns what the researcher learns and can utilise from prior research and knowledge building efforts to help shape their own research journey. However, when evaluating such prior efforts, the researcher will make supporting judgments involving all three realisation meta-criteria, to decide whether they are convinced by the research outcomes conveying such prior efforts (the researcher essentially becoming the audience for those prior efforts). The researcher should also consider the effectiveness of the contextualisations of those prior efforts. As the researcher examines the research outcomes for those prior efforts, they will find several of the explication meta-criteria useful: value for learning, change and impact to evaluate the contributions and take-away messages from each effort; fertilisation of new ideas and actions to evaluate the future directions for research put forward in each prior effort (which the researcher may build up/follow-up on in their own journey); and presentational character to evaluate the effectiveness of the research outcome in meeting the researcher's expectations as a relevant audience member. Thus, in this sub-story, the researcher builds up arguments and a basis for their own research journey through the critical evaluation of what has come before, and those critical evaluations can themselves be assisted by the meta-criteria.

'Conceptualisation' Sub-story

A 'conceptualisation' sub-story may appear as a discrete sub-story or distributed and integrated, as appropriate, with other sub-stories, especially the 'prior research/knowledge building' sub-story, that focus on the early parts of the researcher's journey. This sub-story allows the researcher to build up a coherent picture of how they have conceptualised relevant aspects of their journey, including, where appropriate, arguing for and presenting theoretical frameworks and anticipations. This sub-story will have a close relationship with evolution of the researcher's questions/propositions/ hypotheses, which may be integrated into the conceptualisation sub-story, or presented separately (e.g., at the end of the 'prior research/knowledge building' story). The conceptualisation sub-story should reflect important aspects of contextualisation. This sub-story needs to be conveyed in such a way that intended audience(s) can follow the logic and argumentation associated with the conceptualising effort. It is a way for the researcher to show how they moved from their starting point(s) to a conceptualisation that underpins their journey and serves as a touchstone for the decisions made about strategies for navigating the 'Data Triangle'. As part of this sub-story, the researcher may wish to make clear the pattern(s) of guiding assumptions, mode of knowledge building, and research frame(s) they have adopted for this journey. These choices may influence how the conceptualisation is shaped, presented, and targeted. However, this is by no means the only sub-story where such revelations can occur. For example, if patterns of guiding assumptions become a critical issue for evaluating prior research/knowledge building efforts, the researcher's decision on what to adopt for their journey may be more usefully developed in the 'prior research/knowledge building' sub-story. If, instead, the researcher wants to make clear that their starting point for making decisions about how to navigate the 'Data Triangle' is the pattern(s) of guiding assumptions and research frame(s) or perhaps the mode of knowledge building they have adopted, then those choices for adoption can be more effectively included in the 'navigating the 'Data Triangle'' sub-story. If the researcher wishes simply to be upfront with where they are coming from, their choice of research frames, mode of knowledge building, and pattern(s) of guiding assumptions and might best be discussed in the 'introductory' sub-story.

The 'conceptualisation' sub-story may not always be required or desirable depending upon the pattern of guiding assumptions the researcher is following and the research frame they are working within. However, if such a sub-story is developed, then the meta-criteria can be of assistance, most especially the contextualisation meta-criteria. Most importantly, when developing a conceptualisation for the research journey (e.g., a conceptual or theoretical framework), the researcher leverages their own positioning, assisted where needed by considerations of relevant prior research/knowledge building efforts, potential participant considerations, and nuances appropriate to the intended research context(s). In addition, the researcher may apply extensional reasoning to explore the potential reach of their conceptual framework and, of course, utilise their developed framework as a focus for new ideas and actions to pursue in their own research journey (evaluated using the fertilisation of ideas and actions meta-criterion).

'Navigating the 'Data Triangle'' Sub-story

A 'navigating the 'Data triangle' sub-story, often referred to as the methodology story, conveys, as completely and transparently as possible given audience expectations and other constraints, how the researcher navigated the 'Data Triangle', the decisions and trade-offs they made, the configuring and learning from any pilot testing/ trialling efforts. Sufficient detail should be provided for relevant audiences to make coherent judgments about aspects of convincingness. This sub-story is one where anticipation of audience expectations and competencies strongly influences the level of detail the researcher provides and the sophistication of the language used. Unfortunately, many research outcomes do not get this level right, either speaking over the heads to a disinterested audience, or omitting too much detail for a critical audience. There is a balance to be achieved in this sub-story and getting that balance wrong can have dramatic impacts on the researcher's capacity to convince. There is no recipe or formulation that can be offered for producing the right balance, the researcher must make their own best judgment, in light of their contextualisations, intentions, knowledge of the audience(s), and expectations of key stakeholders and gatekeepers. The researcher's journal can be a key resource to draw upon when fleshing out this sub-story. There should be a natural and smooth transition between the 'prior research/ knowledge building sub-story, the 'conceptualisation' sub-story (if the researcher has crafted one), the research questions/propositions/ hypotheses, and the 'navigating the 'Data Triangle'' sub-story. The 'Navigating the 'Data Triangle'' sub-story has a number of interconnected sub-sub-stories embedded within it, to be discussed below. However, the precise composition and ordering of these sub-sub-stories will at least partly depend upon the researcher's choice of research outcome to craft, audience considerations, research frame(s), adopted pattern(s) of guiding assumptions, and the 'Data Triangle' navigation strategies they will be implementing.

Choice of Pattern(s) of Guiding Assumptions

If not discussed or indicated in another sub-story (e.g., the 'introductory' or 'conceptualisation' sub-story), the researcher's adopted pattern(s) of guiding assumptions should be signalled in the 'navigating the 'Data Triangle' story. It may not always necessary to include such a sub-sub-story if the pattern is clearly conveyed by the researcher's choices (i.e., tacit adoption, which is often the case in certain disciplines where positivist inquiries are the norm). However, it is becoming an increasingly important aspect of transparency for researchers to openly state their pattern choices and their reasons for making them (part of such arguments may make reference to the mode of knowledge building the researcher is undertaking). Moving from tacit adoption to explicit adoption signals, to relevant audiences, what they can expect to see and what pattern-specific research quality criteria are relevant. If the researcher's adoption choices vary from what has been the case in prior research or knowledge building efforts, this creates an obligation to present explicit arguments. For certain research outcomes, such as a manuscript for submission to an academic journal or a thesis, dissertation, or portfolio, explicit arguments for pattern adoption can facilitate appropriate inferences about convincingness from reviewers and examiners.

Research Context(s)

Here, the researcher provides some details on the context(s) in which they conducted their research, even if it was laboratory-based research, along with their reasoning for the choice(s). If the researcher conducted research in the field, then they should convey a sub-sub-story about the groups/organisations/institutions in which they gathered data. They should also detail any relevant historical events that impacted on that context. Irrespective of the researcher's adopted pattern(s) of guiding assumptions, the researcher should include details on their positioning as a researcher with respect to the research contexts.

Research Frame(s)

In this sub-sub-story, the researcher would argue for and describe the research frame (or synergistic combination of frames) that they adopted for their research. The choice of frame(s) should logically emerge from the development of the researcher's research goals and questions/propositions/ hypotheses in conjunction with research contexts, pattern(s) of guiding assumptions, key stakeholder interests and involvement, the researcher's positioning, the positioning of participants and other data sources and the audience(s) they intend their research to be speaking to. The precise ordering of above three sub-sub-stories will depend on the relationship between research frames, guiding assumptions, and research context(s). Choice of research frame may lead the researcher to adopt a particular pattern of guiding assumptions; choice of research context(s) may demand a certain pattern of guiding assumptions be adopted; choice of research frame may suggest the types of research contexts that the researcher is interested in, and so on.

Research (MU) Configuration

This sub-sub-story conveys the exact structural composition of MU activities and how they are configured in time and space. Here, audiences learn how MU-related activities unfolded during the navigation of the 'Data Triangle' in terms of data source choices, data form(s) gathered, and contextualised data gathering. The MU configuration employed may have been informed by research frame(s), pattern(s) of guiding assumptions, the centrality of time for connecting data gathering strategies or events, the centrality of an intervention or change event, and the breadth or depth of learning pursued. In short, specific scoping decisions may have influenced the choice of MU configuration and these should be unpacked to some degree for the audience. If a mixed methods approach, involving both quantitative and qualitative data forms, is employed, then different MUs will focus on the different data forms (via a simultaneous, sequential, hierarchical, case-based, hybrid or evolutionary configuration), unless a hybrid/merged data form is gathered. If an explicit intervention or change event is of interest, the researcher should discuss any manipulations of context (e.g., in the physical, psychological or social environment) and control procedures (such as a placebo condition, the use of random assignment of participants to treatment groups or a double-blind control procedure), implemented (typically associated with a positivist inquiry), types of group or occasion comparisons to be made, and the number of occasions on which data will be gathered. If a case-based MU configuration is employed, discussion of the choice and contextualisation of the case(s) will be required.

Choices of and Engagement with Participants and Other Data Sources

In this sub-sub-story, the researcher would describe their data choice strategies and implementation processes (involves choices of people, documents, organisations, communities, meetings to observe, events, and so on, as appropriate), the nature of the sample(s) sought, how they gained access to their sample(s) (e.g., how they passed relevant gatekeepers), and how they dealt with any attrition in participants. In research guided by the pattern of guiding assumptions for a positivist inquiry or intended to explore/understand external causation, the researcher would also describe their intended population, as this provides the audience with one target for any generalisations. The researcher may also include a report on the nature of their obtained sample(s) and response rates and, where appropriate, make some comments on the suitability of their sample size(s) and/or composition(s). If the researcher is undertaking one or more case studies, they should spell out how they selected the case study organisations as well as how they chose the participants (or non-human data sources, such as documents, photos/pictures) inside each organisation. In all cases, the researcher should discuss the positioning of participants and other data sources with respect to the research context and problem. For some types of research (e.g., grounded theory), the researcher may need to signal that their choices of data sources were distributed in phases, interlaced with data gathering and analysis activities (e.g., using the theoretically-driven, negative case, and/or sequential/saturation data source choice strategies). Another critical aspect of this sub-sub-story story concerns the issue of ethical constraints and ethical approval(s) for the research, should such be required. Here, the researcher should describe the steps they went through to gain approval from their university, professional organisation, community, and/or another institution to conduct their research. For Indigenous/First Nations research, the pathways taken to gain access to relevant communities, via community Elders or other key individuals or groups, should be discussed as well as the assurances given with respect to reciprocity and protection/appropriate sharing of culturally-privileged knowledge.

Instruments or Measures

This sub-sub-story is essential for any research conducted under the pattern of guiding assumptions for a positivist inquiry or that uses a Measurement data-shaping strategy in some form (e.g., within a survey, descriptive, or explanatory research frame). Here, the researcher would operationally define their constructs, clearly identify the various roles to be taken by their measures (i.e., independent, dependent, moderating, mediating, or extraneous variables, perhaps displayed in a structural relations or network model) and describe how they constructed any instruments or measurement processes. Alternatively, if the researcher uses another researcher's published instrument, they should briefly review any prior research that demonstrated the validity, reliability, and perhaps cultural suitability of that instrument. If the researcher constructed their own questionnaire or other type of instrument, they need to set out the logical or empirical processes by which they constructed the instrument, ordered its sections, designed its scales, validated its measurement properties, and organised its layout, instructions, appearance and delivery/collection systems (e.g., mail, web-based, face-to-face, group-based). If the researcher conducted a cross-cultural comparison study, they should argue for, or demonstrate the equivalence of, their instruments in each cultural setting in which they were used.

Procedures

In this sub-sub-story, the researcher should describe: how the data were gathered (including all instructions to participants; details of any training undertaken by people involved in data gathering, such as interviewers or questionnaire administrators); how the data were recorded (including any mechanical, computerised, or web-based support for data gathering/recording) or created (if the Generative data-shaping strategy is used), in what contexts (including, for example, details on physical layout of interview rooms, procedures for dealing with interruptions or withdrawals from participation); and how long interviews took to conduct or how long questionnaires took to complete, on average. This is the 'nuts and bolts' sub-sub-story, where the researcher explains how they actually obtained the data they end up working with in their analyses. If the researcher conducted a pluralist or mixed methods investigation, they should fully unpack the procedures associated with each of their data gathering strategies/MUs. Transparency and attention to relevant details are very important in this sub-sub-story. Where appropriate, the researcher should also include any discussion on how they adapted their data source strategies or data gathering procedures 'on the fly' to handle emergent issues (a more common occurrence in interpretivist/constructivist or other kind of non-positivist inquires). If they carried out participant observations, they would need to set out their procedures for organising and recording their observations, including how they structured any field notes. Any gaps in this sub-sub-story could leave the audience wondering how the researcher obtained specific types of information.

Pilot Testing/Trialling of 'Data Triangle' Strategies (Where Relevant)

In this sub-sub-story, assuming the researcher conducted a pilot testing/trialling phase, they should summarise how they organised it, who was involved and how they were chosen/selected, what information was gathered, what they learned, and what changes were made to their approach and its procedures and/or instruments as a consequence. What is important in this sub-sub-story is not necessarily a full analysis of the pilot testing/trialling data gathered, but a story about the insights the researcher gained about the viability of the 'Data Triangle' navigation choices they have made and any resulting changes or adaptations they implemented. If the researcher was pilot testing, refining, and validating a new or heavily adapted measurement instrument, then this sub-sub-story would be extended by conveying the validation evidence (reliability evidence would normally be reported as well).

Approach(es) to Data Analysis

In this sub-sub-story, the researcher should detail: their planned approach(es) to analysing the data, including what software support systems they used, if any; how they prepared/transformed their data for analysis (e.g. questionnaire coding for a quantitative survey, transcription for recorded interviews, conversations, meetings); what statistical procedures they planned on conducting if they gathered quantitative or hybrid/merged data (including preliminary analyses for data screening, assumption checking and construct validation and comments on the suitability of their obtained sample for planned analyses), or the approach(es) to analysis of qualitative or hybrid/merged data which they used to address specific research questions; and the reasoning behind important data display choices they made. In research guided by pattern of assumptions for non-positivist inquiries, the fact that the researcher may commence data analysis at the same time or nearly the same time as data gathering means that they will have to unfold this story for the audience and show how results from earlier analyses fed into later data gathering and analysis. This will typically be a nonlinear story and may require the researcher to cross-refer to their 'choices of and engagement with participants and other data sources' and 'procedures' sub-sub-stories.

The 'navigating the 'Data Triangle' sub-story should be evaluated primarily against the three realisation meta-criteria. This is where the researcher makes and defends their choices for navigating the 'Data Triangle' and such choices need to be consistent with quality criteria expectations relevant to the pattern(s) of guiding assumptions they have adopted associated with their researcher positioning. Researcher positioning should also be reflected in any discussions about the insider/outsider status of the researcher, as this will influence choices and implementation procedures for data source and gathering strategies. Above and beyond this, the realisation meta-criteria help the researcher to gauge the quality of their various strategy implementations, irrespective of choice of pattern of guiding assumptions, with respect to convincingness. Since navigating the 'Data Triangle' considers data source selection and engagement, the positioning of participants and other data sources becomes a useful meta-criterion to apply. Prior research may be drawn upon to support certain choices made, which makes juxtapositioning with other research a relevant meta-criterion for aspects of the 'navigating the 'Data Triangle' sub-story. Contextual sensitivity needs to be effectively demonstrated when making choices for data gathering strategies to be implemented in specific contexts as well as in adaptations arising from pilot testing/trialling. Certain choices or trade-offs may have been made during the researcher's journey and any limitations those adaptive choices impose need to be acknowledged. This can especially be the case when the researcher is discussing what was learned and what was done following pilot testing/trialling. Presentational character may also play a role, especially if the research outcome being produced is accompanied by externally-imposed expectations, guidelines, or rules for format, content, and sub-sub-story ordering.

'Learning from/Meaning of Data' Sub-story

The 'learning from/meaning of data' sub-story, sometimes referred to as the results and discussion story, conveys what has been learned from the data that have been gathered and how. This sub-story must be closely linked to and logically follow from the 'conceptualisation' and 'navigation of the 'Data Triangle' sub-stories. In this sub-story, the balance between narrative text, numbers, tables, and figures/diagrams, and other types of data displays needs to be carefully thought through; such balance depending upon a number of considerations such as the form and setting for experiencing the research outcome, audience expectations and competencies, and stakeholder/gatekeeper expectations and competencies. Getting the balance wrong risks losing the audience, which adversely impacts on convincingness. Getting the balance right makes it much easier for audience(s) to get on board with the researcher's intentions and interpretations. This balance is much more challenging to achieve in cases where the researcher intends to speak to multiple audiences of differing levels of expectations and competencies (e.g., academics, practitioners, workers, lay people or members of the community/public). This sub-story may comprise several interconnected sub-sub-stories, each focusing on learnings from specific analytical efforts.

It is now common practice to combine results and discussion of immediate interpretation of those results, whereas many years ago it was more common to present results and discussion in separate sub-stories (this is still the practice in some disciplines such as psychology). The discussion aspect of this sub-story should focus on interpretations, and not on the drawing of conclusions, generalisations, speculations, or implications. What the researcher wants to accomplish is to display the meanings of what they have learned to relevant audiences, consistent with the pattern of guiding assumptions under which the data were gathered. The focus should be on the research questions/propositions/hypotheses with minimal tangential or side excursions; audiences should be able to clearly connect them to what is being conveyed. For research gathering qualitative or hybrid/merged data, the researcher needs to ensure that their sub-story stays close to the data; it must be transparent so that audiences can see how the researcher has moved from the data to their interpretations. As the research questions will tend to be more general, it may take a range of analyses and data displays, at various levels, for the researcher's story to emerge in ways that address their questions. For research gathering quantitative data, the researcher needs to ensure that their story draws upon appropriate statistical methods, reports appropriate statistical and/or graphical evidence, and that it draws appropriate conclusions from that evidence (see, for example, Cooksey, 2020). Here, the researcher would likely use a sequence of sub-sub-stories, analysis by analysis, discussing the outcomes for each. Here, the research propositions/hypotheses will tend to be more specific, and it will often be convenient to organise the story on a question-by-question basis. If quantitative, qualitative, and/or hybrid/merged data are gathered, the discussion of analyses may be sequential or integrated and the researcher would need to use their best judgment to decide which (the MU configuration, whether simultaneous, sequential or hierarchical, will also have a bearing on this decision).

Where quantitative data are concerned, the reporting of outcomes from statistical analyses is relatively formulaic (see, for example, Miller, 2005). One aspect of the sub-story should address processes undertaken to screen data for errors, anomalies, congruence with assumptions, and missing data patterns as well as, where appropriate, how specific variables were coded for analysis (see examples in Cooksey, 2020, Procedure 8.2 and Fundamental Concepts III and VI). Another aspect is transparently communicating the criteria for judging statistical significance and effect sizes that the researcher will utilise (such criteria must then be consistently applied throughout the sub-story). Numbers, tables, and graphs need to be carefully chosen or crafted to communicate precisely what the researcher wants to convey (refer to Sect. 2.3.2; see also, for example, Knaflic, 2015; Krum, 2013; Zaumanis, 2021, for some guidance here). Swamping the audience(s) with unnecessary numbers and tables or overly cryptic graphs (i.e., 'overkill') runs a strong risk of over-whelming, thereby obscuring, the researcher's message. This is especially a danger if the researcher simply cuts-and-pastes tables and graphs directly from the statistical software package that produced them (which must be identified, but which invariably includes much more detail than any researcher needs), without critically evaluating exactly what needs to be included and what can be excluded from such presentations. Not using the right numbers or a correctly configured table or graph (i.e., 'underkill') risks rendering the researcher's interpretations (perhaps even their competence) open to question. The critical part of the 'learning from/meaning of data' sub-story, based on quantitative analyses, is the narrative the researcher weaves around the statistics, tables, and graphs they employ (Cooksey, 2020 provides concrete illustrations in his Appendix C). If the researcher engages in model building/testing, the sub-story should have a part that addresses how well the data met the assumptions required by the modelling procedure (see, for example, discussions in Cooksey, 2020, Procedure 7.13, Fundamental Concepts V and VI). Statistics, numbers, tables, and graphs should be seen simply as vehicles for short-hand representations of patterns and meaning in the data. They should support the narrative of the sub-story and be actively discussed rather than simply displayed.

Where qualitative data are concerned, the reporting of analysis outcomes is much less rule-driven and formulaic compared to quantitative analyses; approaches are very diverse and often innovative (see, for example, discussions and illustrations in Bazeley, 2021; Grbich, 2013). Qualitative analyses are typically iterative in nature, often supported by qualitative analysis software (which must be identified, but which do not do the analyses, they support the analyses developed by the researcher). The researcher can develop coding systems (see Saldaña, 2021), create memos, build tables, maps, and displays using such software (refer to Sect. 2.3.3; see also, for example, Jackson & Bazeley, 2019; Kuckartz & Rädiker, 2019; Silver & Lewins, 2014; see also Knaflic, 2015, Krum, 2013, and Zaumanis, 2021, for some tips on the creation/construction of graphics outside of specific qualitative analysis software packages), but these then need to be judiciously evaluated for their usefulness in helping the researcher convey meanings and interpretations. How the data were prepared and checked should also form part of the sub-story. The researcher also needs to decide which kinds of interpretive pathways their data might support or inform, such as focusing on or emphasising narrative accounts, events, codes or themes, participant roles, time, contexts, relationships, meanings, or cases. Any data display created needs to remain as close to the data as possible, so that audiences can see the links and reasoning that went into creating the display. The researcher needs to decide on the proper balance between narrative text and data displays. Data displays are very useful devices for conveying particular aspects of data interpretation but require discussion so that their meaning and intent is clear (see, for example, the numerous illustrations in Miles et al., 2020). Judicious choice of quote material to support interpretations is also required; such materials may become part of the narrative text or be embedded in tables but serve the specific purpose of illustrating particular interpretations or displaying particular meanings to support the emergent analytical sub-story.

Where mixed data types are concerned (typically arising from MU configurations that many researchers would characterise as 'mixed methods' approaches), the researcher needs to clearly spell out the reasoning behind the approach they are taking and explain how the various data types are being brought together in a coherent analytical framework (e.g., how qualitative data were quantitised (i.e., the transformation of qualitative data or analytical outcomes (such as data source attributes and codes) into a quantitative form (see, for example, Onwuegbuzie & Leech, 2005, p. 287; Sandelowski et al., 2009); how quantitative data were qualitised (i.e., qualitative labelling, description, and/or characterisation of quantitative patterns (see, for example, Happ, 2009; Onwuegbuzie & Leech, 2005, p. 287); also see discussions and illustrations in Tashakkori & Teddlie, 2010, Chap. 16 through 19; Poth, 2018, Chap. 9). Sometimes, sequential reporting of analysis outcomes (e.g., qualitative results first, then quantitative results, or vice-versa, depending upon the sequential or hierarchical MU configuration involved) is called for, in which case, each type of data will be the subject of its own sub-sub-story. At other times, a more integrative approach, where hybrid/merged data are available (see, for example, discussions/

illustrations in Bazeley, 2018 or Gobo et al., 2022) is necessary, especially in simultaneous, hierarchical, and case study MU configurations. Hybrid/merged data displays may involve the creation of matrix displays with quantitative counts represented numerically or symbolically, cause or concept maps with quantitative strengths of linkages reflected in the thickness of links, or quantitative graphs overlaid with qualitative and clarifying text. Certain hybrid/merged data forms may be accompanied by their own specialised and innovative data displays (refer to Sect. 2.3.4; see also, for example, Bazeley, 2018, with respect to social network analysis; Gobo et al., 2022, with respect to the Symlog method and the event calendar and intervey methods; Kane & Trochim, 2007, with respect to participant-led concept mapping). While such hybrid displays often involve quantitising, it is also possible to produce qualitised displays. Again, attention to the proper balance between narrative, numbers, tables, graphs/diagrams, and other data display devices is essential in the sub-story. Unless audience(s) can see how the quantitative, qualitative, and/or hybrid/merged data were brought together, any interpretations put forward by the researcher will not be convincing.

The 'learning from/meaning of data' sub-story is all about the data analysis vertex of the 'Data Triangle', what was done, and what was learned. The analytical integrity meta-criterion is extremely useful in evaluating this sub-story. Internal coherence is also key, as it is internal coherence that influences what can/should be inferred or understood from analyses. Extensional reasoning becomes relevant if the researcher tries to extend what analyses are telling them to other contexts and participants. When the researcher tries to make sense of what the data are telling them, the contextualisation meta-criteria become useful to apply, i.e., the researcher's own positioning (or reconsidering or withholding their own positioning) becomes relevant, as do the positionings of participants and other data sources. Analyses should follow logical pathways to addressing research questions/propositions/ hypotheses in context and interpretations should be sensitised to the context(s) in which they emerge. Analyses may reveal unexpected outcomes and anomalies that should be explicitly handled, perhaps with additional or revised analyses. Presentational character becomes relevant where the research outcome being produced is accompanied by externally-imposed format and content expectations. Presentational character is also relevant to judgments about how well the 'learning from/meaning of data' sub-story connects with the needs, competencies, and expectations of its intended audiences, in terms of linking and communicating analysis results and displays to interpretations.

'Pulling It All Together' Sub-story

The 'pulling it all together' sub-story, sometimes referred to as the conclusions and implications story, conveys the overall conclusions for the researcher journey in terms of what has been learned, what the findings mean, and the implications the findings may have for other people, places, times, circumstance, conceptualisations and so on. If this sub-story is conveyed well, even research beset by many problems and surprises can be rendered convincing. This sub-story depends heavily upon the researcher's journal for getting interpretive and contextual details right. As well, this sub-story will refer to and critically reflect on other relevant sub-stories within the research outcome and may be accompanied by researcher reflexivity with respect to how they have been impacted/changed by their research journey. This sub-story itself comprises several interconnected sub-sub-stories, which may focus on broad conclusions, surprising or unexpected findings, implications for specific purposes/stakeholders (e.g., implications for other researchers, theory, methodology, practitioners, social or institutional policy and practices, workplaces, and other settings, future research journeys), as well as the limitations that attend the current research effort. It is via this sub-story that bridges between the researcher and intended audience(s) are either built or burned. Based chiefly on this sub-story, the audience decides whether the journey has been worth the effort, whether they have been convinced by what has been claimed/concluded, and whether they can clearly see the take-away messages for themselves, significant others, and/or their constituencies/communities. In this sub-story, the goal is less about what the researcher thinks about their own research and more about whether they have done enough to allow intended audience(s) to see and decide for themselves; it is more about making meaning for others than making meaning for themselves. What is said and done in this sub-story strongly influences the ultimate impact of the researcher's journey and the knowledge they have built by undertaking that journey.

The 'pulling it all together' sub-story is really a three-part series of linked sub-stories. In the first sub-sub-story, the researcher operates at the highest conceptual level to draw major conclusions from their research, focusing on their research questions/propositions/hypotheses. They want to characterise the main contributions to knowledge and learning and identify their key take-home messages for audiences and they want to do this in a way that acknowledges their contextualisations, research frame(s), pattern(s) of guiding assumptions, positionings, scoping, and configuration. The researcher should also juxtapose what they have learned with what had or had not been learned in prior research/knowledge building efforts. While new, perhaps very recent, literature may be introduced in this sub-sub-story, it is not reviewed in the same way as for the prior research/knowledge building sub-story. Rather, the researcher should use such relevant prior research as touchstones to connect their research with a pre-existing body of knowledge.

In the second sub-sub-story, the researcher spells out how the conclusions, learning achieved, and knowledge produced might carry meaning or be useful for other contexts and audiences. Here is where the researcher sets out what they see as the implications of their research for those target contexts and audiences. Depending upon the specifics of the research journey, the researcher may want to discuss implications for theory, for practice, for training, education and development, for methodology, for policy, for stakeholders, for social justice, equality, and/or equity, for innovation diffusion and further development, for communities, for the environment at large, and so on. Some speculation is permissible in this sub-sub-story, but the researcher should be very careful that they do not extend too far beyond what their data or contextualisations could reasonably support.

In the third sub-sub-story, the researcher takes a step back to discuss the caveats, constraints and limitations, and future potentials associated with their research (i.e., engaging in critical reflection, reflexivity, and anticipation). Here, the researcher wants to be reflexive and self-critical. Despite common beliefs to the contrary, the researcher does not 'shoot themselves in the foot' by pointing out the limitations of their research. Instead, they are displaying maturity and critical reflection in understanding the limits to what they were able to learn and therefore limits to what relevant audiences might take away from the research. Part of this discussion should be to remind audiences of any steps that the researcher took to offset or mitigate some limitations; in short, to explore any trade-offs made to make the research feasible whilst remaining convincing. There is another purpose for this type of critical self-disclosure and that is to set up springboards for the researcher to identify future directions for research. Once they have discussed the limitations and caveats associated with their research, the researcher can set out what they see as potential future directions for research and the limitations that attended their current research journey may point to some concrete future directions. Here, the researcher is thinking forward to address the question 'where to from here?'.

The pulling it all together' sub-story is where challenge to achieve convincingness in the eyes of relevant audience(s) is won or lost. All the meta-criteria have a role to play in evaluating this sub-story and its interconnected/interwoven sub-sub-stories. The contextualisation meta-criteria come into play when conclusions are made, and implications are drawn. There may be references to prior research/knowledge building that are used to help make interpretive comparisons and contrasts and these should be assessed for their quality. Positioning of both the researcher and participants and other data sources are highly relevant to drawing defensible conclusions and implications, and all such conclusions and implications need to reflect contextual sensitivity. When shaping conclusions and implications, the researcher must reflect on how effective their choices were for navigating the 'Data Triangle', and those choices can be evaluated using all three realisation meta-criteria. The entire sub-story should be subject to judgment under all five explication meta criteria, which means that various sub-sub-stories should reflect value for learning, change and impact. Unexpected and anomalous outcomes and surprises should be fully and openly discussed and related to the conclusions and implications. Limitations must also be openly acknowledged and discussed and related back to conclusions and implications. In both cases, the issue is what qualifications or constraints might attend the researcher's learning claims. As with other sub-stories, presentational character is uniformly relevant, especially if the research outcome being produced is accompanied by externally-imposed guidelines and expectations. Presentational character is also relevant to judgments about how well the 'pulling it all together' sub-story connects with the needs, competencies, and expectations of its intended audiences, in terms of conveying meaning and linking audiences to implications.

6.3.4 The Uniqueness of Proposals or Funding/resource Requests as Research Outcomes

When anticipatory/prospective outcomes (e.g., research proposals or funding/ resource requests/ applications) are considered (see, for example, Locke et al., 2013; Punch, 2016), what is produced may follow directly from one research journey but is less of a research outcome itself and more of an anticipated outcome if the new research journey is supported. The pattern of sub-stories is somewhat abbreviated and restricted to illuminating a researcher's plan for a research journey. The sub-stories focus on how what is proposed relates to what has been done before and to what will hopefully be learned, from whom or what, in what contexts, for what purposes, and, importantly with what resource requirements. 'Learning from/meaning of data' sub-story will not make an appearance, except as anticipatory statements in other sub-stories; the 'navigating the 'Data triangle' story will be significantly terse and typically relatively light on implementation/procedural considerations. Where research funding applications are concerned, there will typically be very strict guidelines for shaping/presenting the overall story as well as certain sub-stories. Two additional sub-stories will also typically be requested: one focusing on budgetary and other resource requirements (the 'request') and one focusing on the researcher(s), their background, expertise, publication history and other prior research outcomes, and their capacities for carrying out the research being proposed (i.e., their competency and capacity positioning). When creating a proposal or request for funding or resources, the contextualisation and realisation meta-criteria come to the fore to help paint a convincing picture of what the researcher plans to do. Value for learning, change, and impact should be anticipated, and presentational character must be attended to. Convincingness will be reflected in approval of a proposal or granting of funding or resource support (in total or in part, and, if in part, this could reflect some doubts as to convincingness or some constraints within the context(s) that the reviewers/decision makers are operating in).

6.3.5 Linearising a Nonlinear Journey

The researcher's journey is inherently nonlinear, full of stops, starts, backtracks, trade-offs, and reconsiderations. However, for research outcomes intended to connect with specific audiences, painting a picture of the journey as it unfolded in real time and space is difficult if not impossible to achieve in certain forms. Whether the research outcome is in a written or an oral form, the sub-stories must unfold in a linear fashion, walking the audience from its starting points to its ending points. In the course of creating such outcomes, the sub-stories can and should interconnect with other sub-stories (which does help the researcher to depict some of the nonlinearities in their journey), however, the overall outcome will follow a linear progressive pathway. The ordering of the pathways typically follows the ordering of the sub-stories depicted in Table 6.1; it is the emphasis, content, and density of the sub-stories that varies according to the researcher's intentions, audience needs, expectations, and competencies and the setting(s) in which the outcome will be experienced (as symbolically represented in Table 6.1).

Theses/Dissertations/Portfolios

In written research outcomes produced to meet some educational or training requirement (such as a thesis, dissertation, or portfolio for a postgraduate degree), the content is expected to be more densely and fully unpacked, so that examiners/assessors can make the most informed judgments about meeting the requirements for the qualification. For theses and dissertations, examiners/assessors, who are typically other academics with research and supervising experience, will be the initial target audience for convincingness and effectiveness in storytelling is critical for the researcher (although their identities may not be known at the time of writing, depending upon particular institutional and cultural rules governing such assessments). For postgraduate portfolios, the initial audience of examiners/assessors will typically be more diverse, drawing upon both academic as well as professional/practice-based/industry-based expertise. While supervisors might be considered the initial audience, they also help to shape the outcome and its content, ensuring it meets institutional requirements, thus playing a role in creating the circumstances for convincingness as well as judging it. If examiners/assessors recommend that changes be made before the final outcome is accepted/passed, conformity with those requests is generally expected, unless solid arguments against doing so can be presented. Generally, in postgraduate research outcomes, other target audiences come into play once the original outcome is reshaped into new outcomes for wider dissemination and sharing (such as conferences papers, seminars, or journal articles). Such reshaping typically sacrifices content density and may recalibrate the emphasis on various sub-stories. In all cases, though, the linearity and ordering of the sub-stories tends to remain largely constant.

If the postgraduate research outcome is expected to be presented/defended, via an oral examination or viva format, in an academic, semi-public, or public setting, then the sub-stories need to be rendered much more concisely and tailored to showcase the researcher's competence and confidence, and capturing and keeping audience interest (especially where more diverse audiences are to be encountered in semi-public or public settings) will be important to pursue (via careful choice of language, balance between narrative talk, tables, and figures/diagrams, typically favouring more visual displays, avoidance of jargon). Such sessions have an expectation that the researcher will field questions from the audience, and it will be important for the researcher to respond in an open and thoughtful manner, rather than by deflecting or being defensive. The researcher must own what they have done and learned. Asking questions is one way that audience members have of gauging/probing convincingness. Note that in an oral examination or viva session, the researcher has the opportunity to inject more of themselves into their sub-stories but needs to attend closely to the amount of time spent on each sub-story. Getting bogged down in the 'introductory' and 'prior research/knowledge building' sub-stories is a sure recipe for being forced to overly compress later sub-stories, which risks damaging convincingness. The 'pulling it all together' sub-story should not be given short shrift.

Submissions for Academic Publication

In written research outcomes intended for submission, review, and publication in an academic journal, the journal itself will set out expectations and guidelines that the researcher must conform with. Such guidelines will tend to reinforce the linearity of the sub-story progression, the formality of language, constrain the density of content expected (through word or page limits and editorial preferences), and constrain the formatting of the submission. The initial audience for submitted manuscripts are the journal's editor/relevant associate editor, and the reviewers they assign. If reviewers/editors suggest changes to the manuscript before final acceptance, conformity with those requests is generally expected, unless solid arguments against doing so can be presented. The challenge will be to effectively incorporate requested changes whist maintaining conformity with journal policies and practices as well as the researcher's goals. This may require the researcher to make trade-off choices between what content to retain and what to remove, when adapting to reviewer and editor requests. The balance between narrative text, tables, and figures/diagrams is important to consider in such submissions because getting the balance wrong (e.g., by including too many figures or tables, including tables with unnecessary detail; trying to rely too much on narrative to present/discuss findings that would be better and more efficiently displayed in a table or figure; using too much or too little quote material in reporting findings from qualitative data) risks damaging convincingness. For journal publications, both efficiency and effectiveness of storytelling is critical. Generally speaking, nonlinearities in the research journey are more difficult to incorporate in journal manuscripts because of the presentational constraints imposed.

Reports

For written research outcomes in report form (generally not intended for publication), how such outcomes are shaped hinges critically upon audience/stakeholder needs, expectations, and competencies and the researcher may be expected to implicitly know these things. Commissioned written reports (such as consulting reports, program/organisational/community reviews, reports to funding bodies) may be expected to include a highly modified 'introductory' sub-story providing an executive summary of the research and listing recommended actions/changes for consideration. Language needs to be less formal and while the linear ordering of the sub-stories tends to remain the same, their content and density will tend to vary significantly. What is found and what is learned by the researcher through their journey and the implications for the group/organisation for whom the research was undertaken will be far more of interest to the audience than how the journey was undertaken and how that journey relates to prior research/knowledge building efforts (unless the report involves specific comparisons to such prior work). Reports presented to one's own organisation/community need to be crafted somewhat differently to reports crafted for other organisations/communities. The former carries insider status and the latter outsider status. For an insider report, certain contextual details may be taken for granted as all members of the organisation/community can be assumed to share them. However, the researcher will need to make clear how they managed the tensions between their role as researcher and their role as organisational/community member; these are contextual details that cannot be neglected, if the report is to have a chance to convince. For an outsider report, getting all contextual details correct about the organisation/community is imperative as the audience will be unconvinced or even become actively hostile if the researcher makes unwarranted/unfounded assumptions about the organisation/community. The researcher must also take steps to ensure that the nature of their relationship with the organisation/community and how it was established is transparently conveyed, lest suspicions about political agendas cloud any judgments of convincingness by the audience.

If reports are expected to be presented in an institutional, community, or public forum, then issues such as capturing and maintaining audience interest in what the researcher is saying, rebalancing and rejigging relevant sub-stories, in favour of offering more visual displays, and clearly spelling out meanings and implications for the audience, should guide the transformation of the report from written to oral form. The researcher would likely be expected to field questions from the audience and the researcher would do well to remember that the audience will likely be comprised of lay people, professionals, and/or practitioners rather than academics when formulating answers to those questions. Overly investing in the use of jargon, being patronising, and talking down runs the run of eroding convincingness and may actively put the audience off (which gives convincingness no chance to shine). The researcher needs to connect with and stay close the audience and such connection and closeness is easier to make and maintain if the researcher downplays their distinctiveness from audience members, speaking to the competencies of the audience rather than reinforcing their own.

Conference Papers

Research outcomes in the form of conference papers, presentations, or posters are subject to rules and guidelines specific to the conference. For academic conferences, where the audience comprises primarily other academic researchers, the content and density of the sub-stories can mirror the expectations for journal articles, except with less density and detail. In short, a conference paper can be likened to a scaled-down journal manuscript. For professional/trade conferences, where the audience comprises primarily industry/business/educational professionals and/or practitioners, the content and density of the sub-stories can mirror the expectations for reports. Submissions to conferences may or may not be reviewed for inclusion; if reviewed, then expectations for adaptation and response to criticisms and requested changes would mirror expectations for journal articles. Research outcomes for conferences, in whatever form, will invariably be much more compressed. Certain sub-stories in the linearised journey may be greatly abbreviated in order to make space for other more relevant (from an audience perspective) sub-stories. It is not uncommon for conference papers, presentations, or posters to present partially completed research journeys, providing interim insights into an on-going journey. Researchers may do this in order to garner early feedback on progress so far, and some of that feedback may be powerful, meaningful, or sensible enough for the researcher to make trade-off choices for research activities yet to come in their journey.

The poster form of conference paper is the most constrained and challenging to organise because it typically must speak for itself. There is no space for extended narrative; reliance on visual displays and short concise statements/paragraphs and dot points is heightened. Space for the sub-stories is highly constrained so use of space must be efficient. Convincingness will be enhanced by emphasising the sub-stories that will build a more effective and efficient (in terms of time taken to read and absorb the essential messages, which could be a minute or less) experience for the passing audience. The linearisation and emphasis/de-emphasis of sub-stories should be captured in the poster in a way that walks the passing audience through them in a simple and timely manner.

For conference presentations, the ordering and density of the various sub-stories may be varied dramatically, in light of audience needs, expectations, and competencies. Time is typically short for a conference presentation, so the researcher needs to be very selective about the sub-stories and details they include. Gray (2021, p. 881) identified four distinct approaches to argument in a conference presentation: the inductive approach (where the researcher presents their facts/findings and then their conclusions); the deductive approach (where the researcher presents their conclusions first, then provides the specific facts/findings that led to the conclusions); the discursive approach (where the researcher presents two or more sides to an argument, then shows using facts/findings, which of the sides is supported); and the 'situation, options, and the way forward' approach (where the researcher explains the situation, presents pros and cons for possible options, and argues for the best way forward). The deductive approach is easier to understand from an audience perspective; it inverts the linear ordering of sub-stories by presenting aspects of the 'pulling it all together' sub-story first, then unfolding the pathways taken to arrive at that sub-story. The discursive approach or the 'situation, options, and the way forward' approach works better if the researcher's goal is to address controversial or contentious issues. The inductive approach, according to Gray, is more interesting to listen to but harder to understand, because the audience must remember what has been presented before in order to correctly judge the conclusions.

Book Chapter or Book

Research outcomes in the form of a book chapter or a book, focusing on part or all of the researcher's journey, can be more open in terms of how the research sub-stories are sequenced and how dense/extensive the content of each is. For book chapters, contributions are generally solicited or invited by the book's editor(s). Book chapters typically do not convey the entirety of a research journey, only the highlights, central messages, and implications for various audiences as well as indicating connections with the overall themes of the book (see, for example, Schuelka & Maxwell, 2016; Sinclair, 2014). A lot of work is done by the 'introductory' sub-story, which is often elaborated to display aspects of a 'prior research/knowledge building', 'conceptualisation', 'navigating the 'Data Triangle', and 'Learning from/meaning of data' sub-stories. The primary work of a typical research-oriented book chapter is carried in the 'pulling it all together' sub-story, which may itself be elaborated by focusing on meanings/implications for readers of book and for other research reported within the book. In short, a book chapter requires extra connective tissue to tie the messages of the chapter in with the overall messages of the book and its other contributors.

A book offers the most open written form for a research outcome, and it need not display a linearised version of the journey. If the entire book is focused on one research journey, the book may take the form of an autoethnography, where the researcher not only presents their research journey, but integrates, with their sub-stories, reflexive observations related to their experiences while undertaking the journey, what they thought, what they felt, how it changed them, and so on (see, for example, discussions in Jones & Adams, 2024). If the book is focused on several, perhaps interconnected, journeys (undertaken by the researcher and/or by others), then snippets of various journeys may be displayed in ways to support discussion/critiques/arguments about some theme or issue, and depending upon the overall thrust of the book, only certain sub-stories from each journey may be included or considered (e.g., for a book focusing on methodological approaches to research on a topic of interest, 'conceptualisation' and 'navigating the 'Data Triangle' sub-stories may primarily be drawn upon, see, for example, Cooksey (1996). For a book reviewing the literature on a topic, the 'prior research/knowledge building', 'conceptualisation', and 'learning from/meaning of data' sub-stories from different journeys may be primarily of interest, see, for example, Hattie (2008). If only their own research journeys are involved, the researcher will need to leave judgments of convincingness to the readership of the book. If the research journeys of others are included, the researcher may render their own convincingness judgments about those research outcomes.

6.3.6 Co-Produced Research Outcomes

If the researcher's journey has been undertaken in partnership with other individuals, a group, or a community, it is often under the guidance of patterns of assumptions for a participatory, Indigenous/First Nations, or systemic othering inquiry, or within an action research, developmental evaluation, Indigenous/First Nations, or pan-disciplinary research frame. In such circumstances, research outcomes may be co-produced by the partnership, rather than by the researcher alone (see, for example, discussions, illustrations, and caveats in Gill et al., 2023; Graham et al. 2023; Martin, 2010; McLean et al., 2023; Turnhout et al., 2020). For Indigenous/First Nations research, co-production occupies a special, ethical, and political place in the building and sharing of culturally-privileged knowledge (see, for example, Kettle, 2019; Maclean et al., 2022; Latulippe & Klenk, 2020; Zurba et al., 2022).

Co-produced research outcomes can be challenging to develop as it involves research where academic researchers and participants/community/lay members shared in the process of creating and sharing knowledge instead of academic researchers working in isolation. With such diversity comes power dynamics and relationship management. For effective, i.e., convincing, co-produced research outcomes to be realised, the co-creators should be motivated to actively (as well as safely) contribute and be on the same page with respect to the goals they want to achieve and the actions they want to promote or take. All voices need to be heard and all inputs considered, even if ultimately not adopted. Part of the 'introductory' sub-story in such work should open up and discuss how dynamics were managed through the co-production process. In the action research frame, participant and contextual needs should take precedence over the researcher's needs, as the goal is to achieve change and improvement at the local level where participants are most closely involved, and this should be reflected in the crafting of the research outcome (Greenwood & Levin, 2007, address some of these considerations, as well as some of political issues discussed below, in their Chap. 17). If multiple research outcomes are to emerge from the same research journey, there may need to be negotiation about credit and ordering of authorship for each outcome produced. If different outcomes take different forms in different settings, different audiences may be involved, and this should also influence the shaping and communicating of outcomes as well as any prioritisation of inputs to the outcome. Creative differences will also need to be resolved, as this can affect the final form of a co-produced research outcome (pattern(s) of guiding assumptions will also exert an influence here).

The co-production processes could be particularly troublesome if there are political overtones to the relationships between the researcher and the partners or between the partners themselves (something that becomes particularly pointed in Indigenous/First Nations research). Those issues will need to be resolved or at least actively attended to as part of the co-production process. In Indigenous/First Nations research, co-produced research/knowledge may be embedded within oral and/or written forms of storytelling (Archibald et al., 2022; Christensen, 2012) and, as Kovach (2021, pp. 244–245) argued, '[s]tories hold knowledge'. Such storytelling comes with certain expectations:

> The teller and the listener are equally active: the listener is not passive. Oral storytelling is a non-directive process, and the story listener is self-determining. (Kovach, 2021, p. 245)

Satisfying different political agendas may require the co-production of different outcomes, each tailored to a specific agenda, a challenging process, especially where political goals are inherently or overtly contradictory. Without appropriate resolution or compromise, the co-produced outcome will be unconvincing for most, alienating for some, and counter-productive for all.

The settings in which co-produced research stories are shared are important considerations, that is, on whose turf/country the stories are being shared. Settings have power and choice of setting can say a lot about the political stance of the research. Furthermore, Maclean et al. (2022) and Latulippe and Klenk (2020) argue that co-production of knowledge is not just about sharing roles in creating research/knowledge outcomes (i.e. integrating Indigenous/First Nations and western knowledges), but for the co-production process to reflect Indigenous/First Nations leadership, values, and traditions, thereby subordinating/decolonising western dominance to Indigenous/First Nations cultural knowledge and knowledge development. It is thus an issue not only of sharing but also of dominance. This is clearly a political issue, one fraught with issues of power, values, ownership, governance, colonisation, and marginalisation, often associated with contested spaces (as Kettle, 2019 put it). Smith (2021, p. 273) went even further when she persuasively argued that research co-production in Indigenous (First Nations) spaces is intimately intertwined with Indigenous (First Nations) activism. She summarises her meaning as relating the 'more generic question about why we do what we do either as researchers and/or activists to questions about the potential ways in which Indigenous activists and Indigenous researchers can collaborate to advance Indigenous interests at local, national and international levels.' Smith is therefore calling for a fundamentally different kind of co-production, one that is driven by larger-scale change imperatives. These issues and intentions must be grappled with, and a way forward evolved in order for Indigenous/First Nations co-produced knowledge to takes its rightful stage (Maclean et al., 2022 provide a clear case example). How these issues are tackled contributes heavily to contextualisation, presentational character, and convincingness.

6.3.7 Creative and Innovation-Oriented Research Outcomes

Under certain patterns of guiding assumptions, such as for an Indigenous/First Nations, systemic othering, postmodern, participatory, or critical inquiry, or within the Indigenous/First Nations, othering/emancipatory, or pan-disciplinary research frames, innovative and creative research outcomes may be produced. Such arts-based research outcomes (e.g., poetry, painting and other artworks, visual images, visual storywork, music, dance, social media postings on platforms such as Tiktok and Instagram, theatrical performance, ethnodrama; see, for example, Archibald et al., 2022; Chilton & Leavy, 2020; Grbich, 2013, Chaps. 10–13; Poth, 2018, Chap. 10; Saldaña, 2024; Siegesmund, 2024) may be intended to reach more diverse audiences for a variety of purposes, including stimulating thought and change; questioning of status quo; imagining new/alternative futures/scenarios; making social observations/commentary in new and creative/inventive, perhaps more impactful, ways; illuminating the complexities and nonlinearities of work/life experiences; exposing/representing/characterising social injustices and oppression; communicating emotions about and interpretations of the world with a wider appeal to the senses, all within some specific context(s). Arts-based research outcomes are creative, inventive, and multi-mode/multi-media in form and their intent may be open, loose, and unfocused, done simply to create something new and different; or focused, to convey specific meanings, emotions, and messages, in a different way. The sub-stories behind arts-based research outcomes may remain largely hidden from the audience, leaving the creative outcome to convey the essential messages and knowledge, or they may accompany the creative outcome in another form (something that Siegesmund, 2024 argues should become a more widespread practice, rather than just leaving creative works to speak for themselves as research outcomes). In some cases, the researcher may work hand-in-hand with participants/communities to produce a creative research outcome (i.e., co-production) or may rework what participants say and experience into a creative form that conveys the emotions and meanings of those experiences, while keeping as close to the participants' original voices as possible (Grbich, 2013, for example, illustrates this latter process in her discussion of poetic inquiry, a process she called 'reframing', p. 130). In all cases, creating an arts-based research outcome is a pathway to achieving much greater, potentially immersive, audience impact than could be achieved by textual/narrative outcomes. Creative research outcomes that present/embed controversial or even threatening ideas can generate impact simply through the discourses and commentaries they stimulate.

For research that develops/refines/evaluates/markets a specific social, policy, or technological innovation (within a developmental evaluation or an action research frame, for example), the innovation itself is a creative outcome and the research that underpins and surrounds that innovation helps to shape it and move it toward wider acceptance and adoption (see, for example, discussions in Cooksey, 2011; Guertler et al., 2020; Patton, 2011; Quintane et al., 2011; Van der Have & Rubalcaba, 2016). In the end, the research goals are likely to be (1) enhancing the chances that the innovation will achieve the goals it is designed to achieve and (2) enhancing the opportunities for, and/or patterns of, adoption. In innovation research then, there are, in fact, dual research outcomes: the innovation itself, and the research story behind the innovation. If done in the context of a professional doctoral program, the latter research outcome may take the form of a portfolio (see illustrations/discussions in Cooksey & McDonald, 2019, Chap. 22). If action research yields an innovation, wide-spread adoption is likely less a goal than effective localised adoption and implementation to achieve a change or improvement objective. Fear et al. (2006) discusses critical engagement as a process of moving from an academic mindset in research to a practice/community-oriented mindset; it is a form of contextualised action research. In such research, producing actions and changes in communities of practice are natural outcomes of a research process designed to engender co-production.

6.3.8 Actions as Research Outcomes

Under certain patterns of guiding assumptions, such as for an Indigenous/First Nations, postmodern, participatory, or critical inquiry, or within the Indigenous/First Nations, othering/emancipatory, developmental evaluation, or action research frames, actions or recommended actions may constitute a research outcome, although typically embedded within a more holistic research story. Action-oriented outcomes reflect the change/emancipation intentions of the research and translate what the researcher has learned into what they think should happen as a consequence of what they learn. Actions may be direct, in the shape of activism and pursuit of social justice (often an intention of Indigenous/First Nations or systemic othering research; Smith, 2021, for example, is strongly in favour of an activist intent for Indigenous/First Nations research) or indirect, in the shape of influencing or directing others to take specific actions, as when the researcher provides a series of recommendations for actions, perhaps to an audience of stakeholders that commissioned the research in the first place. In either case, the researcher is creating (or co-creating, depending upon circumstances and contexts, such as would occur within the action research frame) a research outcome that is explicitly value-centric and intended to motivate change. Sub-stories may guide the actions/recommendations but remain largely behind the scenes or only tersely explored, except perhaps for the 'introductory' and 'pulling it all together' sub-stories which may be more concretely represented. This, of course, depends heavily upon the context(s) in which the research journey was undertaken and facilitated. Presenting recommendations in a research report is one major reason why an executive summary of the research may be a required sub-story.

6.4 Enhancing the Impact of Research Outcomes

Correctly balancing the various aspects of research stories to better meet the needs and expectations as well as competencies, of target audience(s) is essential. That balance is where impact resides. Of course, what impact means depends upon which audience perspective the researcher attends to. For an academic or university perspective, impact may refer to enhanced citation rates for journal articles, enhanced research reputation, or enhanced opportunities for obtaining research funding. For a professional or practitioner audience, impact may mean improvements in practices and work processes. For a community or the general public, impact may mean that the research influences public policy and community circumstances in beneficial ways. For government departments/officials, impact may mean that the research informs sound/beneficial public policy creation and change. For business leaders, CEOs, managers, principles and the like, impact may mean that the research provides or signals pathways to realising enhanced prosperity and, perhaps, social/environmental as well as fiscal responsibility for the organisation s/institutions they lead. For an Indigenous/First Nations community, impact may mean that Indigenous/First Nations culture, traditions, and knowledge are prioritised, respected, protected, and self-determined (i.e., not shaped or dictated by the colonising culture) or that actions are signalled or taken to rectify injustices and colonial oppression; impact is therefore embedded in and engendered by reciprocity. For the researcher, impact may mean career progression/promotion, new/different research opportunities, or access to new networks of colleagues. For mentors or supervisors, impact may mean satisfaction with their effect on the researcher's capabilities and career. For funding bodies, impact may mean that sound/convincing return is achieved for the money/support invested and provided.

There is, however, an important caveat to be mentioned here. For some perspectives on research (held, for example, by universities, governments, and research funding bodies), enhancing research impact means enhancing measurable quantitative indicators of impact, such as citation rates, impact ratings, and assessment of research outlet rankings. These indicators may actually drive funding and promotion policies as well as reputational/quality rankings but they will likely also have the unfortunate effect of stifling creativity in research outcomes. Doing 'good' research gets reduced to doing research that is seen, by significant institutional stakeholders, to conform to a limited perspective on what counts as convincing and meaningful. However, such indicators typically apply only to research published in academic/scholarly journals and effectively reduces judgments of convincingness to decontextualised numbers. All sense of research context is lost by using such indicators. While there has been a recent increase in more qualitative assessments of research impact, those assessments are typically done by individuals and groups who were not necessarily part of the researcher's intended audiences. The point being made here is that, while convincingness is in the eye of the audience, it may not be judged by the appropriate audience.

If the researcher intends to target multiple audiences, it means there are likely to be multiple and potentially contradictory impacts, all of which may not be achievable within the scope of a single research outcome. Thus, the researcher needs to use their best judgment to decide how best to orient their research outcome(s) toward achieving/demonstrating impact and pluralist systemic thinking can be employed to inform and support such judgments. Achieving the most impactful balance can be facilitated via a number of potentially interconnected pathways for creating a convincing research outcome.

6.4.1 'Nuts and Bolts'

First and foremost, the researcher needs to choose the right audience(s) to target because nothing convinces less than 'speaking to the wrong crowd'. As well, the researcher needs to carefully choose the setting(s) in which audience(s) will encounter and experience the research outcome to ensure that their experience is congruent with their expectations. Research outcomes should avoid overuse of esoteric language and jargon; even in the most academic forms of outcome, excessive reliance on esoteric language and jargon can tend to obscure rather than support convincingness. The researcher should strategically consider the best blend of narrative, tables, graphs, figures and/or diagrams to employ in their research outcome, especially if it is in a textual or narrative form. Complex graphs and tables in analysis reporting can be taxing for an audience, especially where audience competencies diverge from what would be needed in order to understand them.

Conforming with expectations about format and content of a research outcome may also enhance the impact of a research outcome if it is congruent with audience or stakeholder expectations. However, such conformity can become a straitjacket that constrains possibilities (e.g., restrictive editorial policies for some journals may, if allowed to affect the research outcome, detract from convincingness) for achieving impact if it requires that essential aspects of the researcher's story be removed, reduced, diluted, restated, or recast. This may move the researcher to consider another outlet or form for their research outcome as an adaptive trade-off response.

6.4.2 *Reflection*

The researcher should consider adopting a reflective approach when creating or co-creating a research outcome, which considers new learning and knowledge against the backdrop of contextualisation and prior learning and knowledge building. Reflection involves demonstrating contextual sensitivity in interpreting and reporting what has been learned/concluded. Reflection should also influence the tailoring of each research sub-story to better suit audience competencies and interest levels, and this involves looking at what is being creating through the eyes of the audience (although not as deeply as a reflexive approach would demand). Finally, being reflective means being open about the limitations associated with the research (i.e., adopting a 'warts-and-all' perspective). This latter aspect can contribute quite effectively to convincingness as long as it is genuine and manifests critical thinking.

6.4.3 *Reflexivity*

In many cases, being reflective in the creation of a research outcome will not go far enough in terms of convincingness. The researcher could, instead or in addition, consider adopting a reflexive approach when creating or co-creating a research outcome, which produces self-awareness, changes in, and new learning for the researcher(s). Such reflexivity provides a way for the researcher to know where they stand so that they can appropriately and fairly gauge/judge/evaluate the stances of others (see, for example, discussions in Bettez, 2015; Hesse-Biber, 2015; Kleinsasser, 2000). Even under patterns of guiding assumptions that pursue an objectivist stance and the understanding of external causality and lawfulness of behaviour, researchers can benefit from adopting a reflexive mindset. Reflexivity has the benefit of allowing the researcher to critical evaluate where they are coming from juxtaposed against/with where others are coming from. The drawback is that reflexivity can feel threatening, even anti-objectivist, to researchers adopting such patterns, and, as a consequence, may lead to non-pluralist non-systemic thinking that is dismissive of other pathways to building knowledge. This fear and the consequences it spawns is counterproductive and antithetical to a truly cumulative human science. Pluralist systemic thinking is the strongest defence against uncritical conformity with disciplinary norms and paradigmatic hubris. Reflexivity should be considered as an embedded and routine aspect of pluralist systemic thinking.

6.4.4 *Capturing the Audience*

Creating a research outcome with the target audience(s) in mind means that the researcher cannot create with only their own needs and preferences driving the process. Audience considerations should always play a role, not only in facilitating convincingness, but also in capturing their attention and commitment to learning from the researcher's journey. For example, the researcher can actively emphasise/demonstrate how interpretations, conclusions, and implications are directly relevant to audience(s) or to relevant stakeholders. This helps audiences to answer the question 'so what?' The researcher can be very clear about what they see as the potential future directions for research following from their own journey. This helps audiences to answer the question 'where to from here?' The researcher can actively invite critical questioning of what has been learned by encouraging dialog. This is also a way of signalling the researcher's reflexive interest.

6.5 Looking to the Future

6.5.1 Where We Are Now

The central purpose of this book was to argue for how pursuing unity in and of knowledge can be achieved by harnessing the diversity of available approaches and stances as well as by creating new approaches and stances. Thus, unity is, quite simply, knowledge. Diversity emerges from the complex interconnectedness of systems associated with human life and exists on multiple fronts (e.g., patterns of guiding assumptions and pattern-specific research quality criteria; research frames;, modes of knowledge building; local, institutional, national, international, cultural, political contexts (to name a few); data sources, data gathering strategies, data analysis strategies for navigating the 'Data Triangle'; trade-offs that can be made to adapt to existing and emergent constraints; varieties of research outcomes) and each front offers pathways of insight into human life. Thus, those pathways of insight constitute ways of approaching social and behavioural research and, therefore, ways for building knowledge. We have seen that knowledge building/social and behavioural research learning cannot occur unless the contexts in which it is embedded and the contexts it is interconnected to are well understood and mapped out. That mapping then provides the background against which the researcher makes their choices for how their research journey will be conceptualised and focused, which instances of prior research/knowledge building efforts they will draw upon, what strategies they will employ for navigating the 'Data Triangle', how they will make sense of what they have learned, and how they will convey what they have (and have not learned) to others. We have reinforced the value of understanding a researcher's identity and of keeping a research journal focused on their research journey. We have explored the concept of convincingness as the primary general goal of any research journey, which is vested in judgments made by relevant audiences for the research outcomes. We have further related judgments of convincingness to twelve meta-criteria for research quality, which are meant to operate above the level of patterns of guiding assumptions. The meta-criteria target three crucial aspects of convincingness: how effectively the research journey has been contextualised (involving four specific meta-criteria); how effectively the data- related activities have been realised (involving three specific meta-criteria), and how effectively the researcher has managed the explication of their journey via a specific research outcome (involving five specific meta-criteria). The meta-criteria are essentially a first-level pathway toward achieving unity from diversity, where diverse patterns of guiding assumptions and the research they underpin can be judged on an equal footing.

Pluralist systemic thinking facilitates the achievement of unity from diversity. We have seen that pluralist systemic thinking embeds a critical process whereby possibilities are evaluated for their potential to addresses the researcher's questions/propositions/hypotheses in a feasible as well as convincing manner against the background of relevant contextualisations. Working through potential choices requires conscious and reasoned consideration of the costs and benefits associated with each. Only then are non-viable choices removed from further consideration, thereby pruning the possibilities into a more manageable set of options. By doing this, the researcher knows what they have ruled in and ruled out and why, without automatically conforming to normative considerations. We have seen that pluralist systemic thinking moves the researcher beyond the constraints imposed by mixed methods thinking without negating the potential benefits that accrue from such thinking. Pluralist systemic thinking prevents premature adoption of research strategies without first thinking about contextualisations, from the researcher's perspective as well as from the perspectives of key others (stakeholders, gatekeepers, participants, downstream considerations) and about which strategies might actually facilitate knowledge building most effectively. It encourages reading and experiencing a more diverse set of prior research learning outcomes. It ensures that any trade-offs made have the best and most realistic chance of facilitating convincing outcomes. It ensures that the forest as well as the trees are examined in order to facilitate and support more effective research journeys. It means that the training experiences need to be much broader than typically encountered and helps to counter the strong gravity exerted by disciplinary norms.

6.5.2 Where We Can Go in the Future

Social and behavioural researchers in the future have access to a much more diverse and continually expanding range of tools and approaches than many disciplines tend to acknowledge. Pluralist systemic thinking therefore opens up pathways for research that researchers (in training, novice, and experienced) may not be aware of or familiar with. It commences within an inclusionary orientation, rather than the exclusionary orientation fostered in many disciplines that favour specific patterns of guiding assumptions. If the inclusionary orientation fails to yield a viable, useful, ethical, or culturally appropriate process for research in some context, this is the stimulus for innovation and creativity, to develop/evolve new methods, new perspectives, and/or new ways of approaching and managing data and creating research outcomes (as evidenced, for example, in recent works by Alhadeff-Jones, 2013; Alvesson & Sandberg, 2021; Archibald et al., 2022; Gobo et al., 2022; Fear et al., 2006; Kovach, 2021; Nind et al., 2013; Poth, 2018; Smith, 2021; Vogt & Williams, 2011). Over time, discipline boundaries tend to become more permeable and open to more diverse approaches. This is the trajectory of evolution that social and behavioural research methods currently travels, albeit at different rates for different disciplines. As well, pan-disciplinary approaches have begun to proliferate as have arts-based, Indigenous/First Nations, and other innovative research approaches. As one example, one can simply looking at the evolution of the content of one of the most respected and useful sources of developments, *The Sage Handbook of Qualitative Research* (now in its 6th edition, see Denzin et al., 2024), through its 6 editions (the first edition being published 30 years ago in 1994, edited by Norman Denzin and Yolanda Lincoln) to see how new and more diversified qualitative approaches have been invented. In the arena of quantitative data-oriented methods, similar diversity in inventiveness can be detected over decades; for example, consider the burgeoning number of 'little green books' in the Sage series of books, *Quantitative Applications in the Social Sciences* (QASS; see https://us.sagepub.com/en-us/nam/qass), which has ballooned to 194 books as at February 2024, commencing with volume 1 in 1976.

In the future, we can only expect greater diversity in approaches developed to support knowledge building in social and behavioural research. We argued earlier that constraints and emergent issues are a primary stimulant for inventiveness and creativity and, given the ever-evolving complexities of human life, the environments we inhabit, the cultures we create, and the technologies we develop/use, the best skills for social and behavioural researchers to develop in order to adapt to and manage such evolutionary diversity are pluralist systemic thinking and flexibility/adaptability. Such thinking allows us to consider the new along with the old and create new where the old does not suffice. Flexibility/adaptability is a natural consequence of engaging pluralist systemic thinking as it focuses on the transformation of systemic feedback about progress thus far as well as anticipation of future problems into feedforward actions and trade-offs to keep the research moving forward in a convincing manner. The end goal will remain the same, building knowledge for learning, for change, for improvement, for emancipation, for innovation, and on the list goes. Getting to that end goal, the unity that is knowledge, will always require contextualisation in order for researchers to make informed choices for their research journeys, understand what they have and have not learned through their journeys, and to envisage what their journeys might mean for others and their contexts. My hope is that this book has been able to move researchers at least one small step toward this future.

References

Alhadeff-Jones, M. (2013). Complexity, methodology and method: Crafting a critical process of research. *Complicity: An International Journal of Complexity and Education, 10*(1/2), 19–44.

Alvesson, M., & Deetz, S. (2021). *Doing critical research.* Sage Publications.
Alvesson, M., & Sandberg, J. (2021). *Re-imagining the research process: Conventional and alternative metaphors.* Sage Publications.
Archibald, J., Xiiem, Q., Lee-Morgan, J., & Santolo, J. (Eds.). (2022). *Decolonizing research: Indigenous storywork as methodology.* Bloomsbury.
Bazeley, P. (2018). *Integrating analyses in mixed methods research.* Sage Publications.
Bazeley, P. (2021). *Qualitative data analysis: Practical strategies* (2nd ed.). Sage Publications.
Bettez, S. (2015). Navigating the complexity of qualitative research in postmodern contexts: Assemblage, critical reflexivity, and communion as guides. *International Journal of Qualitative Studies in Education, 28*(8), 932–954.
Chilisa, B. (2020). *Indigenous research methodologies* (2nd ed.). Sage Publications.
Chilton, G., & Leavy, P. (2020). *Arts-based research: Merging social research and the creative Arts.* In P. Leavy (Ed.), *The Oxford handbook of qualitative research* (2nd ed., pp. 602–632). Oxford Academic.
Christensen, J. (2012). Telling stories: Exploring research storytelling as a meaningful approach to knowledge mobilization with Indigenous research collaborators and diverse audiences in community-based participatory research. *The Canadian Geographer, 56*(2), 231–242.
Cooksey, R. (1996). *Judgment analysis: Theory, methods, and applications.* Academic Press.
Cooksey, R. (2011). Yours, mine or ours: What counts as innovation? *The Journal of Agricultural Education and Extension, 17*(3), 283–295.
Cooksey, R. (2020). *Illustrating statistical procedures: Finding meaning in quantitative data* (3rd ed.). Springer Nature.
Cooksey, R., & McDonald, G. (2019). *Surviving and thriving in postgraduate research* (2nd ed.). Springer Nature.
Denzin, N. (2024). The elephant in the living room, or extending the conversation about the politics of evidence, Part 2. In N. Denzin, Y. Lincoln, M. Giardina, & G. Cannella (Eds.), *The Sage handbook of qualitative research* (6th ed., pp. 549–566), Sage Publications.
Denzin, N., Lincoln, Y., Giardina, M., & Cannella, G. (Eds.). (2024). *The Sage handbook of qualitative research* (6th ed.). Sage Publications.
Fear, F., Rosaen, C., Bawden, R., & Foster-Fishman, P. (2006). *Coming to critical engagement: An autoethnographic exploration.* University of America Press.
Gill, J., Arthur, M., & Leng, R. (2023). What is co-production? Conceptualising and understanding co-production of knowledge and policy across different theoretical perspectives. *Evidence & Policy, 19*(2), 275–298.
Gobo, G., Fielding, N., La Rocca, G, & van der Vaart, W. (2022). *Merged methods: A rationale for full integration.* Sage Publications.
Graham, I., Rycroft-Malone, J., Kothari, A, & McCutcheon, C. (Eds.). (2023). *Research coproduction in healthcare.* Wiley-Blackwell.
Gray, D. (2021). *Doing research in the real world* (5th ed.). Sage Publications.
Greenwood, J. (2019). Arts-based research. *Oxford Research Encyclopedia of Education.* https://doi.org/10.1093/acrefore/9780190264093.013.29
Greenwood, D., & Levin, M. (2007). *Introduction to action research: Social research for social change* (2nd ed.). Sage Publications.
Grbich, C. (2013). *Qualitative data analysis: An introduction* (2nd ed.). Sage Publications.
Guertler, M., Kriz, A., & Sick, N. (2020). Encouraging and enabling action research in innovation management. *R&D Management, 50*(3), 380–395.
Happ, M. (2009). Mixed methods in gerontological research: Do the qualitative and quantitative data "touch"? *Research in Gerontological Nursing, 2*(2), 122–127.
Hattie, J. (2008). *Visible learning: A synthesis of over 800 meta-analyses relating to achievement.* Routledge.
Hasim, Z., & De Luca, R. (2011). Issues and challenges of using web blogs as a medium for research communication. *Waikato Journal of Education, 16*(1), https://doi.org/10.15663/wje.v16i1.76.

Hesse-Biber, S. (2015). Introduction: Navigating a turbulent research landscape: Working the boundaries, tensions, diversity, and contradictions of multimethod and mixed methods inquiry. In S. Hesse-Biber, & R. Johnson, (Eds.), *The Oxford handbook of multimethod and mixed methods research inquiry* (pp. xxxiii–liii). Oxford University Press.

Jackson, K., & Bazeley, P. (2019). *Qualitative data analysis with NVivo* (3rd ed.). Sage Publications.

Jones, S., & Adams, T. (2024). Autoethnography as becoming-with. In N. Denzin, Y. Lincoln, M. Giardina, & G. Cannella (Eds.), *The Sage handbook of qualitative research* (6th ed., pp. 421–435), Sage Publications.

Kane, M., & Trochim, W. (2007). *Concept mapping for planning and evaluation.* Sage Publications.

Kleinsasser, A. (2000). Researchers, reflexivity, and good data: Writing to unlearn. *Theory into Practice, 39*(3), 155–162.

Knaflic, C. (2015). *Storytelling with data: A data visualisation guide for business professionals.* Wiley.

Koranteng, F., & Wiafe, I. (2019). Factors that promote knowledge sharing on academic social networking sites: An empirical study. *Education and Information Technologies, 24*(2), 1211–1236.

Kovach, M. (2021). *Indigenous methodologies: Characteristics, conversations, and contexts* (2nd ed.). University of Toronto Press.

Krum, R. (2013). *Cool infographics: Effective communication with data visualization and design.* Wiley Australia.

Kuckartz, U., & Rädiker, S. (2019). *Analyzing qualitative data with MAXQDA: Text, audio, and video.* Springer Nature.

Latulippe, N., & Klenk, N. (2020). Making room and moving over: Knowledge co-production, Indigenous knowledge sovereignty and the politics of global environmental change decision-making. *Current Opinion in Environmental Sustainability, 42*, 7–14.

Locke, L., Spirduso, W., & Silverman, S. (2013). *Proposals that work: A guide for planning dissertations and grant proposals* (6th ed.). Sage Publications.

Maclean, K., Woodward, E., Jarvis, D., Turpin, G., Rowland, D., & Rist, P. (2022). Decolonising knowledge co-production: Examining the role of positionality and partnerships to support Indigenous-led bush product enterprises in northern Australia. *Sustainability Science, 17*(2), 333–350.

Martin, S. (2010). Co-production of social research: Strategies for engaged scholarship. *Public Money & Management, 30*(4), 211–218.

McLean, R., Carden, F., Aiken, A., Armstrong, R., Bray, J., Cassidy, C., Daub, O., Ruggiero, E., Fierro, L., Gagnon, M., Hutchinson, A., Kislov, R., Kothari, A., Kreindler, S., McCutcheon, C., Reszel, J., Scarrow, G., & Graham, I. (2023). Evaluating the quality of research co-production: Research Quality Plus for Co-Production (RQ + 4 Co-Pro). *Health Research Policy and Systems, 21*(51), https://doi.org/10.1186/s12961-023-00990-y.

McIntyre, A. (2008). *Participatory action research.* Sage Publications.

Miles, M., Huberman, A., & Saldaña, J. (2020). *Qualitative data analysis: A methods sourcebook* (4th ed.). Sage Publications.

Miller, J. (2005). *The Chicago guide to writing about multivariate analysis.* University of Chicago Press.

Kettle, N. (2019). Knowledge co-production in contested spaces. *Arctic, 72*(1), 43–57.

Nind, M., Wiles, R., Bengry-Howell, A., & Crow, G. (2013). Methodological innovation and research ethics: Forces in tension or forces in harmony? *Qualitative Research, 13*(6), 650–667.

Onwuegbuzie, A., & Leech, N. (2005). On becoming a pragmatic researcher: The importance of combining quantitative and qualitative research methodologies. *International Journal of Social Research Methodology, 8*(5), 375–387.

Patton, M. (2011). *Developmental evaluation: Applying complexity concepts to enhance innovation and use.* Guilford Press.

Patton, M., McKegg, K., & Wehipeihana, N. (2015). *Developmental evaluation exemplars: Principles in practice.* Guilford Press.

Poth, C. (2018). *Innovation in mixed methods research: A practical guide to integrative thinking with complexity*. Sage Publications.
Punch, K. (2016). *Developing effective research proposals* (3rd ed.). Sage Publications.
Quintane, E., Mitch Casselman, R., Sebastian Reiche, B., & Nylund, P. (2011). Innovation as a knowledge-based outcome. *Journal of Knowledge Management, 15*(6), 928–947.
Saldaña, J. (2021). *The coding manual for qualitative researchers* (4th ed.). Sage Publications.
Saldaña, J. (2024). Performance shapes for qualitative inquiry. In N. Denzin, Y. Lincoln, M. Giardina, & G. Cannella (Eds.), *The Sage handbook of qualitative research* (6th ed., pp. 437–452), Sage Publications.
Sandelowski, M., Voils, C., & Knafl, G. (2009). On quantitizing. *Journal of Mixed Methods Research, 3*(3), 208–222.
Schuelka, M., & Maxwell, T (Eds.) (2016). *Education in Bhutan: Culture, schooling, and gross national happiness*, Springer Science.
Siegesmund, R. (2024). The arts as research: Nomadic materiality and possible futures. In In N. Denzin, Y. Lincoln, M. Giardina, & G. Cannella (Eds.), *The Sage handbook of qualitative research* (6th ed., pp. 453–465), Sage Publications.
Sinclair, M. (Ed.). (2014). *Handbook of research methods on intuition*. Edward Elgar Publishing.
Silver, C., & Lewins, A. (2014). *Using software in qualitative research: A step-by-step guide* (2nd ed.). Sage Publications.
Smith, B., & McGannon, K. (2024). Co-production and impact: Challenges and opportunities. In N. Denzin, Y. Lincoln, M. Giardina, & G. Cannella (Eds.), *The Sage handbook of qualitative research* (6th ed., pp. 535–548), Sage Publications.
Smith, L. (2021). *Decolonizing methods: Research and Indigenous peoples* (3rd ed.). Zed Books.
Smith, S. (2006). Encouraging the use of reflexivity in the writing up of qualitative research. *International Journal of Therapy and Rehabilitation, 3*(5), 209–215.
Tashakkori, A., & Teddlie, C. (Eds.). (2010). *Sage handbook of mixed methods in social & behavioral research* (2nd ed.). Sage Publications.
Turnhout, E., Metze, T., Wyborn, C., Klenk, N., & Louder, E. (2020). The politics of co-production: Participation, power, and transformation. *Current Opinion in Environmental Sustainability, 42*, 15–21.
Van der Have, R., & Rubalcaba, L. (2016). Social innovation research: An emerging area of innovation studies? *Research Policy, 45*(9), 1923–1935.
Vogt, W., & Williams, M. (2011). *The Sage handbook of innovation in social research methods*. Sage Publications.
Zaumanis, M. (2021). *Research data visualisation and scientific graphics* (Peer Recognized series, Book 2). Kindle Direct Publishing.
Zurba, M., Michael, A. Petriello, M., Madge, M., McCarney, P., Bishop, B., McBeth, S., Bodwitch, H., & Bailey, M. (2022). Learning from knowledge co-production research and practice in the twenty-first century: Global lessons and what they mean for collaborative research in Nunatsiavut. *Sustainability Science, 17*(2), 449–467.

Index

R. W. Cooksey, *Unity from Diversity*, https://doi.org/10.1007/978-981-97-3462-7

T

U

V

SPRINGER NATURE

GPSR Compliance

The European Union's (EU) General Product Safety Regulation (GPSR) is a set of rules that requires consumer products to be safe and our obligations to ensure this.

If you have any concerns about our products, you can contact us on ProductSafety@springernature.com

In case Publisher is established outside the EU, the EU authorized representative is:

Springer Nature Customer Service Center GmbH
Europaplatz 3
69115 Heidelberg, Germany

The manufacturer's authorised representative in the EU is Springer Nature Customer Service Centre GmbH, Europaplatz 3, 69115 Heidelberg, Germany. If you have any concerns regarding our products, please contact ProductSafety@springernature.com

Printed and bound by CPI Group (UK) Ltd, Croydon, CR0 4YY
30/09/2026
02225298-0016